新 蝶類生物学英和辞典

岩野秀俊　監修
鍛治勝三　著

北 隆 館

New English - Japanese Dictionary of Butterflies' Biology

Written by

Katsuzo Kaji

Supervised by

Dr. Hidetoshi Iwano

Published by

はじめに

本書「新蝶類生物学英和辞典」は、従来の単なる生物系の英和辞典とはひと味内容の異なった辞典と考えています。蝶類学、鱗翅学、昆虫学、応用昆虫学の各学問分野を主体として、それらの分野に関連する生物学や分子生物学なども包含した幅広い学問分野で、主に使用される頻度が高いと思われる英単語や英用語などを集録してあります。旧書「蝶類生物学英和辞典」での集録用語数は約 8,800 語にすぎず、グリーンブックスシリーズ（ニューサイエンス社）の一つとして刊行しましたが、本書では、特に昆虫学や分子生物学関連の語句や用語を大幅に充実させましたので、集録した語句や用語数は 4,700 語以上も増えて、約 13,500 語にまで及んでいます。さらには、内容もリニューアルして、新たな辞典として生まれ変わりました。

かつて昆虫学に関する辞典としては、私自身が学生時代から愛用していた素木得一博士が書かれた「昆虫学辞典(1962)、北隆館」が古くから著名な好書として知られていましたが、現在、復刻版はあるものの高価ということもあり、これに代わる昆虫学関係の普及版の新辞典が望まれていました。さらには、蝶や蛾を主体とした昆虫学に関する英和辞典となると、皆無といって過言でないほど良書が見あたらず、その必要性を痛感していたこともありました。そんな折り、私が代表を務める「相模の蝶を語る会」の会員でもあった鍛治勝三氏より出版企画の話を伺い、是非私に監修をお願いしたいという依頼がありましたので、快諾した次第です。

いざ実際に校閲作業を行ってみると、予想以上に作業が難航して進みませんでした。一字一句ずつ、根気よく徹底的に点検するというのは、精神的にも疲れて負担が大きく、その上で莫大な時間を要する作業となってしまいました。特に、今回の校閲では、凡例にも書いてあるように、できるだけ単なるカタカナ表記は避ける方針で臨んではいたのですが、それでも昨今の分子生物学関連の語句や化学物質名などでは、英語の読みをそのままカタカナ読みして日本語に当てはめている場合が多いために、適当な和訳語が見あたらず、結構苦労しました。例えば、「genome」や「primer」などは、やむなくそのままカタカナ読みとしてあります。

かつて旧書が刊行された後、英和辞典だけでなく、次は是非とも蝶類生物学に関連した和英辞典も作って下さいという多くの読者からの声を伺いました。しかしながら、和英辞典の作成は、日本語の語彙（ごい）の豊富さなどを考えると、障壁が多く、作成にはより多くの困難が予想されました。それでも、発行元である北隆館編集部のご協力により、本書の中に、「日本語用語索引」という形で集録英用語の和訳を、アイウエオ順に並び替えて掲載していただいたことで、日本語から該当する英単語や英用語を検索することができるようにしました。例えば、「幼虫」や「幼生」、「若虫」を索引で探すと、それに該当する英単語として「larva」が見つかります。さらに、「若虫」に関しては「nymph」も見つかります。その点では、本書は「英和辞典」でありながら、「和英辞典」的な使用ができる画期的な特徴を持った新しいスタイルの辞典になったと自負しています。さらには、日本産蝶類全275種の英名についても掲載してあります。

掲載すべき語句や用語はまだたくさんあるかと自覚していますが、ページ数の制約などもあり、今回はご了承いただければ幸いです。現代はインターネットで容易に何でも検索できる時代ではありますが、本書のような小型サイズの辞典であっても、読者の皆様方にとって存在感のある座右の書となって活用されることを祈っております。

2018年6月

6月4日（ムシの日）の執筆に合わせて　　岩野 秀俊

目　次

はじめに（岩野秀俊） …… 1 ～ 2
目次 …… 3
凡例 …… 4 ～ 10

新蝶類生物学英和辞典
■本編：新蝶類生物学英和辞典 …… 11 ～ 237
■付録：日本産蝶類名称の英和／和英編
英和編 …… 238 ～ 243
和英編 …… 243 ～ 248

日本語用語索引 …… 249 ～ 375

おわりに（鍛治勝三） …… 376 ～ 379
監修者・著者略歴 …… 380

〔表紙写真〕
1段目：スジグロシロチョウ／スジボソヤマキチョウ／アサギマダラ
2段目：ミドリヒョウモン／ツマベニチョウ／エルタテハ
3段目：クジャクチョウ／ミスジチョウ／シータテハ
4段目：イシガケチョウ／オナガアゲハ／ミカドアゲハ

裏表紙：アオスジアゲハ

凡　例

1. 本編：新蝶類生物学英和辞典

●編纂方針

・本英和辞典は、「蝶類の生物学」の学問範囲である「分布、分類、生理、生態、形態、遺伝、発生、病理、行動、保全、採集」の各領域をできる限り網羅し、かつ、現代の研究潮流である最新の分子生物学の専門英用語もかなり網羅している。また、関連する昆虫学分野での使用に供するために、（基礎）昆虫学及び応用昆虫学の専門英用語についても補追した。収載見出し語（英用語）数は、約 13,500 語に及ぶ。

・従来の専門分野の英和辞典と異なり、「専門英用語の充実」は勿論のこと、「生物学や医学などの分野での専門語の造語で基盤となる（古典）ラテン語や（古典）ギリシャ語の接頭辞」、「専門語以外の適切な英単語の選定」、「実際によく使われる英単語やフレーズ」にも力点を置いている。

●特記事項

・学術用語では常用漢字表に基づいた表記が基本になっているので、漢字の使用にあたっては常用漢字の使用を原則とした。

①「跗節」:「跗節」の「跗」は常用漢字表にないので、「ふ節」と表記した。ただし、基本語である「tarsus」に関してのみ、「跗節」と表記した。

［例１］ tarsus :（複 . -si）、跗節（ふせつ）、ふ節〔ラテン語〕

［例２］ pretarsus : 前ふ節（ぜんふせつ）

②「嚢」:「嚢」は常用漢字表にないので、「のう」と表記した。

［例］ bladder : 香嚢 → 香のう

③「楯」と「盾」:「楯」は常用漢字表にないので、「盾」を使用することを原則とした。

［例１］ scutum :（複 . -ta)、盾板(じゅんばん)、楯板（胸部）〔ラテン語〕
→（複 . -ta)、盾板(じゅんばん)(胸部）〔ラテン語〕
［例２］ clypeus :（複 . -pei)、唇基部、頭楯(とうじゅん)、頭楯板、頭盾〔ラテン語〕
→（複 . -pei)、唇基部、頭盾(とうじゅん)、頭盾板〔ラテン語〕

④「棘」と「刺」:「棘」は常用漢字表にないので、「刺」を使用することを原則としたが、機械的に変換すると違和感を感じる用語もあるので、両論併記した上で、送り仮名を付けた。
［例１］ spine : トゲ、刺、棘(とげ)
［例２］ seta :（複 . -ae)、棘毛(きょくもう)、剛毛〔ラテン語〕
→ 刺毛(しもう)、棘毛(きょくもう)、剛毛〔ラテン語〕
［例３］ dorsal spine : 背棘 → 背棘(はいきょく)、背刺(はいし)

・「sequence」のカタカナ表記：「シークエンス」と「シーケンス」

約 60 年前に作成が開始された文部（科学）省の『学術用語集』（または『オンライン学術用語集』）では「シーケンス」が圧倒的に多いが、本辞典では最新の「分子生物学」にウエイトを置いており、かつ、原語の発音に近いこともあり、「シークエンス」を採用した。
［例１］ sequence : 配列、シーケンス → 配列、シークエンス

同様に、その進行形である「sequencing」を「シーケンシング」ではなく、「シークエンシング」を採用した。
［例２］new generation sequencing : 新世代シーケンシング、新世代シーケンサー
→ 新世代シークエンシング、新世代シークエンサー

・略語に関しては、全綴りを明記した。また、略称を含む項目については、略称の頭文字を大文字で表記した正式な綴りを項末にカッコで示した。
［例１］ DNA : デオキシリボ核酸（DeoxyriboNucleic Acid）
［例２］ deoxyribonucleic acid : デオキシリボ核酸（DNA）
［例３］ UV-C-irradiated plant : UV-C（遠紫外線）照射された植物
（UV-C ＝ UltraViolet-C）

・同じ見出し語で語義が異なる語（同形異義語〈homograph〉）は、２番目の見出し語の直後に「[hg.]」を付与して別項目（別見出し語）とした。

［例１］ bold : はっきりした

BOLD [hg.] : バーコード・オブ・ライフ・データ（生物種同定システム）（Barcode of Life Data）

［例２］ F : 前翅（ぜんし）（Forewing）

F [hg.] : F 値、検定統計量、t 統計量（分散分析）

・化合物名中の幾何異性体を表す「(*Z*)-」や「*cis*-」の接頭語、結合位置を示す「*N*-」では、英文字のみをイタリック体で表記した。

［例１］(*Z*)-3-hexenol : (Z)-3-ヘキセノール、シス-3-ヘキセノール（*cis*-3-Hexenol）

［例２］ 11-*cis*-retinal : 11-シス-レチナール、11-シス型レチナール

［例３］ *N*-β-alanyl dopamine synthase : *N*-β-アラニルドーパミンシンターゼ、*N*-β- アラニルドーパミン合成酵素

・見出し語の配列はアルファベット順としたが、見出し語中の先頭の括弧文字「(」は無視して配列した。例えば、

(*Z*)-3-hexenol : (*Z*)-3- ヘキセノール、シス-3- ヘキセノール（*cis*-3-Hexenol）

は、「Z」の項にある。

・見出し語（英用語）が実際に使用される時に複数形で多用される場合でも、基本は「単数形表記」で統一した。ただし、複数形に関しては、（　）内に示し、ハイフンを用いて見出し語と共通の部分を省略したものもある。

［例１］ antenna :（複 . -nae）、触角〔ラテン語〕

［例２］ census datum :（複 . -ta）、調査データ

・蝶類学や生物学分野などでは、２語以上から成る英用語（複合語）で語間の空白文字やハイフン文字が省略されるときがある（米国では空白文字を用いない書き方、英国では空白文字を用いる書き方が好まれる傾向があり、また、両者には意味〈ニュアンス〉面での微妙な違いがある）が、基本用語の場合には、両方を見出し語とした。

［例１］ fore wing : 前翅（ぜんし）、上翅（じょうし）

forewing (FW) : 前翅（ぜんし）、上翅（じょうし）

［例２］ hind-gut : 後腸（こうちょう）

hindgut : 後腸

・蝶の科や亜科の表記は、「複数形表記」とした。

［例１］Parnassians and Swallowtails : アゲハチョウ科

［例２］Longwings : ドクチョウ亜科

・重要な用語で「英語」と「ラテン語」の表記がある場合には、両方を併記した。

［例］palp : 鬚（ひげ、しゅ）〔英語〕

palpus : （複 . -pi）、鬚（ひげ、しゅ）〔ラテン語〕

・冗長な重複の英用語は、どちらかを削除した。

［例］bursa copulatrix

bursa copulatrix cell → 削除

・見出し語の綴りは、英語式より米語式を優先した。

［例１］behaviour → behavior

［例２］colour → color

・蝶の種名表記を統一した。

［例］Apollo → Apollo butterfly （「butterfly」を加筆した）

・蛾の種名表記を統一した。

［例］Sloane's urania → Sloane's urania moth （「moth」を加筆した）

・できるだけ単なるカタカナ表記は回避した。そのままで日本語として通用する場合や他に適語がない場合は、カタカナ表記とした。

［例］black list → ブラックリスト

・学名や蝶類学分野で必修の遺伝子名／遺伝子略名（記号）などはイタリック体で表記した。

［例１］*bithorax* : 双胸遺伝子

［例２］*Hox* gene : ホメオティック遺伝子、相同異質形成遺伝子

［例３］*in vivo* : インビボ、生体内で（ラテン語）

・遺伝用語に関して、2017 年 9 月に、「日本遺伝学会」が新たに遺伝学用語の和訳を提案した。その中でも、蝶類学分野でも最重要語である表現型を対象する

「variation」を「変異」ではなく、「多様性」と和訳するように変更した。例えば、「variation」にいきなり「多様性（diversity）」の和訳を当てると、すぐには理解しがたいので、従来の和訳を残して、その後ろにカッコ書きで改訂した和訳を示す「両論併記」を採用した。

［例１］dominant : 優性の、優勢形質（顕性形質）、優占種、顕性

［例２］recessive : 劣性の、劣勢形質（潜性形質）、不顕性

［例３］variation : 変異（多様性）、変動

・一部の和名（アレチコビトシジミ）には差別用語が残っているため、あえてそのまま掲載すると共に、今回改訂した新和名も並記掲載した。

○ western pygmy blue butterfly : アレチコビトシジミ

（アレチコシジミ：改訂新名称）

2. 付録：日本産蝶類名称の英和／和英編

●編纂方針

・日本産の蝶類に関しては、最新の日本産蝶類標準図鑑（学習研究社発行）（白水，2006）の和名に準拠した。その図鑑に記載された迷蝶類は削除して、275 種の蝶類を選定し、その名称（和名）の英名（common English name）を記載した。

・日本産蝶類名称の英名に関しては、場合によっては亜種水準まで勘案して、英名を選定・採用した。

・「The」の表記は不要と考え、統一して削除した。

・１つの和名に対して複数の英名表記がある場合には、そのまま列記した。

［例］カラスアゲハ : Bianor Peacock

カラスアゲハ : Chinese Peacock

・「和名に対して該当する英名がない種」について

「和名に対して該当する英名がない種」に関しては、以下の２種があり、その英名を著者が創作した。このことを明示するために、英名の後に「***」を付加した。

○ソテツシジミ : Cycad Butterfly ***

○ヤエヤマカラスアゲハ : Yayeyama Peacock ***

●特記事項

・「学名に由来した英名」について

蝶の名称には、学名（属名 種名）に由来した英名もある。

[例] ギフチョウ : Japanese Luehdorfia（属名を流用）

ヒメウラナミジャノメ : Argus Rings（種名を流用）

・「ミスジチョウ類を表す英名」について

ミスジチョウ類を表す英名には「Sailer」と「Sailor」、さらには「Glider」があるが、本付録では「Sailer」に統一して使用した。

[例] コミスジ : Common Sailer

「Glider」に関しては、広く普及していると想定される場合には、それも採用した。「フタスジチョウ」がその例である。

○フタスジチョウ : Hungarian Glider

・「オオイチモンジ」について

「オオイチモンジ」に関しては、「Popular Admiral」という英名が過去の記載誤りでそのまま通用してしまったと考えられるので、ここでは「Poplar Admiral」のみを採用した。

○オオイチモンジ : Poplar Admiral

・「ムラサキオナガウラナミシジミ」について

「ムラサキオナガウラナミシジミ」は、以下のような英名が付されている。

Catochrysops strabo strabo（インド方面産）→ Oriental forget-me-not

C. starabo luzoneris（フィリピン方面産）→ Forget-me-not

日本に飛来する個体群は、フィリピン方面産と考えられているので、ここでは「Forget-me-not」を採用した。

○ムラサキオナガウラナミシジミ : Forget-me-not

・「ヤマトスジグロシロチョウ」について

「ヤマトスジグロシロチョウ」は該当する英名がなく、「エゾスジグロシロチョウ」が北海道（道東）産の別種ではあるが、（両種は、分子系統的な差異があるが、半種状態にあると考えるので、）ここでは「エゾスジグロシロチョウ」と同名の英名を採用した。

○ヤマトスジグロシロチョウ : Green-veined White

○エゾスジグロシロチョウ : Green-veined White

・「リュウキュウムラサキ」について

「Jacintha Eggfly」は、大陸亜種（学名：*Hypolimnas bolina jacintha*）の英名であるが、日本には大陸亜種も飛来するので、削除せずにそのまま残した。

○リュウキュウムラサキ : Common Eggfly

○リュウキュウムラサキ : Great Eggfly

○リュウキュウムラサキ : Jacintha Eggfly

○リュウキュウムラサキ : Varied Eggfly

以上

新蝶類生物学英和辞典

-（記号）

-aceae : - 科（植物の場合）
-idae : - 科（動物の場合）
-inae : - 亜科（動物の場合）
-ini : - 族（動物の場合）
-logy : - 学、- 論〔ギリシャ語〕
-oideae : - 亜科（植物の場合）
-phagy : - 食性、- 食〔ギリシャ語〕
-pter : - 翅〔ギリシャ語〕

0 – 9（数字）

1 ha : １ヘクタール（100m × 100m の正方形の面積）（hectare）
11-*cis*-retinal : 11- シス - レチナール、11- シス型レチナール
2-way ANOVA : 二元配置分散分析（ANalysis Of VAriance）
20-HE : 20- ヒドロキシエクジソン（20-HydroxyEcdysone）
20-hydroxyecdysone : 20- ヒドロキシエクジソン、20- ハイドロキシエクダイソン
25 bp overlap : 25 塩基オーバーラップ
3'UTR : 3' 側非翻訳領域、3' 非翻訳領域（three prime UnTranslated Region）
3-hydroxykynurenine : 3- ヒドロキシキヌレニン
35-cycle PCR : 35 サイクル（数）での PCR（Polymerase Chain Reaction）
3D structure : 三次元構造（three Dimensional structure）
3rd stadium : ３齢
4-methylsulfinylbutyl : 4- メチルスルフィニルブチル
5'UTR : 5'側非翻訳領域、5'非翻訳領域（five prime UnTranslated Region）
72-well rotor : 72 穴ローター
7TMP : ７回膜貫通型タンパク質（seven TransMembrance Protein）
7TMR : ７回膜貫通型受容体（seven TransMembrance Receptor）
96 well plate : 96 穴プレート

a

A : 臀脈（でんみゃく）、肛脈（Anal vein）
a new genus and a new species : １新属１新種
a prior selection of branch : 分岐の先験的選択
a prior specification of lineage : 系統の先験的明確化
a tenth : 10 分の 1、10 回につき 1 回
a- : 無 -、非 -〔ギリシャ語〕
a.s.l. : 海抜（Above Sea Level）
AAT : アスパラギン酸アミノ基転移酵素、アスパラギン酸アミノトランスフェラーゼ（Aspartate AminoTransferase）
ab- : 離れて -、外側の -、反対側の -〔ラテン語〕
ab. : 異常型（aberrant form）
abandonment : 放棄、中止
abaxial : 反軸側（はんじくそく）の、背軸側の
abbreviation : 略記
abdomen : （複 . -mina）、腹部〔ラテン語〕
abdomen mass : 腹部重量
abdominal : 腹部の
abdominal defensive gland : 腹部防御腺、腹性防御腺
abdominal ganglion : （複 . -glia）、腹部神経節〔ラテン語〕
abdominal proleg : 腹脚（ふくきゃく）、腹部の脚
abdominal tergite : 腹部背板（ふくぶはいばん）
aberrant : 異常型、異常
aberrant form : 異常型
aberrant whitish form : 白化異常型

aberration : 変体、変性種、異常
ABGD : 自動バーコードギャップ発見、バーコードギャップの自動発見（Automatic Barcode Gap Discovery）
abide : 守る、遵守する、とどまる、待つ
ability of to-and-fro movement : 往復移動の能力
abiotic constraint : 非生物的制約
abiotic environment : 非生物的環境
abiotic factor : 非生物的要因
abiotic stress : 非生物的ストレス
abiotic stress response : 非生物的ストレス応答
ablative operation : 除去手術
abnormal activity : 異常行動
abnormal climate : 異常気候
abnormal wing : 異常な翅
abolish : 抑制する、機能しなくなる
abort : 流産する
about a day later : 約 1 日後
abrupt change : 突然の変化
absolute tautonymy : 完全同語反復
absolute term : 絶対項
absorb : 吸収する
absorbance spectrum : 吸収スペクトル
absorbance spectrum maximum : 吸収スペクトルの極大波長、最大吸収スペクトル
absorbent paper : 濾紙
absorbent tissue : 吸収性のティッシュ
absorption : 吸収
absorption efficiency : 吸収効率
absorption maximum : 吸収極大
absorption peak : 吸収ピーク
absorption spectrum : 吸収スペクトル
absorption value : 吸収値
absorption wavelength region : 吸収波長域
abstract : 要旨、概要、摘要
abundance : 多量、多数、豊富、存在量、個体数、個体数量
abundance of a species : 種の個体数
abundance of nectar source : （吸）蜜源の量
abundant : 豊富な、豊富にある、大量の
abundant evidence : 豊富な証拠
abundant grass : 豊富な草、たくさんの草
abuse of species diversity index : 種多様性指標の誤用、種多様性指数の誤用
Acanthaceae : キツネノマゴ科
accelerated metamorphosis : 加速化された変態
accept : 受理、アクセプト
acceptability : 受容性
acceptable threshold : 容認可能な限度
access road : アクセス道路
accessibility : アクセス容易性
accessible : アクセスが容易、アクセスが可能な
accession code : 受入番号、受入コード
accession number : 受入番号
accessory genital gland : 生殖器付属腺
accessory gland : 付属腺
accessory medulla : 副視髄
accessory pigment cell : 補助色素細胞
accessory reproductive gland : 生殖付属腺（ARG）
accessory reproductive gland secretion : 生殖付属腺分泌物
accidental acquisition of novel gene : 新規遺伝子の偶然的な獲得
acclimation : 順応、順化、馴化(じゅんか)
acclimation period : 馴化期間
acclimatization : 順化、馴化(じゅんか)
acclimatization process : 順化過程
accompanying commentary : 付随コメント、付随論評
accrue : 生じる、手に入れる
accumulated temperature : 積算温度
accumulated value : 積算値
accumulation : 蓄積

accumulation curve : 累積曲線
accumulation of neutral mutation : 中立的突然変異の累積
accurate answer : 正確な解答
accurate datum : (複 . -ta)、正確なデータ
acetamiprid : アセタミプリド(「ネオニコチノイド系」の殺虫剤)
achene dispersal distance : 痩果の分散距離
achondroplasia : 軟骨異栄養症、軟骨発育［形成］不全症
achromatic vision : 明暗視、全色盲
acknowledgements : 謝辞
ACO : アリコロニー最適化法(Ant Colony Optimization)
acoustic : 聴覚の、音響の
acoustic stimulus : 音刺激、音響刺激
acquired ammonia : 獲得されたアンモニア
acquisition of novel trait : 新規形質の獲得
Acraeinae : ホソチョウ亜科
acrotergite : 端背板、端背片
Act on Conservation of Endangered Species of Wild Fauna and Flora : 絶滅のおそれのある野生動植物の種の保存に関する法律
act synergistically : 相乗的に作用する、協力的に働く
actin polymerization : アクチン重合
actin staining : アクチン染色
acting in concert : 共同行為、共同して行動する
actinomycin D : アクチノマイシン D
action effect : 作用効果
action spectrum : 作用スペクトル
activate : 活性化する、活動的にする
activation stimulus : 活性化刺激
active brain : 活性脳
active ingredient : 有効成分(農薬製剤の中で効き目を持つ物質)
active movement : 自発的移動
active phase : 活動相、活動位相
active season : 活動時期、活動季節
active stage : 活動期
activity budget : 活動時間配分、行動の時間配分
activity locus : 活動点、活動場所
activity radius : 行動半径、行動範囲
activity range : 行動範囲
activity schedule : 活動スケジュール
actogram : アクトグラム
actual mating probability : 実際の交尾確率、実効的交尾確率
aculea : (複 . -ae)、アクレア(微小な刺毛)、微毛、微刺、微小針状体
acute : 急性の、鋭角
acylation reaction : アシル化反応
adaptation : 適応、順応
adaptative learning : 順応学習、適応学習
adaptive convergence : 適応的収斂
adaptive differentiation : 適応的分化
adaptive dispersal : 適応的分散
adaptive dynamics : 適応ダイナミクス
adaptive evolution : 適応進化
adaptive hypothesis : 適応的仮説
adaptive introgression : 適応的遺伝子移入
adaptive management : 順応的管理
adaptive mutation : 適応的突然変異
adaptive novelty : 適応的新奇性、適応的新規性
adaptive radiation : 適応放散
adaptive response : 適応的反応
adaptive search behavior : 適応的探索行動
adaptive seasonal polymorphism : 適応的季節多型
adaptive significance : 適応的意義、適応上の意義
adaptive strategy : 適応戦略
adaptive system : 適応様式
adaptive trait : 適応形質

additional factor : 追加要因
additional note : 追加された知見
additional record : 追加記録
additional study : 追加研究、付随研究、追加調査
additive : 加法的な、付加的な、相加的な
additive genetic variance : 相加的遺伝分散
additive genotypic value : 相加的遺伝子型値
additive tree : 相加的樹形図
additively : 加法的に、相加的に
adenosine deaminase-related growth factor : アデノシンデアミナーゼ関連増殖因子（ADGF）
adenosine triphosphate : アデノシン三リン酸
ADGF : アデノシンデアミナーゼ関連増殖因子（Adenosine Deaminase-related Growth Factor）
adhesion of the barbed and sticky bristle : 棘(とげ)のある粘着性剛毛の粘着力
adhesive tape : 粘着テープ
adipokinetic hormone : 脂質動員ホルモン
adjacent distribution : 隣接分布地域
adjacent districts : 周辺地域
adjacent lane : 隣接レーン
adjoin : 接合する
admiral butterfly : アカタテハ属の蝶
Admirals and Relatives : イチモンジチョウ亜科
admit : 認める
admixture : 混合
adopt : 採用する
adoption : 採用
adult : 成虫、成体
adult butterfly : 成虫の蝶、蝶成虫
adult diapause : 成虫休眠
adult eclosion : 羽化した成虫、成虫の羽化
adult emergence : 成虫の羽化
adult emergence season : 成虫出現期、成虫羽化時期
adult form : 成虫形状
adult life history stage : 成虫（成長した動物）の生活史段階
adult life span : 成虫寿命
adult longevity : 成虫の寿命
adult mass : 成虫の体重
adult maturation : 成虫の成熟
adult morphology : 成虫の形態
adult overwintering : 成虫での越冬、成虫越冬
adult population : 成虫個体群、成虫個体数
adult sex-specific plasticity : 成虫の性特異的可塑性
adult size : 成虫のサイズ
adult stage : 成虫期
adult weight : 成虫体重
adult wing : 成虫翅
adult-female : 成虫雌
adulthood : 成虫期
adventitious bud : 不定芽
adversity : 逆境
adversity selection : 逆境淘汰
adzuki bean borer moth : アズキノメイガ
aedeagus (ae) : 交尾棍、エデアグス、挿入器
aedoeagus (aedeagus) (ae) : エデアグス、挿入器
aequorin : エクオリン、イクオリン（クラゲの発光タンパク質）
aequorin gene : エクオリン遺伝子、イクオリン遺伝子
aerial dispute : 空中戦闘、空中戦
aerial hawking bat : ヒメホリカワコウモリ
aerial interaction : 空中相互作用
aerobic respiration : 酸素呼吸
aerodynamics : 空気力学
aeropyle : 気孔
aestivate : 夏眠する〔米語〕
aestivation : 夏眠、夏休眠〔米語〕
Aethionema : タイリンミヤコナズナ属

afar : 遠くに
aff. : の近似(affinis)、類似、近似種
affected native species : 影響を受ける在来種
affinity : 密接な関係、親和性
affinity chromatography : 親和性クロマトグラフィー
afforestation program : 植林プログラム、造林プログラム
afforested area : 植林地、造林地
AFLP : 増幅断片長多型 (Amplified Fragment Length Polymorphism)
aforementioned : 前述の、前記の
African migrant butterfly : ミズアオシロチョウ
Afrotropical region : 熱帯アフリカ区、エチオピア区
afternoon peak : アフタヌーンピーク、A ピーク
afternoon territoriality : 午後のナワバリ制
agarose gel : アガロースゲル
agarose gel electrophoresis : アガロースゲル電気泳動
age distribution of gene duplicate : 遺伝子重複の年齢分布
age of adult : 成虫の年齢
age specific life table : 齢別生命表、年齢別生命表
age-related sexual receptivity : 日齢に関係する交尾受容性
age-related site fidelity : 日齢に関係する出生場所固執性、日齢に関係する生息場所執着性
Agency for Cultural Affairs : 文化庁(日本)
aggravate : さらに悪化させる
aggregate : 集群、集合(体)
aggregation : 集団、集まり
aggregation pheromone : 集合フェロモン
aggregative feeding : 集団摂食
aggressive caterpillar : 攻撃的な幼虫
AGH : 造雄腺ホルモン (Androgenic Gland Hormone)
agility : 機敏、機敏性
agonist : アゴニスト、作動筋、作動物質
agri-environment scheme : 農業・環境スキーム、農業環境計画(英国の政策)
agricultural intensification : 農業集約化
agricultural landscape : 農業景観
agricultural pest : 農業害虫
agricultural pesticide : 農業用殺虫剤
agriculture : 農耕、農業
AIC : 赤池情報量基準 (Akaike's Information Criterion)
air hole : 通気孔
air temperature : 気温
airborne communication : 空気伝播コミュニケーション(空気中の香りを介したコミュニケーション)
Akaike's information criterion : 赤池情報量基準
aktionsraum : 行動圏
alarm pheromone : 警報フェロモン
alarming rate : 警告率
alata : (複 . -ae)、有翅型
alate species : 有翅種(ゆうししゅ)
albatross : アホウドリ
albeit : ～にもかかわらず、～ではあるが、～であろうとも
albinic : 白化の
albino form : 白化型
alcohol : アルコール
alcoholic Bouin's fixative : アルコールブアン固定液
aldehyde : アルデヒド
aldoxime product : アルドキシム産物
Alexander Koenig Museum : アレクサンダー・ケーニヒ博物館
algorithm : アルゴリズム、問題解決のための段階的手法

algorithmic approach : アルゴリズム的アプローチ
alien species : 外来種、移入種、帰化種
alighting : 着陸
align : 並べる、整列させる
aligned ambiguous region : 整列された曖昧領域
alignment : 整列、アラインメント
alignment coverage : 整列被覆(率)、整列網羅(率)
alignment length : アラインメント長
alignment with sequence : 配列のアラインメント
alimentary canal : 消化管(しょうかかん)
aliquot of medium, an : 培養液の小分け(分取分)
alkaloid : アルカロイド
all year : 年中、一年中
all-against-all alignment : 完全総当りの整列
all-female brood : すべて雌の同腹仔
all-female-producing matriline : すべて雌を産生する母系群
all-male cage : すべて雄が入っているケージ
allantoic acid : アラントイン酸
allantoin : アラントイン
allatostatic hormone : アラトスタチンホルモン(昆虫脳で分泌されるアラタ体抑制ホルモン)
allatostatin : アラトスタチン
allatotropin : アラトトロピン
Allee effect : アリー効果
Allee's effect : アリー効果
allele : 対立遺伝子、アレル
allele difference : 対立遺伝子の違い
allele frequency : 対立遺伝子頻度
allele-specific expression : 対立遺伝子特異的な発現
allelic change : 対立遺伝子の変化
allelic divergence : 対立遺伝子(の)分岐
allelic richness : アレリックリッチネス
allelic size variation : 対立遺伝子のサイズ変異
allelic turnover : 対立遺伝子転換、対立遺伝子の代謝回転
allelic variant : 対立遺伝子変異体
allelic variation : 対立遺伝子変異
allelo- : 相互の -、お互いの -〔ギリシャ語〕
allelochemical : アレロケミカル、他感作用物質、他感物質
allelopathy related compound : アレロパシー関連化合物
allied species : 近縁種
alligate : 帯蛹
allo- : 異 -、別 -〔ギリシャ語〕
allochronic : 異時性、異時的な
allochronic isolation : 時間的隔離、異時的隔離
allochthonous element : 他生的な要素
allochthonous species : 他生的な種、外来種
allomelanin : アロメラニン
allometry : アロメトリー、相対成長
allomimesis : 隠蔽的異物擬態
allomone : アロモン
allopatric : 異所性の、異域性の
allopatric and genetic speciation : 異所的な遺伝的種分化
allopatric speciation : 異所的種分化
allopatry : 異所性
allopolyploid : 異質倍数体
allopolyploidy : 異質倍数性
allotment : 市民菜園、割り当てられたもの
allotype : アロタイプ、別模式標本
allozyme : アロザイム
allozyme allele frequency : アロザイム対立遺伝子頻度
allozyme diversity : アロザイム多様性
allozyme electrophoresis : アロザイム電気泳動

allozymic differentiation : アロザイム分化
ally : 近縁種
allylisothiocyanate : アリルイソチオシアネート
alongside : 並んで、一緒に
alorium : 爪間盤(板)、アロリウム
alpha-diversity : アルファ多様性、α-多様性(「種多様性」のことで、「ある１つの環境における種多様性」の意味)
alpha-tubulin (α-tubulin) : アルファチューブリン、α-チューブリン
alpine : 高山性、高山性の
alpine butterfly : 高山蝶
alpine meadow : 高山性草原
alpine organism : 高山性生物
Alpine ringlet butterfly : ベニヒカゲ
alpine tundra : 高山ツンドラ
alpine tundra grassland : 高山ツンドラ草地
alpine zone : 高山帯
alteration : 変更、修正
altered regulation : 変化した調節
alternate phenotype : 交互に発現する表現型
alternating season : 交互に入れ替わる季節
alternating seasonal environment : 交互に入れ替わる季節環境
alternative adult phenotype : 交互に発現する成虫の表現型
alternative allele : 代替の対立遺伝子
alternative model : 代替モデル
alternative phenotype : 選択的表現型、二者択一の表現型
alternative splicing : 選択的スプライシング、択一的スプライシング
alternative temperature environment : 交互に入れ替わる温度環境
altitude : 高度、標高
altitudinal cline : 高度的クライン、標高クライン、標高傾度
altitudinal optimum : 高度的最適
altitudinal shift : 高度シフト、高度変化
altruistic behavior : 利他(的)行動
Alu sequence : Alu 配列、アル配列(制限酵素 Alu I〈*Arthrobacter luteus* 由来の制限酵素〉の認識部位を有することから名付けられた短い配列)
am- : 周囲 -〔ラテン語〕
amateur naturalist : アマチュアの自然科学研究者、アマチュア研究者
Amazon basin : アマゾン川流域
Amazonian butterfly : アマゾンの蝶
amb- : 周囲 -〔ラテン語〕
ambient condition : 周囲条件、環境条件
ambient temperature : 周囲温度、外気温
ambiguity : 多義性、不確かさ
ambiguous character : 曖昧な特性、多義的な特性
ambiguous electromorph : 多義的電気泳動パターン
ameliorate : 改善する、良くなる
amensalism : 片害共生、片害作用
American copper butterfly : アメリカベニシジミ
American snout butterfly : アメリカテングチョウ
ametaboly : 無変態
amibient temperature : 周囲温度
amino acid : アミノ酸
amino acid change : アミノ酸置換、アミノ酸変異
amino acid difference : アミノ酸配列の違い
amino acid replacement : アミノ酸置換
amino acid residue : アミノ酸残基
amino acid secretion : アミノ酸の分泌物
amino acid site : アミノ酸部位
amino acid substitution : アミノ酸置換
amino acid substitution frequency : アミノ酸置換頻度
amino acid substitution model : アミノ酸置

換モデル
amino acid tryptophan : アミノ酸トリプトファン
amino acid-lipid conjugate : アミノ酸型脂質共役体
amino-acid sequence : アミノ酸配列
ammonia : アンモニア
ammonia ingestion : アンモニア摂取
ammonia uptake : アンモニアの吸収、アンモニアの摂取
ammonium : アンモニウム
ammonium chloride : アンモニウム塩化物、塩化アンモニウム
amnion : 羊膜
Amniote genome : 羊膜類のゲノム
Amniotes : 羊膜類
among-taxa component : 分類群間の成分
amount of assimilation : 同化量
amount of blue bilin : 青色ビリン量
amount of light : 光量
amount of light energy : 光エネルギー量
amount of seminal protein : 精液中のタンパク質の量
AMOVA : 分子分散分析
(Analysis of MOlecular VAriance)
amphibian : 両生類
amphibian genome : 両生類のゲノム
amphipathic amino acid : 両親媒性アミノ酸
ample anatomical datum : (複 . -ta)、十分な解剖データ
ample artificial nectar : たっぷりな人工果汁
ample opportunity : 十分な機会
amplicon : アプリコン、単位複製配列、増幅産物
amplicon length variation : アンプリコン長の変異
amplification of short fragment : 短断片の増殖、短断片の増幅
amplification product : 増幅産物
amplification reaction : 増幅反応
amplified DNA fragment : 増幅 DNA 断片
amplified fragment : 増幅断片
amplified fragment length polymorphism marker : 増幅断片長多型マーカー
ampula : 膨大部
Amur type : アムール型
anabolism : 同化
anadromous Pacific salmon : 遡河性の太平洋サケ
anagenesis : 前進進化
anagram : アナグラム
anal : 肛門(こうもん)、臀部、臀脈(でんみゃく)、A 脈、肛側
anal angle : 肛角部
anal area : 臀域(でんいき)
anal dash : 後角縦線(蛾類)
anal fan : 臀扇(でんおう)
anal fold : 肛門褶
anal lobe : 尾葉、臀葉、葉片
anal margin : 後翅内縁、臀縁
anal proleg : 肛門部の脚、尾脚
anal pyramid : 肛錐
anal segment : 肛節、尾節
anal vein : 臀脈、A 脈
analogous rapid evolutionary change : 類似の進化面の急速変化
analysis of covariance : 共分散分析
analysis of fixed effect : 固定効果の分析、母数効果の分析
analysis of flight : 飛翔の解析
analysis of marker recombination rate : マーカー組換え率の解析
analysis of molecular variance : 分子分散分析
analysis of stage-structured population : ステージ別個体群解析
analysis of variance : 分散分析
anapleural cleft : 背側板溝

anapleural suture : 背側板溝状線(胸部)
anatomically : 解剖学的に
anatomy : 体の構造、解剖
ancestral amino acid substrate : 祖先(型)アミノ酸基質
ancestral bat detector : 先祖伝来のコウモリ探知器
ancestral compound : 祖先伝来の化合物、祖先型化合物
ancestral heliconiine group : 祖先のドクチョウグループ
ancestral larval foodplant : 祖先的な幼虫の寄主植物
ancestral node : 祖先ノード
ancestral state : 祖先状態
ancestral state reconstruction : 祖先状態再構成、祖先状態復元、先祖状態再構築
anchor : 固定する、アンカー、アンカー配列
anchor locus : アンカー遺伝子座、固定遺伝子座
ancient asexual : 古代無性の(種)
ancient DNA : 古代 DNA
ANCOVA : 共分散分析
(ANalysis of COVAriance)
andesitic : 安山岩質の
androconial organ : 発香器官
androconial scale : 発香鱗
androconial secretion : 発香器官分泌物
androconial system : 発香器官系
androconium : 香鱗、発香鱗
androgenic gland : 造雄腺
androgenic gland hormone : 造雄腺ホルモン
anellus : 陰茎支持片、陰茎包隔膜(蛾類)〔ラテン語〕
anemotaxis : 走風性
anepimeron : 上後側板
anepisternal suture : 上前側板溝状線
anepisternum : 上前腹板
anesthetize : 麻酔をかける、麻痺(まひ)させる
angiosperm : 被子植物
angiosperm flower : 被子植物の花
angiosperm host-plant : 被子植物の寄主植物、寄主被子植物、宿主被子植物
angular : とがった、角張った、鋭角的な
angular spot : 角張った斑点
anhydrobiosis : 乾燥休眠、アンヒドロビオシス、乾眠
animal : 動物
animal behavioral decision : 動物の行動に関する意思決定
animal droppings : 獣糞
animal genome : 動物のゲノム
animal movement : 動物の移動
animal navigation : 動物のナビゲーション
animal speciation : 動物の種分化
animal trajectory : 動物の移動軌跡
animal's current position : 動物の現在位置
anion : アニオン、陰イオン
anion channel : アニオンチャ(ン)ネル、陰イオンチャ(ン)ネル
anion superoxide : 超酸化物アニオン
anisogamy : 異型配偶
anisotropy : 異方性、不均等性
ankyrin repeat : アンキリンリピート
ANN : 人工ニューラルネットワーク
(Artificial Neural Network)
annealing temperature : 焼きなまし温度、アニーリング温度
annelifer : アネリフェル
annellus : 薄膜の鞘
annonaceous acetogenin : アノナセウスアセトゲニン
annotated gene : 注釈付きの遺伝子
annotation : 注釈、注解
annual change : 周年変化
annual cycle of cool-warm climate : 寒暖気候の周年サイクル
annual cycle of wet-dry climate : 乾湿気候

の周年サイクル
annual fluctuation : 年次変動
annual grass : 一年生の草
annual herbaceous plant : 一年生草本
annual life cycle : 周年経過
annulate : 環のある、環紋をつけた .
annulated : 環化した、環のある
anonymous : 匿名の、名のない
anonymous reviewer : 匿名の査読者
anonymous work : 匿名の著作物
Anopheles gambiae : ガンビエハマダラカ
ANOVA : 分散分析（ANalysis Of VAriance）
ant : アリ、蟻
ant colony optimization : アリコロニー最適化法
ant-enriched habitat : アリが豊富な生息地
antagonism : 拮抗作用
antagonistic pathways, two : 二種の拮抗的な生産経路
antagonistic pleiotropy : 拮抗的多面発現
Antarctic region : 南極区
ante- : 前の -、前に - 〔ラテン語〕
antecostal suture : 前縁溝状線（胸部）
antemedial area : 内側中央部（蛾類）
antemedial line : 内横線（蛾類）
antenna : （複 . -nae）、触角〔ラテン語〕
antenna shape : 触角の形状
antennal : 触角の
antennal flagellomere : 触角の鞭小節
antennal socket : 触角挿入孔
antennal type : 触角形態、触角形状（蛾類）
antennapedia : アンテナペディア遺伝子
anterior : 前、前部の、前側の、前方の〔ラテン語〕
anterior basalare : 前方の基翅節片、前方の基翅甲
anterior branch : 前分枝
anterior cross vein : 前横脈
anterior cubitus vein : 前肘脈、Cu 脈、中脈（ちゅうみゃく）
anterior edge : 前縁、前端
anterior eyespot : 前方の眼状紋
anterior location : 前方の位置
anterior margin : 前縁（部）
anterior tentacle organ : 前部伸縮突起、前方伸縮突起
anterior-posterior axis : 前後軸、AP 軸、前縁 - 後縁軸
antero- : 前部 - 〔ラテン語〕
anterolateral scutal suture : 前外側盾板溝状線
anteroposterior axis : 前後軸、AP 軸、前縁 - 後縁軸
anteroposterior scale ring : 前縁 - 後縁方向に取り巻く鱗粉リング
anteroposteriorly : 前縁 - 後縁方向へ、前後方向に
anthesis : （複 . -ses）、開花
Anthocharidini : ツマキチョウ族
anthophyta : 被子植物門
anthoxanthin : アントキサンチン
anthropic disturbance : 人為的攪乱
anthropogenic climate warming : 人為的な気候温暖化
anthropogenic habitat : 人為的に改変された生息地
anthropogenic habitat change : 人為的な生息地変化
anthropogenic impact : 人為的な影響
anthropogenic pressure : 人為的圧力
anti- : 反 -、不 - 〔ギリシャ語〕
anti-aphrodisiac compound : 抗催淫性化合物
anti-freezing substance : 凍結防止物質
anti-herbivore defense : 抗植食者防御
antiaphrodisiac : 抗催淫剤
antibiosis : 抗生作用
antibiotic : 抗生物質
antibiotic treatment : 抗生物質処理
antibiotic-containing diet : 抗生物質入り

の飼料
antibody : 抗体
anticodon : アンチコドン（コドンに相補的な３個のヌクレオチドの配列）
antifeedant : 摂食阻害物質
antifreeze protein : 不凍タンパク質
antigen : 抗原
antimicrobial peptide cecropin B : 抗微生物ペプチドであるセクロピン B
antimicrobial property : 抗菌性、抗微生物性
antioxidant enzyme : 抗酸化酵素
antisense RNA : アンチセンス RNA
antiviral action : 抗ウイルス作用
antiviral defense : 抗ウイルス防御
antiviral response : 抗ウイルス応答
Antp : アンテナペディア遺伝子（*Antennapedia* gene）
anucleate spermatozoon : 無核精子
any time diapause : 随時的休眠
aorta : 大動脈（だいどうみゃく）
AP axis : 前後軸、AP 軸、前縁 - 後縁軸（Anterior-Posterior axis）
Apaturina erminia : パプアコムラサキ
ape : 類人猿
apex : （複 . apices）、頂部、翅頂部、（翅）端部、翅端、棲、先端（方）の
aphid : アブラムシ
aphrodisiac : 催淫剤
aphrodisiac pheromone : 催淫性フェロモン
apical : 頂室、翅頂室、（翅）端室、頂端の、翅頂部帯、翅端部、先端（方）の〔ラテン語〕
apical cell : 頂室、端室、末端細胞
apical side : 頂端側
apical spot of cell : 中室端斑
apico- : 先端 -〔ラテン語〕
apiculture : ミツバチ飼育、養蜂業
apiculus : ふ先、先端の細い毛
apo- : 離れて -、別れて -〔ギリシャ語〕
Apollo butterfly : アポロウスバシロチョウ
apolysis : アポリシス、クチクラ遊離（「apo-」は「離れる」の意味で、「lysis」は「溶解」の意味）
apomixis : アポミクシス、無融合生殖
apomorph : 後生的進化形質、後天的新形質
apomorphy : 派生形質
apoptosis : アポトーシス、プログラム細胞死
apoptosis-like cell death : アポトーシス様細胞死
Aporia nabellica : ウスグロミヤマシロチョウ
Aporini : ミヤマシロチョウ族
aposematic : 警告する、警戒色の
aposematic caterpillar : 警戒色の幼虫
aposematic coloration : 警告色、警戒色
aposematic insect : 警告性昆虫
aposematism : 警告誇示、警告戦略、警告擬態
apparency : 顕示性（度）
apparent competition : 見かけの競争
apparent decline : 明白な衰亡
apparent fusion of eyespot : 眼状紋の外見上の融合
apparently dead pupa : 外見的には死んだ蛹
apparently fixed mitochondrial DNA introgression : 見かけ固定しているミトコンドリア DNA の遺伝子浸透
apparently viable population : 見たところ存続できる個体群
appearance : 外見、発現、外観
appendage : 付属肢（ふぞくし）、付属器、付属突起
appendix bursa : （複 . -ae）、交尾突起（蛾類）〔ラテン語〕
apple maggot fly : リンゴミバエ
applied entomology : 応用昆虫学
appreciate : 正しく認識する、～の真価（性質、差異）を認める
approach : 接近

approach probability : 接近確率
approx. : 約(approximate)
approximate : 近づける、近似の
approximately : おおよそ
aptera : (複 . -ae)、無翅型
apterous : 無翅の
apyrene sperm : 無核精子
apyrene spermatozoon : 無核精子
AQS : 単為生殖による女王位継承システム(Asexual Queen Succession)
aquatic invertebrate assemblage : 水生無脊椎動物群集
aqueous cell wall : 水溶性の細胞壁
Arabidopsis lineage : シロイヌナズナ属の系統
Arabidopsis thaliana : シロイヌナズナ
arable farming : 耕作農業
aradid bug : ヒラタカメムシの一種
arbitrarily made : 任意に設定された
arbitrary : 任意の
arbitrary combination of letter : 文字の任意組合せ
arbitrary threshold : 任意の閾(いき)値、任意閾
arbor : あずまや、日よけの場所
Archaea : 古細菌
archaeabacteria : 古細菌
Archaeognatha : イシノミ目、古顎目
Archean : 始生代(しせいだい)の
archival preservation : アーカイブ保存、公文書史料保管
Archosaurs : 主竜類(しゅりゅうるい)(絶滅したハ虫類と鳥頸類〈恐竜や翼竜など〉からなる)
arcsine transformed : 逆正弦変換
arcsine transforming : 逆正弦変換
arctic : 北極(圏)の
arctic species : 北極圏種
arctic zone : 北寒帯、北極帯
arctiid moth : ヒトリガ科の蛾
Arcto-Tertiary element : 第三紀北極要素、第三紀周北極要素、第三紀周極要素
area apicalis : 端三角部(たんさんかくぶ)
area of overlap : 重複地域、重複する区域
area-based accumulation curve : 面積に基づく累積曲線
area-wide integrated pest management : 広域的総合的害虫管理、広域総合的害虫管理
areal extent : 空間範囲、範囲
areola : (複 . -lae)、小室〔ラテン語〕
areolate : 小室の
areole : 小室〔英語〕
areolet : 鏡胞(きょうほう)、小翅室(しょうししつ)
ARG : 生殖付属腺(Accessory Reproductive Gland)
ARG reservoir : ARG 貯蔵器、生殖付属腺の貯蔵器(Accessory Reproductive Gland)
ARG secretion : 生殖付属腺分泌物(Accessory Reproductive Gland)
ARG supernatant : ARG 上清、生殖付属腺上清(Accessory Reproductive Gland)
Argentine ant : アルゼンチンアリ
arguably : ほぼ間違いなく、おそらく
arid grassland : 乾燥地帯
arid region : 乾燥地
arid tropical : 乾燥熱帯性
arid zone : 乾燥地帯
arise behaviorally : 行動結果として生じる
aristate antenna : (複 . -nae)、芒形(のぎがた)の触角(針のような突起)
Aristolochiaceae : ウマノスズクサ科
arithmetic progression : 等差数列
arms race : 軍拡競争
army : 群れ、大群
arolium : 爪間盤(そうかんばん)、アロリウム
arrangement : 配列、並び方
arrangement pattern : 配置パターン
arrested development : 発育停止、発育遅

滞、発育遅延
arrested embryogenesis : 阻害された胚形成、胚発達停止
arrival of a new species : 新たな種の出現
arrow-shaped marking : 矢印状の斑紋
arthropod : 節足動物(せっそくどうぶつ)
arthropod transcription start site : 節足動物の転写開始点
arthropoda : 節足動物門
article : 条項、論説、品目
artifact : 創造的所産、人工産物、人為構造
artificial barrier : 人工障壁
artificial chromosome : 人工染色体
artificial diet : 人工飼料
artificial environment : 人工的な環境
artificial food : 人工飼料
artificial intelligence approach : 人工知能的アプローチ
artificial landscape : 人工的な景観、人為的な景観
artificial nectar : 人工果汁
artificial neural network : 人工ニューラルネットワーク(ANN)
artificial selection : 人為選択
artificial short-day : 人工短日
artificial substrate : 人工基質、人工的な培養基
artificial transfer : 人工的な転換
artificially induced species : 人為導入種
arylphorin : アリルフォリン、アリルホリン遺伝子
ascertain : 確かめる、突きとめる
ascidian genome : ホヤのゲノム
aseasonal migration : 非季節的移住
aseasonal summer frost : 季節外れの夏に降りた霜
asexual queen succession : 単為生殖による女王位継承システム
Asia-Pacific area : アジア・太平洋地域
Asian mainland and its adjacent island : アジア大陸とその属島
aside from : ～は別にして、～はさておき
ASL : 海抜(Above Sea Level)
asparagine-to-serine substitution : アスパラギンからセリンへの置換
aspartate aminotransferase : アスパラギン酸アミノ基転移酵素、アスパラギン酸アミノトランスフェラーゼ
aspect ratio : アスペクト比、縦横比
assay plate : アッセイプレート、評価プレート
assemblage : 群集、集団
assemblage-level thinning hypothesis : 群集水準の間伐説、群集レベルの間伐説、群集レベルの間引き説
assembled contig : アセンブリしたコンティグ、アセンブルコンティグ
assembled genome scaffold : アセンブルされたゲノム上の足場
assessment criterion : (複 . -ria)、評価基準
assessment protocol : 評価プロトコール、評価計画案
assimilate : 同化する
assimilation : 同化、同化作用
assimilation rate : 同化率、同化速度
association : 群集
association between color and preference : 色と選好性の関連
association mapping : 連合定位法
association study : 関連解析、関連研究
associative learning : 連合学習
assortative mate preference : 同類交配選好
assortative mating : 同類交配、選択交配
asymmetric : 非対称の
asymmetric gene flow : 非対称遺伝子流動
asymmetric skeleton photoperiod : 非相称枠光周期、非相称スケルトン光周期
asymmetrical color pattern : 左右非対称の色彩パターン

asymmetrical topology : 非対称的樹形、非相称的樹形
asymptote : 漸近線
asymptotic estimator : 漸近的推定量
asymptotic richness : 漸近的種数、漸近種数
asymptotic richness estimator : 漸近種数推定量
asymptotically : 漸近的に
asymptotically equivalent : 漸近的に等価
asynchronous between sexes : 性間の(発生の)非斉一化
asynchronous emergence : 不斉一発生、発生の不斉一性
Atlantic Forest : 大西洋岸森林(南米)
atlas moth : ヨナクニサン
atmosphere : 大気、空気、雰囲気
atmospheric phenomenon : 気象現象
atomic coordinate : 原子座標
ATP : アデノシン三リン酸(Adenosine TriPhosphate)
ATP synthase : ATP 合成酵素、アデノシン三リン酸合成酵素(Adenosine TriPhosphate)
attach distally : 末端側で結合する、末梢側で結合する
attack : 攻撃(する)
attacker : 捕食者
attain : 達成する
attenuate : 減衰させる、減衰する、弱める
attenuated feminization : 減衰された雌化
attenuated feminizing activity : 減衰された雌化行動
attenuation : 減衰、抑制
attract : 誘引する
attractant : 誘引物質
attraction of carnivorous natural enemy : 肉食性天敵の誘引
attraction of herbivore enemy : 植食者天敵の誘引
attraction of natural enemy of herbivore : 植食者の天敵の誘引
attraction of pollinator : 花粉媒介者の誘引、送粉者の誘引
attributable : 帰すことができる、基因する
atypical : 異常な、定型的でない
auctorum : 著者たちの〔ラテン語〕
audible sound : 聞き取れる音
audiogram : オージオグラム、聴力図
audiogram experiment : 聴力図実験、オージオグラム実験
audiogram procedure : 聴力検査手順、聴力図作成手順
auditory : 聴覚(の)
auditory nerve : 聴覚神経
auditory response : 聴覚応答、聴覚反応、聴性反応
auditory sense : 聴覚
auditory threshold : 聴覚閾(いき)値、聴覚閾、最小可聴値
augment : 増大(する)
augmentation : 放飼増強法
augmented ultraviolet color vision : 増強された紫外線光に対する色覚
auricular : 耳の、聴覚の
Australian region : オーストラリア区
autapomorphy : 固有派生形質
autecology : 個生態学
author : 命名者、著者
authorisation procedure : 承認手続き
auto- : 自己 -、同種 -、自動 -〔ギリシャ語〕
autoapomorphy : 固有派生形質
autoclaved : 高圧滅菌された
automated purification instrument : 自動精製装置
Automatic Barcode Gap Discovery : 自動バーコードギャップ発見、バーコードギャップの自動発見(バーコード間のギャップを計算することで配列を自

動的に振り分ける方法)、ABGD
automimicry : 種内擬態、自己擬態
automixis : オートミクシス、自家生殖
autonomous driving robot : 自動運転ロボット
autonomous oscillator : 自律振動体
autonomous process : 自律過程
autopolyploid : 同質倍数体
autosomal gene : 常染色体遺伝子
autosomal inversion : 常染色体逆位
autosomal locus : 常染色体の遺伝子座
autotomy : 自切(じせつ)
autotomy position : 自切位置(じせついち)
autotrophy : 独立栄養、自家栄養
autumn form : 秋型
autumn morph adult : 秋型成虫
autumn rain : 秋雨
availability : 適格性、利用価値
availability of habitat : 生息地の利用可能性
available name : 適格名、適用名
available nomenclatural act : 適格な命名法的行為
available work : 適格な著作物
average : 平均、平均値
average density : 平均密度
average length : 平均長(塩基の)
average number of allele per locus : 遺伝子座あたりの平均対立遺伝子数
average sound pressure level : 平均音圧レベル
avian predator : 鳥捕食者
avid nectar feeder : 熱烈な吸蜜者、蜜の愛飲者
avoidance : 忌避作用、忌避性
avoidance behavior : 忌避行動
AW-IPM : 広域的総合的害虫管理、広域総合的害虫管理(Area-Wide Integrated Pest Management)
await : 待ち構える
awakening : 目覚め、覚醒
axial : 軸側(じくそく)
axiality : 軸性
axilla : (複 . -ae)、三角節片、三角板〔ラテン語〕
axillar : 三角板の
axillary sclerite : 腋節片
axis : (複 . axes)、軸、中心線〔ラテン語〕
axon : 軸索
axon terminal : 軸索末端

b

BA : 基底面積、胸高断面積(Basal Area)
BAC : 細菌人工染色体、バクテリア人工染色体(Bacterial Artificial Chromosome)
BAC clone identification : BACクローン同定
Bacillus thuringiensis : バチルスチューリンゲンシス(昆虫病原細菌の一種)
back : 背、背面
back and side : 背部と体側
back mutation : 復帰突然変異
back out of eye : 目の後ろあたりで
back-transformed : 逆変換(対数を取った値を元の値に戻すような場合は、「逆変換」と呼ばれる)
backcross : 戻し交配、戻し交雑
backcross brood : 戻し交配の同胞
backcross family : 戻し交配の家系
backcross male preference : 戻し交雑の結果の雄の選好性
backfill : バックフィル
background activity : 背景活動
background extinction : 背景絶滅、バックグラウンド的絶滅
background noise : 背景雑音、暗騒音
backward model selection : 後退的モデル選択法
backward trajectory analysis : 後退流跡線解析
bacmid : バクミド(baculovirus と plasmid をくっつけた造語とされる)

bacmid DNA : バクミド DNA
bacmid-derived virus : バクミド由来ウイルス
bacterial artificial chromosome : バクテリア人工染色体(BAC)
bacterial cell : バクテリア細胞、細菌細胞
bacterial endosymbiont Wolbachia : 細胞内共生細菌ボルバキア
bacterial genome : 細菌のゲノム
bacterial infection : 細菌感染
bacterium : (複 . -ia)、細菌、バクテリア
baculovirus : バキュロウイルス
baculovirus infection : バキュロウイルス感染
baculovirus infectivity : バキュロウイルスの感染性
baculovirus motility : バキュロウイルスの運動性
baculovirus multiplication : バキュロウイルス増殖
bagworm : ミノムシ
balance of nature : 自然のバランス
balancing selection : 平衡選択
ball-like cluster of hair : ボールのようなかたちをした毛束
bamboo : 竹
bambusoid plant : 竹類植物
band : 帯、模様、縞状バンド
band region of adult wing : 成虫翅の帯領域、成虫翅の帯部分
band string : 帯糸(たいし)
banded orange butterfly : オビモンドクチョウ
banding pattern : バンドパターン
bandwidth : 帯域幅
banker plant system : バンカー法
barcode analysis : バーコード解析
Barcode Index Number system : バーコード索引番号システム(種名とは別にバーコード塩基配列から認識されたクラスターに一意な番号をつけ、暫定的な分類情報として用いる方法)
bare ground : 草木のない土地、裸地
barely significant reduction : まれに有意となる減少
barrier resistance : バリアー抵抗
basad : 基部側へ
basal : 基部帯、基部(方)の
basal area : 基部(蛾類)
basal band : 基部の縞状バンド
basal cell : 基底細胞
basal dash : 基縦線、(剣状紋〈ensiform〉)(蛾類)
basal group : 基底グループ、基部群
basal line : 基線、(亜基線)(蛾類)
basal node : 基底ノード、基部の結節
basal ring : 基環
basal species : 原始的な種、基部種
basalar cleft : 基翅節片溝
basalar muscle : 基翅節片筋
basalare : 基翅甲、前翅基骨(胸部)
base : 基角、塩基、基部
base change : 塩基置換
base frequency : 塩基頻度、塩基配列の出現頻度
base of forewing : 前翅基部
base of tympanal chamber : 鼓膜室の基部
base pair : 塩基対
base sequence : 塩基配列
base substitution : 塩基置換
base substitution number : 塩基置換数
base thigh section : 腿節基部(たいせつきぶ)
base- : 基部 - 〔ラテン語〕
based on the fact : 事実に基づいている
basement membrane : 基底膜
basic information : 研究基礎資料、基礎情報
Basic Local Alignment Search Tool : BLAST 検索
basic plan : 基本プラン
basic skeleton : 基本骨格
basin : 盆地、流域、入り江

basis of the shift : 移動の契機
basisternal-parepisternal suture : 基腹板 - 側前側板溝状線
basisternum : 基腹板（胸部）
basitarsus : 基ふ節
basking : 日光浴
bat detection : コウモリ探知
bat evasion : コウモリからの逃避
batch of egg : 卵塊
Bateman's principle : ベイトマンの原理
Batesian mimicry : ベイツ型擬態
battle : 闘争
Bayes posterior probability : ベイズの事後確率
Bayesian algorithm : ベイズのアルゴリズム、ベイズの問題解決手法
Bayesian inference tree : ベイズ推定法による系統樹、BI 法による系統樹
Bayesian information criterion : ベイズ情報量基準（BIC）
Bayesian method : ベイズ法
Bayesian phylogenetic analysis : ベイズ系統解析、ベイズ法による系統解析
BC : 戻し交配（BackCross）、シアン化ベンジル（Benzyl Cyanide）
BCAA : 分岐鎖アミノ酸（Branched Chain Amino Acid）
bcd gene : ビコイド（*bicoid*）遺伝子
beaded : 数珠玉状の
beak mark : ビークマーク、鳥の嘴型、咬み跡
bean weevil : マメゾウムシ
bearing strain : 保有（菌）株
beat hypothesis : ビート仮説
Beaufort Wind Scale : ビューフォート風力階級
bee : 蜜蜂
beech band : ブナ帯
beet armyworm : シロイチモジヨトウ
beetle horn : 甲虫の角（つの）、カブトムシの角
beg : 避ける、回避する、請う
beginning of light period : 明期開始
behavior : 行動、活動
behavior level : 行動レベル
behavior of female refusing male : 雌の交尾拒否行動
behavioral : 行動によるもの、行動の
behavioral and neurophysiological response : 行動的・神経生理的応答
behavioral approach : 行動学的アプローチ
behavioral change : 行動変化
behavioral characteristic : 行動特性
behavioral ecology : 行動生態学
behavioral incompatibility : 行動的不一致
behavioral isolation : 行動的隔離
behavioral observation : 行動学的観察、行動観察
behavioral protandry : 行動学的プロタンドリー、行動面での雄性先熟
behavioral response : 行動的応答、行動的反応
behavioral syndrome : 行動シンドローム
behavioral-based model : 行動ベースモデル
bellows : 蛇腹
below surface : 裏翅、裏面
belt shaped area : 帯状地域
Ben gene : *Ben* 遺伝子、ベン遺伝子、ベンドレス遺伝子（*Bendless*）
BEN protein : BEN タンパク質、ベンタンパク質、ベンドレスタンパク質
benchmark dataset : ベンチマークデータセット、標準データセット
benchmark seedbank dataset : 基準種子銀行データセット、基準シードバンクデータセット
beneath : 真下に
beneficial effect : 有益な効果
beneficial insect : 益虫
beneficial trait : 有益な形質

benefit : 利益、収益
benign introduction : 保全的導入
benthic invertebrate assemblage : 河口底生無脊椎動物群集
benthic marine mass collection : 海洋底生の大量採取品
benzyl : ベンジル
benzyl cyanide : シアン化ベンジル（BC）
berry : 木の実、果実
best linear unbiased prediction : 最良線形不偏予測量（BLUP）
best threshold : 最適な閾(いき)値、最適閾値
best-fit partitioning scheme : 最良適合分割法、最適な分割法
best-studied class : 最もよく研究された分野
bet-hedging : 危険分散、二股賭け戦略、両掛け戦略
bet-hedging adaptation : 危険分散適応、両掛け適応、両掛け戦略適応
bet-hedging effect : 危険分散の効果
bet-hedging strategy : 危険分散戦略、両掛け戦略
beta globin : β- グロビン
beta-alanine (β-alanine) : β- アラニン、ベータアラニン
beta-carotene (β-carotene) : β- カロチン、ベータカロチン
beta-diversity : ベータ多様性、β- 多様性（「生息地間多様性」のことで、「別々の環境間での種多様性の違い」の意味）
beta-glucosidase (β-glucosidase) : β- グルコシダーゼ
between and within years : 年間および年内
between-morph courtship : 異色の翅間での求愛
between-sex genetic correlation : 性間の遺伝的相関
BI analysis : BI 法の解析（Bayesian Inference）
BI tree : BI 法による系統樹、ベイズ推定法による系統樹
bi- : 2-、二 -、双 -、二倍 -〔ラテン語〕
bias : バイアス、先入観、偏り
bibliographic reference : 参考文献
BIC : ベイズ情報量基準（Bayesian Information Criterion）
bicuspid apex of valva : 交尾弁二尖、交尾弁縁棘、交尾弁縁刺
bidirectional SSH : 双方向 SSH 法（Suppression Subtractive Hybridization）
bidirectional subtracted library : 双方向サブトラクションライブラリー
biennial : 二年ごとの、二年生の、越年生の
bifurcating : 二分岐
bilin : ビリン
bilin-binding protein : ビリン結合タンパク質
biliverdin IX : ビリベルジン IX
biliverdin type : ビリベルジンタイプ
billion : 十億の
bimodal : 双峰
bimodal activity periods : 双峰性の活動期（本来の「bimodal」は統計学用語で、モードが２つあることを意味する用語であり、頻度分布曲線が２つの山をもつことを表す）
bimodally : 二峰的に、二方式に
BIN : バーコード索引番号システム（Barcode Index Number）
binary variable : 二値変数
binned raw datum : （複 . -ta）、値域ごとにまとめた生データ
binocular : 双眼鏡
binocular microscope : 双眼顕微鏡
binomen : （複 . -mina）、二語名〔ラテン語〕
binomial datum : （複 . -ta）、二項データ
binomial error distribution : 二項誤差分布
binomial nomenclature : 二名法、二名式命名法、二名式用語体系

binomial regression model : 二項回帰モデル、二項分布回帰モデル
binomial response : 二項反応
binominal name : 二語名
binominal nomenclature : 二名法
bio- : 生物 -〔ギリシャ語〕
bioassay : 生物検定、バイオアッセイ
bioateral symmetry : 左右相称
biochemical evolution : 生化学的進化
biochemical level : 生化学レベル
biochemistry : 生化学
biocoenosis : 群集
biodiversity : 生物多様性、生物学的多様性
biodiversity assessment : 生物多様性評価、生物の多様性評価
biodiversity conservation : 生物多様性の保全
biodiversity study : 生物多様性研究
bioeconomics : 生物経済学
biogeographic study : 生物地理学的研究
biogeographical region : 生物地理区
biogeography : 生物地理学
bioindicator : 生物指標
bioinformatic analysis : バイオインフォマティクス解析、生物情報解析
bioinformatic study : バイオインフォマティクス研究、生物情報研究解析
biological characteristic : 生物学的特徴
biological clock : 生物時計、体内時計
biological control : 生物(学)的防除
biological difference : 生物学的差異
biological diversity : 生物多様性
biological factor : 生物学的要因、生物学的因子
biological infrared imaging : 赤外光による生体イメージング
biological infrared sensing : 赤外光による生体センシング
biological invasion : 生物学的侵入
biological isolation : 生物的隔離
biological reason : 生物学的理由
biological replicate : 生物学的反復実験、生物学的反復(実験サンプルとなる生物の個体間差などに起因する実験誤差を小さくするために行う反復実験)
biological species : 生物種、生物学的種
biological species concept : 生物学的種概念
biological specimen : 生物試料、生物標本
biologically : 生物学的に
biologically important expansion : 生物学的に重要な拡大
biology : 生物学
bioluminescent jellyfish : 発光クラゲ
biomass : 生物(体)量、バイオマス
biomass density : 植物量の密度
biome : 生物群系、バイオーム
biomimetics : バイオミメティク、バイオミメティックス、生物情報科学
biordinal crochet : 二様長短交互型の鉤爪、二様長短交互型の鈎爪
biosphere : 生物圏
biosynthesis : 生合成
biosynthetic pathway : 生合成経路
biosystematics : 生物系統学、生物分類学
biota : 生物相
biotic factor : 生物的要因
biotic interaction : 生物間相互作用
biotic interrelationship : 生物的相互関係、生物間相互関係
biotic inventory : 生物目録、生物資源目録
biotic stress : 生物的ストレス
biotic survey : 生物調査、生物資源調査
biotope : 小生活圏、ビオトープ
biotype : バイオタイプ、生物型、遺伝因子型
bipartition tree : 二分割系統樹
bipectinate : 両櫛歯状の(蛾類の触角)
biped : 二足動物、二足を有する
bird : 鳥

bird droppings : 鳥の糞
bird vocalization : 鳥のさえずり、鳥の鳴き声
birdcall : 鳥の鳴き声
birdwing butterfly : トリバネアゲハ
birth : 誕生
birth and death dynamics : 生死動態、新生・消失の動態
biserial crochet : 双列型の鉤爪
biserrate : 両鋸歯状(りょうきょしじょう)の(蛾類の触角)、重鋸歯状(じゅうきょしじょう)の(葉縁にそれぞれの鋸歯に小さな鋸歯がある)
bistability phenomenon : 双安定現象、二相安定現象
bithorax : 双胸遺伝子
biting : (虫が)咬(か)む
biting mouthparts : 咬み型口器(かみがたこうき)
bivariate scatter plot : 2変量散布図
bivoltine : 二化性の、二化性
bivoltine population : 二化性集団
bivoltine zone : 二化性地域、二化地帯
bizarre : 奇異な、奇怪な、風変わりな
black aberrant form : 黒化異常型
black and buff eyespot : 暗黄褐色の眼状紋
black annulus : 黒環
black dorsal hindwing : 黒色の背側後翅
black dot : 黒丸、黒点
black flanking orange : 橙色で囲まれた黒色
black list : ブラックリスト
black marking pen : 黒色のマーキングペン
black mustard : クロガラシ
black patch : 暗色斑
black pigment : 黒色色素
black pupation board : 黒色の蛹化台紙
black ring : 黒環
black scale : 黒色鱗粉
black streak : 黒条
black stripe on outer edge : 外縁黒帯
black stripe size : 黒帯幅
black swallowtail butterfly : 黒色系アゲハ
black-ringed : 黒く縁どられた
black-vein : 黒条
blackish brown : 黒味をおびた褐色
blackish upperside : 黒っぽい翅表
bladder : 香のう
BLAST comparison : BLAST 比較(Basic Local Alignment Search Tool)
BLAST search : BLAST 検索
blastoderm : 胚盤葉、胞胚葉
Blattodea : ゴキブリ目、網翅目
blemish : 傷、傷つける
BLL gene : *BLL* 遺伝子、ブラコウイルス様レクチン遺伝子(*Bracovirus-Like Lectin* gene)
block : 阻止する
blood : 体液、血液
blooming season : 開花季節、開花期
blow fly : クロバエ
blue bilin accumulation : 青色ビリンの蓄積
blue bilin pigment : 青色ビリン色素
blue bilin production : 青色ビリン色素の生成
blue jay : アオカケス
blue light : 青色光
blue metalmark butterfly : シジミタテハ
blue morpho butterfly : メネラウスモルフォ
blue pigment : 青色色素
blue poppy : 青いケシ
blue spectral shift : 青色のスペクトル領域へのシフト、青色スペクトル範囲へのシフト、青色スペクトルシフト
blue-banded female model : 青色翅の雌モデル
blue-green coloration : 青緑色化
blue-shift : 青色側へのシフト、青方偏移、ブルーシフト
blue-shifted lineage : 青色シフト系統
blue-type larva : 青色型の幼虫

Blues : ヒメシジミ亜科、シジミチョウ科
blunt : 鈍い、先のとがっていない
BM : 大英博物館(British Museum)
BmN4 : カイコ(*Bombyx mori*)の卵巣由来の樹立培養細胞株
BMNH : ロンドン自然史博物館(British Museum (Natural History))、英国自然史博物館
body : 胴、体、胴体
body color : 体色
body color change : 体色変化
body component : 体の器官
body fluid : 体液
body length : 体長
body mass : 体重
body morphology : 体部の形態、体の形態
body of female : 雌の体
body of lepidopteran host : 鱗翅類宿主の体
body plan development : ボディプランの発育
body resource : 体部の資源
body size : 体の大きさ、体サイズ
body size indicator : 体の大きさの指標、体サイズ指標
body structure : 体構造
body surface area : 体表面積
body temperature : 体温
bog : 湿原、沼地
bog butterfly : 湿原性蝶
boiling point : 沸点
bold : はっきりした
BOLD [hg.] : バーコード・オブ・ライフ・データ(生物種同定システム)(Barcode of Life Data)
bold line : 太線
Bombykol : ボンビコール(カイコの雌成虫が出す性フェロモン)
Bombyx mandarina : クワコ、クワゴ
Bombyx mori : カイコ
Bombyx sequence : カイコ類の塩基配列
bombyxin : ボンビキシン
bona fide scientific purpose : 本物の科学的目的
bonanza : 幸運、大当たり
bone : 骨
Bonferroni adjustment : ボンフェローニの調整
Bonferroni correction : ボンフェローニ(の)補正
book review : 書評
boom and bust cycle : 好景気 - 不景気サイクル
boost : 増加する
bootstrap : ブートストラップ
bootstrap probability : ブートストラップ確率
bootstrap replicate : ブートストラップ複製
bootstrap support : ブートストラップサポート
bootstrap value : ブートストラップ値(BV)
bootstrapping : ブートストラップ法
borax solution : ホウ砂溶液
border : 外縁、縁どり、国境、境界、接する、と境界を成す
border control : 国境検査
border ocellus : 辺縁部の単眼
boreal species : 北方系種、寒帯種
boreo-alpine taxa : 北方高山性の分類群、寒帯高山性の分類群
Borneo : ボルネオ
both flanking marker : 両側の隣接マーカー
both sexes : 雌雄両方、両性
both vasa deferentia : 両輪精管(りょうゆせいかん)
bottleneck : ボトルネック、瓶首
bottleneck effort : ボトルネック効果、瓶首効果
bottom-up effect : ボトムアップ効果
bottomland hardwood forest : 低地広葉樹林
bottomland-butterfly : 低地性の蝶
bottomland-flower : 低地性の花
boundary : 境界、境界線

boundary layer : 境界層
boundary of territory : テリトリーの境界
boundary region : 境界領域
bourgeois strategy : ブルジョワ戦略
bovine rhodopsin : ウシロドプシン
bovine template : ウシテンプレート
bowl shaped : 椀状の、椀型の
bp : 塩基対(単位)(base pair)
bp long : 塩基長
brachypter : 短翅型
brachypterism : 短翅
brachypterous form : 短翅型
braconid family : コマユバチファミリー、コマユバチ科
Braconidae : コマユバチ科
bracovirus gene : ブラコウイルス遺伝子(BV)
bracovirus insertion : ブラコウイルスの挿入、ブラコウイルスの注入
bracovirus life cycle : ブラコウイルスの生活環
bracovirus particle entry : ブラコウイルス粒子の侵入
bracovirus protein : ブラコウイルスタンパク質
bracovirus sequence : ブラコウイルス遺伝子配列
bracovirus virulence protein : ブラコウイルス毒性タンパク質
bracovirus-associated wasp : ブラコウイルス関連の蜂
bracovirus-lectin like protein : ブラコウイルス - レクチン様タンパク質
bracovirus-like *Ben* sequence : ブラコウイルス様 *Ben*(遺伝子)配列
bracovirus-like lectin gene : ブラコウイルス様レクチン遺伝子、*BLL* 遺伝子
brain : 脳〔ラテン語〕
brain hormone : 脳ホルモン
brain, prothoracic gland, corpus allatum, corpus cardiacum complex : 脳 - 前胸腺 - アラタ体 - 側心体の複合体
brainless animal : 除脳動物
brainless diapausing-pupa : 除脳休眠蛹
brainless pupa : 除脳蛹
branch : 枝
branch length : 枝長
branch-site model of selection : 自然選択の分岐点モデル
branch-site test of selection : 選択の分岐点テスト
branched chain amino acid : 分岐鎖アミノ酸(BCAA)
branched spine : 枝分れしたトゲ
branching process : 分枝過程
brand : 性標、性斑
Brassica oleracea : キャベツ
Brassicaceae : アブラナ科
brassicaceous plant : アブラナ科植物(例えばキャベツ)
Brassicales : アブラナ目
Brassicales-feeding Pieridae butterfly : アブラナ目を摂食するモンシロチョウ科の蝶
breakthrough finding : 画期的発見
breathing pore : 気門、気孔
breed : 産む、繁殖する
breeding area : 繁殖領域、繁殖する地域
breeding ground : 繁殖場、飼育場
breeding habitat : 繁殖地、繁殖場所
breeding season : 繁殖期
breeding value : 育種価
bricolage : ブリコラージュ、あり合わせ製作
brief flight : 短時間の飛翔
brief light pulse : 短時間の光パルス
bright wing pattern : 鮮やかな翅の色彩
bright yellowish green pupa : 明黄緑色の蛹
brightness : 明度
brightness processing : 明度処理、輝度処理
brightness-contrast vision : 明暗対比型視覚
bring about : 引き起こす、もたらす

brink of extinction : 絶滅の瀬戸際
bristle : 剛毛
British Museum (Natural History) : ロンドン自然史博物館(BMNH)、英国自然史博物館
British Museum : 大英博物館(BM)
broad sense, in a : 広義で、広い意味で
broad-scale population distribution : 大規模な個体群分布
broadcast : 放送、ブロードキャスト
broadleaved woodland : 広葉樹林地帯
broadly speaking : 大ざっぱに言えば、概して
broken band : 不連続な帯
broken line : 不連続な帯、破線
brood : 同腹の仔、ブルード、同腹、同腹仔
brood comparison : 同腹仔の比較、ブルード比較
brood-dependent variation : 同腹仔依存の変動
brother-sister mating : 兄妹交尾、同胞交配
brown butterfly : ジャノメチョウ
brown leaf litter : 褐色の腐葉(層)
brown marking : 褐色斑紋
brown pigment : 褐色色素
brown planthopper : トビイロウンカ
brownish : 褐色がかった
brownish necrotic tissue : 褐色がかった壊死組織
Browns : ジャノメチョウ科
browsing : 若葉を食べる
bruchid beetle : ヨツモンマメゾウムシ
Bruchidae : マメゾウムシ科
brush : 雑木林、低木林、やぶ
brush-footed : 刷毛足の
Brush-footed Butterflies : タテハチョウ科
Brussels sprout : 芽キャベツ
BSC : 生物学的種概念(Biological Species Concept)
buckeye butterfly : アメリカタテハモドキ
budded virus : 発芽型ウイルス、出芽型ウイルス、出芽ウイルス(BV)
buddleia : フジウツギ、ブッドレア
buffer strip : 緩衝帯
built-up area : 建て込んだ地区、市街地
bulla : (複 . -ae)、胞器、水胞(水疱)状の構造物〔ラテン語〕
Bulletin of Zoological Nomenclature : 動物命名法紀要
bumblebee : マルハナバチ
bump : こぶ、瘤
burned longleaf pine : 野焼きされたダイオウショウ
burned stand : 野焼きされた林分
burned treatment : 野焼き処理
burnin : バーンイン(ベイズ法関連用語)
bursa copulatric : 交尾のう
bursa copulatrix : 交尾のう
bursicon : ブルシコン
burst of diversification : 多様化爆発
butter-colored fly : バター色のハエ・アブ
butterflies : 蝶類
butterfly : 蝶
butterfly activity : 蝶の活動
butterfly appearance : 蝶の出現
butterfly atlas : 蝶の地図帳
butterfly bush : 蝶の木(フジウツギ科ブッドレアの別称)
butterfly collecting report : 蝶採集記
butterfly community stability : 蝶群集の安定性
butterfly detoxification mechanism : 蝶の解毒機構
butterfly distribution : 蝶の分布
butterfly diversity hotspot : 蝶類多様性のホットスポット
butterfly eclosion : 蝶の羽化
butterfly fauna : 蝶相、チョウ相

butterfly fossil : 蝶化石
butterfly garden : バタフライガーデン、蝶園
butterfly gardening : 蝶の来る庭作り、バタフライガーデニング
butterfly genome : 蝶のゲノム
butterfly lineages colonized plant : 蝶系統群が定着した植物
butterfly maintenance : 蝶の維持
butterfly morphology : 蝶の形態(学)
butterfly of the forest : 森の蝶
Butterfly of the South East Asian Islands : 東南アジア島嶼の蝶
butterfly pigment : 蝶の色素
butterfly plant arms-race : 蝶と植物の軍拡競争
butterfly population : 蝶の個体群
butterfly sampling : 蝶採集、蝶採取
Butterfly Science Society of Japan, The : 日本蝶類科学学会
Butterfly Society of Japan, The : 日本蝶類学会
butterfly status : 蝶の状態
butterfly tour : 蝶の旅
butterfly transect : 蝶の観察路、蝶の観察地
butterfly-co-rotating : 蝶の共回転
butterfly-flower interaction : 蝶と花との相互作用
BV : 発芽型ウイルス、出芽型ウイルス、出芽ウイルス(Budded Virus)、ブートストラップ値(Bootstrap Value)
BV gene : BV 遺伝子、ブラコウイルス遺伝子(BracoVirus)
bx : 双胸遺伝子(*bithorax* gene)
by-product : 副産物
by-product of ecological selection : 生態(学的)選択の副産物
by-product of ecological shift : 生態転換の副産物
by-product of sexual selection : 性選択の副産物
bya : 十億年前(billion years ago)
Byasa alcinous : ジャコウアゲハ
Byasa hedistus : ヘディストスジャコウアゲハ
bylaw : 条例、内規

C

C : 前縁脈(Coastal vein)
C value paradox : C 値パラドックス
C-lectin domain : C 型レクチンドメイン
C-terminal part : C 末端部
C-type lectin gene : C 型レクチン遺伝子
CA : アラタ体(昆虫の頭部にある内分泌器官)(Corpus Allatum)(複 . -ta)〔ラテン語〕
ca. 60% : 約 60%(「*circa*」の省略形で、「キルカー」と読む)〔ラテン語〕
cabbage : キャベツ
cabbage crop variety : キャベツの作物変異(多様性)
cabbage white butterfly : モンシロチョウ
cactophilic *Drosophila* : 好サボテン性ショウジョウバエ、サボテン好性ショウジョウバエ
caeliferin : エリシターの一種(トウモロコシに揮発成分を放出させるエリシター)
Caenorhabditis elegans : エレガンス線虫、線虫
Caesalpinia sp. : ジャケツイバラ属の一種
caffeine-sensitive neuron : カフェイン感受性ニューロン、カフェイン感受性神経細胞
cage : ケージ、籠(かご)
cage complex : 籠複合体、飼育ケージ複合体
Cairns birdwing butterfly : メガネトリバネアゲハ(旧称)、プリアムストリバネアゲハ(ケアンズはオーストラリア北東部の地名で、生息地の一つ)
calcareous grassland : 石灰土壌の草地
calcareous soil : 石灰質土壌
calcium concentration : カルシウム濃度

calcium imaging assay : カルシウムイメージング(検定)法
calcium-dependent luminescent protein : カルシウム依存的発光タンパク質
calcium-dependent protein kinase : カルシウム依存性タンパク質キナーゼ
calculated migration : 目標移動
calibrate : 調節する、調整する
California dog-face butterfly : カルフォルニアイヌモンキチョウ
calling sound of bird : 鳥の呼ぶ音
calm weather : 穏やかな気象、風がない天候
Camberwell beauty butterfly : キベリタテハ
Cambrian explosion : カンブリア爆発
camouflage : 擬態、隠蔽、カムフラージュ
camouflaged ventral side of wing : 擬態した翅の腹側
campaniform sensillum : 鐘状感覚子
canalize : 向ける
candidate gene : 候補遺伝子
candidate gustatory receptor gene : 味覚受容体候補遺伝子、味覚受容体遺伝子候補
candidate marker : 候補マーカー、マーカー候補
candidate normalization gene : 候補基準化遺伝子
candidate spectral tuning site : スペクトル調整部位候補
cannibalism : 共食い(ともぐい)、カニバリズム、種内捕食
canopied tree : 天蓋に覆われている木
canopy : 林冠、樹冠、天蓋
canopy closure : 林冠閉鎖(樹冠どうしが隙間なく埋まったような状態にあること)
canopy hardwood : 上層の広葉樹林、林冠層の広葉樹林
canopy pine : 上層のマツ、林冠層のマツ
canopy tree : 上層の木
cap : 帽鞘(ぼうしょう)、キャップ
capillary : キャピラリー、毛細管、毛管
capital breeder : キャピタル型ブリーダー(産卵におけるエネルギー投資戦略の一つで、産卵前の蓄積エネルギーに依存する繁殖パターン；エネルギーの獲得・配分パターンの一つで、繁殖期間中に餌を食べないため繁殖行動に必要なエネルギーは事前に蓄える繁殖パターン)
capitate antenna : (複 . -nae)、球桿状(きゅうかんじょう)の触角(触角先端部が丸く膨らむ)(蛾類)
capitulum : 蓋帽(がいぼう)
Capparaceae : フウチョウソウ科
capping : キャップ形成
capsule : 朔(さく)【植物学】、卵殻(らんかく)、朔殻(さくかく)
captivating hobby : 魅力たっぷりな趣味
captive breeding : 飼育環境下での繁殖、人工繁殖
captive condition : 飼育条件、捕獲下条件
captive-bred rare species : 飼育環境で繁殖された希少種
captive-bred stock : 飼育環境で繁殖されたストック
captivity, in : 飼育されている
capture for identification : 同定ための捕獲
capture-by-circularization : 環状化捕獲法
capture-by-hybridization : ハイブリダイゼーション捕獲法
car-based recorder : 自動車を利用した記録者
carbohydrate : 炭水化物、糖質
carbohydrate metabolism : 炭水化物代謝
carbohydrate-binding protein : 炭水化物結合タンパク質
carboxylesterase : カルボキシルエステラーゼ
cardenolide : カルデノリド、強心配糖体
cardenolide concentration : カルデノリド濃縮
cardenolide content : カルデノリド含有

cardenolide poison : カルデノリド毒
cardiac glycoside : 強心配糖体
cardiac glycoside insensitivity : 強心配糖体に対する非感受性
cardiac valve : 噴門弁
Cardinium : カルディニウム
cardo : (複 . -dines)、軸節〔ラテン語〕
caribou : カリブー(トナカイの一種)
carina : (複 . -nae)、稜線、隆起線
carinate : 隆起線
caring for young : 育仔(いくし)
carnivore : 肉食(性)哺乳類、肉食性動物
carnivorous : 肉食性の
carotenoid : カロチノイド
carrion : 腐肉、死肉
carrying capacity : 環境収容力、牧養力
case : 案件、格、事例
case study : 事例研究、ケーススタディ
Cassia sp. : カワラケツメイ属の一種
caste : カースト(階級制度)
casual release : 思いつきの放蝶
catabolism : 異化
catalyze : 引き起こす、接触作用を及ぼす
catching : 採集
categorical fixed factor : 質的固定要因、質的確定要因
category name : 分類名、分類階級名
category--subcategory ratio : カテゴリ - サブカテゴリの比、分類項目 - 細分類項目の比
category--subcategory taxonomic ratio : カテゴリ - サブカテゴリの分類数比
catenate flight : 連結飛行、連結飛翔
caterpillar : 幼虫(蝶/蛾の幼虫)、イモムシ、ケムシ
caterpillar host immune defense : 幼虫宿主の免疫防御
caterpillar host immune response : 幼虫宿主の免疫反応(応答)
caterpillar stage : 幼虫期
cation : カチオン、陽イオン
caudal : 後端にある、尾(状)の
caudal gill : 尾鰓
causal factor : 成立要因
causal link : 因果関係
causative agent : 原因病原体、起因子
cautery : 焼灼法(しょうしゃくほう)
caution, with : 用心深く、慎重に
caveat : 但し書き、断り書き
cavity area among cell : 細胞間腔領域
CBD : 生物多様性条約(Convention on Biological Diversity)
CCR5 : ケモカイン受容体5(Chemokine Receptor 5)
CCYV : ウリ類退緑黄化ウイルス、ウリ類退緑黄化病(Cucurbit Clorotic Yellows Virus)
cDNA : 相補的DNA(complementary DNA)
CDS : コード配列(CoDing Sequence)
CDS coding : CDS コーディング(CoDing Sequence ; コーディング領域で、アミノ酸に翻訳される領域)
cease : 中止する、停止する
cecropia moth : セクロピアサン
cecropin : セクロピン
celestial compass : 天体コンパス
celestial navigation : 天体航法
cell : (翅)室、中室、細胞
cell autonomous : 細胞自律的
cell bar : 中室の棒状紋
cell body : 細胞体
cell count : 細胞計数
cell culture : 細胞培養
cell culture medium : 細胞培養培地
cell end : 中室端斑
cell entry mechanism : 細胞侵入機構
cell free : 無細胞、セルフリー
cell homogenate : 細胞磨砕物

cell image : 細胞像
cell interaction : 細胞間相互作用、細胞相互作用
cell line : 細胞株、細胞系統
cell periphery : 細胞周辺(部)
cell proliferation : 細胞増殖
cell wall : 細胞壁
cell-autonomous manner : 細胞自律的な方法
cell-to-cell spread of infection : 細胞間の感染拡大
cellular actin : 細胞アクチン、細胞内アクチン
cellular arrangement : 細胞の配列
cellular compartment : 細胞区画
cellular component : 細胞成分
cellular cytoplasm : 細胞質
cellular cytoskeleton dynamics : 細胞骨格の細胞内動態
cellular localization : 細胞(内)局在性
cellular machinery : 細胞機構、細胞内装置
cellular pattern formation mechanism : 細胞の配列パターン形成機構
cellular response : 細胞応答
cellulose : セルロース
Celsius degree : 摂氏度
cement layer : セメント層
cenchrus : (複 . -ri)、背瘤、背粒〔ラテン語〕
Cenozoic era : 新生代(地質時代の一時代)
census : 調査を行う、調査、センサス
census datum : (複 . -ta)、調査データ
census of butterfly : 蝶類調査
center of distribution : 分布の中心
center of frequency : 頻度の中心
center of gravity : 重心
center of maximum variation : 最大変異の中心
center of occurrence : 発生の中心
center of origin : 起源の中心
center of speciation : 分化の中心地、分化の中心

centiMorgan : センチモルガン(遺伝学的な距離の単位)
central : 中央部
central band : 中央部の縞状バンド
Central China : 中国中部
central complex : 中心複合体
central dogma : 中心教義、セントラルドグマ
central focus : 中心フォーカス
central fusion : 中央融合型
Central Laos : 中部ラオス、ラオス中部
central nervous system : 中枢神経系
central neurosecretion cell : 中央神経分泌細胞
central place foraging : 中心点採餌(餌を巣に持ち帰る採餌)(CPF)
Central Sakhalin : 中部サハリン
central signalling region : シグナル伝達の中心部
central silk girdle : 中央の帯糸
Central Thailand : タイ中部
Central Vietnam : ベトナム中部
centrifugation : 遠心分離
centrifuge : 遠心分離機にかける、遠心分離を行う
centromere : セントロメア、動原体
cercal : 尾角(びかく)の(鱗翅類などの幼虫の尾端にある突起)
cercus : (複 . -ci)、尾角、尾毛、尾葉
cereal : 穀物、禾穀類(かこくるい)
certain record : 確実な記録
certain state variable : 特定の状態変数
certainly true : 疑いなく正しい、確実に正しい
cervical sclerite : 頸節片
cervical shield : 頸部シールド
cessation : 休止、中断
cf. : 「ラテン語の confer の略号で、『参照する』、『比較する』 といった意味を持つ記号」、「『おそらくこの種であろ

う』と推測される場合、『属名 cf. 種名』の形式で表記」
CFC : 隠れた雌による選り好み、雌による密かな性選択（Cryptic Female Choice）
CG : 強心配糖体（Cardiac Glycoside）
chaetosema : （複 . -ata）、毛隆
chaetotaxy : 刺毛相（しもうそう）、刺毛式、毛式（刺毛の数、位置を記録したもの）
chaetotaxy of thorax : 胸部の刺毛相
chalaza : カラザ（卵の中で卵黄を安定させるひも）
challenging question : 挑戦的な質問
change of photoperiod : 日周変化、光周期変化、日長変化
chaparral : チャパラル（米南西部の矮性カシの木の密林）、イバラのやぶ
chaperone : シャペロン
Chapter : 章
character : 形質、性質、特性、特徴
character displacement : 形質置換
character map : キャラクターマップ
character mapping : キャラクターマッピング
character state : 形質状態、標徴
characteristic : 特性
characteristic distribution : 特徴的分布
characteristic protective coloration : 特徴的な保護色
characteristically logarithmic trend : 典型的な対数傾向
Charaxinae : フタオチョウ亜科
charaxine butterfly : フタオチョウ
charcoal : 木炭
chase : 追飛行動
chasing : 追跡飛翔、追飛
chauvinism : 偏重主義、対外強硬主義
cheater : チーター
checkered pattern : 格子縞、まだら模様
checkerspot butterfly : ヒョウモンモドキ
cheek : 頬（ほほ）
chemical : 化学物質、化学薬品
chemical arms race : 化学的軍拡競争
chemical change : 化学的変化、化学変化
chemical control : 化学的防除
chemical cue : 化学刺激、化学的な刺激
chemical defense : 化学防御
chemical ecology : 化学生態学、ケミカルエコロジー
chemical evolution : 化学進化
chemical group : 化学群
chemical halo : ケミカルハロー、化学的な暈（かさ）
chemical mimicry : 化学擬態
chemical modification : 化学的変形
chemical sense : 化学的感覚、化学感覚
chemical structure : 化学構造
chemically mediated interaction : 化学的媒介による相互作用
chemically mediated signal : 化学的媒介による信号
chemo- : 化学 -〔ギリシャ語〕
Chemokine Receptor 5 : ケモカイン受容体5（CCR5）
chemoreception : 化学受容
chemoreceptor : 化学受容器、化学受容体
chemosensillum : 化学感覚子
chemosensory gene : 化学的感覚遺伝子
chemosensory hair : 化学感覚毛
chemosensory neuron : 化学感覚ニューロン、化学感覚神経細胞
chemosensory protein : 化学感覚タンパク質
chemosensory regulation of feeding : 化学感覚による摂食行動制御
chemotactile receptor : 接触化学受容器
chemotaxis : 走化性
chestnut band : クリ帯
chevron : 山形紋
chewing mouth : 咀嚼口（そしゃくぐち）
chewing-type insect : 咀嚼型昆虫

chi-square of homogeneity : 等質性のカイ二乗検定、等質性のχ二乗検定
Chi-square test : カイ二乗検定、χ二乗検定
chickadee : チッカディ（シジュウカラの仲間の小鳴鳥）
chicken genome : ニワトリのゲノム
child term : 子用語、下位概念用語（GO 用語〈Gene Ontology term〉で、その用語には親子関係がある）
chilling : 低温化、冷たい
chilling period : 冷蔵期間
chimera : キメラ（同一個体中に遺伝子型の違う組織が互いに接触して存在する現象）
chimpanzee : チンパンジー
China mountain : 中国山地
Chinese type : 中国型
ChIP-seq : クロマチン免疫沈降シークエンシング（Chromatin ImmunoPrecipitation sequencing）
chitin : キチン（質）
chitin-binding protein : キチン結合性タンパク質
chitinous ring : キチン環
Chlie : チリ（チリ共和国）
chloramphenicol : クロラムフェニコール
chloroplast : 葉緑体
chloroplast DNA : 葉緑体 DNA（cpDNA）
chloroplast genome : 葉緑体ゲノム
choice of flower : 花の選択
cholesterol : コレステロール
chordata : 脊索動物門
chordate : 脊索動物
chordates : 脊索動物
chordotonal organ : 弦音器官
chordotonal sensory organ : 弦音感覚器官
chorion : 卵殻
chosen period : 選定期間
chrom- : 色 -〔ギリシャ語〕
chromatic adaptation : 色彩適応
chromatin : クロマチン、染色糸
chromatin immunoprecipitation sequencing : クロマチン免疫沈降シークエンシング（ChIP-seq）
chromatin modeling : クロマチンモデリング
chromatin organization : クロマチン構成
chromatogram : クロマトグラム
chromo- : 色 -〔ギリシャ語〕
chromophore-binding pocket : 発色結合ポケット
chromosomal breakage : 染色体切断
chromosomal crossover : 交叉
chromosomal evolution : 染色体進化
chromosomal fusion : 染色体融合
chromosomal genome : 染色体ゲノム
chromosomal interval : 染色体間隔
chromosomal inversion : 染色体逆位、染色体並びが逆向き（逆位）
chromosomal location : 染色体位置
chromosomal map : 染色体地図
chromosomal organization : 染色体の構成
chromosomal rearrangement : 染色体再配置
chromosomal structure : 染色体（の）構造
chromosome : 染色体
chromosome banding : 染色体分染法、染色体バンド法
chromosome containing *ras* : *ras* 遺伝子を含有する染色体
chromosome level : 染色体レベル
chromosome mapping : 染色体マッピング
chromosome survey : 染色体調査
chronic : 慢性の
chrono- : 時 -〔ギリシャ語〕
chronobiology : 時間生物学
chronological order : 時代的順序
chronology : 年代（推定学）、年表
chrysalis : 蛹（さなぎ）、クリサリス（ギリシャ語で「金」の意味）

Chugoku District of Japan : 中国山地（日本）
CI : 細胞質不和合（Cytoplasmic Incompatibility）
Ci [hg.] : 肘脈中断遺伝子（翅脈が途切れている）（gene Cubitus-interruptus）
CI, 95% : 95% 信頼区間（95% Confidence Interval）
cicada : セミ
cichlid : シクリッド、カワスズメ
ciliary opsin : 繊毛型オプシン
ciliary photoreceptor cell : 繊毛型光受容細胞
ciliary-type : 繊毛型
ciliate : 繊毛状(せんもうじょう)の（糸状の一形状）（蛾類の触角）、毛縁(もうえん)（葉縁に縁毛がある）
cilium : （複 . -ia）、繊毛〔ラテン語〕
circa : およそ、概〔ラテン語〕
circadian : 概日性、サーカディアン
circadian clock : 概日時計
circadian hypothesis : 概日仮説
circadian oscillation : 概日振動
circadian rhythm : 概日リズム、日周リズム
circannual rhythm : 概年リズム
circle reintegration : 環の再組み込み
circled region : 丸で囲った領域
circular flight : 旋回飛翔、回転飛翔
circular flight area : 旋回飛翔領域、旋回飛翔圏
circular-flying individual : 旋回飛翔個体
circularization of linear molecule : 直線状分子の円形化、直線状分子の環状化
circulating level of JH : 幼若ホルモン（JH）の血中濃度、JH の循環濃度
circulatory system : 循環系
Circum Japan Sea Area : 周日本海地域
circum- : 周囲 -、周辺 -〔ラテン語〕
circumpolar : 極周辺の、極域周辺の
circumpolar species : 環北極種
circumstantial line of evidence : 一連の情況証拠、一連の状況証拠
cis effect : シス効果（ある配位子〈ligand〉が主に立体的な効果により、シスの位置で起こる配位子交換反応の速度などへ影響を及ぼすこと ;「trans effect〈トランス効果〉」を参照）
cis regulatory change : シス調節転換
cis regulatory element : シス調節要素、シス調節エレメント
cis-element : シスエレメント、シス因子
CITEs : 絶滅のおそれのある野生動植物の種の国際取引に関する条約、ワシントン条約（Convention on International Trade in Endangered species of Wild Fauna and Flora）
citrus swallowtail butterfly : アフリカオナシアゲハ、オナシアゲハ
citrus tree : 柑橘類の木
clade : 分岐群、クレード、単系列、完系列、単系統
clade selection : 分岐選択
clade support : 系統分岐の支持度、クレードの支持度
cladistic analysis : 分岐解析
cladistics : 分岐論、分岐分類主義、分岐学
cladogenesis : 分岐進化、クラドジェネシス
cladogram : 分岐図、分岐関係図
clasper : 把握器(はあくき)、クラスパー、交尾器
class : 綱（分類階級の「こう」）、級【統計学】
classic biogeography : 古典的生物地理学
classic example : 古典的な例、古典的事例
classic theory : 古典仮説
classical biological control : 伝統的生物的防除
classical hybrid inviability : 古典的な雑種不和合性
classification : 分類、類別
classification unit : 分類単位
classified object : 分類された対象群

clavate antenna :（複 . -nae）、棍棒状(こんぼうじょう)の触角（アゲハチョウ上科）
claviform spot : 楔状紋(けつじょうもん)（蛾類）
claw : 爪
clay treatment : 粘土処理
clean environment : クリーンな環境
cleaning : 洗浄
clear asymptote : 明らかな漸近線、完全な漸近線
clear-water country : 水清き国
clearance of JH : 幼若ホルモン（JH）の除去、JH の浄化
clearing : 開拓地、空地
clearing population : 開拓地に適応した個体群
clearly proven : 明確に証明された
Clearwing Butterflies : トンボマダラ亜科
Cleavase fragment length polymorphism : CFLP 法、クリーバーゼ断片長多型
cleaved amplified polymorphic sequence : CAPS 法
cleft : 溝、裂け目、裂片
cleft of rock : 岩の裂け目
Cleomaceae : フウチョウソウ科
cliff : 崖(がけ)、絶壁
cliff-dwelling : 崖を住居とする
climate : 気候
climate chamber : 人工気象室
climate change : 気候変動
climate difference : 気候的差異
climate regulation : 気候調節
climate warming : 気候の温暖化
climatic adaptation : 気候適応
climatic condition : 気候条件
climatic cooling : 気候の冷涼化
climatic datum :（複 . -ta）、気候データ
climatic lethal limit : 気候的な致死限界
climatic variable : 気候変数
climatic zone : 気候帯
climax : 極相
climber : つる植物、つる性の植物
climber species : ツタ類などの植物種
clinal : クライン的
clinal variation : クライン的変異
cline : クライン、連続変異、勾配、傾斜
clock-shifting experiment : 時間移動実験、時間偏移実験
clone : クローン、栄養分枝系
cloned primer : クローン化プライマー
cloning efficiency : クローニング効率、クローン生存効率
close causal association : 密接な因果関係
close ecological relationship : 密接な生態関係
close proximity on the chromosome : 染色体上で近接した
close relation : 近縁
close relative : 近縁種
close-up : クローズアップ、大写し
close-up cross-section : 拡大断面図
closed population : 閉鎖個体群
closely related sequence : 近縁配列
closely related species : 近縁種
closely related sympatric pair : 近縁で同所的に生育するペア
closely related taxon : 近縁な分類群
closely related virus : 近縁のウイルス
closely-interrelated life-history trait : 密接に相互関連した生活史形質
closest relative : 最近縁種
clothianidin : クロチアニジン（「ネオニコチノイド系」の殺虫剤）
cloud :（雲のような）大群
club : 棍棒、棍棒状部、触角先端部、クラブ、触角の先端の太い棒
cluster : クラスター
cluster structure : クラスター構造
clustered gene : クラスターを構成する遺伝子

clustering : 集合化、クラスタリング
clustering of flower : 花の群生
clutch : 卵塊
clypeal : 頭盾の
clypeus : (複 . -pei)、唇基部、頭盾(とうじゅん)、頭盾板〔ラテン語〕
cM : センチモルガン(遺伝学的な距離の単位)(centiMorgan)
CMS : 細胞質雄性不稔(Cytoplasmic Male Sterility)
CNE : 保存されたタンパク質非コード領域(Conserved Noncoding Element)
CNV : 遺伝子コピー数変異、コピー数多型(Copy Number Variation)
co- : 共同 -、相互 -、共通 -〔ラテン語〕
co-author : 共著者、共同執筆者
co-effect of biotic and abiotic stress : 生物的ストレスと非生物的ストレスとの共同効果
co-mimic : 共通擬態種、相互擬態種
co-occur : 同時に起こる
co-worker : 共同研究者
coalescence : コアレセンス、合祖
coalescence time : 合祖時間
coalescent part : 合着した部分(共通の祖先に行き着く部分)
coalescent theory : 合祖理論、合体理論
coalescent tree prior : 合着した系統樹の事前情報、合着した系統樹の事前分布、合祖系統樹の事前情報、合祖系統樹の事前分布
coarctate : 囲蛹(いよう)(最終齢の幼虫の脱皮殻が硬化し外皮となること)、囲蛹殻に包まれている
coarectate pupa : 囲蛹
coarse : 粗い
coastal : 沿岸の
coastal area : 海岸域
coastal back-swamp : 砂丘後背湿地
coastal vein : 前縁脈(ぜんえんみゃく)、C 脈
coat : 覆う
coauthorship : 共著者、共同執筆者
cobalt lysine : コバルトリジン
cobalt-coloring method of axon : 軸索のコバルト染色法
cocoon : 繭(まゆ)
cocoonase : コクナーゼ
code : 規約、暗号、コード
coding region : コード領域、コードしている領域
coding sequence : コード配列
codominant : 共優性
codominant molecular marker : 共優性分子マーカー
codon : コドン
codon evolution : コドン進化
codon position : コドン位置
codon-based maximum-likelihood analysis : コドンに基づく最尤解析
coecum penis (coe) : 陰茎盲のう
coefficient of gene differentiation : 遺伝子分化係数
coefficient of variation : 変動係数(母集団の母標準偏差と母平均との比率)(CV)
coevolution : 共進化
coevolutionary interaction : 共進化の相互作用
coevolutionary partner : 共進化の共同者、共進化のパートナー
coexist sympatrically : 同所的に共存する
coexistence : 共存
coexisting species : 共存種
coexisting strain : 共存系統
cognitive map : 認知地図
cohesive framework : 整合的な枠組み、整合性のある枠組み
cohort : 区(補助的な分類階級の「く」)、同時出生集団、コホート、同類

cohort analysis : コホート解析
COI : チトクロム酸化酵素サブユニット 1 （Cytochrome Oxidase subunit 1）
coin : 造り出す
coincide with : 一致する
coinfect : 重複感染する、共感染する
coinfecting Spiroplasma strain : 重複感染しているスピロプラズマ系統
Col-0 ecotype : 生態型コロンビア（Columbia）
cold acclimatization : 寒冷順化
cold blooded : 変温の
cold chain : 低温系列
cold hardiness : 耐寒性
cold Nagano prefecture : 寒冷地の長野県
cold paralysis : 寒冷麻痺(かんれいまひ)
cold patroller : 体温の低い巡回者、低体温巡回型
cold region : 寒帯地域
cold shock : 低温ショック、寒冷ショック
cold-hardiness : 耐寒性
cold-season phenotype : 寒候期の表現型
Coleman method : コールマン法
Coleoptera : コウチュウ目、鞘翅目
colinearity : 線形一致、共直線性
collapse : 衰弱する
collar : えり、カラー、襟状部（蛾類）
collate : 集めて分析する
collect : 採集する、収集する
collect randomly : ランダムに採取する、ランダムに採集する
collecting butterfly : 蝶採集
collecting pressure : 採集圧
collecting trip : 採集旅行
collection : コレクション
collection efficiency : 採集効率
collection flask : 収集フラスコ
collection point : 採集地
collective group : 寄集群
collector : 収集家、コレクター
colleterial gland : 粘液腺
collinearity : コリニアリティ、共線性
colonisation : 移住
colonisation bottleneck : 定着ボトルネック
colonisation distance : 繁殖地間の距離
colonised site : 集団繁殖地
colonising history : 定着の歴史
colonist : コロニスト、定着者
colonization : 移住、移入、入植、定着
colonization ability : 定着力
colonization dynamics : 定着動態
colonization event : 定着イベント
colonization probability : 定着確率
colonization rate : 定着率、移入率
colonizing ability : 定着力、定着能力、移住能力
colonizing population : 定着しつつある個体群
colony : 集団繁殖地、コロニー
colony growth : コロニー成長
colony health : コロニーの健康状態
color : 色彩
color band : 色帯
color change : 体色変化
color constancy : 色（彩）恒常性
color field : カラーフィールド
color form : 色彩型
color information : 色彩情報
color marking : 色斑
color morph : 色彩型
color pattern : 色彩パターン、色彩斑紋
color pattern diversification : カラーパターンの多様性
color pattern shift : カラーパターン変動、カラーパターン変化
color pattern-specific expression : 色彩パターンの特異的発現
color polymorphism : 色彩多型

color preference : 色選好性
color preference map : 色彩選好地図
color tone : 色調
color variation : 色彩のバリエーション
color vision : 色覚
color-opponent response : 色対比型応答、色対立型応答、反対色反応
color-pattern region : カラーパターン領域
coloration : 色彩化、着色
coloration of predetermined wing pattern : 前もって決まっている翅の紋様の色彩
Colore naturali edita - Lepidoptera : 原色蝶類検索図鑑〔ラテン語〕
colore naturali edita : 原色版、天然色版〔ラテン語〕
colored illustration : 色図版、カラー図解
Colored illustrations of the butterflies of Japan : 原色日本蝶類図鑑
Colored illustrations of the insects of Japan : 原色日本昆虫図鑑
colored lateral filtering pigment : 外側の着色フィルタリング色素
colorful flower : 色鮮やかな花
coloribus naturalibus : 原色〔ラテン語〕
columnar neuron : カラムを構成しているニューロン、円柱のニューロン
combination : 結合
combination of "hot-cold" and "dry-wet" : 寒暖と乾湿の組合せ
combination of mate choice experiment : 配偶者選択の組合せ実験
combinatorial probability : 組合せ確率
combinatoric theory : 組合せ(理)論
combined action : 組合せ操作
Comet : 縞模様
comma butterfly : シータテハ
comma-shaped red marking : 赤い半月紋
commencing : 開始
commensal : 片利共生的
commensal interaction : 片利共生的相互作用
commensalism : 片利共生(へんりきょうせい)
comment : 論評、批評、コメント
commercial hybrid of maize : トウモロコシの市販交雑種
Commission : 審議会
commitment : コミットメント
common ancestor : 同一祖先種、共通祖先種、共通祖先
common ancestor of primate : 霊長類の共通祖先
common ancestor of wasp : 蜂の共通祖先
common ancestral group : 共通の祖先群
common blue butterfly : ウスルリシジミ、イカロスシジミ
common blue morpho butterfly : ペレイデスモルフォ
common bluebottle butterfly : アオスジアゲハ
common cutworm : ハスモンヨトウ
common environment : 共通環境、共有環境
common garden condition : 共通庭園条件、同一環境条件
common garden experiment : 共通庭園実験、同一環境実験(異なる産地の生物を同じ土地で育成しての比較研究；生物の差異と生育環境の相関を調べる実験)
common garden mark-recapture study : 同一庭園で飼育した蝶による標識 - 再捕獲研究
common grackle : オオクロムクドリモドキ
common imperial blue butterfly : エバゴラスヒスイシジミ
common name : 通称名、俗名
common opal butterfly : アフリカサバクシジミ
common oviduct : 総輸卵管(そうゆらんかん)、中央輸卵管
common pathogen : 共通の病原体、普通にみられる微生物
common phenomenon : 共通の現象

common physical condition : 共通の物理的条件
common species : 普通種
common viral pathogen : 共通のウイルス病原体
common, diurnally active, mute butterfly : 通常、昼行性で音を出さない蝶
common-environment rearing : 共通環境での飼育
commonly used plant : 一般的に使用されている植物、自生する植物、日常的に利用される植物
commonly visit : よく訪問する
commonness and rarity of species : 種の普通性と希少性、普通種と希少種
communal foraging behavior : 共同採餌行動
communal rooster : 集団で塒(ねぐら)を形成する鳥
community : 群集
community composition : 群集(群落)構成
community diversity : 群集多様性
community ecology : 群集生態学
community ecology study : 群集生態学的研究
community module : 群集モジュール
community structure : 群集構造
community/ecosystem genetics : 群集・生態系遺伝学
comparable density : 比較可能な密度
comparative Ct method : 比較 Ct 法
comparative genomics : 比較ゲノム学、比較ゲノム解析
comparative mapping : 比較マッピング
comparative morphological method : 比較形態学的方法
comparative morphological study : 比較形態学的研究
comparative study : 比較研究
comparison : 比較
compartment : 区画、コンパートメント、領域
compartmentation : 構造化、区画化
compartmentation of substrate and enzyme : 基質と酵素の区画化
compass navigation : コンパスナビゲーション
compatibility method : 適合法
compatible solute : 適合溶質
compensatory growth : 補償成長
competing theory : 競争理論
competition : 競争
competition theory : 競争理論、競合説
competitive ability : 競争力
competitive interaction : 競争的相互作用
competitor : 競争種
competitor relationship : 競争者との関係
complementary DNA : 相補的 DNA
complementary sex determination : 相補的性決定
complete darkness : 暗黒
complete metamorphosis : 完全変態
complete wing expansion : (羽化直後の)完全な開翅、完全な翅の伸長
completed adult body formation : 完全な成体形成
completed adult morphogenesis : 完全な成虫の形態形成
completely bifurcating tree search method : 完全２分岐樹探索法
completely melanic variant : 完全黒化型
complex : 複合体
complex issue : 複雑な論争点
complex local orography : 複雑な局所的山地地形、複雑な局所的オログラフィー
complex of gene : 遺伝子複合体
complex peak : 複合ピーク
complex situation : 複雑な状況
complex statistical correction : 複雑な統計的補正
component : 構成分子、構成要素

component species : 構成種
Compositae : キク科
composite interval mapping : 複合区間マッピング法
composition of insect : 昆虫の構成
compositional similarity : 構成的類似性
compound : 複合(の)
compound action potential : 複合活動電位
compound eye : 複眼
compound microscope : 複合顕微鏡、複式顕微鏡
compound name : 複合名
comprehensive geographical sampling : 広範な地理的サンプリング
comprehensive modeling framework : 包括的なモデリング枠組み、モデリングの包括的枠組み
compressed : 側圧した、圧縮した
comprising about 50% : 約 50％を含んでいる
compromise : 危うくする、妥協する
computationally intensive : 計算集約的、計算性を強化したもの
computer simulation : コンピュータシミュレーション
Comstock-Needham system : コムストック - ニードハム(ニーダム)の体系
con-specific : 同種、同種の個体
concatenated protein : 連環型タンパク質、コンカテマー化したタンパク質、連鎖状タンパク質、連結したタンパク質
concave : くぼんだ、凹む
concavity : 凹面、くぼみ
conceal : 隠す
conceivable : 考えられる、想像できる
conceivably : ひょっとすると、おそらく
concentration : 濃度
concentric circle : 同心円
concentric eyespot pattern : 同心円状の眼状紋パターン
concentric ring : 同心円状の環
conceptual supergene : 概念的超遺伝子、仮説的超遺伝子
concise guide for the identification : 手軽な同定法
conclusion : 結論
conclusive prediction : 決定的な予測
concordant : 調和した、一致した
concordant pattern : 一致したパターン
condensed sex chromatin body : 凝縮性染色質体
condition dependence : 条件依存性
condition dependence hypothesis : 条件依存仮説
conditional : 条件付きの
conditional proposal : 条件付きの提案(書)
Condolence : 弔辞
cone : 円錐形
cone photopigment : 錐体視物質
cone photoreceptor cell : 錐体視細胞
cone tweeter : 円錐ツイーター
confer : 与える
confidence interval : 信頼区間
confidence limit : 信頼限界
confirm : 確かめる、裏付ける
confirmed population : 確認された個体群
conflict avoidance : 衝突回避
confocal analysis : 共焦点分析
confocal microscopy : 共焦点顕微鏡
confocal observation : 共焦点顕微鏡観察、共焦点観察
confound : 混同する
confronting behavior : 対峙する行動
congener-like larval form : 幼虫の新たな色彩型
congeneric : 同属の
congeneric coexistence : 同属の共存
congeneric species : 同属種

congregate : 集まる
conical morphogen gradient : モルフォゲンの円錐形濃度勾配
conifer : 針葉樹
connected population : 孤立化していない個体群
connecting vowel : 結合母音
connection between individual level movement and population level distribution : 個体水準の移動と個体群水準の分布との結合
connectivity : コネクティビティ、連続性、連結性【景観生態学】
consecutive in time : 時間に連続した
consecutive minutes, 30 : 連続 30 分間
consensus sequence : コンセンサス配列、共通配列
consensus tree : 合意系統樹
consent : 同意、承諾、許可
conservation : 保護、保存、保全
conservation biogeography : 保全生物地理学
conservation biology : 保全生物学
conservation concern : 保護に対する懸念
conservation decision : 保全決定
conservation ecology : 保全生態学
conservation effort : 保護努力、保全努力
conservation expenditure : 保護支出、保護費用
conservation introduction : 保全的導入
conservation measure : 保全措置、保全対策
conservation of biodiversity : 生物多様性の保全
conservation of butterfly diversity : 蝶類多様性の保全
conservation of plant diversity : 植物多様性の保全
conservation of population : 個体群の保全、集団の保全
conservation organization : 保護団体、保護機関
conservation policy : 保全方針、保護方針
conservation strategy : 保全戦略
conservationist : 自然保護者、環境保全主義者
conservative hypothesis : 保守的仮説
conservative selection : 保守的選択
conservatively : 保守的に
conserve : 保全する
conserved developmental pathway : 保存された発生経路
conserved domain : 保存されている領域(機能が共通しているタンパク質で保存されている領域)
conserved name : 保全名
conserved noncoding element : 保存されたタンパク質非コード領域(CNE)
conserved PCR primer : 保存された PCR プライマー
conserved polymerase chain reaction primer : 保存されたポリメラーゼ連鎖反応用プライマー
conserved supergene locus : 保存されたスーパー遺伝子座
conserved viral site : 保存されているウイルス性部位
conserved work : 保全された著作物
consistency : 一致性
consistent amplification : 一貫した増幅
conspecific : 同種の、同種類、同じ種類
conspecific attraction : 同種的誘引、同種他個体に対する誘引
conspecific communication : 同種間コミュニケーション、同種間通信
conspecific female : 同種の雌
conspecific individual : 同種の個体
conspecific mating : 同種間の交配
conspecific phenotype : 同種の表現型、同種表現型

conspecific recognition : 同種認知
conspecific wing : 同種の翅
conspicuous : 目立つ、目立った
conspicuous biological difference : 顕著な生物学的差異
conspicuous coloration : 目立つ色、顕著な色
conspicuous eyespot : 目立つ眼状紋
conspicuous territorial behavior : 目立つテリトリー行動
conspicuous ventral wing marking : 腹側の翅の目立つ斑紋
conspicuousness : 目立つこと、顕著さ
constant short photoperiod : 一定短日日長
constant temperature : 恒温
constant temperature regime : 恒温条件、一定温度条件
Constitution : 審議会規則、制定、制度、法令
constitutive heterochromatin : 構成的ヘテロクロマチン
consume : 食べ尽くす、消費する
consumer : 消費者
consumer level : 消費者水準
consumption of fat : 脂肪消費
consumptive effort : 消費効果
contact chemoreceptor : 接触化学感覚器、接触化学受容器
contact pheromone : 接触フェロモン
contact zone : 接触帯、接触域
containment : 封じ込め
containment area : 封じ込め地域
contaminate : 汚染する
contamination of non-crop plant : 非作物植物の汚染
contamination of water : 水の汚染、水質汚染
contest tactics of contender : 競争相手の競争戦術
context dependence : 位置関係の依存性、文脈依存性、背景依存性、生息環境依存性
context-independent : 状況に独立な
context-specific : 状況特異的な、環境特異的な
context-specific fashion : 状況特異的なやり方、環境特異的なやり方
contig : コンティグ(DNA 配列断片群を重ね合わせてできるコンセンサス配列や、それを構成する配列断片群)、整列群、連結断片
contig length : コンティグ長
continent : 陸地
continental : 大陸産
continental characteristic climate : 内陸性気候
continental climate : 大陸的気候、大陸性気候
continental scale : 大陸規模
continental-scale estimate : 大陸規模での推定
contingency plan : 緊急時対応計画
contingent : 依存する
continuation of the report : 報告の続編
continuous distribution : 連続分布
continuous expenditure : 継続的な支出
continuous flight : 連続飛翔、継続飛翔
continuous flight behavior : 連続飛翔行動
continuous habitat : 連続した生息地
continuous line : 実線、連続線
continuous nucleotide : 連続した塩基、連続したヌクレオチド
continuous parental antibiotic treatment : 親の継続的な抗生物質処理
continuous ring : 連続したリング
continuous variation : 連続的変異、連続変異、連続した変異
continuously flying type : 連続飛翔型
contradict each other : 相矛盾する
contradiction : 反論
contrary effect : 逆効果

contrast : コントラスト、対比
contrasting flight path : 対比して目立つ飛翔経路
contribution ratio : 寄与率
control : 防除、制御、管理、対照、統制
control larva : (複 . -ae)、対照幼虫、コントロール幼虫
control measure : 防除措置
control mechanism : 調節機構
control of flight : 飛翔調節、飛翔制御
control PCR : コントロール(ポジコン／ネガコン)PCR、対照 PCR
control plant : 対照植物
control site : 対照場所、統制群の場所
control stand : 対照林分、統制群の林分
control threshold : 要防除密度
control treatment : 対照処理、統制処理(統制群対象の処理)
control virus : 対照ウイルス、コントロールウイルス
controlling factor : 支配要因、決定要因
controlling mechanism : 調節機構、制御機構
controversial : 議論の余地がある
controversial topic : 議論の余地のある話題
controversy : 論争
Convention on Biological Diversity : (CBD)、生物多様性条約
Convention on International Trade in Endangered species of Wild Fauna and Flora : (CITEs)、絶滅のおそれのある野生動植物の種の国際取引に関する条約
Convention on International Trade in Endangered species of Wild Fauna and Flora Convention : (CITEs)、ワシントン条約
conventional sodium dodecyl sulfate-proteinase K digestion : 従来のデシル硫酸ナトリウム - プロテイナーゼ K による消化
converge : 収斂する、収束する
convergence : 収斂(しゅうれん)、収斂現象
convergence behavior : 収斂現象
convergent amino acid change : アミノ酸の収束的変化
convergent change analysis : 収束的変化解析
convergent evolution : 収斂した進化、収斂進化、収束進化
convergent functional-site evolution : 機能的部位の収斂進化
convergent mimicry : 収斂的擬態
convergent phenotype : 収斂した表現型、集束した表現型
convergent result : 収斂的な結果
convergently : 収斂的に
conversely : 逆に言えば
convex bubble : 凸状の泡、凸状の気泡
convex inner membrane : 凸状の内膜
convincing example : 確証例、確信させてくれる例
cool place : 冷涼な場所
cool temperature : 低温、冷温
cool temperature treatment : 低温処理
cool winter : 寒い冬
cool- and warm-temperate zone : 冷温帯域
cool-temperate deciduous broadleaf forest : 冷温帯性落葉広葉樹林
cool-temperate deciduous forest : 冷温帯性落葉樹林
cool-temperate habit : 冷温帯性
cool-temperate species : 冷温帯性の種
cool-temperate zone : 冷温帯
cooled pupa : 冷却蛹
cooling : 寒冷化
cooling treatment : 冷却処理
coordinated evolution : 同調した進化
copper-zinc superoxide dismutase domain : 銅／亜鉛超酸化物不均化酵素領域

Coppers : シジミチョウ亜科
coppice : 雑木林
coppicing : 雑木林の伐採、コピス
copulated individual : 交尾個体
copulation : 交尾(こうび)
copulatory : 交尾の
copulatory aperture : 交尾孔
copy number variation : 遺伝子コピー数変異、コピー数多型
coral reef fish : サンゴ礁に生息する魚、サンゴ礁魚
core area : 活動中心部
core biosynthetic pathway : 中核的生合成経路、コア生合成経路
coremata : 発香総
corn bollworm : アメリカタバコガ
corn field : トウモロコシ畑
cornea : 角膜
corneal lens : 角膜レンズ
corner : かど(角)、辺縁部
cornutus (cor) : (複 . -ti)、射精のう棘、射精のう刺〔ラテン語〕
corolla depth : 花冠深度
corona : 冠状部(蛾類)、副花冠(副冠)
corpus allatum : (複 . corpora allata)、アラタ体〔ラテン語〕
corpus bursa : (複 . corpora bursae)、交尾のう体
corpus cardiacum : (複 . corpora cardiaca)、側心体〔ラテン語〕
correct original spelling : 正しい原綴(つづ)り
correction and additional information : 訂正と追加の情報
correlate : 相関がある
correlate with : 相関関係がある
correlated random walk : 相関ランダム歩行、相関をもつランダム歩行
correlated turn : 相関旋回
correlation : 相関、相関関係
correlation angle : 相関角度
correlation coefficient : 相関係数
correlative : 相関関係がある、相関的な
corresponding author : 連絡著者、連絡先となる著者、コレスポンディングオーサー、通信担当者、責任著者
correspondingly large and significant peak : 対応する大きく有意なピーク
corridor : コリドー、回廊
corrigendum : (複 . -da)、正誤表
corroborate : 裏付ける、確証する、補強する
cosegregation : 共分離
cosmopolitan : 普遍種、汎存種(はんぞんしゅ)、広汎種
cosmopolitan distribution : 世界各地の分布
cosmopolitan species : 普遍種
cosmopolite : 普遍種
cost : 出費、コスト、損失、費用
costa : 前縁、前縁脈、C 脈〔ラテン語〕
costal : 前縁部、前縁室
costal fold : 前縁脈のひだ、前縁褶
costal margin : 前縁
costal vein : 前縁脈、C 脈
costly : 費用がかかる、犠牲が多い
Cotesia congregata : コマユバチ科の一種
Cotesia glomerata : アオムシサムライコマユバチ(コマユバチ科の一種)(= *Apanteles glomerata*)
Cotesia sesamiae : コマユバチ科の一種
cotton bollworm moth : アメリカタバコガ、オオタバコガ
cotton field : 綿畑
cotype : コタイプ、副基準標本
coumarin : クマリン
counter adaptation : 対抗適応
counter behavior : 対立行動
counteract : 阻止する
counteracting power : 消去力
countermeasure : 対策、対抗手段

counterpart : 代置種
countertactics : カウンター戦術、対抗戦術
countless garden : 無数の菜園
countless site : 無数の場所
countryside : 田園地帯
countryside butterfly : 里山の蝶
court : 求愛する、誘う
courtesy : 好意
courtship : 求愛行動
courtship behavior : 求愛行動
courtship dance : 求愛ダンス
courtship display : 求愛誇示
courtship element : 求愛の行動要素
courtship pheromone receptor : 求愛フェロモン受容体
courtship preference : 求愛選好
covariance analysis : 共分散分析
covariance structure : 共分散構造
covariate : 共変量、共変数、共分散分析
covary : 共変動する
cover : 被覆(ひふく)
cover photograph : 巻頭写真
cover scale : カバースケール
cover site : 隠れ場所
cover slip : カバースリップ、カバーグラス(スライドグラスに載せた試料の上に載せるための薄いガラス板のこと)
coverage : 包括度、カバー率
coverage depth : 被覆されている領域の(読み)深度
coverslipped : 覆った
coxa : (複 . -ae)、基節(きせつ)〔ラテン語〕
coyness : はにかみ、内気
cpDNA : 葉緑体 DNA(chloroplast DNA)
CPF : 中心点採餌(餌を巣に持ち帰る採餌)(Central Place Foraging)
CPV : 細胞質多角体病ウイルス(Cytoplasmic Polyhedrosis Virus ; CyPoVirus)
cracker butterfly : カスリタテハ属の蝶
crackling sound : パチパチという音
Cramer's blue morpho butterfly : レテノールモルフォ
cranial : 前(ぜん)
craw : 爪(つめ)
cream : 乳白色、クリーム
creased appearance : ひだ模様
creature : 生物、創造物
credibility value : 信頼性の値
cremaster : 懸垂器、尾鉤
crenulate : 小鈍鋸歯状の
crepuscular : 薄暮活動性、薄明、薄暮時の
crescent-shaped : 三日月形の
Cretaceous period : 白亜紀
crevice : 割れ目
cricket : コオロギ
criminal offence : 犯罪行為
crimson tip butterfly : ダナエツマアカシロチョウ
CRISPR/Cas9 technology : CRISPR-Cas9 技術(遺伝子改変技術の名称)、クリスパーキャスナイン技術(Clustered Regularly Interspaced Short Palindromic Repeats / CRISPR associated protein 9)(規則的な間隔をもってクラスター化された短鎖反復回文配列)
critical day-length : 臨界日長
critical illuminance : 限界照度
critical light intensity : 限界光量
critical mate preference cue : 配偶者選好の臨界刺激
critical night-length : 臨界夜長
critical number of day : 臨界日数
critical photoperiod : 臨界日長、臨界光周期
critical species : 危篤種
critical threshold : 臨界閾(いき)値
critical value : 臨界値

critically endangered native species : 危機的な在来の絶滅危惧種
crocale form : (ウスキシロチョウの)無紋型、クロカレ型
crochet : 鉤爪(かぎづめ)、鈎爪
crop : 素のう
crop field : 作物畑
crop tissue : 作物組織
cross : 交雑種、交雑、断面、クロス、交配
cross even twice : さらに二回交雑する
cross habituation : 交差慣化
cross once : 一回交雑する
cross rib : 横梁
cross section : 断面、断面図
cross tolerance : 交差耐性
cross vein : 横脈(おうみゃく)
cross-band : 網目模様
cross-hatched area : 格子囲い領域、クロスハッチング領域
cross-linking of cuticle protein : 表皮タンパク質の架橋
crossbreed : 雑種、異種交配種
crossing : 種間交雑
crossing scheme : 掛け合わせ様式、異種交配様式
crossing-over rate : 交叉率(こうさりつ)、交差率
crosstalk between phytohormone pathway : 植物ホルモン経路間での相互作用
crossvein : 横脈
crosswind drift : 横風ドリフト、横風偏流
crowding : 群がり、込み合い、群集
crowding condition : 混み合う条件、混雑条件
crucial role : 決定的な役割
crucial stage in speciation : 種分化の決定的な段階
crucifer : アブラナ科植物
crumpled wing : しわしわの翅
crustacean lectin : 甲殻類レクチン
CRY : クリプトクロム(概日リズムをつかさどる時計遺伝子の1つ)、クライ(CRYptochrome)
cryophase : 低温期
cryoprotectant : 冷凍保存、低温障害防御物質
crypsis : 隠蔽(いんぺい)
cryptic behavior : 擬態行動
cryptic biodiversity : 隠蔽的な生物多様性
cryptic color : 隠蔽色、保護色
cryptic coloration : 保護色、隠蔽色
cryptic diversity : 隠蔽的多様性
cryptic female choice : 隠れた雌による選り好み、雌による密かな性選択
cryptic fraction of butterfly diversity : 蝶類多様性の隠蔽的な部分
cryptic mimicry : 隠蔽型擬態
cryptic species : 隠蔽種
crypticity : 隠蔽性、隠蔽色
cryptobiosis : 乾燥休眠、クリプトビオシス(「隠された生命活動」の意味)
cryptochrome : クリプトクロム(時計遺伝子の1つ)
crystal formation : 結晶形成
crystalline cone : 円錐晶体
CSD : 相補的性決定(Complementary Sex Determination)
CSP : 化学感覚タンパク質(ChemoSensory Protein)
CT : 要防除密度(Control Threshold)
Ct value : Ct 値(Threshold Cycle value)(PCR 増幅産物がある一定量に達したときのサイクル数)
Cu : 肘脈(ちゅうみゃく)(Cubitus)
Cu stress : 銅(重金属)ストレス
CuA : 肘脈前枝、前肘脈(Cubitus Aanterior)(蛾類)
cube-root : 立方根
cubital : 肘脈の

cubital crossvein : 肘横脈
cubital vein : 肘脈、Cu 脈
cubitus : 肘脈、Cu 脈
cubitus anterior : 肘脈前枝、前肘脈（CuA）（蛾類）
cubitus interruptus : キュビタスインターラプタス（転写制御因子）、肘脈中断
cubitus posterior : 肘脈後枝、後肘脈（CuP）（蛾類）
Cuc Phuong National Park : クックフォン国立公園
cucullus : 交尾弁端部〔ラテン語〕
cue signal : 刺激信号
cultivar : 栽培品種
cultivate : 開墾する、養殖する、栽培する
cultivated crucifer : 栽培されたアブラナ科植物
cultivated land : 耕作地、耕地
cultivation : 耕作、開墾
cultural control : 耕種的防除
culture medium : 培養基、培地
cultured cell : 培養細胞
cumulative heat unit : 積算温量
cumulative mortality : 累積死亡率
cumulative number of species : 累積種数
cumulative temperature : 積算温度
cumulative weight : 累積荷重、累積重量
CuP : 肘脈後枝、後肘脈（Cubitus Pposterior）（蛾類）
curiosity : 珍奇なこと、物珍しさ
curiously : 不思議なことに
curled : 曲がった
curtail courtship : 求愛行動を抑制する
curve-fitting extrapolation method : 曲線当てはめ型外挿法、カーブフィッティング外挿法
curved wing-tip : 鉤状の前翅先端
cuspis : 指突起、尖突起
cut flower : 切り花
cut *Lantana* sp. flower : ランタナ種の切り花
cuticle : クチクラ、表皮、キューティクル
cuticle composition : クチクラ組成
cuticle formation : 表皮形成
cuticle integrity : 表皮の強固さ
cuticle protein : 表皮タンパク質
cuticle scale : 表皮鱗粉
cuticle-forming tissue : 表皮形成組織
cuticular hydrocarbon : 体表炭化水素
cuticular surface : 皮膚表面
cuticular ultrastructure : 表皮の超微細構造
cutoff level : カットオフ水準、切り捨て水準
cutting of dry leaf: 枯れ葉の切断
CV : 変動係数（Coefficient of Variation）
cv. : 園芸品種（cultivar）、栽培変種
cyanide release : シアン化物の放出
cyanobacteria : シアノバクテリア
cyanogenesis : シアン生成、青酸生成能
cyanoglucoside : シアノグルコシド
cyclic cicada : 周期ゼミ
cyclops : 単眼体
Cydno longwing butterfly : シロオビドクチョウ
cypovirus : 細胞質多角体病ウイルス（CPV）
cyrus : サイラス型（シロオビアゲハの非擬態型雌）
cyst : シスト、シスト細胞、のう胞、包のう
cysteine arrangement pattern : システイン配置パターン
Cytb : *Cytb* 遺伝子、チトクロム b 遺伝子（cytochrome b）
cytochrome : チトクロ（ー）ム
cytochrome oxidase : チトクロム酸化酵素
cytochrome oxidase subunit 1 : チトクロム酸化酵素サブユニット 1
cytochrome oxidases I and II : チトクロムオキシターゼ I と II
cytochrome P450 : チトクロム P450
cytokine : サイトカイン、シトキン

cytological observation : 細胞学的観察
cytologically : 細胞学的に
cytoplasmic incompatibility : 細胞質不和合
cytoplasmic incompatibility microorganism : 細胞質不和合微生物
cytoplasmic male sterility : 細胞質雄性不稔
cytoplasmic polyhedrosis virus : 細胞質多角体病ウイルス（CPV）
cytoplasmic sex ratio distorter : 細胞質因子による性比歪曲因子、細胞質性比歪曲因子
cytoplasmically inherited gene : 細胞質遺伝性遺伝子
cytoskeletal reorganization : 細胞骨格の再組織化、細胞骨格の再構築
cytoskeleton dynamics : 細胞骨格動態
cytoskeleton mediated baculovirus motility : 細胞骨格媒介性バキュロウイルスの運動性
cytoskeleton rearrangement : 細胞骨格再配置

d

D : 暗期（Dark）、表面（Dorsal）、上面、背側
D-statistic : D 統計量
d.f. : 自由度（degree of freedom）
D.P.R. Korea : 北朝鮮、朝鮮民主主義人民共和国（Democratic People's Republic of Korea）
Da : ダルトン、ドルトン（分子量の単位）（Dalton）
dagger-like : 短刀状の
daily abundance pattern : 日周の個体数パターン
daily activity : 日周活動、日周行動
daily movement : 日常飛翔、日常的移動
daily periodicity : 日周サイクル
daily visit : 日々の訪問
damage-associated molecular pattern : 傷害関連分子パターン、傷害由来分子パターン、傷害関連分子構造（DAMP）（被傷害者の表面に存在し、植物によって認識される分子で、植物はこれを認識することで免疫反応を起こす）
damaged area : 損傷領域
damaging invasive species : 有害な侵入種
DAMP : 傷害関連分子パターン（Damage-Associated Molecular Pattern）
damp area : 湿地
danaid butterfly : タテハチョウ科の蝶
danaidone : ダナイドン（性行動刺激物質で、雄のヘアペンシル成分／性フェロモン成分）
Danaina subtribe : マダラチョウ亜族
Danio rerio : ゼブラフィッシュ
DAPI : ダピ（ジアミジノフェニルインドール）（DNA 結合性で核染色に用いられる蛍光色素の一種）（4',6-DiAmidino-2-PhenylIndole）
dark box : 暗箱
dark brownish color : こげ茶色
dark form : 暗色型、暗化型
dark mimic : 暗化擬態
dark period : 暗期
dark phase : 暗期、暗相
dark reddish-brown color : 暗赤褐色
dark time : 暗期
dark treatment : 暗黒処理
dark-adapted live insect : 暗順応性の生きた昆虫
dark-time measurement : 暗期測定
darken light wavelength : 減光波長
darker, shadier place : より暗く、より薄暗い場所
darkness : 暗所
darting flight : 矢のようにすばやく飛ぶ
Darwin's finch : ダーウィンフィンチ（キンパラ科の鳥）

Darwin, Charles : ダーウィン、チャールズ・ダーウィン
Darwinian selection : ダーウィン流淘汰
dashed gray line : 破線の灰色の線、灰色の破線
dashed line : 破線
data logger : データロガー、データ計測器
data mining : データマイニング
data preparation : データ準備
data quality : データの質
data suitability : データの適切性
database : データベース
date : 日付
date of publication : 公表の日付、刊行日
dating of the common ancestor : 共通祖先の年代決定
daughter cell : 娘細胞、嬢細胞
dawn : 薄明、夜明
day degree : 日度
day length : 日長
day-flying : 昼行性の
day-length : 日長、日長時間
day-old : 日齢
day-to-day variation : 日々(の)変化
daylight hour : 日照時間
days post transfection, six : 形質移入6日後
dazzle : 目がくらむ
dB : デシベル(decibel)
DD : 全暗黒、恒暗(continuous Dark あるいは Dark-Dark)
DDBJ : 日本 DNA データバンク(DNA Data Bank of Japan)
DDC : ドーパ脱炭酸酵素、ドーパデカルボキシラーゼ(DopaDeCarboxylase)
de novo : 初めから、新たに〔ラテン語〕
de novo genome assembly : デノボゲノムアセンブリー
de novo mutation : 新生突然変異
de novo sequencing : デノボシークエンシング
de novo transcriptome generation : デノボトランスクリプトーム生成、新しく発生したトランスクリプトームの生成
de- : 下降 -、否定 -〔ラテン語〕
de-efferented : 神経遮断の
dead feigning : 擬死
dead insect : 死んだ昆虫
dead leaf : 枯れ葉
dead mimicry : 擬死
deadly pathogen : 死の病原体、致死性微生物、致死的病原体
deal with : 対処する、対応する
dearth of experimental test : 実験的検定の不足
dearth of genomic resource : ゲノム資源の欠乏、ゲノム資源の不足
death mimicry : 擬死(ぎし)
debilitation : 衰弱化
decapentaplegic : デカペンタプレジック遺伝子(シュウジョウバエの形態形成遺伝子)
decapitation : 断頭
decaying : 腐敗しかけた、朽ちる
decaying fruit : 腐敗している果実、腐果
decent rule of thumb : 厳しくない経験則
deceptive flower : だまし花
deciduous : 落葉性の
deciduous broadleaf forest : 落葉広葉樹林
deciduous forest : 落葉樹林
deciduous species : 落葉性の種
deciduousness : 落葉性
decision making : 意志決定、意思決定
decision-making : 意思決定
decisive step : 決定的なステップ
deck of shuffled card : シャッフルされたカードの束
Declaration : 布告書
decline : 衰亡、消長、傾斜、衰退
decline of genetic diversity : 遺伝的多様性の低下、遺伝的多様性の衰亡

decline of grassland butterfly : 草原の蝶の衰退
decline stage : 滅亡の段階
declining abundance : 個体数の減少
declining butterfly : 衰亡しつつある蝶
decomposer : 分解生物、分解者
decomposing fruit : 腐敗している果実、腐果
decomposition : 分解、腐敗、腐朽
decorate : 飾る
decoy apparatus : おとり装置
decrease : 低下する、減少する
decreasing abundance : 減少する存在量
decreasing similarity : 低下する類似度
deem : 見なす
deep : 深(しん)、深みのある
deep-red in upperside : 小豆色の地色
deeply diverged lineage : 深く分岐した系統
deer : シカ
defend territory : なわばりを防御する
defending behavior : 防御行動
defense activity : 防衛行動
defense compound : 自己防衛物質
defense gene : 防御遺伝子、防衛遺伝子
defensin : ディフェンシン
defensive flavor : 防衛的芳香
defensive glucosinolate : 防衛的グルコシノレート
defensive property : 防衛特性
defensive regurgitation : 防御的吐き戻し
defensive response : 防御応答、防御反応
defensive substance : 防御物質
definition : 定義
deflect : そらす、曲折させる
deflection of predator attack : 捕食者攻撃のぶれ
deforestation : 森林破壊、森林伐採
deformed testis : 奇形の精巣
deformed wing : 奇形の翅
degenerate : 退化する、悪化する
degenerate forward primer : 順方向縮重プライマー、縮重フォワードプライマー
degenerate primer : 縮重プライマー
degenerative : 退化的
degradation : 低下
degree of change : 変化程度
degree of cover : 網羅度合
degree of crowding : こみあいの程度、混み具合
degree of error : 誤りの度合、誤差の程度
degree of occurrence : 出現程度
degree-day : 日度
dehydrate : 脱水する
dehydration : 脱水
delayed larva : 遅れた幼虫、遅発幼虫
delaying emergence : 遅延発生、遅延出現
deleterious allele : 有害対立遺伝子
deleterious effect : 有害作用、有害効果
deleterious genetic effect : 有害な遺伝的影響
deleterious recessive : 有害な劣性種
deletion : 欠失
deliberate introduction : 故意の導入
deliberate release of butterfly : 故意の放蝶
delimitation : 境界設定
delineation : 境界
Delonix sp. : ホウオウボク属(マメ科)の一種
deme : 地域集団、ディーム
demographic character of metapopulation : メタ個体群の人口学的特性
demographic concern : 人口学的関心
demographic inertia : 人口学的慣性
demographic parameter : 人口学的パラメータ、人口学的母数
demographic scenario : 人口学的筋書き、人口学的シナリオ
demographic stochasticity : 人口学的確率性
demographic unit : 人口統計上の単位
demographically : 人口統計的に
denaturing high-performance liquid

chromatography : DHPLC 法、変成高速液体クロマトグラフィー分析法
dendrite : 樹状突起
dendritic tip : 樹状突起先端部
dengue fever : デング熱
dense : 稠密(ちゅうみつ)な、濃密な
dense genus : 稠密な属
dense green covered area : 密集した緑に覆われた地域
dense lowland rainforest : 密集した低地雨林
dense patch : 密集したパッチ
dense shade : 稠密な日陰
dense species : 稠密な種
dense vegetation : 密集した植生
dense zone : 濃密な地帯
dense, shady wood : 密集した日陰の多い森
densely distributed marker : 密に分散されたマーカー
denser : より濃い
density board : 密度板
density effect : 密度効果
density of individual : 個体数密度
density of microtrich : 微毛密度、微毛の密度
density-dependent dispersal : 密度依存的分散
density-dependent process : 密度に依存する過程
density-independent process : 密度に依存しない過程
density-mediated indirect effect : 密度の変化を介する間接効果
dentate : 歯のある
dentation : 歯列
deoxyribonucleic acid : デオキシリボ核酸（DNA）
dependent pattern : 依存パターン、翅脈依存パターン
dependent variable : 従属変数
deploy : 配置に付かせる
deposit : 置く、預ける
depressed : 平圧される、抑圧された
depression : 衰弱
derived group : 派生グループ
dermal gland : 皮膚腺
Dermaptera : ハサミムシ目、革翅目
descendant family : 子孫ファミリー
descending flight : 相手を下方に抑え込もうとする飛翔行動
describe : 記載する
description : 記載
description of a new subspecies : 新亜種記載
desert : 砂漠、荒原
deserve : 受けるに足る、値する
desiccating : 乾燥化
desiccation : 乾燥
designate : 指定する
designation : 指定
designation of lectotype : 後模式指定
desk lamp : デスクランプ
detach : 取り外す、分離する
detectable association : 検出可能な関連
detectable enzymatic activity : 検出可能な酵素活性
detected transcript : 検出された転写物、検出された転写産物
deter : 防ぐ、抑止する、やめさせる
deteriorate : 衰える、悪化する、劣化する
deteriorated food condition : 悪化した食物条件
deteriorating environment : 悪化しつつある環境
deterioration : 悪化
determinant factor : 決定要因
determination : 決定機構
deterministic model : 決定論的モデル
deterrent : 抑止力、抑制因子
detoxication enzyme : 解毒酵素
detoxification gene : 解毒遺伝子

detoxify : 解毒する、無毒にする
detrimental high noon temperature : 有害な真昼の温度
detritophagy : 腐泥食性、腐食性
detrivore : 腐泥食(性)生物、腐食(性)生物
devastating attack : 壊滅的な打撃
developed country : 先進国
developing forewing : 前翅の発生
developing leg : 脚発生
developing wing : 翅発生
development : 発生、発育、成長
development construction of ～ : ～の造成、～開発
development history : 形成史、発展史
development induction : 発育誘導
development phase : 発育相
development program : 発育プログラム
development rate : 発育速度、生長速度、発育率、発生率、成長率、発達率
development stage : 発育相、発育段階
development time : 発育期間、発育時間
development zero : 発育零点
developmental advantage of female : 雌発育の有利点
developmental biology : 発生生物学
developmental choice : 発生的選択
developmental constraint : 発生拘束、発生上の制約(突然変異体のある種の存続を防ぐ発生機構の性状)
developmental delay : 発育遅延、成長遅延
developmental endemism : 発展的固有
developmental gene : 発生遺伝子
developmental genetic mechanism : 発生遺伝機構
developmental heterochrony : 発生の異時性、発育異時性
developmental homeostasis : 発育安定性、発生的恒常性、発育恒常性
developmental homologue : 発生ホモロジー、発生相同
developmental mechanism : 発生機構
developmental pathway : 発生経路
developmental process : 発育過程
developmental programming : 発生プログラミング、発生プログラム
developmental regulator Acheron : 発生調節因子アケロン、発育調節因子アケロン
developmental season : 発育季節
developmental stage : 発育期、発育ステージ
developmental switching mechanism : 発生上のスイッチ機構
developmental time : 発育時間、発達時間
developmental zero point : 発育零点
developmentally arrested pupa : 発育が抑制された蛹、発育が阻害された蛹
devote : 充てる
devour : むさぼり食う
DH : 休眠ホルモン(Diapause Hormone)
DHPLC : DHPLC 法、変成高速液体クロマトグラフィー分析法(Denaturing High-Performance Liquid Chromatography)
di- : 2-、二-、双-〔ギリシャ語〕
dia- : 通して-、間の-〔ギリシャ語〕
diachronicity : 通時性、経時性(単系統種)
diacritic mark : 区別的発音符
diagnosable : 診断可能な
diagnosis : 診断
diagnostic character : 診断形質、識別形質
diagnostic PCR detection : 診断 PCR 検出法、診断 PCR 法による検知
diagonal bar : 斜めの条線、斜め線
diagonal broken line : 対角破線
dialect of genetic code : 遺伝暗号の方言
diamond pattern : ひし形
diamond symbol : ひし形記号
Diana butterfly : ダイアナヒョウモン
diapause : 休眠(主に動物の)、発生休止状態の休眠、休眠期

diapause ability : 休眠能力
diapause bioclock protein : 休眠関連の生物時計に関わるタンパク質
diapause characteristic : 休眠特性
diapause condition : 休眠条件
diapause destined larval period : 休眠になる場合の幼虫発育期間、休眠性幼虫発育期間
diapause developing individual : 休眠をして発育した個体
diapause development : 休眠成長、休眠発育
diapause development individual : 休眠をして発育した個体
diapause development period : 休眠発育期、休眠期
diapause duration clock : 休眠持続時間時計
diapause egg : 休眠卵
diapause entry : 休眠突入
diapause factor : 休眠因子
diapause hormone : 休眠ホルモン
diapause induction : 休眠誘起
diapause induction ratio : 休眠誘起率
diapause instar stage : 休眠齢期
diapause intensity : 休眠深度、休眠の深さ
diapause larva : 休眠幼虫
diapause pathway : 休眠経路
diapause phase : 休眠相
diapause phenotype : 休眠表現型
diapause program : 休眠プログラム
diapause ratio : 休眠率
diapause response : 休眠反応
diapause site : 休眠場所
diapause stage : 休眠期
diapause state : 休眠状態
diapause syndrome : 休眠シンドローム、休眠症候群
diapause termination : 休眠覚醒、休眠消去
diapause termination rate : 休眠解除率
diapause-blocking effect : 休眠阻止効果
diapause-destined larva : 休眠予定幼虫、休眠性幼虫
diapause-destined pupa : 休眠予定蛹
diapause-inducing photoperiod : 休眠誘起の日長、休眠誘起の光周期
diapause-maintaining mechanism : 休眠維持機構
diapausing adult : 休眠成虫、休眠型成虫
diapausing pupa : 休眠蛹
diaphragm : 隔膜、横隔膜
dichotomous key : 二分岐のキー
dichotomy : 二項対立、二分法、二分
dichromatic : 二色型色覚の、二色性の
dicotyledon : 双子葉植物
dideoxy method : ジデオキシ法
dideoxy terminator : ジデオキシターミネーター
diet : 食料、飼料
diet quality : 食料の質、餌の質
diet quantity : 食料の量
diet shift : 食性転換
diet type : 餌の型
dietary nitrogen : 食餌性窒素、食物の窒素
dietary requirement : 食物要求
differ : 異なる
differance : 差延(「空間的差異と時間的遅延」を意味する短縮語；フランス語の造語)
difference in fitness : 適応性の相違
difference of development stage : 発育段階の差
different climatic region : 異なる気候の地域
different genus : 別属
different species : 異種
differential gene expression : 識別的遺伝子発現
differential scanning calorimetry : 示差走査熱量測定(DSC)
differentially expressed gene : 異なる表現

型の遺伝子、発現量が異なる遺伝子、発現変動遺伝子
differentiate : 識別する、分化する
differentiated morphologically and genetically : 形態的かつ遺伝的に分化している
differentiation : 分化
differentiation of food habit : 食性分化
differently-pigmented scale : 異なる色素で着色された鱗粉
differing architecture of pine and angiosperm : マツ（裸子植物の一種）と被子植物との異なる構成
difficult condition : 悪条件
difficult-to-identify species : 同定困難な種
diffuse coevolution : 拡散共進化
diffuse coevolutionary interaction : 拡散共進化の相互作用
diffusion : 拡散、放散
diffusion and freeze-back : 拡散と寒冷押し戻し
diffusion equation : 拡散方程式
diffusion process : 放散過程、拡散過程
diffusion-based random walk : 拡散に基づくランダム歩行
diffusion-reaction model : 拡散反応系モデル
digestibility : 消化性、消化吸収率
digestion : 消化
digestive system : 消化系
digestive tract : 消化管(しょうかかん)
digital camera : デジタルカメラ
digital oscilloscope : デジタルオシロスコープ
digital photograph : デジタル写真
digital signal processor : デジタル信号処理装置、デジタルシグナルプロセッサー
digital video : デジタルビデオ
digitus : 端指
diglyceride : ジグリセリド
diglyceride-carrying protein : ジグリセリド運搬タンパク質
dihydroxy-phenyl-alanine : ジヒドロキシフェニルアラニン
dil- : その溶液が希薄であることを示す接頭語
dilute : 弱める、希釈する
diluted honey : 薄めたハチミツ
diluted solution : 希釈溶液
dilution : 希釈液、希釈物
dilution effect : 希釈効果
dilution series : 希釈系列（一定の割合で希釈したシリーズ）
Diluvium : 洪積世
dim light : 薄暗い光
dim monochromatic flash set : 薄明かりの単色閃光セット
dimethylsulphoxide : ジメチルスルホキシド
diminish : 減らす
diminished JH titer : 減少した幼若ホルモン力価
diminutive black larval mutant : 小さい黒色幼虫の突然変異体
dimorphic body color : 二型の体色、性的二型の体色
dimorphism : 二形性、二型
dioecy : 雌雄異体
dip : （水平線の磁針の）伏角
diploid embryo : 二倍体胚
Diptera : ハエ目、双翅目(そうしもく)
dipteran insect : 双翅目の昆虫、ハエ目の昆虫
dipteran species : 双翅目の種、ハエ目の種
direct cloning : 直接クローニング法
direct coevolutionary interaction : 直接的な共進化の相互作用
direct contact : 直接接触
direct defense : 直接防衛
direct developing individual : 休眠をしないで発育した個体、不休眠で発育した個体
direct development individual : 休眠をし

ないで発育した個体、不休眠で発育した個体
direct driver : 直接の影響要因
Direct Repeat Junction : 直接反復結合(DRJ)
direct repellent of herbivore : 植食者に対する直接的な反発
direct selection : 直接的な選択
direct sequencing : 直接シークエンス法
direct sequencing of cDNA size fraction : cDNA のサイズ画分の直接シークエンシング法、cDNA のサイズ画分の直接塩基配列決定法
direct sunlight : 直射日光
directed manner : 一定方向を目指して、定位性
Direction : 告示書
direction of transfer : 伝播の方向
direction ratio : 方向比
directional effect : 方向性効果
directional flight : 方向移動飛翔
directional flight ability : 方向移動能力
directional selection : 方向性選択、定方向性選択、定方向選択、方向性淘汰
dirt : 泥
dis- : 分離 -、不 -、非 -、無 -〔ラテン語〕
disassortative mating : 異類交配
disc : 原基、(翅)円盤
disc equation : 円盤方程式
discal : 中央帯、中央部
discal cell : 中室
discal lacuna : 横脈くぼみ、横脈欠落部
discal lunule : 横脈月状紋、横脈紋(蛾類)
discal spot : 横脈紋
discern : 見つける、気づく、見分ける
Disclaimer : 棄権宣言
discontinuous distribution : 不連続分布
discovery : 発見
discrepancy : 不一致、相違、矛盾
discrete alternative morphology : 離散的な選択的形態
discrete focus : 分離した焦点
discrete sampling : 離散サンプリング
discrete wing pattern element : 不連続な翅のパターン要素
discrimen : 分離線、ディスクリーメン〔ラテン語〕
discriminate : 識別する、区別する
discrimination : 識別
discrimination between known neighbor and stranger : 既知の隣者とよそ者との間の識別
discussion : 考察、議論
disease myxomatosis : 疾患粘液腫病
disease outbreak : 病気の大流行、病気の発生
disease symptom : 病気の症状、病徴
diseased hybrid brood : 病気の雑種同腹仔
disentangle : もつれをほどく、解決する、見出す、選り分ける
disguise : 擬態する
disinfection : 消毒
disinhibit : 脱抑制する、脱抑制をきたす、抑制解除する
disjunct distribution : 隔離分布、分離分布
disorientate : 方向感覚を失う
disparity : 異質性、不均衡
dispersal : 分散、拡散、移動、放散
dispersal ability : 分散力、分散能力
dispersal behavior : 分散行動
dispersal capacity : 分散能力
dispersal morph : 分散型
dispersal polymorphism : 分散多型
dispersal potential : 分散の潜在能力
dispersal propensity : 分散傾向
dispersal rate : 分散率
dispersal tendency : 分散性向、分散傾向
dispersal trait : 分散形質
dispersant : 分散剤

dispersed distribution : 分散分布
dispersion ability : 分散能力
dispersive movement : 分散移動
display : 誇示、表示
display behavior : 誇示行動
disproportionate influence : より大きすぎる影響
disproportionately large : 不釣り合いなほど多くの、かなり多くの
disprove : 誤りを実証する、反証をあげる
dispute : 反論する、争い
disputed whether : ～かどうかを議論する
disregarding abundance : 無視できる程度の個体数
disrupt : 崩壊させる、分裂させる
disruption : 崩壊、分裂、分断
disruptive coloration : 分断色（偽装戦略の一種）
disruptive ecological selection : 分断的生態選択、分断化生態選択、分断性生態選択
disruptive natural selection : 分断自然選択、分断自然淘汰
disruptive selection : 分断選択、分断淘汰
disruptive sexual selection : 分断性選択、分断性淘汰
dissect : 解剖する、詳細に調べる、切断する
dissected wing : 切り取った翅
dissection : 解剖体、解剖
dissection technique : 解剖手法、解剖技法
dissemination : 播種、増殖、拡散、伝播
dissimilar behavior : 似てない行動、異なる行動
dissimilarity between butterfly community : 蝶群集間の非類似性
dissimilarity matrix : 非類似度行列
dissimilation : 異化
dissolution : 崩壊
dissolve : 溶かす
dissolved mineral : 溶解した無機物、溶解したミネラル
distal : 末端の、末梢の、端部の、遠位
distal ablation : 末端部位の切除
distal band : 末端側の縞状バンド（外横線）
distal region : 末端領域、末梢領域、遠位領域
distal tip : 外縁の先端、付属肢の先端
Distal-less expression : 末端部のない遺伝子の発現
distal-less gene : 末端部のない遺伝子
distal-less protein : 末端部のないタンパク質
distally : 末端側に、末梢側に
distance effect : 距離効果
distance matrix method : 距離行列法
distance of flight bout : 飛翔バウト（一続きの行動期間）の距離【行動生態学】
distance parsimony method : 距離節約法
distance scale : 距離尺度
distance threshold : 距離閾（いき）値
distance Wagner method : 距離ワグナー法
distant lineage : 離れた系統
distant relative : 遠縁
distant target : 遠くの目標、遠く離れた目標
distantly related : 遠縁の
distantly related co-mimetic species : 遠縁の相互擬態種
distantly related organism : 遠縁生物
distantly related species : 遠縁種
distasteful : 味の悪い、まずい
distasteful model : 味の悪いモデル
distasteful species : 味の悪い種
distilled water : 蒸留水
distinct clade : 離れた分岐群、異なった分岐群
distinct color pattern : 明瞭なカラーパターン
distinct species : 別種
distinct subspecies : 別亜種
distinct substrate : 異なる基質
distinct west-east gradient : 西から東にか

けての明確な勾配
distinctive allele : 明確に区別できる対立遺伝子
distinguishing point : 識別点
distort : ゆがめる
distortion : ゆがみ
distribute : 分布する、分配する
distribution : 分布
distribution and abundance of species : 種の分布と個体数
distribution change : 分布の変化
distribution element : 分布要素
distribution frequency : 分布頻度
distribution of genus : 属分布
distribution of organism : 生物分布
distribution of velocity : 速度分布
distribution pattern : 分布パターン、分布様相
distribution region : 分布域
distribution route : 分布経路
distribution survey : 分布調査
distribution type : 分布型
distributional change : 分布変化
distributional expansion : 分布拡大
distributional range of genus : 属分布圏
distributional range of species : 種分布圏
disturbance : 攪乱(かくらん)
disturbance frequency : 攪乱頻度
disturbance intensity : 攪乱強度
disturbance treatment : 攪乱処理
disturbed woodland : 攪乱された疎林、人為的攪乱を受けた林分
ditrysia : 二門亜目
ditto underside : 同裏面
ditto, UN : 同裏面(ditto UNderside)
diuresis : 利尿
diuretic hormone : 利尿ホルモン
diurnal : 昼行性の、昼間の
diurnal activity : 日周活動、日周行動
diurnal behavior : 日周活動、日周行動、昼行活動
diurnal butterfly : 昼行性蝶
diurnal moth : 昼行性蛾
diurnal periodicity : 日周性
diurnal rhythm : 日周リズム
diurnal wood nymph : 昼行性の森林性若虫
diurnally active : 日中に活動的な、日中に活動する、昼行性の
diurnally active and mute : 昼行性で音を出さない
diverge : 分岐する、分散する
diverged population : 分岐した個体群
divergence : 分岐、相違、分化、多様化
divergence between higher classification : 高次分類間の分岐
divergence time : 多様化期間、分岐時間、分岐年代
divergent clade : 多様な分岐
divergent evolution : 分岐進化
divergent function : 異なる機能、多様な機能
divergent mate preference : 多様な配偶者選好
divergent pattern : 異なるパターン、分岐パターン
divergent wing pattern : 多様な翅パターン
diverse age structure of tree : 多様な樹齢構造
diverse array : 分岐配列、多様な列挙
diverse group : 多様なグループ
diverse herbaceous community : 多様な草本群落
diverse insect order : 多様な昆虫目
diverse species : 多様な種
diverse template source : 多様なテンプレートソース
diverse variation : 多様な変異
diversification : 多様化
diversification of the larval feeding habit : 多様な幼生期食性

diversification rate : 多様化速度、多様化率
diversifying selection : 多様化選択
diversion of plant resource : 植物資源の転換
diversity : 多様性、分岐
diversity hot spot : 多様な種のホットスポット
diversity index : 多様性指標、多様性指数、多様度指数
diversity of consumer : 消費者の多様性
diversity of life : 生活の多様性、生命の多様性
diversity of resource : 資源の多様性
diversity on earth : 地球上の多様性
divert : 惑わせる
division : 区分、区画、分裂
DLM : 背縦走筋、背側縦走筋（Dorsal Longitudinal Muscle）
DmelGr43a : キイロショウジョウバエ（*Drosophila melanogaster*）の味覚受容体 43a
DMSO : ジメチルスルホキシド（Dimethylsulphoxide）
dN : 非同義置換率（Non-synonymous substitution rate）
DN [hg.] : 背方ニューロン（Dorsal Neuron）
dN/dS : 非同義・同義塩基置換数比（Non-synonymous/Synonymous nucleotide substitution ratio）
DNA : デオキシリボ核酸（DeoxyriboNucleic Acid）
DNA amplification : DNA 増幅
DNA barcode : DNA バーコード
DNA barcode record : DNA バーコード記録
DNA barcode reference library : DNA バーコード参照ライブラリー、DNA バーコードリファレンスライブラリー
DNA barcoding : DNAバーコーディング、DNA バーコード法（特定の遺伝子領域の短い塩基配列〈DNAバーコード〉を用いて簡便に生物種の同定を行う方法）
DNA barcoding performance : DNAバーコーディングの性能、DNAバーコード法の性能
DNA chip : DNA チップ
DNA Data Bank of Japan : 日本 DNA データバンク（DDBJ）
DNA degradation : DNA 分解
DNA exchange : DNA 交換
DNA extraction : DNA 抽出
DNA fingerprint : DNAフィンガープリント法
DNA fragment-length polymorphism : DNA 断片長多型（DFLP）
DNA marker : DNA マーカー
DNA marker assisted selection : DNA マーカー利用選抜
DNA polymorphism : DNA 多型
DNA repair : DNA 修復
DNA transposon : DNA トランスポゾン
DNA-based assessment : DNAに基づく評価
DNA-based identification : DNA に基づく同定
DNA-based specimen identification : DNA に基づく標本同定
DNA-binding motif and dimerization domain : DNA 結合のモチーフ・二量体形成の領域
DNase treated : DNase 処置、DNase 処理（DeoxyriboNuclease ; デオキシリボヌクレアーゼ）
dNTP : 4 種類のデオキシリボヌクレオチド三リン酸（dATP, dCTP, dGTP, dTTP）を混合したものを表記する
dodder : ネナシカズラ
dodging : 肩透かし、素早く逃げる、ジグザグ飛翔
domain : ドメイン（インターネット上で用いるコンピュータのグループ名や住所）、上界、変域、領域

dome-like shape : ドーム状の形
domestic alien species : 国内由来の外来種、国内外来種
domestic animal : 家畜
domestic native species : 国内在来種
domesticate : 順化する、順応する
domesticated animal : 家畜
domesticated *Ben* gene : 順化(順応)した(昆虫の)*Ben* 遺伝子
domestication of bracovirus sequence : ブラコウイルス配列の順化
dominance : 優性、顕性、優占度
dominance of melanic element : 黒化要素の優性
dominance relationship : 優性関係
dominance series : 優位系列
dominant : 優性の、優勢形質(顕性形質)、優占種、顕性
dominant mutation : 優性突然変異
dominant white and recessive yellow allele : 優性白色／劣性黄色の対立遺伝子
dominate and wide-distributed species : 優占な広域分布種
dominate species : 優占種
donate : 提供する、寄付する、寄贈する
donor : 供与体、供与菌、ドナー、寄贈者
donor circle : ドナーサークル
donor plant : 供与植物
donor population : 供与者個体群
donor site : 提供側生息地
donor species : ドナー種
donor stock : ドナーストック
doom : 運命づける
door scope : 覗(のぞ)き見る
dopa : ドーパ(アミノ酸の一種)
DOPA decarboxylase : ドーパデカルボキシラーゼ、ドーパ脱炭酸酵素(DOPA = DihydrOxyPhenylAlanine)(DDC)
dopadecarboxylase : ドーパ脱炭酸酵素、ドーパデカルボキシラーゼ
dopamine : ドーパミン(脳内の神経伝達物質)
dopamine derivative : ドーパミン誘導体
dopamine quinone : ドーパミンキノン
dormancy : 休眠(主に植物の)、活動休止状態の休眠
dormant phase : 休止相
dormant state : 休止状態
dorsal : 背部の、背面の、上面の、表面の、背側(はいそく)、背(方)の
dorsal appendage : 背側突起、背部の付属肢
dorsal basking : 背面日光浴、開翅日光浴
dorsal direction : 背面方向
dorsal dissection approach : 背側解剖法
dorsal eyespot : 背部眼状斑点、背眼状斑点
dorsal forewing : 背側の前翅
dorsal lateral line : 背側の側線
dorsal longitudinal muscle : 背縦走筋、背側縦走筋(DLM)
dorsal margin : 背側縁、背縁
dorsal nectary organ : 背部蜜腺
dorsal neuron : 背方ニューロン
dorsal part of pars intercerebralis : 脳間部背面
dorsal plane : 背断面、背面図
dorsal retina : 背側網膜
dorsal section : 背断面、背面図
dorsal seta : 背毛、背部の刺毛、背部の剛毛
dorsal side : 背側
dorsal side of wing : 翅背側
dorsal sinus : 背腔(はいこう)
dorsal spine : 背棘(はいきょく)、背刺(はいし)
dorsal surface : 背断面、背面図
dorsal surface of forewing : 前翅背面
dorsal vessel : 背脈管(はいみゃくかん)、背管
dorsal view : 背側から見た図(写真)
dorsal-basking : 背面日光浴
dorsal-ventral axis : 背腹軸、DV 軸

dorsally : 背側に
dorsoventral axis : 背腹軸、DV 軸
dorsoventral flight muscle : 背腹方向に走る飛翔筋、背腹飛翔筋
dorsoventral flight musculature : 背腹方向に走る飛翔筋組織
dorsum : 後縁、後縁帯〔ラテン語〕
dorsum of forewing : 前翅後縁
dose-dependent manner : 用量依存的手法、濃度依存的な様式
dot : 小斑点
double crossing over : 二重交叉
double digested : ２種類の制限酵素で切断した、二重に切断した（２種類の制限酵素を同時に用いてDNAを切断すること）
double helix : 二重らせん
double-strain-infected all-female brood : 二系統に感染したすべて雌の同腹仔、二重感染のすべて雌の同腹仔
double-stranded DNA : 二本鎖 DNA、二重鎖 DNA
double-stranded RNA : 二本鎖 RNA、二重鎖 RNA
doublesex : 両性
doublesex [hg.] : 両性遺伝子、*dsx* 遺伝子
doublesex isoform : 両性のアイソフォーム
doubleton : ２個体だけ現れた種、ダブルトン
doubling time : 倍増時間
doubly infected female : 重複感染した雌
doubly infected insect line : 重複感染の昆虫系統
down wind direction : 追い風の方向、風下の方向
downgrade : 優先度を下げる
downstream of sex determination cascade : 性決定カスケードの下流
downstream of sex determination system : 性決定システムの下流
downstream of the stop codon : 終止コドンの下流
downstream pathway : 下流経路
downstream production of defense compound : 防御化合物の下流の産出機構
downward selection : 下方選択
downwind : 風下に
Dpp : デカペンタプレジック遺伝子（*Decapentaplegic* gene）
drab little wing : 茶褐色の小さな翅
draft genome : 概要ゲノム
draft genome assembly : ドラフトゲノムのアセンブリー
draft sequence : 概要塩基配列
dragonfly : トンボ
drainage : 排水、排水路、排水法
draining : 乾燥化
dramatic alteration : 劇的な改変
drawing : 線画
dried specimen : 乾燥標本
drift : 浮動
drift selection : 浮動選択
drilling of dressed seeds : 粉衣された種子の条播き
drink : 吸水する
drinking : 吸水中
drinking water : 吸水
driver of change : 変化の駆動因子
driver of movement : 移動の駆動因子
driving force : 推進力
drizzle : 霧雨
DRJ : 直接反復結合（Direct Repeat Junction）
drooping tail : はかまの裾のような後翅（尾端）
drop back : 後退する、減少する
drop of saline : 塩水の滴下
drop sharply : 急降下する
drop-feeding method : 滴注式給餌法

dropping-off : 脱落化
droppings : 糞
Drosophila : ショウジョウバエ属
drought : 旱魃(かんばつ)、渇水、日照り、干ばつ
drumming : 連打、ドラミング
dry climate : 乾燥気候
dry forest : 乾燥森林
dry forest region : 乾燥森林地域、乾燥森林地帯
dry season form : 乾季型、乾型
dry thorn-scrub habitat : 乾燥有刺低木林の生息地
dry upland grass : 乾燥高地の草、乾地草
dry zone : 乾燥地帯
drying : 乾燥、乾燥化
Ds : 横脈紋(Discal spot)
dS [hg.] : 同義置換率(Synonymous substitution rate)
DSC : 示差走査熱量測定(Differential Scanning Calorimetry)
dsDNA circle : dsDNA の環
dsf. : 乾季型(dry season form)
dsRNA : 二本鎖 RNA、二重鎖 RNA(double-stranded RNA)
dsx : *dsx* 遺伝子、両性遺伝子(*doublesex*)
dual role : 二つの役割
dubious : 疑わしい、怪しげな
ductus : (単複同形)、管〔ラテン語〕
ductus bursa : (複 . -ae)、交尾管(蛾類)〔ラテン語〕
ductus ejaculatorius : (複 . -rii)、射精管(しゃせいかん)〔ラテン語〕
ductus seminalis : (複 . -les)、受精管〔ラテン語〕
dull-colored : 鈍い色の
dung : 糞(動物の)
dung fly : クソバエ、糞蝿
Dunnett's test : ダネット(の)検定
duplicate gene : 重複遺伝子
duplicate ultraviolet opsin : 重複型紫外線オプシン
duplication : 重複
duration of larval instar : 幼虫期の齢期間
duration of light exposure : 照射される光の期間
duration of response : 応答期間、反応継続期間
dusk : 薄暮、薄暗がり、たそがれ
dusky gene : 暗色化遺伝子
dust : 散在する
dv : 背腹方向に走る(dorsoventral)
DV axis : 背腹軸、DV 軸(Dorsal-Ventral axis)
dwarf bamboo thicket : 矮性の竹藪
dwell : 棲息する
Dyar's law : ダイアーの法則
dynamic airflow : 動的空気流、動的気流
dynamic environment : 動的環境
dynamic equilibrium : 動的平衡
dynamic escalation process : 動的増大過程
dynamic expression profile : 動的発現プロファイル
dynamic landscape : 動的景観
dynamic programming : ダイナミックプログラミング
dynamic range : ダイナミックレンジ
dynamic state : 動的状態

e

E-value cutoff : E 値のカットオフ値、期待値のカットオフ値
E. coli : 大腸菌(*Escherichia coli* の省略形)
E. coli colony : 大腸菌コロニー
EAG : 触角電図、触角電図法(ElectroAntennoGram)
eagle owl : ワシミミズク
ear : 耳
ear drum : 鼓膜

earliest stage of speciation : 種分化の最初期
early Cenozoic era : 新生代初頭
early detection : 早期発見
early embryonic event : 胚の初期発生イベント
early embryonic stage : 胚発生の初期段階
early iconography : 初期図像学
early instar larva : 若齢幼虫
early male-killing : 初期型雄殺し
early morning : 早朝
early phase of invasion : 侵入の初期
early pupa : 蛹初期
early pupal epidermis : 蛹初期の表皮
early pupal hindwing : 蛹初期の後翅、蛹期前期の後翅
early pupal stage : 蛹初期、蛹期初期
early pupal wing : 蛹初期の翅
early spring : 初春、早春
early stage : 幼生期
early stages extend over 100 days : 幼生期100日をこえる
early warning of change : 変化の早期警報
early-summer form : 初夏型
earnest : 本気の、熱心な
earth : 地球
earth's magnetic field : 地磁気、地球の磁場
East Asia : 東アジア
eastern Bhutan : ブータン東部
EC : 酵素分類(Enzyme Classification)
ecdysial droplet : 脱皮顆粒
ecdysial gland : 脱皮腺(だっぴせん)
ecdysial line : 脱皮線(だっぴせん)
ecdysis : 脱皮(だっぴ)、脱け殻(ぬけがら)
ecdysis triggering hormone : 脱皮刺激ホルモン
ecdysone : エクジソン、エクダイソン(脱皮ホルモンの一種)
ecdysone receptor isoform : エクジソン受容体のアイソフォーム
ecdysteroid : エクジステロイド(脱皮ホルモンの一種)
ecdysteroid hormone : エクジステロイドホルモン、脱皮ホルモン
ecdysteroid release : エクジステロイドの放出
ecdysteroid titer : 脱皮ステロイド力価、エクジステロイド量
ecdysterone : エクジステロン(エクジソンに似た脱皮ホルモン)
eclose : 羽化する
eclosed female : 羽化した雌
eclosion : 羽化、脱蛹
eclosion hormone : 羽化ホルモン
eclosion rate : 羽化率
eclosion season : 羽化時期
eclosion termination : 羽化消去
eco-chemical : 生態化学
Ecocene : 始新世
ecoevolutionary dynamics : 生態進化ダイナミクス、生態進化の動態
ecological balance : 生態的バランス、生態的平衡、生態系のバランス
ecological character : 生態(学)的特性
ecological color variation : 生態的色彩変異
ecological community : 生態(学)的群集
ecological condition : 生態的条件
ecological consequence : 生態的結末、生態学的結果
ecological context : 生態的環境
ecological disturbance : 生態系の攪乱、生態学的攪乱
ecological divergence : 生態的多様化、生態(学)的分岐
ecological engineering : エコロジカルエンジニアリング、生態工学
ecological factor : 生態的要因、生態的因子
ecological forecasting : 生態学的予測
ecological function : 生態的機能
ecological geography : 生態地理学

ecological model : 生態(学)モデル、生態系モデル
ecological niche : 生態(学)的地位、エコロジカルニッチ
ecological opportunity : 生態的機会
ecological photograph : 生態写真
ecological population dynamics : 生態的個体群動態、生態で定義した個体群動態
ecological region : 生態学的地域
ecological release : 生態的解放
ecological requirement : 生態的要求、生態学的要求事項
ecological resurgence : 生態的誘導多発生
ecological role : 生態的役割
ecological significance : 生態的意義
ecological space : 生態的空間、生物相表現空間
ecological specialization : 生態的特殊化
ecological trait : 生態的特性
ecological variable : 生態学的変数
ecology : 生態、生態学
ecology of association between herbivorous insect and plant : 植食性昆虫と植物間の関係に関する生態学
economic balance of deciduous forest : 落葉林の経済収支
economic injury level : 経済的被害許容水準(EIL)
ecosystem : 生態系、エコシステム
ecosystem engineer : 生態系エンジニア
ecosystem functioning : 生態系(の)機能
ecosystem service : 生態系サービス
ecosystem-led approach : 生態系主導のアプローチ
ecotype : 生態型
ectoparasite : 外寄生者
ectopic eyespot : 転位した眼状紋、異所的眼状紋
ectopic white pattern : 転位した白色紋様
ectotherm : 外温生物
ectothermic : 変温の、外温の
Ecuador : エクアドル
edge effect : 林縁効果、エッジ効果、周縁効果
edge of unpaved road : 未舗装道路の路端
edge of wing : 翅端
edge-following : 周縁に沿う(飛翔)
edible part for the larva of butterfly : 蝶の幼虫の餌となる部分
edible part of plant : 植物の食べられる部分
edit distance : 編集距離
editor : 編集者
EditSeq program : EditSeqプログラム(塩基配列の編集用プログラム)(DNASTAR 社製)
EDTA : エチレンジアミン四酢酸(EthyleneDiamineTetraacetic Acid)
education : 教育
EF1α : 伸長因子 1α(Elongation Factor 1-alpha)
EF1α gene : EF1α 遺伝子、伸長因子 1α 遺伝子(Elongation Factor-1α)
effect of abundance : 個体数効果
effect of acclimation period : 馴化期間の効果
effect of climate change : 気候変動の影響
effect of continuous development : 発育継続の効果
effect of fire : 野焼きの効果、野焼きの影響
effect of inoculation : 接種効果
effect of light interruption : 光中断効果
effect of population age : 個体群が発生してからの年数の効果(影響)
effect of prior residence : 先住効果
effect of seasonal change : 季節変化の効果
effect of spatial aggregation : 空間的集団効果
effect of starvation : 絶食の影響
effect of stem density : 樹幹密度効果、幹数密度効果、本数密度効果

effective cumulative heat : 有効積算温量
effective growth season : 効果的な成長期、効果的な発育時期
effective number of allele : 有効アリル数、有効対立遺伝子数
effective number of locus : 遺伝子座の有効数
effective number of nucleotide : 実効塩基数
effective population number : 有効な集団の大きさ
effective population number of size : 有効集団サイズ、有効個体群サイズ
effective population size : 有効集団サイズ、有効個体数
effective spectrum : 有効スペクトル
effective temperature : 有効温度
effective wavelength : 有効波長
effector : 効果器
effector molecule : エフェクター分子
efficiency of conversion of food : 食料変換効率
efficiency of conversion of food into offspring : 次世代への餌の転化効率
efficiency of foraging behavior : 採餌行動の効率
EGF : 表皮成長因子、上皮細胞成長因子、上皮細胞増殖因子、上皮成長因子(Epidermal Growth Factor)
egg : 卵(らん、たまご)
egg cannibalism : 卵の共食い、卵食
egg cluster : 卵塊
egg clutch : 卵の一群、卵塊
egg collection : 卵収集、卵回収
egg deposition : 卵堆積、産卵
egg diapause : 卵休眠
egg diploidization : 卵の二倍体化
egg encapsulation : 卵の包囲化
egg formation : 卵形成
egg hatch : 卵の孵化
egg hatching rate : 卵孵化率
egg hatching success : 卵の孵化成功率
egg mortality : 卵死亡率、卵死亡数
egg overwintering : 卵越冬
egg parasitoid : 卵捕食寄生者(宿主の卵に産卵する寄生生物)
egg parasitoid-attracting volatile : 卵寄生蜂を誘引する揮発性物質
egg production : 卵の生産
egg protein : 卵タンパク質
egg recognition : 卵認識
egg stage : 卵期
egg-associated secretion : 卵由来分泌物
egg-derived elicitor : 卵由来エリシター(エリシターとは「植食者誘導性植物揮発性物質〈HIPV〉」生産を誘導する化学物質をいう)
egg-diapausing species : 卵休眠種
egg-induced plant response : 卵で誘導された植物応答
egg-infested plant : 卵が産み付けられた植物
egg-killing defense : 卵殺し防御、殺卵防御
egg-killing direct defense : 殺卵の直接防御
egg-killing parasitoid : 殺卵性捕食寄生者
egg-laden Brussels sprout leaf : 卵をたくさん付けた芽キャベツの葉
egg-larval parasitoid : 卵捕食寄生者 - 幼虫捕食寄生者
egg-laying : 産卵
egg-load assessment : 卵数の評価
egg-shell : 卵殻(らんかく)
egg-to-adult developmental time : 卵から成虫までの発育時間
eggs in mass : 卵塊
eggshell : 卵殻
EIA : 環境影響評価(Environmental Impact Assessment)
EIF4α : 真核生物型(翻訳)開始因子 4α(Eukaryotic Initiation Factor 4-alpha)

EIL : 経済的被害許容水準（Economic Injury Level）
ejaculate : 射精する、射精、精液
ejaculatory duct : 射精管
El Nino : エルニーニョ現象〔スペイン語〕
elaborate hypothesis : 精巧な仮説
electric balance : 電気計量器、電気バランス
electroantennogram : 触角電図、触角電図法
electrode : 電極
electromorph : 電気泳動パターン、電気泳動型特異性
electrophoresis : 電気泳動（法）
electrophoretic analysis : 電気泳動解析、電気泳動分析
electrophysiological assay : 電気生理学的方法、電気生理学的測定
electrophysiological response : 電気生理学的反応
electroporated : 電気穿孔処理された
electroporation : 電気穿孔法、エレクトロポレーション法
electroporation-mediated small interfering RNA incorporation : 電気穿孔法を用いた低分子干渉 RNA の取込み実験、エレクトロポレーション法を用いた低分子干渉 RNA の取込み
elementary color-coded neuron : 色情報をコードする要素的なニューロン
elevated mortality : 上昇した死亡率
elevated risk of population extinction : 個体群絶滅の高まっているリスク
elevated threshold : 上昇した閾(いき)値
elevation : 高度、標高
elevational shift : 高度変動、標高変動
elfin butterfly : コツバメ
elicit : 解発する、誘起する、顕在化する
elicitation : 誘発、顕在化
eliciting effect : 誘引効果、誘発効果
eliciting function : 誘引機能、誘発機能
elicitor : エリシター
elide : 脱字を施す、取り除く
elimination of anterior eyespot : 前方眼状紋の切除
ellipsoid : 楕円体
ellipsoidal eyespot : 楕円体状の眼状紋
elliptical : 長円形の
elongate : 長く伸びる
elongate photoreceptor cell : 細長い光受容細胞
elongation : 伸長部、伸長
elongation factor 1-alpha : 伸長因子 1α（EF1α）
elongation factor : エロンゲーションファクター、伸長要素
elucidate : 解明する
elucidation : 解明
elusive quantity : 捉えどころのない数量
elytron : （複 . -tra）、上翅(じょうし)、翅鞘(ししょう)、鞘翅(しょうし、さやばね)
em- : 中に入れる -、にする -〔ギリシャ語〕
emanate : 生じる、発する、出る
emarginate : 湾入形（円く窪んでいる形状）（触角の形状）、凹形、凹頭(おうとう)（葉先端の形状）
emarginated : 切れ込んでいる、凹形
Embioptera : シロアリモドキ目、紡脚目
EMBL : 欧州分子生物学研究所（European Molecular Biology Laboratory）
emblematic group : 象徴的なグループ
embryogenesis : 胚形成、胚発生
embryology : 発生学
embryonic development : 胚発育
embryonic mortality : 胚死亡率、胎児死亡率
embryonic sex determination : 胚の性決定
emendation : 修正、修正名
emerge : 羽化する
emerge from an egg : 孵化する
emergence : 羽化(うか)、孵化(ふか)、発生

emergence equalization : 発生時期の斉一化
emergence pattern : 発生パターン、羽化パターン
emergence rate : 羽化率
emergence time : 出現時間、発生時刻、羽化時間
emergence timing of cell row : 細胞列の出現時期
emergency : 創発、創発性
emergent : 巨大木、抽水植物
emergent population pattern : 創発される個体群パターン、突発的な個体群パターン
emerging adult : 羽化成虫
emerging form : 新生の型
emerging species : 新生種
emigrant : 移出種
emigrate : 移住する
emigration : 移出、移住して行くこと、渡り、回遊
emigration rate : 移出率
emperor butterfly : コムラサキ属の蝶
Emperors : コムラサキ亜科
empirical animal tracking datum : (複 . -ta)、実験的な動物追跡データ
empirical asymptote : 実証的漸近
Empirical Bayes approach : 経験ベイズ法
empirical evidence : 実証的証拠
empirical landscape : 実験的景観
empirical movement : 実験的移動
empirical research : 実証研究
empirical result : 実証的な結果、実験に基づいた結果
empirical study : 実証的研究
empirical test : 実証試験
empodium : エムポディウム、爪間突起〔ポルトガル語〕
empty case : 脱け殻(ぬけがら)
empty habitat : 空き生息地

en : エングレイルド遺伝子(*engrailed* gene)
en- : 中に入れる -、にする -〔ギリシャ語〕
enact : 制定する
encapsulation : 包囲作用
encoded Dsx protein : コードされた Dsx タンパク質
encoding C-type-lectin : エンコーディング C 型レクチン
encounter : 遭遇する、出会う
encounter rate : 遭遇率
encroachment of woody vegetation : 木本植生の侵食
end hook : 端鉤(たんこう)
end of light period : 明期終了
end sequence : 末端配列
endanger : 衰亡の危機にさらす
endangered species : 絶滅危惧種
endemic : 固有の
endemic organism : 固有生物
endemic species : 固有種
endemic species to Japan : 日本特産種、日本固有種
endemic to Japan : 日本特産
endemicity : 固有性、特有性
endemism : 固有性、固有種
ending : 語尾、終了
endo- : 内 -〔ギリシャ語〕
endocellular bacterium : (複 . -ia)、細胞内細菌
endocrine factor : 内分泌要因
endocrine gland : 内分泌腺(ないぶんぴつせん)
endocrine mechanism : 内分泌機構
endocrine organ : 内分泌器官
endocrine regulation : 内分泌調節
endocrine substance : 内分泌物質
endocrine system : 内分泌系
endocuticle : 内クチクラ、内表皮(ないひょうひ)、内原表皮、エンドクチクラ
endogeneous circannual clock : 内因的な

概日時計、内因性リズム
endogenous chromosomally transmitted virus : 染色体に組み込まれて伝達された内在性ウイルス
endogenous octopamine receptor : 内在性のオクトパミン受容体
endogenous stimulus : 内因性刺激
endogenous viral element : 内在性ウイルス由来配列(EVE)
endogenous virus : 内在性ウイルス
endomitosis : 核内有糸分裂
endoparasite : 内部寄生者
endoparasitism : 内部寄生
endoparasitoid egg : 内部捕食性寄生(蜂の)卵、内部捕食寄生者の卵
endoparasitoid wasp : 内部捕食性寄生蜂、内部捕食寄生性の蜂
endophagous : 内食性の
endosperm : 内胚乳
endosymbiont : 細胞内共生細菌、細胞内共生微生物、内部共生微生物
endosymbiotic microorganism : 細胞内共生微生物
endothermic : 内温性の、温血の
enemy attraction : 天敵の誘引
enemy-free space : 天敵不在空間、天敵低密度空間、天敵真空空間
energetic constraint : エネルギー的な制約
energetic equivalence rule : エネルギー等価則
energy demand : エネルギー需要
energy demand of diapause period : 休眠期間のエネルギー要求量
energy intake : エネルギー摂取(量)
enforcement : 補強、強化
English aurelian : 英国人蝶類採集家(「オーレリアン〈aurelian〉」は、ギリシャ語で「黄金」を意味し、金色の蛹にたとえられ、広い意味で蝶を愛する人たちのことを言う。)
engrailed : エングレイルド遺伝子、波形縁の
enhanced stress tolerance : 強化されたストレス耐性
enhancement : 強化
enhancer : エンハンサー
enlarged region : 拡大した領域
enlarged trachea : 伸張した気管、拡張した気管
enlarged ventral eyespot : 腹側の伸長した眼状紋列
enlargement : 拡大図
enocytoid : エノシトイド(昆虫血球細胞の一種)
enrich : 豊かにする、強化する
entail : 課す
enteric caeca : 盲のう(もうのう)
enthusiast : 愛好家
entire : 全縁(翅縁に切れ込みがない)
entire daylight period : 全日中期間
entire experimental period : 全実験期間
entire population's yearly range : 個体群全体の年間分布範囲
entomo- : 昆虫 -〔ギリシャ語〕
entomogenous : 昆虫寄生性(の)
Entomological Society of Japan : 日本昆虫学会
entomologist : 昆虫学者
entomology : 昆虫学
entomopathogenic : 昆虫病原性(の)
entomophagous : 食虫性(の)
entomophilous flower : 虫媒花
entrainment : 同調
entrenched : 確立した
envelope : 三角紙、エンベロープ、包み
environment : 環境
environment cue : 環境刺激
environmental change : 環境変化、環境変動
environmental condition : 環境条件

environmental conservation : 環境保護、環境保全
environmental covariate : 環境共変量
environmental determinant : 環境決定要因
environmental experience : 環境体験、環境経験
environmental extreme : 過酷な環境
environmental factor : 環境要因
environmental fluctuation : 環境変動
environmental genomics : 環境ゲノム学
environmental gradient : 環境傾度、環境勾配
environmental health : 環境保健
environmental impact assessment : 環境影響評価
environmental manipulation : 環境操作、環境変数の操作(例えば、温度操作)
environmental movement : 環境保護運動
environmental parameter : 環境パラメーター、環境因子
environmental period : 環境周期
environmental protection : 環境保全
environmental rearing condition : 環境的飼育条件
environmental regulation : 環境支配
environmental resistance : 環境抵抗
environmental risk : 環境リスク
environmental sensitivity : 環境(的)感受性、環境的敏感性
environmental stimulus : 環境刺激
environmental stochasticity : 環境的確率性、環境変動の確率性
environmental variable : 環境変数
environmental variation : 環境変動、環境多様性
environmentally conscious population : 環境意識が高い国民
environmentally sound : 環境への影響が少ない、環境にやさしい
envisage : 考察する、予想する、心に描く
envision : 描く、想像する
enzymatic activation : 酵素的活性化
enzyme : 酵素
enzyme classification : 酵素分類(EC)
enzyme code equivalent : 酵素コード等価物、酵素コード同等物
enzyme esterase : エンザイムエステラーゼ
enzymes tyrosine hydroxylase : エンザイムチロシンヒドロキシラーゼ
Eocene series : 始新統
eon : 累代、イーオン
ephemeral food source : 短命な餌源
Ephemeroptera : カゲロウ目、蜉蝣目
epi- : 上 -〔ギリシャ語〕
epicnemial : 前腹板の、前側片の
epicnemial carina : 前腹板隆起線、前腹板隆縁
epicnemium :(複 . -mia)、前腹板、前側片
epicranial suture : 頭蓋縫合線(とうがいほうごうせん)
epicuticle : 上クチクラ、外表皮(がいひょうひ)、エピクチクラ
epicuticular wax : 上表皮ワックス
epidemic : 伝染病、流行病
epidermal cell : 表皮細胞、上皮細胞
epidermal cell layer : 表皮細胞層
epidermal cell sheet : 表皮細胞シート、表皮細胞層
epidermal growth factor : 表皮成長因子、上皮細胞成長因子
epidermal organ : 表皮器官
epidermal response : 表皮反応
epidermis : 表皮、真皮、表皮細胞
epigamic behavior : 求愛行動
epigenetic : エピジェネティック、後成的な
epigenetic mechanism : 後成的機構
epigenetic regulation : エピジェネティック制御(遺伝子の塩基配列〈ACGT の順序〉によらない遺伝子発現の制御)

epigenetics : エピジェネティクス、後成学
epimeral : 後側板の
epimeral suture : 後側板線
epimeron : (複. -ra)、後側板(こうそくばん)(胸部)〔ラテン語〕
epimicrospectrophotometry : エピ顕微分光測光法
epimorphic field : 全射域
epiphysis : 葉状片
epiproct : 肛上板、肛上片
epistasis : エピスタシス、上位(性)、上下位性
epistatic incompatibility : 上位不和合性
epistatic sex determinant locus : 上位性決定遺伝子座
epistatic shutter allele : 上位シャッター対立遺伝子
episternal : 前側板の
episternal groove : 上腹板溝
episternal suture : 上腹板線
episternum : (複. -na)、前側板(ぜんそくばん)、上腹板、前腹板(胸部)〔ラテン語〕
epistomal groove : 頭額溝
epithelial cell : 上皮細胞
epizootic : 流行性、流行病
epizootiology : 流行病学(伝染病学)
epoch : 世(地質時代の「せい」)
epomia : (複. -ae)、前胸側斜架〔ラテン語〕
epsilon-diversity : イプシロン多様性、ε- 多様性(「地域多様性」のこと)
equality : 斉一、均質
equalization : 斉一化、均質化
equally : 等しく
equator : 赤道
equatorial : 赤道直下の
equilibrium density : 平衡密度
equipositional value line : 等位置価で描かれた線
equivalent chromosomal location : 等価な染色体位置
era : 代(地質時代の「だい」)
eradicate : 撲滅する
eradication : 撲滅、根絶
eradication campaign : 根絶キャンペーン
erase : 消失する
erode : 侵食する
erosion : 低下
erosion of genetic diversity : 遺伝的多様性の低下、遺伝的多様性の喪失、遺伝的多様性の衰退
erratic suitability : 変動性
error : 過誤、ミス、誤り
error of multiple comparison : 多重比較の誤り
eruciform : 毛虫状の、イモ虫型
erythropterin : エリスロプテリン
escalation : 増大、拡大、上昇
escape behavior : 逃避行動
escape mimicry : 逃避擬態
escape response : 回避反応
Escherichia coli : 大腸菌
esophagus : 食道
ESS : 進化的安定戦略、進化的に安定な戦略(Evolutionarily Stable Strategy)
essential insight : 本質的な洞察力
Essex skipper butterfly : カラフトセセリ(「Essex」〈エセックス〉は英国の州名)
EST : 発現遺伝子配列断片、発現配列タグ(Expressed Sequence Tag)
EST database : EST データベース、発現遺伝子配列断片データベース(Expressed Sequence Tags)
EST sequence : EST 配列(Expressed Sequence Tag sequence)
establish : 設立する、確立する
established invasive alien species : 定着してしまった侵略的外来種
established method : 確立された手法

establishment : 定着、定着性、確立
establishment ability : 定着力
ester : エステル
esterifying acid : エステル化する酸
estimated density : 推定密度
estimated distance : 推定距離
estimated probability : 推定確率
estimated tree : 推定系統樹
estimated value : 推定値
estimation of amino acid substitution : アミノ酸置換数の推定
estimation of diversity index : 多様性指標の推定、多様性指数の推定
estivation : 夏眠、夏休眠〔英語〕
estuary : 河口
ESU : 進化的に重要な単位、進化的重要単位(Evolutionary Significant Unit)
ESU delineation : ESU の境界設定、進化的に重要な単位の境界設定
ET : エチレン(Ethylene)(信号伝達物質の一種)
ETH : 脱皮刺激ホルモン (Ecdysis Triggering Hormone)
ethanol (EtOH) : エタノール(EtOH)(「Et」がエチル〈基〉で、「OH」はアルコールを指し、「エチルアルコール」とも呼ばれる)
ethanol : エタノール
ethanol-precipitated DNA sample : エタノール沈殿させた DNA サンプル
ethanol-preserved muscle tissue : エタノール保存されていた筋肉組織
ether : エーテル
ethical : 倫理的
Ethiopia region : エチオピア区
ethogram : エソグラム、行動目録(動物行動学において用いられる言葉で、野生動物の見せる一連の行動要素をリスト化し、それを時間的・空間的・順序的にどのように発現するかを細かく記録・解析したもの)
ethyl acetate : 酢酸エチル
ethyl ethanoate vapor : エタン酸エチルの蒸気
ethylene : エチレン(ET)(信号伝達物質の一種)
ethylene biosynthesis : エチレン生合成
etymology : 語源学
Eubacterium : (複 . -ia)、真正細菌
eucaryote : 真核生物
euchromatin : ユークロマチン、真正染色質
eukaryote : 真核生物
eukaryote genome : 真核生物のゲノム
eukaryotic biodiversity : 真核生物の生物多様性
eukaryotic initiation factor 4-alpha : 真核生物型(翻訳)開始因子 4α(EIF4α)
eumelanin : 真性メラニン、真正メラニン
euplantula : (複 . -ae)、ふ節盤(ふせつばん)、真褥盤葉
eupyrene sperm : 有核精子
eupyrene spermatozoon : 有核精子
Eurasian continent : ユーラシア大陸
Eurasian origin : ユーラシア起源
European arctic-alpine lepidoptera : ヨーロッパの北極圏の高山性鱗翅類
European butterfly : ヨーロッパの蝶
European colonization : ヨーロッパ諸国による植民地化
European map butterfly : アカマダラ
European Molecular Biology Laboratory : 欧州分子生物学研究所(EMBL)
eusociality : 真社会性
Euthalia djata : ベニオビイナズマ
Eutheria : 真獣類
evaginate : 外転する、裏返しにする
evaluation indicator : 評価指標
evasive flight maneuver : 逃避の飛翔行動作戦

evasive mimicry : 逃避擬態
evasive response : 逃避反応
EVE : 内在性ウイルス由来配列（Endogenous Viral Element）
evenness : 均等度、均衡度
evenness of relative abundance distribution : 相対的個体数分布の均等度
eventual result : 終局の結果
evergreen broadleaf forest : 常緑広葉樹林
evergreen forest : 常緑樹林
evergreenness : 常緑性
eversible tubercle : 反転性伸縮突起
every year : 毎年
everywhere : いたるところに
evident : はっきりわかる、明らかな
evo-devo : 進化発生学、エボ - デボ
evolution : 進化
evolution of butterfly foodplant : 蝶の食性の進化
evolution of dispersal : 分散の進化
evolution of diversity : 多様性の進化
evolution of genetic code : 遺伝暗号の進化
evolution of key innovation : 重大な革新の進化、重要な進化的革新
evolution of migration rate : 移動率の進化
evolutionarily stable optimal harvesting strategy : 進化的に安定な最適収穫戦略（ESOHS）
evolutionarily stable strategy : 進化的安定戦略
evolutionary arms race : 進化的軍拡競争
evolutionary biogeography : 進化生物地理学
evolutionary biology : 進化生物学
evolutionary branching : 進化的分岐
evolutionary change : 進化的変化
evolutionary consequence : 進化的影響、進化的結果
evolutionary convergence : 進化的収斂、進化的収束
evolutionary dead-end : 進化の袋小路
evolutionary distance : 進化距離
evolutionary divergence : 進化的分岐
evolutionary dynamics : 進化動態
evolutionary genomics : ゲノム進化学、進化ゲノム学
evolutionary history : 進化の歴史
evolutionary inertium : 進化的慣性
evolutionary novelty : 進化の新規性、進化的新奇性
evolutionary optimum : 進化的最適条件
evolutionary perspective : 進化的見方、進化的な観点、進化的視点
evolutionary plasticity : 進化的可塑性
evolutionary point of origin : 進化上の起点
evolutionary process : 進化過程
evolutionary process of butterfly : 蝶の進化過程
evolutionary radiation : 進化的放散
evolutionary rate : 進化速度、進化率
evolutionary relationship : 進化的類縁関係、進化的関係
evolutionary rescue : 進化的救済、進化的レスキュー
evolutionary response : 進化的応答、進化的反応
evolutionary scenario : 進化シナリオ
evolutionary sequence : 進化的配列
evolutionary significance : 進化的意味、進化的意義
evolutionary significant unit : 進化的に重要な単位、進化的重要単位（ESU）
evolutionary spiral : 進化的スパイラル
evolutionary stable strategy : 進化的安定戦略
evolutionary strategy : 進化戦略
evolutionary tendency : 進化傾向、進化的傾向
evolutionary trajectory : 進化軌跡、進化軌道

evolutionary trapping : 進化上のトラップ、進化的トラップ、進化的罠
evolve neutrally : 中立的に進化する
ex vivo assay : 生体外での試験〔ラテン語〕
ex- : 外へ -、無 -、非 -〔ラテン語〕
ex-situ conservation : 生息域外保全
exacerbate : 激化させる、悪化させる
exact probability : 正確確率
examination of lectotype : タイプ標本(レクトタイプ)の調査
exarate pupa : 裸蛹(らよう)(繭を作らないタイプの蛹)、自由蛹
excess larval developmental time : 幼虫の過剰な発育時間
excess, the : 超過分
exchange of information : 情報交換
excluded : 除外された
excluded name : 除外名
exclusion probability : 排除確率
exclusive preserve : 排他的分野
excrete meconium : 蛹便(ようべん)を排泄する
excretion : 排泄
excretory system : 排泄系
excurved : 湾曲状の
exemplar species : 典型種
exemplify : 例証する
exert : 使う、働かせる、及ぼす、引き起こす
exhaustive search method : 網羅的探索法
exhibit : 示す、提示する
existence of diversified intermediate form : 多様な中間型の存在
existence of territorialism : ナワバリ制の存在
existing population : 既存の個体群
exo- : 外 -〔ギリシャ語〕
exocrine secretion : 外分泌、外分泌物
exocrine substance : 外分泌物質
exocuticle : 外クチクラ、外原表皮、エキソクチクラ
exocuticle formation : 外原表皮形成、外クチクラ形成
exodus : 移住
exogenous ammonia : 外生アンモニア
exogenous β-glucosidase : 外因性 β- グルコシダーゼ
exon : エクソン
exon shuffling : エクソンかきまぜ説
exoskeleton : 外骨格
exotic : 外来の
exotic garden plant : 移入種の園芸植物
exotic species : 外国種、外来種
expanding northern population : 北方に拡大する個体群
expansion of the distribution : 分布拡大
expected heterozygosity : ヘテロ接合度の期待値
expected tree : 期待系統樹
expedition report : 遠征報告
expense of host fitness : 宿主の適応度犠牲
experimental and correlative studies : 実証的相関研究、実証的相関解析
experimental arena : 実験地、実験アリーナ
experimental attention : 実証的関心
experimental confirmation : 実験的確証、実験的確認
experimental design : 実験計画
experimental evidence : 実験的証拠
experimental group : 実験群、実験グループ
experimental introduction : 実験的導入
experimental manipulation : 実験的操作
experimental period : 実験期間
experimental population : 実験個体群
experimental procedure : 実験手順
experimental productivity gradient : 実証的生産性勾配
experimental screen : 実験的スクリーン、実験的スクリーニング
experimental standardization : 実験の標準化

experimental study : 実験的研究、実験研究
explanatory variable : 説明変数
exploitation : 搾取作用
exploitative competition : 消費型競争
explorative walk : 探索的歩行
explosive adaptive radiation : 急激な適応放散
exponential rise of genomic datum : (複. -ta)、ゲノムデータの指数的上昇、ゲノムデータの指数的増加
exposed place : 露出した場所
exposure : 曝(さら)すこと、晒すこと
exposure to, in : 暴露、暴露時の
express barely : かろうじて発現する、まれに発現する
expressed sequence tag : 発現遺伝子配列断片(EST)
expression : 発現、表現
expression level : 発現量、発現レベル
expression level of *RpL3* : *RpL3* 遺伝子の発現水準、リボソームタンパク質 L3 遺伝子(Ribosomal Protein L3)
expression of female phenotype : 雌の表現型発現
expression of seasonal form : 季節型発現
expression pattern : 発現パターン
expression plasticity : 表現可塑性
expression profile : 発現プロフィール、発現プロファイル
expression regulation : 発現調節
expression signature tag : EST 解析
expression stability : 発現安定性
expression system : 発現系
expression tendency : 発現傾向
expression verification : 発現検証
expressional regulation : 発現調節
exquisite spatial and temporal regulation : 鋭敏な時空調節
extant : 現存している、現生、現存
extant colony : 現存の集団繁殖地
extant population : 現存個体群
extant species : 現生種
extant variety : 在来種
extend laterally : 外側に伸長する
extend posteriorly : 後方に伸長する
extended point : 突出部
extensive local knowledge of the flower : 花に関する広範な局所的知識
extensive observation : 広範囲にわたる観察
extensive survey : 広範囲な調査、大規模な調査
extent of wing melanization : 翅の黒化の程度
external : 外(がい)、外部の
external anatomy : 外部組織、外部の解剖学的構造、外部構造
external body feature : 外部構造
external character : 外部形質
external coincidence model : 外的符合モデル
external genitalia : 外部生殖器
external morphology : 外部形態(学)
external node : 外部結節
external stimulus : 外部刺激、外界の刺激
extinct : 絶滅した、絶滅
extinct species : 絶滅種
extinct Taiwan race : 絶滅した台湾亜種
extinction : 死滅、絶滅
extinction and colonisation dynamics : 絶滅 - 定着動態、絶滅 - 移入動態
extinction rate : 絶滅率
extinction risk : 絶滅リスク
extirpated : 根絶した
extra molt : 過剰脱皮
extra- : 領域外の -、範囲外の - 〔ラテン語〕
extracellular environment : 細胞外環境
extracellular matrix : 細胞外基質
extracellular recording : 細胞外記録
extrachromosomal genome : 染色体外ゲノム

extract : 抽出物、エキス
extracted DNA : 抽出 DNA
extrafloral nectary : 花外蜜腺
extraocular photoreceptor : 視覚外光受容器
extraordinary number : 非常に大きな個数、途方もない個数
extraordinary proportion : 異常な割合、非常に大きな割合
extrapolation : 外挿
extraretinal photoreceptor : 網膜外光受容器
extreme condition : 極限状態
extreme fluctuation : 極端な変動
extreme unevenness : 極端な不均等、極端な不均一
extremity : 先端、最端
extrinsic : 外因的、(環境的)
exuded tree sap : 滲み出た樹液
exuviae : 脱け殻(ぬけがら)、脱皮殻
exuvium : 脱皮殻(だっぴから)
eye : 眼
eye cap : 目蓋
eye-antennal disc : 眼 - 触角原基
eye-opening : 開眼
eyeshine : 暗視眼(動物の目が暗闇の中で光る効果)
eyespot : 眼状紋(がんじょうもん)、目玉模様、眼状斑点
eyespot absent : 眼状紋の消失
eyespot focus : 眼状紋フォーカス
eyespot formation : 眼状紋形成
eyespot size : 眼状紋サイズ
eyespot-associated gene : 眼状紋関連遺伝子

f

F : 前翅(ぜんし)(Forewing)
F [hg.] : F 値、検定統計量、t 統計量(分散分析)
F value : F 値(二元配置分散分析)
f. : 品種(form/forma)、「『属名 種名 f. 品種名』の形式で表記」
FA : 変動非対称性(Fluctuating Asymmetry)
Fabaceae : マメ科
fabric of nature : 自然の構造
face : 顔面
facet : 個眼(面)
facial : 顔面の
facies : 外観、外見、顔面〔ラテン語〕
factorial design : 要因配置計画
facultative : 外因性、条件的、機会的、任意の
facultative association : 任意共生的関係
facultative diapause : 外因性休眠
facultative migration : 随意移動
facultative mutualism : 任意的相利共生、条件的相利共生、不偏性相利共生
facultative parthenogenesis : 機会的単為生殖
facultative pupal diapause : 条件的蛹休眠
FAD : フラビンアデニンジヌクレオチド(Flavin Adenine Dinucleotide)
falcate : 鎌状の(セセリチョウ上科の触角形状、蛾類の前翅翅頂部の形状)
fall : 落ちる
fall armyworm : ツマジロクサヨトウ
fall form : 秋型
fall in : 分類される
fallen fruit : 落果
falling titer : 低下する力価
falling together : 共倒れ
fallow : 休耕地
false ally : 似而非(えせ)近縁種、にせの近縁種
false alpine butterfly : 偽の高山蝶
false discovery rate : 偽発見率(陽性判定となる遺伝子の中に含まれる偽陽性の割合を表わす値)(FDR)
false eye : 擬眼、擬目
false head : 擬頭
false negative : 偽陰性
false positive : 偽陽性

falsification : 反証
fam. : 科(family)
fam. nov. : 新科名(family novus)
familiarity : 精通性
family : 科(分類階級の「か」)
family group : 科階級群
family name : 科名
family--order ratio : 科数 - 目数の比
family-group name : 科階級群名
Far Eastern Russia : 極東ロシア (Russian Far East)
far field pressure component of sound : 遠距離場の音の圧力成分
Far-Eastern Russia : 極東ロシア
Faraday cage : ファラデーケージ、ファラデー箱
fare : 暮らす
farmer-selected landrace : 農家が選択した農産物
farming : 耕作化、養殖
farmland : 農地、農村地帯、農村地域
farmland habitat : 農地性生息場所
farmland site : 農地の場所
fasciculate : 毛束状の(糸状の一形状)(蛾類の触角)
fascinating : 興味をそそられる
fast larval growth : 早い幼虫成長
fast-changing environment : 急速に変化する環境
fast-flying : すばやく飛ぶ
fast-slow continuum : (成熟が)早いものと遅いものを両極に持つ連続体
fat : 太らす
fat body : 脂肪体、肥満体
fatally damage : 致命的に損傷する
fate of the mimetic phenotype : 擬態表現型の最終結果、擬態表現型の運命
fate-determined wing : 運命的に決定された翅
fatty acid : 脂肪酸
fatty acid-amino acid conjugate : 脂肪酸 - アミノ酸縮合物、脂肪酸 - アミノ酸接合体(FAC)
fauna : (複 . -ae)、動物相、ファウナ
Fauna Europaea : ヨーロッパ動物相データベース(ヨーロッパに産する陸上及び淡水産の全動物のデータベース)
faunal diversity : 動物相の多様性
faunal region : 動物地理区
favea : 窩、中心窩
favor : 好む
favorable breeding condition : 有利な繁殖条件、良好な繁殖条件
favorable host plant : 好都合な寄主植物
favored : 気に入られている
FBLRH : FBLRH プロジェクト(ババリアの動物相の DNA バーコード構築プロジェクトの 名 称)(Fauna Bavarica - Lepidoptera RHopalocera)
FBS : ウシ胎児血清(Fetal Bovine Serum)
FDR : 偽発見率(False Discovery Rate)
feather-like marking : 羽毛状の斑紋
feathered spine : 羽毛のようなトゲ
feature : 特徴
fecundity : 蔵卵数、産卵数、繁殖力、産卵力、生殖力、妊性、稔性
fecundity selection on female : 雌に対する妊性選択、雌における妊性選択
federally endangered butterfly : 連邦政府によって絶滅危惧種に指定されている蝶
feed : 食事をする
feed conversion efficiency : 餌の転換効率
feed on : 食する
feeding : 吸蜜中、摂食中
feeding activity : 摂食活性
feeding attractant : 摂食誘引物質
feeding behavior : 摂食行動、摂食活動
feeding cut off : 摂食停止

feeding damage : 食害、摂食傷害
feeding deterrent : 摂食阻害物質
feeding differentiation : 食性分化
feeding habit : 食性
feeding herbivore : 摂食する植食者
feeding on the elm : ニレを食餌植物とする
feeding period : 摂食期
feeding preference : 摂食選好性
feeding site : 摂食場所、餌場
feeding stimulation : 摂食刺激
female : 雌(めす)、♀、メス
female butterfly : 母蝶
female choice : 雌の選択
female defense polygyny : 雌防衛型の一夫多妻
female eclosion : 雌の羽化
female embryo : 雌(の)胚
female emergency schedule : 雌の出現予定表、雌の羽化スケジュール
female genitalia : 雌性生殖器
female heterogametic : 雌ヘテロ型、雌異型配偶子型
female income : 雌の収入
female model : 雌モデル
female monogamy : 雌性単婚性
female parent : 母親
female phenotype : 雌の表現型
female progeny : 雌の子孫
female pupa : 雌蛹
female reproductive tract : 雌の生殖器官
female sexual differentiation : 雌の性分化
female sexual receptivity : 雌の性的交尾受容力
female specific chromosome : 雌特異的染色体
female type larva : 雌型幼虫
female-biased SSD : 雌のほうが雄よりも体格が大きい性的二型、雌に偏ったSSD(Sexual Size Dimorphism)
female-heterogametic chromosomal constitution : 雌ヘテロ型性染色体構成
female-heterogametic sex chromosomal system : 雌ヘテロ型性染色体システム、雌ヘテロ型性染色体様式
female-informative : 雌に有益な
female-limited Batesian mimicry : 雌に限定されたベイツ型擬態
female-limited isoform : 雌に限定されたアイソフォーム
female-limited mimicry : 雌に限定された擬態
female-limited polymorphic mimicry : 雌に限定された多型擬態
female-limited polymorphism : 雌に限定された多型
female-specific *dsx* isoform : *dsx* 遺伝子から生じた雌特異的なアイソフォーム
female-specific molecular mechanism : 雌に特有の分子機構
female-specific organ : 雌特有の器官
feminization : 雌性化、雌化
feminize : 雌化する
femoral : 腿節の
femur : (複 . -ra)、腿節(たいせつ)〔ラテン語〕
fermenting : 発酵した
fermenting fruit : 発酵した果実
fertile : 繁殖力のある、妊性、稔性の
fertile egg : 有精卵、受精卵
fertilise : 肥沃にする、受精させる
fertility : 妊性、稔性、生殖能、受精率
fertilization : 受精
fertilization rate : 受精率
fertilize : 受精させる
fetal bovine serum : ウシ胎児血清(FBS)
Feulgen's reaction : フォイルゲン反応
few days later, a : 数日後、２・３日後
FGFR3 : 繊維芽胞成長因子受容体タイプ３(Fibroblast Growth Factor Receptor-3)

FH : 両翅(Forewing and Hindwing)
fibrinopeptide : フィブリノペプチド
fibroblast growth factor receptor-3 : 繊維芽胞成長因子受容体タイプ３(FGFR3)
fibular : 腓(ひ)側、腓骨の
FID : 水素炎イオン化型検出器(Flame Ionization Detector)
fidelity : 忠実性、適合度
field : 野外、フィールド、圃場、分野
field assay : 野外検定
field assistance : フィールド支援、野外支援
field condition : 野外条件
field evidence : フィールド証拠
field expedition : 野外遠征、フィールド遠征
field experiment : 野外実験
field margin : フィールド周縁部、圃場縁
field memory recorder : フィールドメモリーレコーダー
field observation : 野外観察、フィールド観察
field observer : 野外観察者
field procedure : 野外での手順
field setting : フィールド実験、野外実験、実地調査
field situation : フィールド状況
field study : フィールド研究、野外研究
field work : 野外観察
field-caught female : 野外採集した雌
field-caught individual : 野外採集個体
field-collected animal : 野外で採集した動物、野外で採取した動物
fig wasp : イチジクコバチ
filament : 触角、触手、糸、花糸、フィラメント
filamentous : 繊維質の
filial : 第○世代の、交配世代の
filiform : 糸状の(蛾類の触角)
filled bar : 黒く塗りつぶされた棒、黒色の棒
filter : 濾過して取り除く
filter paper : 濾紙(ろし)
filtering procedure : フィルタリング手順
fimbriate : 縁毛状の(糸状の一形状)(蛾類の触角)
final dose : 最終投与、最終投与量
finch : フィンチ(スズメ科の小鳥)
fine : 罰金、細かい、微細な
fine brush : 微細ブラシ
fine forceps : (ハサミムシなどの)微細なハサミ、微細な鉗子、先細のピンセット
fine spatial scale : 詳細空間スケール、空間の詳細スケール、微細空間スケール【景観生態学】
fine structure : 微細構造
fine weather : 良い天候、晴天
fine-grained : 微粒、細粒
fine-scale genetic mapping : 詳細な遺伝子地図
fine-scale mapping : 詳細なマッピング、細密なマッピング
fine-scale spatial mosaic : 微細スケールの空間(的)モザイク
fine-tuned response of individual : 個体の微調整された反応
fire : 野火(自然火災)、野焼き
fire regime : 野焼き体制、野焼きの管理体制
fire-maintained community : 野焼きで維持されている群落
first author : 筆頭著者、第一著者
first comer : 最も早く飛来した個体
first discovery : 初発見
first filial generation : 雑種第一代目
first generation : 第一世代
first generation adult : 第一世代成虫
first instar : 初齢、一齢
first line of defence : 防衛の最前線
first passage time : 初回の継代時間、初通過時間、初到達時間
first record : 初記録
first reviser : 最初の改訂者、第一校訂者
first segment : 第一節

first-generation hybrid : 第一代雑種、第一世代の雑種
FISH : 蛍光インサイチュー・ハイブリダイゼーション法（Fluorescent *In-situ* Hybridization）
fish genome : 魚類のゲノム
Fisher's exact test : フィッシャーの正確確率検定
fishing line : 釣糸
fitness : 適応度、フィットネス
fitness benefit : 適応度向上の利益
fitness criterion : 適応度基準
fitness parameter : 適応度因子、適応度パラメータ、適応パラメータ
fitted model value : 当てはめられたモデル値
fitted value : あてはめ値、当てはめられた平均値
five prime untranslated region : 5' 非翻訳領域（5'UTR）
fixation : 固定
fixation by elimination : 消去法による固定
fixation index : 固定指数
fixation of mimicry gene : 擬態遺伝子の固定
fixation probability : 固定確率
fixative : 固定剤、定着剤、色留め剤
fixed day : 定日型（ていじつがた）
fixed divergent : 固定、固定分岐
fixed instar number : 一定の齢数、固定の齢数
fixed nonsynonymous substitution : 非同義性固定置換
fixed synonymous substitution : 同義性固定置換
fixed time diapause : 定時的休眠、定時休眠
flabellate : 扇状の（触角の形状）〔英語〕
flagellar : 鞭節の
flagellomere : 鞭小節
flagellum : 鞭節（べんせつ）〔ラテン語〕
flagship group : 象徴グループ
flagship insect : 象徴昆虫
flagship species : 象徴種
flame ionization detector : 水素炎イオン化型検出器
flameshift : フレームシフト
flange : 突縁（蛾類）
flank : 側面を接する
flanking region : フランキング領域（遺伝子上流側に隣接する領域）
flanking sequence : フランキング配列
flapping wing : バタバタさせている翅、羽ばたきをする
flash coloration : 閃光色
flat cover : 扁平な袋
flat shaped : 平坦状の
flattened club : 平べったい先端
flattened spheroidal surface of wing : 扁平な楕円体状の翅表面
flavin adenine dinucleotide : フラビンアデニンジヌクレオチド（FAD）
flavone : フラボン
flavonoid : フラボノイド
flavonoids : フラボノイド系色素
flavonol : フラボノール
flavonol glycoside : フラボノールグリコシド、フラボノール配糖体
fleeing point : 逃避点
flesh fly : ニクバエ
fleshy : 肉質の
fleshy spine : 肉質の突起
flexor : 筋、筋肉、屈筋
flier form : 移動型
flight : 飛翔、移動飛翔
flight ability : 飛翔能力、飛行能力
flight activity : 飛翔活動
flight area : 飛翔領域、飛翔する領域
flight behavior polymorphism : 飛翔行動多型
flight boundary layer : 飛行境界層
flight capacity : 飛翔能力

flight control : 飛翔調節、飛翔制御
flight distance : 移動距離
flight morphology : 飛翔関連形態
flight muscle : 飛翔筋
flight muscle mass : 飛翔筋重量
flight muscle polymorphism : 飛翔筋多型
flight path : 移動経路
flight period : 発生期間
flight polymorphism : 移動多型
flight season : 飛翔の季節
flight sound : 飛翔音
flight sound of avian predator : 鳥捕食者の飛翔音
flight space : 飛翔空間
flight speed : 飛翔速度
flight time : 飛翔時間
flight-related behavior : 飛翔関連行動
flood mitigation : 洪水緩和
flora : (複 . -ae)、植物相、フローラ
floral belt : 植生帯、植物帯
floral chemistry : 花化学
floral composition : 花の構成
floral nectar : 花蜜
floral organ formation : 花器形成
floral scent : 花の香気成分、花の発散香気成分
floral scent emission : 花の香りの発散、花の香りの放散
floral vegetation : 花植物
floral volatile : 花の揮発性成分、花の揮発性物質
floral volatile benzyl acetone : 花が放散する揮発性ベンジルアセトン、花由来の揮発性ベンジルアセトン
floristic element : 植物相的要素
floristically : 植物相的に
flour beetle : 小麦粉につく甲虫、コクヌストモドキ
flourish : 繁茂する
flower abundance : 花の個体数
flower and blossom fascinating butterfly : 蝶を呼ぶ花
flower constancy : 定花性、花選好性、花種選好性
flower garden : 花壇、花園
flower phenology : 開花フェノロジー、花歴学
flower preference : 花選好性
flower shape : 花の形状、花形
flower visitor : 花の訪問者
flower-like spider : ハナグモ
flower-visiting behavior : 訪花行動
flowering crop : 開花作物
flowering individual : 顕花個体
flowering plant : 顕花植物
fluctuate : 変動する
fluctuating asymmetry : 変動非対称性、(左右)対称性のゆらぎ、ゆらぎの非対称性
fluctuating temperature treatment : 変温処理
fluctuation : 変動、ゆらぎ
fluid : 流動体、液体
fluid secretion : 液体分泌、水分泌
fluorescence : 蛍光
fluorescent bulb : 蛍光灯電球
Fluorescent *In-situ* Hybridization : 蛍光インサイチュー・ハイブリダイゼーション法(FISH)
fluorescent mounting medium : 蛍光染色用封入剤
fluorescent paint : 蛍光塗料
fluorescent-labelled primer : 蛍光標識したプライマー
flutter : 羽ばたきをする、飛翔する
flutter response : はばたき反応
flutter wing : 翅をパタパタと開閉する
fluttering speed : 飛翔速度
fly : 飛ぶ、ハエ
fly vertically from low to high altitude : 平地と高標高地間を垂直に移動する

fly way : 蝶道、飛行経路
flycatcher : ヒタキ(小鳥の一種)
flying off : 飛び散る
flying pattern : 飛翔パターン
flying type : 飛翔型
focal ablation : フォーカスの切除
focal butterfly : 注目の蝶
focal graft : フォーカスの移植
focal group : 焦点をあてたグループ、主要なグループ
focal period : 注目の期間
focal signal : フォーカスシグナル、焦点信号
focus : (複 . -ci)、フォーカス、焦点
foe : 敵、かたき、敵対者
fold : 折りたたむ、ひだ、褶曲、重なり
fold difference in expression : 発現量の倍数差(2 fold difference ; 2倍差)
fold-change difference : 発現比の差、倍数変化の差
folded : よじれた
foliage : 群葉、葉、葉物
foliage density : 群葉密度
foliage type for oviposition : 産卵に適した葉の種類
folivore : 葉食動物
follicle : 濾胞
follicular cell : 濾胞細胞
following day : 次の日、翌日
following procedure : 次の手順
following project : 後続プロジェクト、後継事業
following protocol : 次の作製手順(書)、次のプロトコール
food abundance : 食物の豊富さ
food chain : 食物連鎖、食物網
food donor plant : 食餌供体植物
food gathering : 採食
food habit : 食性
food intake : 食餌摂取(量)
food limitation : 食料制限、食料制約
food plant : 食草、食餌植物
food recruitment : 食物調達、食物確保
food requirement : 食物要求、食餌要求
food resource : 食餌資源
food shortage : 食物不足
food supply : 食物供給
food web : 食物連鎖、食物網
foodplant change : 食草転換
fool : 騙(だま)す
forage : 食糧をあさる、食糧を入手する
foraging : 採餌
foraging behavior : 採餌行動
foraging preference : 採餌選好性
foraging site : 食料を入手する場所
foraging strategy : 採餌戦略
foraging success : 採餌成功(率)
foraging trip : 採餌旅行
foramen : 後頭孔、孔、窩(か)
foramen magnum : 後頭孔、大窩
fore gut : 前腸(ぜんちょう)
fore leg : 前脚(ぜんきゃく)、前肢(ぜんし)
fore wing : 前翅(ぜんし)、上翅(じょうし)
fore wing sheath : 前翅芽
fore- : 前 - 〔ラテン語〕
forefront : 最前線
foregut : 前腸
foreign pest : 外来性害虫
foreleg : 前脚(ぜんきゃく)、前肢(ぜんし)
foreleg folded on its thorax : 胸部に折り畳まれている前脚
foreleg magnified : 前脚の拡大
forest : 森林
forest and relictual genus : 森林遺存属
forest and wide-distributed butterfly : 森林性広域分布蝶
forest and wide-distributed genus of Southeast Asian origin : 東南アジア森林広域分布属

forest butterfly : 森林性蝶
forest canopy habit : 林冠性
forest clearing : 森林開拓地
forest dweller : 森の住人
forest edge : 林縁部、森林の周縁域
forest environment : 森林的環境
forest fire : 森林火災
forest floor : 林床
forest floor habit : 林床性
forest gap : 森林ギャップ
forest margin : 林縁
forest margin area : 森林の周縁域
forest matrix : 森林マトリックス
forest non-gap : 林冠
forest remnant : 森の残部、残っている森林
forest species : 森林性種
forest structure : 森林構造
forest tent caterpillar : オビカレハ幼虫、森林天幕毛虫
forest type : 森林型
forest, savanna and wide-distributed genus : 森林サバンナ広域分布属
forest-dwelling species : 森林性種
forested ecosystem : 森林生態系
forested habitat : 森林性の生息地
forested landscape : 森(林)に覆われた景観
forested protected area : 植林保護地域
forestland : 森林地、森林
forestry : 林業
foretarsal receptor : 前脚ふ節感覚器、前脚ふ節受容器
foretarsus : 前脚ふ節
forewing (FW) : 前翅(ぜんし)、上翅(じょうし)
forewing band : 前翅の帯
forewing bud of pupa : 蛹の前翅芽、蛹期の前翅原基
forewing color : 前翅の色
forewing color gene : 前翅の色彩遺伝子
forewing fringe color : 前翅縁毛色
forewing length : 前翅長
forewing overlying hindwing : 後翅を覆っている前翅
forewing patch : 前翅の大斑点
forewing structure : 前翅の構造
forewing upperside : 前翅表面
forewing's apex : 前翅の最上端部
form : 型、形、態
form ectopically : 正常でない位置に起こる形
formal demonstration : 形式的論証
formalin solution : ホルマリン溶液、ホルマリン水溶液
formation mechanism : 形成機構
formation of ectopic eyespot : 転位した眼状紋の形成
formation of garter : 帯糸形成
formation process : 形成過程
formerly-declining butterfly : 以前に衰亡しつつあった蝶
fortify : 防備を強化する
forward and reverse primers : フォワードプライマーとリバースプライマー、順方向プライマー及び逆方向プライマー
forward or reverse reading : 順方向または逆方向の読み込み
fosmid : フォスミド
fosmid clone : フォスミドクローン
fosmid library : フォスミドライブラリー
fosmid sequencing : フォスミド配列、フォスミドシークエンシング
fossil : 化石
fossil calibration : 化石記録に基づく年代較正
founder : 創始者
founder effect : 創始者効果
founder event : 創始者事象
founder selection : 創始者選択
four degenerate site : 四重縮退サイト

FPKM value : FPKM 値(Fragments Per Kilobase of exon per Million mapped sequence fragments または Fragments Per Kilobase of transcript per Million mapped reads)(全サンプルの中央値に対する百万単位読み取り断片あたりの千単位エクソン断片の正規ログ比)
fragmented landscape : 分断された地形、分断化された景観
fragmented population : 分断化された個体群、断片化された個体群
fragrance composition : 芳香組成物
frame shift mutation : フレームシフト突然変異
frass : 糞粒、糞
free glycine : 遊離グリシン、遊離型グリシン
free progress : 自由進行
free-run : 自由継続する
free-running period : 自由継続周期
free-running rhythm : 自由継続リズム
freeze intolerance : 非耐凍性
freeze susceptible : 非耐凍性、凍結感受性
freeze tolerance : 耐凍性
freeze-intolerant insect : 非耐凍型昆虫、非耐凍性昆虫
freeze-intolerant species : 非耐凍型種、非耐凍性種
freeze-tolerant species : 耐凍性種
freezing injury : 凍傷
French Guiana : 仏領ギアナ
frenulum : 翅鉤(はねかぎ)、翅棘(しきょく、はねとげ)、翅刺(しし)
frequency : 周波数、振動数、頻度
frequency demultiplication hypothesis : 周波数積算仮説
frequency dependent selection : 頻度依存選択
frequency discrimination : 周波数弁別
frequency distribution : 頻度分布
frequency of adult : 成虫の頻度
frequency of flight bout : 飛翔期間(一続きの行動期間)の頻度、飛翔バウト回数
frequency of the yellow allele : 黄色対立遺伝子頻度
frequency-dependent competition : 頻度依存的な競争
frequency-dependent predation : 頻度依存的な捕食、頻度依存性捕食
frequency-dependent selection : 頻度依存選択
frequent : しばしば行く
frequently cut : しばしば伐採されている
fresh cutting : 新鮮な切断部分
fresh medium : 未使用の培養液、未使用培地
fresh mutation : 新生突然変異
fresh nectar : 新鮮な果汁
fresh weight : 生体重、新鮮重
freshly emerged female : 新鮮な羽化したばかりの雌
freshly emerged specimen : 生き生きと出現した標本
freshwater fish : 淡水魚
freshwater invertebrate : 淡水産無脊椎動物
fright tactic : 脅かし戦術
fringe : 縁毛、縁毛帯、フリンジ
fringe color : 縁毛色
Fritillary : ヒョウモンチョウ
fritillary marking of black spot on an orange ground : オレンジ色の地色に黒斑のあるヒョウモンチョウの翅模様
frog : カエル
from autumn through next spring : 秋から翌春にかけて
from winter diapause onward : 冬眠以降、冬眠以後
frons : (複 . frontes)、額、ひたい、額板、前頭〔ラテン語〕
front pair of leg : 前脚、前肢
frontal carina : 額線、額隆起線(頭部)

frontal line : 額線(頭部)
frontal tubercle : 額伸縮突起(蛾類)、額瘤(がくりゅう)
frontispiece : 口絵
frontoclypeal suture : 額頭盾縫合線(がくとうじゅんほうごうせん)、前頭 - 頭盾縫合線(頭部)
frost : 氷点下、氷結
frozen sample : 冷凍標本
fruit fly : ミバエ、ショウジョウバエ
fruit mimic : 果実に擬態
fruit orchard : 果樹園
fruit-feeder : 果物食者
fruitfully environment : 多産な環境、実り豊かな環境
fuel : たきつける、あおる
full brother : 同父母の兄(弟)、実の兄(弟)
full ORF : 全長ORF、全長オープンリーディングフレーム(Open Reading Frame)
full protection : 完全保護
full sibling : 全部の近縁種、全同胞種
full-grown larva : 完全に成長した幼虫、十分成長した幼虫
full-length : 全長
full-length sequence : 全長配列
fully grown : 終齢
fully protected species : 完全保護種、完全に保護されている種
fully-formed larva : 成体が完成した幼虫
fumigant : 燻蒸剤(くんじょうざい)
function downstream of the SA signaling pathway : SA(サリチル酸)の信号伝達経路の機能性下流
functional analysis : 機能解析
functional annotation : 機能注釈
functional difference : 機能的差異
functional divergence : 機能的分岐、機能分化、機能多様性、機能発散
functional female : 機能的な雌、機能性雌
functional insect ear : 昆虫の機能的な耳
functional interrelationship : 機能分担
functional locus : 機能的遺伝子座
functional morphology : 機能的形態学、機能形態学
functional organization : 機能的機構、機能的構成、機能的構築
functional performance : 機能的パフォーマンス
functional response : 機能の反応、機能的反応
functional role : 機能的役割
functional study : 機能研究、機能解析研究
functionally related gene : 機能的に近縁な遺伝子
fundamental niche : 基本ニッチ、基本生息場所
fundatrix : 幹母
fungal genome : 真菌類のゲノム
fungal infection : 糸状菌感染、菌類感染
fungus : (複 . -gi)、糸状菌、カビ、菌類
funicle : 繋節
funicular : 繋節の、糸節の
funiculus : 鞭節、触角鞭状部
furcal pit : 叉甲孔(胸部)
furcasternum : 叉甲腹板(胸部)
further afield : より遠く離れた
further experiment : 継続実験、追加実験
fuse : 融合する
fused eyespot doublet : 融合して一つになった眼状紋の対
fused microvillar membrane : 融合した微絨毛膜
fusiform : 紡錘状(ぼうすいじょう)の(触角の形状)
fusion : 癒合(ゆごう)
fusion gene : 融合遺伝子
fusion of gradients : 勾配群の融合
future direction : 将来の方向性、今後の方向性

future study : 将来の研究、今後の研究

g

G protein-coupled receptor : Gタンパク質結合受容体、Gタンパク質共役型受容体
G-protein : G タンパク質
g. : 属（genus）
g. nov. : 新属名（genus novus）
G6PDH : グルコース-6-リン酸デヒドロゲナーゼ（Glucose-6-Phosphate DeHydrogenase）
GA : 遺伝的アルゴリズム（Genetic Algorithm）
gain of function : 機能亢進、機能獲得
galea : （複 . -ae）、外葉〔ラテン語〕
gall : ゴール、虫癭(ちゅうえい)、虫こぶ
gallery forest : ガレリア森林（サバンナなどの川沿いの帯状林）、拠水林
gamete : 配偶子
gamete duplication : 生殖核倍加型
gametic isolation : 配偶子隔離、配偶子単離
gamma distribution : ガンマ分布
gamma irradiation : γ 線照射、ガンマ線照射
gamma model of rate heterogeneity : 置換速度の不均質性のガンマモデル
gamma rate category : ガンマ率のカテゴリ数、ガンマ速度のカテゴリ数
gamma-distributed rate : ガンマ分布比
gamma-diversity : ガンマ多様性、γ- 多様性（「景観多様性」のことで、「対象とするすべての環境での種多様性」である）
ganglion thoracicum primum : 第一胸節神経球
gap : ギャップ、割れ目、隔たり
gap habitat : ギャップ生息地
gap penalty : ギャップペナルティ
GAPDH : GAPDH 遺伝子（グリセルアルデヒド -3- リン酸脱水素酵素）（Glyceraldehyde-3-phosphate dehydrogenase）
garden tiger moth : ヒトリガ
garden warbler : ニワムシクイ
gardening : ガーデニング、園芸、庭造り
gas chromatography : ガスクロマトグラフィー
gas exchange : ガス交換
gastrocoel : 原腸のう
gate : ゲート
gatekeeper butterfly : ゲートキーパー（ジャノメチョウ科の一種）
GC : ガスクロマトグラフィー（Gas Chromatography）
GC-biased gene conversion : GC 含量に偏りのある遺伝子変換（Guanine and Cytosine ; グアニンとシトシン）
GC-MS : ガスクロマトグラフィー - マススペクトロメトリー（Gas Chromatography Mass Spectrometry）
gel : ゲル〔ドイツ語〕、ジェル〔英語〕
gen- : 遺伝の -、遺伝子の -〔ギリシャ語〕
gen. : 属（genus）
gen. nov. : 新属名（genus novus）
gena : 頬〔ラテン語〕
genal : 頬の
GenBank accession no. : 遺伝子銀行受入れ番号、ジーンバンクアクセス番号
gender : 性別、性
gender agreement : 性の一致
gender ending : 性語尾
gender-specific expression : 性別特異的な発現、性差発現
genders : 雄雌、雄と雌、ジェンダー
gene : 遺伝子
gene acquisition : 遺伝子獲得
gene annotation : 遺伝子注釈
gene birth : 遺伝子の新生
gene controlling pigmentation : 色素沈着制御遺伝子
gene conversion : 遺伝子変換
gene Cubitus-interruptus : 肘脈中断遺伝子（翅脈が途切れている）

gene death : 遺伝子の消失
gene diversity : 遺伝子多様性
gene duplication : 遺伝子重複
gene expression : 遺伝子発現
gene expression analysis : 遺伝子発現解析
gene expression cascade : 遺伝子発現カスケード
gene expression study : 遺伝子発現研究
gene family : 遺伝子ファミリー、遺伝子族
gene flow : 遺伝子流動、遺伝子の流れ、遺伝子交流、遺伝的交流
gene flux : 遺伝子流動
gene genealogy : 遺伝子系図、遺伝子の系譜、遺伝子系統学、遺伝子系図学
gene knockout : 遺伝子欠損、遺伝子破壊
gene level : 遺伝子レベル
gene loss : 遺伝子喪失、遺伝子欠失
gene mutation : 遺伝子突然変異
gene ontology : 遺伝子オントロジー、遺伝子概念体系（遺伝子の機能の記述に関して、生物学分野における共通語彙を策定するプロジェクト）（GO）
gene order : 遺伝子配列順、遺伝子順序
gene pool : 遺伝子プール
gene product : 遺伝子産物
gene region : 遺伝子領域
gene sequence : 遺伝子配列
gene specific primer : 遺伝子特異的プライマー
gene transfer agent : 遺伝子導入体、遺伝子伝播因子（GTA）
gene tree : 遺伝子系統樹
gene *wingless* : 無翅遺伝子
gene-encoding region : 遺伝子のコーディング領域
gene-family dynamics : 遺伝子ファミリー動態
gene-for-gene coevolution : 遺伝子対遺伝子共進化
gene-specific reverse primer : 遺伝子特異的逆転写用プライマー
genealogical change : 系統変化
genealogical tree : 系統樹
genealogy : 系統、系譜
general epidermal cell : 一般的な表皮細胞
general functional category : 一般的な機能分類
general primer : 普遍プライマー、普遍的なプライマー、ゼネラルプライマー
general protocol : 一般的なプロトコル、汎用プロトコール
general sampling method : 一般的なサンプリング法
general term : 一般用語
general time reversible model : 一般時間反転可能モデル
general transcription factor : 基本転写因子（GTF）
generalist : ジェネラリスト、万能家
generalist butterfly : 広食性の蝶、ジェネラリストの蝶
generalist feeding : 雑食性
generalist species : 普遍性の種
generalization : 一般化、普遍化
generalized linear model : 一般化線形モデル
Generalized Mixed Yule-Coalescent model : GMYC モデル、一般化融合（混合）Yule-Coalescent モデル（調査地域に生息する種数を推定する手法で、種分化の分岐に Yule モデルを、種内変異の分岐に Coalescent モデルを当てはめたモデル）
generation : 世代
generic name : 属名
generic placement : 属の所属
generic richness : 属数
genesis : 発生
genetic algorithm : 遺伝的アルゴリズム（GA）

genetic analysis : 遺伝子解析、遺伝子分析
genetic and biochemical approach : 遺伝生化学的手法
genetic architecture : 遺伝的構成
genetic association : 遺伝相関
genetic background : 遺伝的基盤、遺伝的背景
genetic basis : 遺伝的基盤、遺伝的基礎
genetic cause : 遺伝的原因
genetic character of metapopulation : メタ個体群の遺伝的形質
genetic code table : 遺伝暗号表
genetic constitution : 遺伝的構成、遺伝学的構成
genetic constraint : 遺伝的制約
genetic control : 遺伝的制御、遺伝的防除
genetic control of plasticity : 可塑性の遺伝的制御
genetic correlation : 遺伝相関
genetic coupling : 遺伝的結合、遺伝的連関
genetic covariance : 遺伝共分散
genetic cross : 遺伝的交雑
genetic diapause : 遺伝的休眠
genetic difference : 遺伝的差異
genetic differentiation : 遺伝的分化
genetic distance : 遺伝(的)距離、地図距離
genetic distinction : 遺伝的相違、遺伝的差異、遺伝的区別
genetic distinctiveness : 遺伝的特殊性
genetic divergence : 遺伝的分化
genetic diversity : 遺伝的多様性
genetic drift : 遺伝的浮動
genetic effect : 遺伝的影響
genetic factor : 遺伝的要因
genetic form : 遺伝型
genetic homology : 遺伝的相同性
genetic incompatibility : 遺伝的不一致、遺伝的不和合性
genetic information : 遺伝子情報、遺伝情報
genetic integrity : 遺伝的完全性
genetic level : 遺伝的レベル
genetic linkage map : 遺伝子連鎖地図、遺伝的連鎖地図
genetic linkage of sexual isolating trait : 性的隔離に関する形質の遺伝的連鎖
genetic load : 遺伝的負荷、遺伝荷重、遺伝的加重
genetic male : 遺伝的雄
genetic mapping : 遺伝子マッピング、遺伝子地図作製
genetic marker : 遺伝マーカー、遺伝子マーカー、遺伝標識
genetic mechanism : 遺伝機構
genetic model : 遺伝モデル
genetic mosaicism : 遺伝的モザイク現象
genetic origin : 遺伝的起源
genetic polymorphism : 遺伝的多型
genetic polyphenism : 遺伝的(表現型)多型
genetic relationship : 遺伝的交流、遺伝的関係
genetic response : 遺伝的応答、遺伝的反応
genetic speciation : 遺伝的種分化
genetic statistics : 遺伝統計学
genetic stochasticity : 遺伝的確率性
genetic structure : 遺伝的構造
genetic task specialization : 遺伝的な役割分業
genetic variability : 遺伝的変異性、遺伝変異性
genetic variance : 遺伝分散
genetic variant : 遺伝的変異体
genetic variation : 遺伝的変異、遺伝性変異、遺伝的変動
genetically associated : 遺伝的に相関した
genetically determine : 遺伝的に決定する
genetically determined preference : 遺伝的に決まっている選好性
genetically determined trait : 遺伝的に決

まっている形質
genetically diverged lineage : 遺伝的分岐した系統
genetically divergent species : 遺伝的に多様な種
genetically female part : 遺伝的雌部位
genetically inherited : 遺伝的に継承された
genetically male individual : 遺伝的雄個体群
genetically male part : 遺伝的雄部位
genetically modified organism : 遺伝子組み換え生物(GMO)
genetically most distant : 遺伝的に最も離れている
genetically similar population : 遺伝的に類似な個体群、遺伝的に類似な集団
genetically-diverse population : 遺伝的に多様な個体群
genetics : 遺伝学、遺伝的特徴
genetics of mimicry : 擬態(の)遺伝学
genetics of pigmentation : 色素沈着の遺伝学
genic selection : 遺伝子淘汰
geniculate antenna : (複 . -nae)、鋭角に屈曲した触角
genital papilla : 側唇、生殖瘤状突起
genital part : 生殖部位
genital photoreceptor : 尾端光受容器
genital valve : 生殖弁(せいしょくべん)、産卵弁(さんらんべん)
genitalia : ゲニタリア、生殖器、交尾器〔ラテン語〕
genitalia vial : ゲニタリアチューブ
genitive ending : 属格語尾
geno- : 遺伝の -、遺伝子の -〔ギリシャ語〕
genome : ゲノム
genome assembly : ゲノムアセンブリー
genome average relationship : ゲノム平均の関係
genome coverage : ゲノムのカバー率
genome destabilization : ゲノム不安定化
genome duplication : ゲノム重複
genome imprinting : ゲノムインプリンティング、遺伝的刷り込み
Genome Information Broker : ゲノム情報ブローカー(GIB)
genome instability : ゲノム不安定性
genome of baker's yeast : パン酵母のゲノム
genome profiling : ゲノムプロファイリング法(生物種がもつ全ゲノムの配列比較による系統解析法)
genome scaffold : ゲノム上の足場
genome scale : ゲノム尺度
genome sequencing : ゲノム配列、ゲノムシークエンシング
genome size : ゲノムサイズ
genome tree : ゲノム系統樹
genome walking : ゲノム配列歩行、ゲノムウォーキング
genome wide SNP datum : (複 . -ta)、ゲノム規模の SNP データ、ゲノム規模の一塩基多型データ(Single Nucleotide Polymorphism)
genome-wide : ゲノム規模の
genome-wide duplication status : ゲノム規模での重複状態、ゲノムワイドでの重複状態
genome-wide introgression : ゲノムワイドの遺伝子移入
genomic BAC : ゲノム BAC、ゲノムバクテリア人工染色体、ゲノム細菌人工染色体(Bacterial Artificial Chromosome)
genomic block : ゲノムブロック
genomic datum : (複 . -ta)、ゲノムデータ
genomic imprinting : ゲノムインプリンティング、ゲノム刷り込み
genomic incompatibility : 遺伝的不和合性、遺伝的不一致
genomic location : ゲノム位置、遺伝子位置
genomic locus : (複 . -ci)、ゲノム遺伝子座

genomic mechanism : ゲノム機構
genomic mutation : ゲノム突然変異
genomic read : ゲノムリード
genomic region : ゲノム領域
genomic resequencing : ゲノム塩基配列の再解析
genomic resource : ゲノムリソース、ゲノム資源
genomic scaffold : ゲノムスキャフォールド、ゲノム骨組
genomic study : ゲノム研究
genomic tool : ゲノミクスの手法
genomics : ゲノミクス(ゲノムと遺伝子についての研究)
genotype : 遺伝子型、遺伝型、遺伝子型を決定する
genotype segregating allele : 遺伝子型の分離対立遺伝子
genotype-environment interaction : 遺伝子型 - 環境相互作用、遺伝子型と環境との相互作用
genotypic difference : 遺伝子型の違い、遺伝型の違い
genotyping : 遺伝子型同定、遺伝子型解析、遺伝子型判定
genotyping by sequencing : GBS 法、ジェノタイピングシークエンシング、塩基配列による遺伝子型解析
Genoveva azure butterfly : オオヤドリギシジミ
genuine transit record : 真正の移動記録
genus : (複 . genera)、属(分類階級の「ぞく」)〔ラテン語〕
genus [hg.] : (単複同形)、膝(ひざ)〔ラテン語〕
genus group : 属階級群
genus name : 属名
genus of neotropical butterfly : 新熱帯区に生息する蝶の属
genus-group name : 属階級群名
geographic area : 地理的地域、地理的な場所
geographic distance : 地理的距離
geographic distribution : 地理的分布
geographic isolation : 地理的隔離
geographic location : 地理的位置、地理的場所
geographic mosaic : 地理的モザイク
geographic pattern : 地理的パターン
geographic population : 地理的個体群
geographic radiation : 地理的放散
geographic range : 地理的分布域
geographic range expansion : 分布拡大
geographic restriction : 地理的制限、地理的制約
geographic scale : 地理的規模
geographic source : 地理的産地
geographic speciation : 地理的種分化、地理的な分化、地理的分化
geographic species : 地理的種
geographic strain : 地理的系統
geographic variation : 地理的変異、地域変異
geographical background : 地理的背景、地誌的背景
geographical cline : 地理的クライン、地理的勾配
geographical coverage : 地理的範囲
geographical distribution type : 地理的分布型
geographical dorsal pattern variation : 背部の模様パターンの地理的変異
geographical gradient : 地理的勾配
geographical race : 地理的系統、地理的品種、地理的亜種
geographical region : 地理区
geographical resolution : 地理的解像度
geographical space : 地理的空間
geographical variation : 地理的変異
geographically close : 地理的に近傍な
geohistorical background : 地史的背景

geological and historical background : 地史的背景
geological succession : 地史的遷移
geological time : 地質時代
geological variation : 地域変異、地史的変異
geomagnetic coordinate : 地磁気座標
geometry of flight path : 飛翔経路の形状、飛翔経路のジオメトリー
geophagy : 土食、土壌食性
geraniol : ゲラニオール（バラ香の化粧品香料）
germ band : 胚帯
germ cell : 胚細胞、生殖系列細胞、生殖細胞
germinate : 芽を出す、発芽する
germination success : 発芽成功
germline cell : 生殖細胞系列の細胞
germline determined sex : 生殖細胞で決定された性
GFP : 緑色蛍光タンパク質（Green-Fluorescent Protein）
ghost authorship : ゴースト著者
ghost moth : コウモリガ
giant glial cell : 巨大グリア細胞
giant redeye butterfly : コウモリセセリ
giant swallowtail butterfly : クレスフォンテスタスキアゲハ、オオタスキアゲハ
Giant-Skippers : イトランセセリ亜科
GIB : ゲノム情報ブローカー（Genome Information Broker）
gift authorship : ギフトオーサーシップ
girdle : 帯、帯糸（たいし）
girdle for support : 支持糸
girdling : 帯糸をかける
gizzard : 砂のう（さのう）
glabrous waxy leaf surface : 無毛の光沢のある葉表面
glacial epoch : 氷期
glacial period : 氷河時代、氷河期
glacial refugium : 氷河期の退避地
glanville fritillary butterfly : グランヴィルヒョウモンモドキ
glass capillary : ガラス製毛細管、ガラスキャピラリー
glass knife : ガラスナイフ
glass microelectrode : ガラス微小電極
glass microscope slide : 顕微鏡用のスライドガラス
glassine envelope : グラシン（紙）の三角紙
gleaning bat : ウサギコウモリ
glial cell : グリア細胞
glial cell layer : グリア細胞層
GLM : 一般化線形モデル（Generalized Linear Model）
glm command : glm コマンド、glm 関数（一般化線形モデル〈generalized linear model〉を扱う関数）
global biodiversity loss : 世界の生物多様性喪失
global dN/dS (ω) ratio : 複数個の非同義置換率（dN）と同義置換率（dS）の比、全体の dN/dS 比
global environmental problem : 地球規模の環境問題
Global Position System : 全地球測位網、全地球測位システム
global positioning system : 全地球測位システム（GPS）
Global Register of Invasive Species : グローバル侵入種登録簿
global warming : 地球温暖化
globalization : グローバリゼーション、世界的規模
globe : 地球
glossa : （複 . -ae）、中舌〔ギリシャ語〕
glossotheca : 蛹鞘、舌鞘
glucose : グルコース
glucose-6-phosphate dehydrogenase : グルコース -6- リン酸デヒドロゲナーゼ

glucosinolate : グルコシノレート、カラシ油配糖体
glucosinolate biosynthesis : グルコシノレート生合成
glucosinolate compound : グルコシノレート化合物
glucosinolate defense system : グルコシノレート防御システム
glucosinolate detoxification : グルコシノレートの解毒化
glucosinolate-containing plant : グルコシノレート含有植物
glue : のり付けする
glutathione-S-transferase : グルタチオン・S・トランスフェラーゼ、グルタチオン S- 転移酵素
glycerin : グリセリン
glycerol accumulation : グリセロールの蓄積
glycine-rich : 高グリシン含有
glycogen : グリコーゲン
glycogen content : グリコーゲン含量
glycolysis : 解糖系、解糖
glycoside : 配糖体、グリコシド
glycosidically-bound volatile : グリコシド結合揮発性物質
glymma : (複 . -ae)、柄側刻〔ラテン語〕
GMO : 遺伝子組み換え生物(Genetically Modified Organism)
gnathos (gn) : 顎(がく)、あご
gnathos : 顎腕(がくわん)(雄の交尾器)(テグメンの後縁と関節した1対の突起で、ウンクスとソキウスの関節部の腹側方に位置する)
GO : 遺伝子オントロジー、遺伝子概念体系(Gene Ontology)
GO term : GO 用語(GO で定義された用語)(Gene Ontology)
goblet cell : 杯状細胞
gold annulus : 金環
gold ring : 金環
gold-drop helicopis butterfly : ミツオシジミタテハ
golden birdwing butterfly : キシタアゲハ
golden piper butterfly : チャオビタテハ
goldpalladium : 金パラジウム、金パラ
golgi body : ゴルジ体
Goliath birdwing butterfly : ゴライアストリバネアゲハ
gonadal development : 生殖腺の発育
gonocoxite : 生殖基節、生殖肢基節
gonotome : 生殖節(せいしょくせつ)
good gene model : 優良遺伝子モデル
Gossamer Wings : シジミチョウ科
Gossamer-wing Butterflies : シジミチョウ科
GPCR : G タンパク質結合受容体(G Protein-Coupled Receptor)
GPS : 全地球測位網、全地球測位システム(Global Positioning System)
Gr : 味覚受容体(Gustatory receptor)
Gr5a : 味覚受容体遺伝子の一つ(Gustatory receptor 5a)
Grace's medium : Grace 培地、グレース昆虫培地
graceful appearance : 美麗種
gradient : 勾配、グラディエント、段階的変化
gradient profile : 勾配プロフィール、勾配プロファイル
gradient variation : 勾配変異
gradual : ゆるやかな
gradual change : 漸進的変化
gradual decrease : 徐々に減少、漸減
gradual evolution : 漸進的進化
gradual tightening of linkage : 連鎖を徐々に密にして行く
graft : 移植組織、移植片、接ぎ木
grafted focus : 移植したフォーカス
grafting experiment : 移植実験
granulocyte : 顆粒細胞(かりゅうさいぼう)

granulose : 粒状の、ザラザラした、細顆粒状の
granulosis : 顆粒病(昆虫ウイルス病の一種)
granulosis virus : 顆粒病ウイルス
granulovirus : 顆粒病ウイルス
Grass Skippers : セセリチョウ亜科
grass with egg : 卵付きの草
grass with oviposition : 産卵された草
grass without egg : 卵なしの草
grasshopper : キリギリス、バッタ、イナゴ
grassland : 草原地帯、草原、草地、牧草地
grassland and relictual genus : 草原遺存属
grassland and wide-distributed genus : 草原広域分布属
grassland butterfly : 草原性蝶
grassland environment : 草原的環境
grassland genus of American origin : アメリカ起源草原属
grassland genus of Eurasian origin : ユーラシア起源草原属
grassland habitat : 草原性生息地
gravid : 抱卵、受胎(じゅたい)
gravity : 重さ、重量、重力
grazing : 放牧
greasy : 油紙のような
great detail : きわめて詳細、非常に詳細
great purple hairstreak butterfly : アメリカヤドリギシジミ(ヤドリギは幼虫の食餌植物)
Greater Sunda Islands : 大スンダ列島
greatest common divisor : 最大公約数
greedy algorithm : 貪欲アルゴリズム、貪欲法
Greek : ギリシャ語
green beard effect : 緑髭効果
green fluorescence : 緑色蛍光
green leaf herbivory : 青葉の植食、緑色葉の植食(性)
green leaf tissue : 青葉組織、緑色葉の組織
green leaf volatile : 青葉の揮発性物質
green light : 緑色光
green revolution : 緑の革命
green-blue marking : 緑青色斑紋
green-fluorescent protein : 緑色蛍光タンパク質(GFP)
green-veined white butterfly : エゾスジグロシロチョウ、ヤマトスジグロシロチョウ
greenhouse : 温室
greenhouse effect : 温室効果
greenhouse gas : 温室効果ガス
gregarious : 群居する、群生する、群居性の、群生性の
gregarious phase : 群生相
gregarious roosting : 集団帰塒(しゅうだんきじ)
gregariously : 集合性があり、群れて
gregariousness : 集合性、群居性、群生、群居
grid square : グリッド四方
grid squares, 10km : 10km グリッド四方
grind : すりつぶす、磨砕する
GRIs : グローバル侵入種登録簿(Global Register of Invasive species)
groove : 溝
ground : 地面
ground color : 地色
ground method : 基礎的手法
ground scale : グランドスケール
ground speed : 対地速度
ground with a pellet and a mortar : 乳棒と乳鉢で磨砕した(「ground」は「grind」の過去形)
group : 集団、個体群、群、階級群
group selection : 群淘汰、集団選択、群選択
grouping behavior : グループ行動、集団行動
grow : 成長
growing period : 生育期、成長期
growing season : 生育期、成長期
growth : 成長、成長量、成長率、生長
growth chamber : 培養室
growth condition : 成長条件、生育条件

growth day : 発育日数
growth inhibitor : 成長･発育を阻害する物質
growth rate : 成長速度、成長率
growth rate difference : 成長速度の差異
growth trait : 成長形質
grub : (甲虫などの)幼虫
Grylloblattodea : ガロアムシ目、欠翅目
GST : グルタチオン･S･トランスフェラーゼ(Glutathione-S-Transferase)
GTA : 遺伝子導入体、遺伝子伝播因子(Gene Transfer Agent)
GTF : 基本転写因子(General Transcription Factor)
Guam Island : グアム島
guava skipper butterfly : シロベリセセリ
guest authorship : ゲストオーサーシップ
guide mark : 花標
Guiding Principles for the Prevention, Introduction and Mitigation of Impacts of Alien Species : 外来種の影響の予防、導入、影響緩和のための指針原則
guise : 見せかけ
gula : 咽喉(いんこう)、喉板
gulching : 待ち伏せ
gustatory neuron : 味覚ニューロン、味覚神経細胞
gustatory organ : 味覚器官
gustatory reception capacity : 味覚受容能力
gustatory receptor : 味覚受容体
gustatory sense : 味覚
gustatory sensory hair : 味覚感覚毛
gustatory stimulus : 味刺激
gut flora : 腸内細菌叢(ちょうないさいきんそう)
gut purge : ガットパージ、脱糞、液状糞
GV : 顆粒病ウイルス(GranuloVirus)
gymnosperm : 裸子植物
gynander : 雌雄型
gynandromorph : 雌雄型、ジナンドロモルフ、雌雄モザイク
gynandromorphism : 雌雄モザイク現象
gypsy moth : マイマイガ

h

H : 後翅(こうし)(Hindwing)
h after, 24 : ～後 24 時間、24 時間後
H locus : *H* 遺伝子座
h old, 24 : 24 時間まで、～後 24 時間
habit : 習性
habit-wise : 習性の面では、習性に関して
habitant : 生息生物
habitat : 生息地、生息環境、生育地、自生地、生息域、ハビタット
habitat abundance : 生息地の個数、生息地数
habitat change : 生息地変化、生息域の変化
habitat destruction : 生息地破壊
habitat deterioration : 生息地の悪化、生息環境の悪化
habitat edge : 生息地の周縁
habitat factor : 生息環境要因
habitat fragmentation : 生息地の分断化
habitat islands : 島嶼生息地
habitat isolation : 生息地隔離
habitat loss : 生息地の減少、生息地の消失、生息地の喪失
habitat management : 生息地管理
habitat manipulation : 生息場所の操作
habitat measure : 生息地の比較測定項目、生息地の測定変数
habitat modification : 生息地の改変、生息地の造成
habitat patch : 生息地パッチ
habitat quality : 生息地の質
habitat range : 生息圏
habitat remnant : 生息地の名残
habitat segmentation : 生息地の分断化
habitat segregation : すみわけ
habitat specialist : 生息場所特定者
habitat specialist butterfly : 生息場所特定

性の蝶
habitat specialist species : 生息場所特定性の種
habitat specialization : 生息地の特殊化
habitat specificity : 生息地の特異性
habitat structure : 生息地の構造、生息地の分布構造
habitat suitability : 生息場適性、生息地適正、生息環境適正、生息地適合性
habitat suitability map : 生息地の好適性地図、生息好適性地図、生息環境適正図
habitat templet hypothesis : 生息場所鋳型説
habitat-patch occupancy : 生息地パッチの占有率
habitus : 習性、行動様式
haemocyte : 血球、血球細胞
haemoglobin : ヘモグロビン
haemolymph : 血リンパ
haemolymph sample : 血リンパのサンプル
Haemophilus influenzae : ヘモフィラス・インフルエンザ菌
hair : 毛
hair tuft : 毛束
hair-like : 毛状の
hairless eye : 毛のない眼、無毛の眼
hairpencil : ヘアペンシル、毛束
hairpencil dihydropyrrolizine : ヘアペンシルから 分 泌 されるジヒドロピロリジン化合物
Hairstreaks : カラスシジミ亜科
hairy eye : 毛がある眼、毛状の眼
Haldane centiMorgan : ホールデンセンチモルガン
Haldane's rule : ホールデインの法則、ホールデンの法則
half-life : 半減期
hallucinogenic : 幻覚作用
haltere : 平均棍(へいきんこん)(ハエ目昆虫における後翅の飛翔機能が退化、変化した可動器官を指す)
hamiform : 鉤状(かぎじょう)の(触角の形状)
Hamilton rule : ハミルトン則、ハミルトンの法則
Hamilton's rule : ハミルトン則、ハミルトンの法則
HAMP : 植食者関連分子パターン(Herbivore-Associated Molecular Pattern)
hamper : 邪魔をする、阻止する
Hampson's classification : ハンプソン 式、ハンプソンの分類
hamulus : (複 . -li)、翅鉤(しこう、はねかぎ)
hand manipulation : 手操作
hand pairing method : ハンドペアリング法
hand-pairing : ハンドペアリング
handful of, a : 少数の
handling : 取り扱い
handling time : 処理時間
hapantotype : ハパントタイプ(標本)
haplo- : 単 -、単一 -
haplodiploid sex determination system : 半倍数性の性決定システム、単数二倍体の性決定様式
haplodiploidy : 半倍数性
haplogroup : ハプログループ
haploid : 半数体(単数体)
haploid effective population size : 半数体集団の有効な大きさ
haploid genotype : 半数体の遺伝子型
haplotype : ハプロタイプ("haploid genotype"の略語)、単相の
haplotype analysis : ハプロタイプ解析、分子系統解析
haplotype diversity : ハプロタイプ多様性
haplotype network : ハプロタイプネットワーク
HapMap : ハプロタイプ地図
harassment : 干渉、ハラスメント
harassment activity : 干渉行動

harbor : 宿る、住みかとなる、ひそむ
hard leaf : 硬い葉
hard rain : 大雨
Hardy-Weinberg equilibrium : ハーディワインベルグ平衡
Hardy-Weinberg expectation : ハーディー - ワインベルグ期待値
harem : ハーレム
harm crop : 作物を傷つける、作物に被害を与える
harmless species : 無害種
harmonic mean : 調和平均
harmonic radar : 高調波レーダー
harmony life : 調和的な生活
harpe : ハルペ、側鉤器
harvest : 採取する、摘出する、収穫する
Harvesters : カニアシシジミ亜科
hatch : 孵化(ふか)する
hatched egg : 孵化した卵
hatching : 孵化(ふか)
hatching size : 孵化サイズ
hatching success : 孵化成功
haustellum : 口吻(こうふん)、吸収管、吸管
hawk moth : スズメガ
Hawk's eyes and ocelli of the Satyridae : ジャノメチョウ科のタカの目と眼状紋、ジャノメチョウ科のタカの目と目玉模様
head : 頭部(とうぶ)
head shell : 頭殻
head truncation : 頭部切断
head width : 頭幅
headspace : ヘッドスペース法
headwaters of the Amazon : アマゾン源流
healthiest population : 最も健康な個体群
healthy environment : 健康的な環境
healthy grass : 健全な草
hearing : 聴覚
hearing function : 聴覚機能
hearing organ : 聴覚器官
heart : 心臓(しんぞう)、背脈管
heat absorption : 熱吸収
heat gain : 熱取得効率
heat paralysis : 熱麻痺(ねつまひ)
heat shock protein gene : 熱ショックタンパク質遺伝子、*HSP* 遺伝子
heat stress : 熱ストレス
heat-shock protein : 熱ショックタンパク質
heated chain : 高温系列
heathland : ヒース地帯、ヒースランド、荒地
heathland vegetation : ヒース地帯の植生
heavily-wooded landscape : うっそうと茂った森の景観、深い森の景観
heavy metal stress : 重金属ストレス
Hebe : ゴマノハグサ
Hebei Province : 河北省(かほくしょう)(中国)
hectare : ヘクタール、ha(面積の単位)
hectographing : ゼラチン版印刷
hedgehog : ヘッジホッグ(遺伝子発現)、ハリネズミ状紋
hedgehog signalling : ヘッジホッグシグナル伝達
hedgerow : 垣根、生け垣
hedgerow management : 低木の列管理、潅木管理、生け垣管理
hedylid moth : シャクガモドキ科の蛾
heliconiine butterfly : ドクチョウ属の蝶
Heliconius butterfly : ドクチョウ、有毒蝶
Heliconius cydno : シロオビドクチョウ
heliotaxis : 走日性
heliotherm : 日光を浴びて体温を上げる変温動物、日光性変温動物
Helitron : ヘリトロン
helper plasmid : ヘルパープラスミド
hemi- : 半 -〔ギリシャ語〕
Hemimetabola : 不完全変態する昆虫類
hemimetabolism : 不完全変態
hemimetaboly : 不完全変態
Hemiptera : カメムシ目、半翅目

hemivoltine : 半化性(生活史が二年)、二年生の化性
hemo- : 血 -〔ギリシャ語〕
hemocoel : 血体腔
hemocyte : 血球、血液細胞
hemoglobin : ヘモグロビン、血色素
hemolin : ヘモリン
hemolymph : 血リンパ、血液
hemolymph ecdysteroid titer : 血液中のエクジステロイド量、血リンパ - エクジステロイド価
hemostatic membrane : 止血膜(しけつまく)
herb : ハーブ、草本
herbaceous cover : 草本被覆率
herbaceous plant : 草本植物
herbicide : 除草剤
herbivore : 草食(性)動物
herbivore community : 草食(性)動物の群集
herbivore derived elicitor : 植食者由来エリシター、植食者由来誘発物
herbivore egg deposition : 植食者の産卵
herbivore induced event : 植食者誘導性イベント
herbivore mediated selection : 草食動物媒介による淘汰
herbivore performance : 植食者のパフォーマンス
herbivore-associated elicitor : 植食者由来エリシター
herbivore-associated molecular pattern : 植食者関連分子パターン、植食者由来分子パターン、植食者関連分子構造(HAMP)(植食者の表面に存在し、植物によって認識される分子で、植物はこれを認識することで免疫反応を起こす)
herbivore-induced plant volatile : 植食者が誘導する植物の揮発成分、植食者誘導性植物揮発性物質
herbivore-induced volatile aldoxime : 草食動物によって誘導された揮発性アルドキシム
herbivore-induced volatile emission : 植食者を誘引する揮発性物質の放出
herbivore-induced within-plant signaling : 植食者誘導性植物内シグナル伝達
herbivore-triggered JA pathway : 植食者引き金性 JA 経路
herbivorous insect : 植食性昆虫、草食(性)昆虫
herbivorous lepidopteran : 植食性チョウ目昆虫
herbivory : 植食性、草食、植食
hereditary wing defect : 遺伝的翅形異常
heritability : 遺伝性、遺伝率(量的形質がどの程度遺伝的に決定されるかを示す尺度)
hermaphrodite : 雌雄同体、雌雄同株、両性動物
hermoperiod : 温度周期
hetero male : 雄ヘテロ型
heterochromatin : ヘテロクロマチン
heterodimeric receptor : ヘテロ二量体を形成する受容体、ヘテロ二量体受容体、ヘテロダイマー受容体
heterodynamic : 異動態的、季節別繁殖動態的、ヘテロダイナミック、([注] 休眠ありの生活環)
heterodynamic type : 異動態的な発生型、周期型
heterogametic sex chromosome : ヘテロな性の染色体、異型配偶子性の染色体、異型性の染色体
heterogamety : 異型性、異型配偶子
heterogeneity : 異種混交性、異質性、不均質性
heterogeneity of butterfly eye : 蝶の眼の異質性、蝶の眼の多様性

heterogeneous environment : 異質な環境
heterogeneous habitat : 異質な生息地、不均一な生息地
heterogeneous selection : 異種選択、異質選択、不均質選択
heterogeneous species : 異質な種
heterogeneously expressed filtering pigment : 不均一に発現したフィルタリング色素
heterologously expressed protein : 異種発現されたタンパク質
heterospecific female : 異種の雌
heterostyly : 異形花柱性
heterotroph : 有機(従属)栄養生物
heterozygosity : ヘテロ接合度、異種接合性、ヘテロ接合性
heterozygosity distribution : ヘテロ接合体個体の分布
heterozygote : ヘテロ接合体、異型接合体
heterozygote genotype : ヘテロ接合体遺伝子型、異種接合体遺伝子型
heterozygous : 異種接合体の、ヘテロ接合体の
heterozygous for the white allele : 白色対立遺伝子のヘテロ接合
heterozygous individual : 異種接合体の個体、対立遺伝子をヘテロで持つ個体、ヘテロ接合個体、ヘテロ接合体の個体
heterozygous male : ヘテロ接合雄
heterozygous region : ヘテロ接合体領域
heuristic search : 発見的探索法
heuristics : ヒューリスティックス、発見的方法
Hewitson's blue hairstreak butterfly : ウラミドリシジミ
hexagon : 六角形
hexamer primer : ヘキサマープライマー
hexapoda : 六脚亜門、六本脚
HGT : 遺伝子の水平伝播(Horizontal Gene Transfer)

hh : ヘッジホッグ(hedgehog)
hibernaculum : (複 . -la)、冬眠場所、越冬生息場所〔ラテン語〕
hibernate : 越冬する
hibernating aspect : 越冬状態
hibernating egg : 越年卵(えつねんらん)
hibernating larva : 越冬幼虫
hibernation : 冬眠、越冬
hibernation form : 越冬形態、越冬型
hidden diversity : 隠れた多様性
hidden layer : 隠れ層
hierarchical dominance : 階層的優性
hierarchical F statistics : 階層的 F 統計
hierarchical G test : 階層的 G 検定、階層的対数尤度比検定
hierarchical likelihood ratio test : 階層的尤度比検定
hierarchical order : 階層的序列、階層的順序
hierarchical state space model : 階層的状態空間モデル
high dimensional model : 高次元モデル
high dose-refuge strategy : 高薬量／保護区戦略、高用量／保護区戦略
high elevation butterfly : 高山蝶
high FDR : 高偽陽性比率、高擬陽性率、擬陽性発見率(False Discovery Rate)
high frequency of backtracking : 来た同じ道を高頻度で引き返す
high humidity : 高湿度
high intensity : 高強度
high level : 高濃度
high light incidence : 高入射角
high magnification : 高倍率、高拡大図
high performance liquid chromatography : 高速液体クロマトグラフィー
high scoring pair : 高スコア分節対(HSP)
high sound pressure : 高音圧
high temperature : 高温
high temperature sensitivity : 高温感受性

high temporal resolution : 高時間分解能
high-dispersal capacity : 広く分散する能力、高分散能力
high-molecular weight : 高分子量
high-quality diet : 高品質な食料
high-quality dispersal habitat : 良質な分散した生息地
high-resolution aerial photograph : 高精度の航空写真
high-rise building : 高層ビル
high-titer stock : 高力価の貯蔵物、高力価のストック
higher classification : 高次分類
higher frequency : 高周波数
higher-order taxa : 高位の分類群
higher-taxon richness : 高次分類数
highest dose : 最高濃度、最高用量
highland : 高原
highland species : 高地性種
highly localized species : 非常に局在化した種
highly polyploid branched nucleus : 多倍数体分岐細胞核
highly unlikely generally : 一般的にはあまりない、一般的にはあり得ない
highly-modified habitat : 高度に改変された生息地
hill topping : ヒルトッピング、山頂占有性
hill-topping behavior : 山頂占有行動
hillside : 丘陵地帯
HIM : 宿主に組み込まれたモチーフ(Host Integration Motif)
Himalayan type : ヒマラヤ型
Himalayas : ヒマラヤ(山脈)
hind leg : 後脚(こうきゃく)、後肢(こうし)
hind thorax : 後胸(こうきょう)
hind wing : 後翅(こうし)
hind wing sheath : 後翅芽
hind-gut : 後腸(こうちょう)
hindgut : 後腸
hindleg : 後脚
hindrance : 障害物
hindwing (HW) : 後翅(こうし)
hindwing coupling : 後翅の結合部
hindwing margin : 後翅縁
hindwing tail : 尾状突起
HIPV : 植食者が誘導する植物の揮発成分、植食者誘導性植物揮発性物質(Herbivore-Induced Plant Volatile)
histological cross section : 組織学的断面
histology : 組織学
histolysis : 解離
histone : ヒストン
historical biogeography : 歴史(的)生物地理学
historical distribution : 歴史的分布
historically : 歴史的に
history : 経緯
hit accession : ヒットした登録番号
hitch-hiking : ヒッチハイキング
hitchhiker species : 付着した外来種
hitchhiking effect : ヒッチハイキング効果
HIV : ヒト免疫不全ウイルス(Human Immunodeficiency Virus)
Hofbauer-Buchner eyelet : H-B アイレット、ホーフバウアー - ブフナーアイレット
hol- : 完全 -、全 -〔ギリシャ語〕
Holarctic region : 全北区(ぜんほっく)
holistic conceptual model : 全体論的概念モデル
holo- : 完全 -、全 -〔ギリシャ語〕
Holometabola : 完全変態する昆虫類
holometabolous : 完全変態の
holometaboly : 完全変態
holophyletic : 完系統的
holopneustic type : 完気門式(かんきもんしき)
holotype : ホロタイプ、完模式標本、正基準標本

home base : ホームベース、居場所
home garden : 家庭菜園
home range : 行動圏、ホームレンジ
homeobox : ホメオボックス
homeosis : ホメオシス、相同異質形成
homeothermic : 定温の
homeotic gene : ホメオティック遺伝子、相同異質形成遺伝子
homeotic mutation : ホメオティック(突然)変異
homing behavior : 帰巣行動
homing capability : 帰巣能力
homing habit : 回帰性
homing navigation : 帰巣ナビゲーション
homo- : 同 - 〔ギリシャ語〕
homodynamic : 同動態的、連続性繁殖動態的、ホモダイナミック、([注] 休眠なしの生活環)
homodynamic type : 無周期型
homoeisis : ホモエイシス
homogametic sex : 同型性、同型配偶子をもつ性
homogenate : ホモジネート、組織粉砕懸濁液
homogeneity : 等質性、均質性
homogeneously distributed : 均一に分布した、均等に分布した
homogenize : 均質化する
homogenized egg : 均質化処理をされた卵、均質化された卵
homogenous subset of sample : サンプルの相同サブセット
homogeny : 相同性
homoiology : ホモイオロジー
homolog : 相同遺伝子、ホモログ
homolog of doublesex : 両性遺伝子の相同体、両性の相同遺伝子、両性のホモログ遺伝子
homologization : 同系化、相同化
homologous : 相同
homologous chromosome : 相同染色体
homologous genetic pathway : 相同遺伝経路
homologous linkage group : 相同連鎖群
homologous marker : 相同マーカー
homologous multichromosomal mimicry architecture : 多相同染色体の擬態構成
homologous nerve branch : 相同(的)神経枝
homologous structure : 相同的構造、相同構造
homologous transcribed sequence : 相同転写配列
homology : ホモロジー、相同
homology model : 相同(性)モデル
homology modeling : ホモロジーモデリング、相同体モデル化
homology region : 相同(性)領域
homology search : ホモロジー検索、相同性検索
homonym : 同名異物、ホモニム、同名
homonymy : 同名関係、同名状態
homoplasy : 成因的相同、類形
homozygote : ホモ接合体、同型接合体
homozygous : ホモ接合型、同型接合体(二倍体生物のある遺伝子座がAA、aaのように同じ対立遺伝子からなる状態のこと)
homozygous for the yellow allele : 黄色対立遺伝子のホモ接合
homozygous locus : ホモ接合型対立遺伝子
honestly significant difference : HSD 法、HSD 検定
honey gland : 蜜腺
honey solution : ハチミツ溶液
honeybee : ミツバチ
honeybee colony : ミツバチのコロニー
honeydew : 蜜、糖液、甘露
honorary authorship : 名誉のオーサーシップ、名誉著者

Honorary chairman : 名誉会長、名誉議長
Honshu population : 本州個体群
hook : 鉤爪(かぎづめ)
hooked : 鉤爪状の
hooked hindwing tail : 鉤爪状の尾状突起
Hopkins host selection principle : ホプキンスの寄主選択則
hopperburn : 坪枯れ
horizontal distribution : 水平分布
horizontal gene transfer : 遺伝子の水平伝播(HGT)
horizontal infection : 水平感染
horizontal layer : 水平層
horizontal transfer of TE : 転移因子の水平伝播(Transposable Element)
horizontal transmission : 水平伝達、水平伝播
hormonal control : ホルモン制御、ホルモン調節、ホルモン支配
hormonal difference : ホルモンの差異
hormonal mechanism : ホルモン機構
hormonal signaling : ホルモンシグナル伝達
hormone : ホルモン
hormone dynamics : ホルモン動態
hormone titer : ホルモン力価
horn : 角状突起
horn length : 角状突起長
horn type : 角状突起の形態
horned head : つのがある頭部
hornet : スズメバチ
horseradish : ワサビダイコン、セイヨウワサビ
horticulturalist : 園芸家
host : 寄主、宿主(しゅくしゅ)
host affiliation : 寄主の協力、寄主起源の認定
host arthropod : 宿主節足動物
host biomass density : 寄主植物量の密度
host body : 宿主の体
host cellular machinery : 宿主の細胞機構
host cyanogenesis : 寄主によるシアン発生、寄主によるシアン形成
host development arrest : 宿主の発生阻止、宿主の発育停止
host domestication : 宿主の順化(順応)
host elongation factor : 宿主由来の伸長因子、宿主伸長因子
host fitness : 宿主の適応度
host generation : 宿主世代
host germ line cell : 宿主の生殖細胞系列細胞
host immune defense : 宿主の免疫防御
host immune suppression : 宿主の免疫抑制
host larva : 宿主幼虫
host plant : 食樹、食草、寄主植物
host plant conspicuousness : 寄主植物の目立ちやすさ、寄主植物の被視認性
host plant recognition : 食草認識
host plant selection mechanism : 食草選択機構
host plant species : 寄主植物種
host population : 宿主個体群、寄主個体群
host preference : 寄主選好性
host preference shift : 寄主選好性転換
host race : ホストレース、寄主品種、寄主系統
host range : 寄主範囲
host recognition protein : 宿主認識タンパク質
host residue : 寄主残留物
host shift : 寄主転換
host specialization : 食草特化適応
host specificity : 宿主特異性
host tissue : 宿主組織
host-marking pheromone : 寄主マーキングフェロモン
host-parasitoid interaction : 宿主 - 捕食寄生者相互作用
host-pathogen arms race : 宿主 - 病原体の軍拡競争、宿主と病原体の軍拡競走

host-plant chemical substance : 寄主植物化学物質
host-plant glucosinolate content : 寄主植物のグルコシノレート含有量
host-plant specialization : 寄主植物特異性
host-plant use : 寄主植物利用
hostile : 外敵に満ちた、相反する、敵対する
hostmarker : 寄主マーカー、寄主標識
hot and humid forest : 高温・湿潤な森林、高温多湿な森林
hot and humid period : 高温湿潤期
hot butterfly species : 今話題の蝶種
hot period : 暑い期間
hot spot : ホットスポット(遺伝子内の突然変異を起こしやすい部分)
hot summer : 暑い夏
hourglass timer : 砂時計型生物時計、砂時計型タイマー
hours post infection : 感染後(経過)時間(hpi)
house built on the sand : 砂上の楼閣
house of card : 砂上の楼閣
house spider : タナグモ
housekeeping gene : ハウスキーピング遺伝子(あらゆる細胞に存在し、特殊な機能は果たさないが、それらの生存に必須な役割を持つ遺伝子の総称)
hovering briefly : 一時的にホバリングして
Hox : ホメオティック遺伝子、相同異質形成遺伝子(Homeotic gene あるいは Homeobox gene)
Hox gene : ホメオティック遺伝子、相同異質形成遺伝子
Hox gene cluster : *Hox* 遺伝子群
HP : ヘアペンシル、毛束(HairPencil)
HPD : 最大事後密度
(Highest Posterior Density)
hpi : 感染後(経過)時間
(hours post infection)
HPLC : HPLC 法、高速液体クロマトグラフィー分析法(High Performance Liquid Chromatography)
HR : 過敏感反応
(Hypersensitive Response)
HR Elicitor : HR エリシター、過敏感反応性エリシター(Hypersensitive Response)
HR-expressing plant : HR 発現する植物、過敏感反応を発現する植物
(Hypersensitive Response)
HR-like necrosis : 過敏感反応様ネクローシス、過敏感反応に類似する壊死
(Hypersensitive Response)
HR-marker : 過敏感反応マーカー
(Hypersensitive Response)
HSD : HSD 法、HSD 検定
(Honestly Significant Difference)
HSP : 熱ショックタンパク質(Heat-Shock Protein)、高スコア分節対(High Scoring Pair)
HSP gene : *HSP* 遺伝子、熱ショックタンパク質遺伝子(*Heat Shock Protein*)
HTU : 仮想的分類単位
(Hypothetical Taxonomic Unit)
huge swathe : 広大な帯状土
human activity : 人間活動
human dwelling : 人間の住居、人里
human genome : ヒトゲノム
human genome project : ヒトゲノム計画
human habitation : 人間の生息地
human immunodeficiency virus : ヒト免疫不全ウイルス(HIV)
human population density : 人間の人口密度
human prosperity : 人間の繁栄
human welfare : 人間の福祉
human-induced disturbance : 人為的攪乱
humeral crossvein : 肩横脈、h 脈
humeral lobe : 肩葉
humeral plate : 翅肩板
humeral vein : 肩脈

humid condition : 湿度条件
humid tropics : 湿潤熱帯
humid zone : 湿潤地帯
humidity : 湿度
hummingbird hawk moth : ホウジャク、ホシホウジャク
hump-shaped diversity curve : こぶ状の多様性曲線
humus : 腐葉土
hunter : 採集者
HvirCR4 : オオタバコガの仲間(*Helicoverpa virescens*)の化学受容体 4
hyaline spot : 透明な斑点
hybrid : 雑種、交雑
hybrid breakdown : 雑種崩壊
hybrid dysfunction : 雑種の機能不全
hybrid egg hatch : 雑種卵の孵化
hybrid exchange of gene : 交雑による遺伝子交換
hybrid genome : 雑種のゲノム
hybrid inviability : 雑種の生存不能、雑種致死、雑種死滅、雑種の生存力低下
hybrid male : 雑種の雄
hybrid mating : 雑種間の交配
hybrid rice : ハイブリッド米
hybrid speciation event : 交雑による種分化イベント、交雑による種形成イベント、雑種種分化イベント
hybrid sterility : 雑種不稔、雑種不妊性
hybrid viability experiment : 雑種の生存力実験
hybrid wing : 雑種の翅
hybrid zone : 交雑帯
hybridization : 交雑、種間交雑
hybridization experiment : 交雑実験
hybridize : 種間交雑
hybridizing species : 交雑種
hydration : 水和
hydrocarbon : 炭化水素
hydrolysis : 加水分解
hydrophilic moiety : 親水性部分
hydrophilic protein : 親水性タンパク質
hydrophobic compound : 疎水性化合物
hydrophobic molecule : 疎水性分子
Hymenoptera : ハチ目、膜翅目
hymenopteran insect : 膜翅目の昆虫、ハチ目の昆虫
hymenopteran lectin : 膜翅目のレクチン、ハチ目のレクチン
hypandrium : 生殖下板
hyper- : 超 -、過度の -〔ギリシャ語〕
hyperdiverse taxa : 超多様な分類、極めて多様な分類
hyperfine structure : 超微細構造
hypergeometric rarefaction curve : 超幾何学的希薄化曲線
hypergeometric sampling distribution : 超幾何学的サンプリング分布
hyperparasitism : 過剰寄生
hypersensitive response : 過敏感反応(HR)(宿主植物が非親和性の病原を認識して、急激に形態的、生化学的変化を起こすこと)
hypersensitive-like response : 過敏感様反応、過敏感型反応
hyphen : ハイフン
HyPhy : 最尤法に基づいて解析を行うソフトウェアの名称
hypodermis : 下皮(かひ)、真皮(しんぴ)
hypopharynx : 下咽頭(かいんとう)
hypopteron : 前前側板(中胸側板の前方にある切片)(胸部)
hypopygial : 尾節の
hypopygium : 尾節
hypostomal bridge : 下口橋
hypostomal carina : 下口隆起線
hypothesis of age and area : 時間と広がり説、年代領域説

hypothesis of neutral evolution : 中立進化説
hypothesis of strong developmental constraint : 強い発生的制約説
hypothesize : 仮定する
hypothetical concept : 仮説的概念、仮説上の概念
hypothetical example : 仮想例、仮想事例
hypothetical protein : 推定（される）タンパク質（名称等が不明なタンパク質）
hypothetical taxonomic unit : 仮想的分類単位（HTU）
hypothetical, sample-based rarefaction curve : 仮想的サンプル数に基づく希薄化曲線
hysterotely : ヒステロテリー（生物体の一部に、通常よりも前の発育段階の形質が現れること）
HZG5 : オオタバコガ（旧学名：*Heliothis zea*）の卵巣由来の樹立培養細胞株

i

IAs : 侵略的外来種、特定外来生物（Invasive Alien species）
IB : 封入体（Inclusion Body）
IBD : 距離による隔離（Isolation By Distance）
Iberian butterfly : イベリア半島の蝶
IBM : 総合的生物多様性管理（Integrated Biodiversity Management）
IBR : 昆虫行動制御剤、昆虫行動制御物質（Insect Behavior Regulator）
ice seeding : 植氷
ice-inoculation avoidance : 植氷凍結回避
ice-nucleating agent : 氷晶核、氷核形成
ICG : 成長休止の時間間隔、成長が停止する期間（Interval to Cessation of Growth）
ichnotaxon : 生痕化石タクソン
ICIPE : 国際昆虫生理生態学センター（International Centre of Insect Physiology and Ecology）
iconographia insectorium Japonicorum colore naturali edita : 原色日本昆虫図鑑〔ラテン語〕
ICZN : 動物命名法国際審議会（International Commission on Zoological Nomenclature）
ideal free distribution : 理想自由分布（IFD）
ideal free pathway : 理想自由経路
ideal population distribution : 理想的な個体群分布
ideal system : 理想的なシステム、理想的体系
idealized spectrum : 理想化されたスペクトル、理想的なスペクトル
identical cumulative number of species : 同一の累積種数
identification : 同定
identification success : 同定成功率
identification system : 同定システム
identity : 同定、識別、正体、同一性
identity threshold : 同一性閾（いき）値
idiosyncratic : 特異的な
if any : もしあれば
IFD : 理想自由分布（Ideal Free Distribution）
igneous rock : 火成岩
ignite : 火を付ける、焼く
IGR : 昆虫成長制御物質、昆虫成長制御剤（Insect Growth Regulator）
IGR [hg.] : 遺伝子間領域（InterGenic Region）
illegal introduction : 違法導入
Illumina technology : イルミナ技術
illuminant difference : 照度差
illuminate : 明らかにする、解明する
illumination : 照度
illustration : 図示、説明図、イラスト
image-resolving eye : 解像能力を有する眼
imaginal disc : 成虫芽、成虫原基、成虫盤
imaginal myrmecophily : 成虫の好蟻性

imago : 成虫、イマーゴ〔ラテン語〕
imago form : 成虫形態、成虫型
imago phenotype : 成虫表現型
imbricate scale : 瓦状に重なった鱗粉、鱗(うろこ)模様状の鱗粉
imidacloprid : イミダクロプリド(「ネオニコチノイド系」の農薬)
immature stage : 幼虫期、未成熟期
immediate consequence : 即座に現れる影響
immigrant : 移入種
immigrant species : 移入種
immigration : 移入、移住して来ること
immobilization : 不動化
immune response : 免疫反応
immunity : 免疫
immunity-related factor : 免疫関連因子、免疫系因子
immunity-related gene family : 免疫系関連遺伝子ファミリー
immunohistochemical localization : 免疫組織化学的局在
impair : 減じる、弱める
impaired immunity : 免疫障害
impaired navigation : ナビゲーション障害
impairment of circle formation : 環形成の障害
impede : 遅らせる、妨害する
impinge : 影響を与える
implicate : 関与している、包み込む
important cue : 重要な刺激
important finding : 重要な発見
imprinting : 刷り込み、インプリンティング
improved grassland : 改良草地
in essence : 本質において、要するに
in line with : と一致して、と合致して
in press : 印刷中
in silico : インシリコ、シリコン内で、コンピュータを用いて〔ラテン語〕
in situ conservation : 生息域内保全、本来の場所での保全
in situ hybridization : インサイチュー・ハイブリダイゼーション法、「その細胞が由来する生物個体内の本来あるべき場所」での交雑実験、原位置標識法〔ラテン語〕
in the sense of : という意味での
in vitro : インビトロ、試験管内で、生体外で〔ラテン語〕
in vivo : インビボ、生体内で〔ラテン語〕
in vivo electroporation : 生体内電気穿孔法、生体内エレクトロポレーション法
in- : 内-、中-、反-〔ラテン語〕
in-depth analysis : 徹底的な分析、網羅的な解析
in-group species : 群内種
inability of eyespot : 眼状紋の発育不全
inability of medial band : 内側の縞状バンドの発育不全、中央部の縞状バンドの発育不全
inadvertent error : 不慮の過誤
inappropriate name : 不適切名
inbred : 近親交配の、同系交配の
inbred larva : 近交系幼虫
inbreeding : 近親交配、インブリーディング、自殖、同系交配、近交
inbreeding coefficient : 近交係数
inbreeding depression : 近交弱勢、近親交配弱勢
inbreeding species : 近親交配種
incertae sedis : (分類学上の)所属位置不明〔ラテン語〕
incidence light : 入射光
incidence of diapause : 休眠率、休眠の発生
incidence of pupal diapause : 蛹休眠率、蛹休眠の発生、蛹休眠の誘起
incipient species : 発端種、初期種
incipient stage of speciation : 種分化の初期段階

inclivous : 内斜
inclusion body : 封入体(IB)
inclusion-body disease : 封入体症、封入体病
inclusive approach : 包括的アプローチ
inclusive fitness : 包括適応、包括適応度
income breeder : インカムブリーダー(産卵時の餌に依存する)、インカム型ブリーダー(産卵におけるエネルギー投資戦略の一つで、産卵エネルギーを産卵時の餌に依存する繁殖パターン)
incoming female : 飛来雌、侵入した雌
incoming virgin female : 飛来してくる無交尾の雌
incommensurate area : 不釣合いな面積
incompatibility : 不和合性、不一致
incompatibility-inducing microbe : 不和合性誘発微生物
incompatible cross : 不和合性交配
incompatible need : 両立しない要求
incomplete metamorphosis : 不完全変態
incongruent : 一致しない
incorporation : 取り込み、結合、合併
incorrect original spelling : 不正な原綴(つづ)り
incorrect subsequent spelling : 不正な後綴(つづ)り
incorrectly assess : 誤認する、不正評価する
increased body mass : 増加した体重、体重増加、過体重
increasing rate : 増加率
increasing stimulus intensity : 増大する刺激強度
increasing titer : 増加するタイター、増加する力価
incubate : 培養する
incubator : 恒温器、定温器、孵化器
incur : 負う
indel : インデル、挿入欠失
indented : くぼんだ
independent infection : 非依存性感染
independent locus : 独立した遺伝子座
independent mating : 自主的な交尾
independent observer : 独立の観測機器
independent species : 独立種
independent variable : 独立変数
independently evolving lineage : 独立の進化系統、独立の進化の系譜
index : 指標、索引
index fossil : 標準(示準)化石
index of vegetation dynamics : 植物動態指標、植生動態指標
index of vegetation productivity : 植物生産性指標
Indian leaf butterfly : コノハチョウ
indication : 指示
indicator : 指標
indigenous : 土着の、原産の
indigenous species : 土着種(どちゃくしゅ)、在来生物種
indirect defense : 間接防衛
indirect effect : 間接効果
indirect evidence : 間接的証拠
indirect interaction web : 間接相互作用網
indirect plant defense : 植物の間接防御
indirect selection : 間接的選択
indirect stabilizing selection : 間接的安定化選択
indispensable : 不可欠な
indistinct brown outer ring : 不明瞭な褐色の外側の環
individual : 個体
individual density : 個体密度
individual difference : 個体差
individual organism : 生物個体
individual stem : 単木
individual variation : 個体変異、個体変動、個体多様性
individual-based accumulation curve : 個体

数に基づく累積曲線
individual-based curve : 個体数に基づく曲線
individual-based dataset : 個体数に基づくデータセット
individual-based protocol : 個体数に基づくプロトコール
individual-based rarefaction : 個体数に基づく希薄化
individual-based rarefaction formula : 個体数に基づく希薄化の計算式
individual-based taxon-sampling curve : 個体数に基づく分類サンプリングの曲線
individual-level behavior : 個体水準の行動、個体レベルの行動
individuals : 個体数
Indo-Australian region : インド・オーストラリア区
Indo-China and adjacent region : インドシナ半島と近隣地域
indole : インドール
indolequinone compound : インドールキノン化合物
indolic glucosinolate : インドールグルコシノレート
indolic melanin : インドールメラニン
induced defense : 誘導防衛、誘導防御
induced plant defense : 植物の誘導性防御応答(植物がエリシターの刺激を受容して立ち上げる後天的な防御応答)
induced plant volatile emission : 誘導性植物揮発性物質の放出
induction condition of autumn morph : 秋型誘導条件
induction factor of summer form : 夏型誘導因子
industrial melanism : 工業黒化型、工業暗化
inert metabolite : 不活性代謝物
inevitable : 避けられない
inevitably : 必ず
infect : 感染する
infected cell : 感染細胞
infected male : 感染雄
infection density : 感染密度
infection frequency : 感染頻度
infectious disease : 感染病
infectious germ : 伝染病菌、感染性細菌
infectious parthenogenesis : 感染性単為生殖
infectivity : 感染性
inferior : 下部の、下(方)の、下(か、した)〔ラテン語〕
inferiority : 劣勢、劣等、不利
infested plant : 被食植物
infiltrate : 浸透する
infinite allele model : 無限対立遺伝子モデル
infinite site model : 無限サイトモデル
inflorescence : 花序
influence of microclimate : 微気象の影響
influence on the beta-diversity turnover : ベータ多様性上での種の入れ替わりの影響(ベータ多様性である「比較する環境間での種の入れ替わり」の影響)
influenza virus : インフルエンザウイルス
influx species : 流入した種
informative site : 情報を持つサイト
informed consent : インフォームド・コンセント(十分な説明に基づく同意)
informed conservation decision : 情報に基づく保全に関する意思決定
infra- : 下 -〔ラテン語〕
infraorder : 下目
infrared region : 赤外域
infrared spectroscopy : 赤外分光
infraspecific name : 種よりも低位の学名
infrasubspecific : 亜種よりも低位の
infrasubspecific name : 亜種よりも低位の学名
infrasubspecific taxon : 亜種よりも低位のタクソン

infrequent : めったに起こらない、まれな
infuscated : すす色の、黒ずんだ
infusion : 注入
ingest : 摂取する
ingredient : 原材料
ingroup : イングループ、内集団
inhabit : 生息する
inhabitant : 生息生物
inheritable variation : 遺伝的変異
inheritance : 遺伝、継承
inhibiting factor : 抑制要因、阻害要因
inhibition of metamorphosis : 変態の抑制、変態の阻害
inhibition of summer-form induction : 夏型の誘導阻害
inhibitory condition : 抑制条件、阻害条件
inhibitory factor : 抑制因子、阻止因子
inhibitory hormone : 抑制ホルモン
inhospitable climate : 住みにくかった気候
initial experiment : 初回実験、初期実験
initial *H* locus screening : *H* 遺伝子座の初期スクリーニング
initial viral entrance : 初期のウイルス侵入
initiation of scale cell development : 鱗粉細胞の発生開始
initiator of speciation : 種分化の開始者
injection : 注射、注入
injection buffer : 注入バッファー
Inka : インカ
innate character : 本質的な性格
innate color preference : 生得的な色選好性
innate immunity : 先天性免疫、自然免疫
inner edge : 内縁
inner epicuticle : 外表皮内層
inner lobe : 内葉片
inner margin : 内縁、後縁
inner membrane : 内膜
inner membrane surface : 内膜表面
inner surface : 内面
innervate : 刺激する、器官を刺激する、神経を刺激する
innervation : 神経刺激伝達、神経支配
innocuous : 無害の
innovative change : 革新的な変化
inoculated SCP : 植氷過冷却点
inoculation : 植氷、接種
inoculative release : 接種的放飼法
inorganic nitrogenous ion : 無機態窒素イオン
insect : 昆虫、インセクト
insect behavior : 昆虫(の)行動
insect behavior regulator : 昆虫行動制御剤、昆虫行動制御物質
insect bioassay : 昆虫の生物学的定量法、昆虫の生物検定、昆虫のバイオアッセイ
insect carrier : 保菌昆虫、媒介昆虫
insect chemoreception : 昆虫化学受容
insect dispersal : 昆虫の分散
insect ecology : 昆虫生態学
insect fitness : 昆虫の適応度
insect growth regulator : 昆虫成長制御剤、昆虫成長制御物質
insect herbivore : 昆虫植食者
insect material : 供試虫、供試昆虫
insect oral secretion : 昆虫の口腔分泌物
insect order : 昆虫目
insect pest : 害虫
insect pest management : 害虫管理
insect phylogenetics : 昆虫の系統解析
insect physiology : 昆虫生理学
insect protection : 昆虫の防御、昆虫の保護
insect saline : 昆虫食塩水
insect tissue : 昆虫組織
insect-tracking technology : 昆虫追跡技術
insect-umbellifer association : 昆虫 - セリ科の関係
Insecta : 昆虫綱、昆虫類
insectarium : 昆虫館

insectary : 昆虫飼育場
insecticidal crystal protein : 殺虫性結晶タンパク質(ICP)
insecticide : 殺虫剤
insecticide resistance : 殺虫剤抵抗性
insecticide resistance management : 殺虫剤抵抗性管理
insectivore : 食虫(性)動物
insectivorous bird : 食虫性(の)鳥
insectivorous bird species : 食虫性の鳥種
inseminated female : 受精した雌、受精雌
insemination : 授精、受精
inseparable : 不可分
insert size : 挿入サイズ
insertion : 挿入
inset : 差込み図、挿入図、インセット
insist : 主張する
insolation hour : 日照時間
insolubility : 難溶性
inspection of electropherogram : 電気泳動図の検査、電気泳動法の検査
instar : 齢(れい)、○齢幼虫、令(れい)〔ラテン語〕
instar number : 齢数
instigate : 扇動する
insulin : インシュリン、インスリン
insulin pathway : インシュリン経路、インスリン信号伝達経路(「インシュリンがシグナルになって起きる連鎖反応」のこと)
insulin signaling pathway : インスリン信号伝達経路
intact : 損なわれていないで、そのままの
intact forest : 手つかずの林
intact plant : 無傷植物
integrase : インテグラーゼ
integrated approach : 統合的アプローチ
integrated biodiversity management : 総合的生物多様性管理
integrated control : 総合防除
integrated form : 組み込み型
integrated pest management : 総合的害虫管理
integrated proviral form : 組み込まれたプロウイルス型
integration ability of circle : 環の組み込み能力
integrative property of endogenous virus : 内在性ウイルスの統合的な性質
integument : 外皮、皮膚(ひふ)、殻
intensification : 強化
intensify : 強める、増感する
intensity of daylight : 太陽光の強さ
intensity of diapause : 休眠深度
intensity of fire : 野焼きの強さ、野焼き強度
intensity of irradiation : 照射強度
intensity of light exposure : 照射される光の強度
intensity of necrosis : 壊死の強度、壊死の重症度
intensity of sound stimulus : 音刺激の強度、音の刺激強度
intensity-response relationship : 強度 - 応答関係
intensive agriculture : 集約農業
intensive chasing : しつこい追跡飛翔、積極的な追飛
intensive research : 徹底的な研究
intensive sampling : 集中(的)サンプリング、集約サンプリング、集中的抽出法
intensively examine : 集中的に調べる
intensively grazed : 集中的に生草が食われた
intensively mown : 集中的に刈られた
intentional introduction : 意図的導入
inter- : 間 -、相互に - 〔ラテン語〕
inter-annual fluctuation : 年間変動、年変動、年次変動
inter-correlated factor : 相関因子、相互相関因子

inter-entity genetic distance : 記載項目間の遺伝距離
inter-individual relocation pattern : 個体間の移住パターン
inter-racial : 品種間の、亜種間の
inter-specific territory : 種間ナワバリ
inter-year predictability : 年ごとの予測可能性、年間の予測可能性
interaction : 交互作用
interaction between fire and habitat : 野焼きと生息地との相互作用
interaction between season and temperature : 季節と温度との間の相互作用
interaction of environmental factor : 環境要因の相互作用
interaction strength : 交互作用の強さ
interaction term : 相互作用項、交互作用項、交差項
interbreed : 交配
intercellular distance : 細胞間の間隔
intercept : 切片【統計学】
intercross : 相互交配、兄妹交配
interested amateur : 愛好家
interesting insight : 興味深い見識
interesting-looking specimen : 興味を釘付けにする標本
interestingly : 面白いことに
interfamily : 科間
interfere : 干渉する、無力化する、阻止する
interference competition : 干渉型競争
interference RNA : RNA 干渉法
interfertile : 異種交配できる
intergeneric cross : 属間交雑
intergeneric hybrid : 属間雑種
intergenic region : 遺伝子間領域
interglacial period : 間氷期
Intergovernmental Panel on Climate Change : 気候変動に関する政府間パネル
intergrade : 中間的段階
intergrade population : 中間型個体群
intergroup variation : グループ内変異
interindividual variation : 個体間変異
interior region : 内陸地方
interior, the : 内陸部、奥地
interlinkage : 連結、結合
interlinked with : 結びついた
intermediate : 中間(ちゅうかん)
intermediate aldoxime compound : アルドキシム中間体化合物
intermediate habitat : 中間的な好適生息地
intermediate measure : 中間的措置
intermediate morph : 中間型
intermediate phenotype : 中間表現型
intermediate photoperiod : 中間的な日長
intermediate rate : 中間的な率
intermediate temperature : 中間的な温度、中間温度
intermediate type : 中間型
intermediate-temperate : 中間温帯(植生帯)
intermediates : 中間型種
internal : 内(ない)、内部の
internal body feature : 内部構造
internal coincidence model : 内的符合モデル
internal control : 内部コントロール、内部標準
internal control gene : 内在性コントロール遺伝子
internal epidermal pouch : 内部表皮の袋
internal lamella : 内板
internal lobe : 内葉片
internal morphology : 内部形態
internal node : 内部結節
internal reproductive organ : 内部生殖器官
International Commission on Zoological Nomenclature : 動物命名法国際審議会(ICZN)
International Plant Protection Convention : 国際植物防疫条約

International Rice Research Institute : 国際イネ研究所、国際稲研究所
interpatch movement : パッチ間移動
interphase nucleus : 間期細胞核
interpolate : 内挿する
interpolated name : 挿入名
interpolation : 内挿、補間
interpopulation variation : 個体群間変異
interpopulational variation : 個体群間変異
interracial contact zone : 亜種間の接触帯、異人種間の接触帯
interracial difference : 亜種間差、亜種間の差異、異人種間差
interracial hybrid zone : 亜種間の交雑帯、異人種間の交雑帯
intersexual defect : 間性欠陥、間性障害
intersexual defect hypothesis : 間性欠陥説、間性障害説
intersexual host trait : 宿主の間性形質
intersexual phenotype : 間性表現型
intersexual selection : 異性間選択、雌雄選択
intersexuality : 間性現象、間性
interspace : 間腔、翅脈間隙
interspecies : 種間
interspecific competition : 種間競争
interspecific crossing : 種間交配
interspecific factor : 種間要因
interspecific gene flow : 種間の遺伝子流動
interspecific horizontal transfer : 異種間の水平移動、異種間の水平伝播
interspecific hybrid : 種間交雑
interspecific hybridization : 種間交雑、種間雑種
interspecific interference : 種間干渉
interspecific mating : 種間の交配
interspecific variation : 種間変異
intersperse : まき散らす
intertropical convergence zone : 熱帯の赤道収斂域、熱帯収束帯
interval mapping : 区間マッピング法
interval to cessation of growth : 成長休止の時間間隔、成長が停止する期間
intervening larval diapause : 休眠を回避した幼虫
intimacy : 深い関係、親しい関係、親密
intimate interaction : 密接な相互作用
intimate relationship : 密接な関係
intra- : 内 -、中 -〔ラテン語〕
intra-individual concordance index : 個体内の一致性指標
intra-plant signaling : 植物内シグナル伝達、植物内情報伝達
intra-specific factor : 種内要因
intra-specific scent mark : 種内の臭跡
intracellular freezing : 細胞内凍結
intracellular signal transduction : 細胞内シグナル伝達系
intrageneric competition : 属内競争
intrageneric speciation of subgenus : 属内の亜属分化
intragenic recombination : 遺伝子内組換え
intragenomic conflict : ゲノム内闘争
intraguild predation : ギルド内捕食
intrasexual selection : 同性内選択、性淘汰、同性内性選択
intraspecific competition : 種内競争
intraspecific diversity : 種内多様性
intraspecific genetic variability : 種内(集団)間の遺伝的変異性
intraspecific horizontal transfer : 種内間の水平移動、種内間の水平伝播
intraspecific mimicry : 種内擬態
intraspecific territory : 種内ナワバリ
intraspecific variation : 種内変異
intrigue : 興味をそそる
intriguingly : 興味をひくように
intrinsic : 内因的、内在的
intrinsic barrier : 固有障壁、内的障壁

intrinsic optimum temperature for development : 内因的な発育最適温度
intrinsic property : 内在的性質
intrinsic rate of increase : 内的増殖率
intrinsic rate of natural increase : 内的自然増加率
intrinsic reproductive power : 内的繁殖力
introduce : 導入する
introduced colony : 移入されたコロニー
introduced plant : 移入植物
introduced species : 導入種、移入種
introduction : 導入、移入、はじめに、序、序論
introduction of paper : 論文紹介、論文の緒言
introduction to the study of the butterfly wing pattern : 蝶の紋様研究入門
introgression : 遺伝子侵入、遺伝子移入、移入交雑、遺伝子浸透
introgressive hybridization : 移入交雑、浸透交雑、浸透性交雑、移入雑種形成
intron : イントロン
intron early/late theory : イントロン前生説／後生説
intron splicing : イントロンスプライシング
intron-exon structure : イントロン - エクソン構造
intruder : 侵入個体
inundative release : 大量放飼法
invade : 侵入する
invader : 侵入生物、侵入者
invalid : 無効な
invalid name : 無効名
invalid nomenclatural act : 無効な命令法的行為
invaluable perspective : 価値のない視点
invariant : 不変
invariant site : 不変部位、変異のないサイト
invasion : 侵入、侵入種
invasion and establishment of unfamiliar land : 未経験の土地への侵入・定着
invasive : 侵略的
invasive alien species : 侵略的外来種、特定外来生物
invasive garden plant : 侵入種の園芸植物
invasive species : 侵入種
invasive species of concern : 問題の侵入種
invasive species release : 侵入種の野外放出
invasive weed seed : 侵入雑草の種子
invasiveness : 侵略性
inverse ratio : 逆比
inversion : 逆位、反転、インバージョン
inversion breakpoint : 逆位のブレイクポイント、逆位の切断点
inversion polymorphism : 逆位多型
invertebrate : 無脊椎動物
invertebrate conservation effort : 無脊椎動物の保全努力
inverted *H* allele : 逆位した *H* 対立遺伝子
inverted orientation : 逆向きの方向、逆方向
investigated season : 調査季節
investigation material : 調査材料
investigator : 調査員、観察員
investment : 投資量、外被、外殼
investment in reproduction : 繁殖への投資量
ionic : イオンの
IPCC : 気候変動に関する政府間パネル（Intergovernmental Panel on Climate Change）
IPM : 総合的害虫管理（Integrated Pest Management）
IPPC : 国際植物防疫条約（International Plant Protection Convention）
IR : 赤外分光（Infrared spectroscopy）
iridescent : 光沢の、虹色の、玉虫色の
iridoid : イリドイド
iridoid glycoside : イリドイドグルコシド、イリドイド配糖体
IRM : 殺虫剤抵抗性管理（Insecticide Resistance Management）

irradiation : 照射
irregular diapause-related phenomenon : 不規則な休眠に関連する現象
irregular projection : 異常な突起部、不規則な突起部、不整形な突起部
IRRI : 国際イネ研究所、国際稲研究所(International Rice Research Institute)
irrigated land : 灌漑(かんがい)地
irrigated meadow : 灌漑された草地、灌漑草地
irrigation : 灌漑
irruption : 大発生、大繁殖
irruptive : 急増した、大発生した、侵入した
ISH : インサイチュー・ハイブリダイゼーション法(*In Situ* Hybridization)
island community : 島の群集
island-hopping : アイランドホッピング、島巡り
islands : 島嶼(とうしょ)
islands of Southeast Asia : 東南アジア島嶼
isochore : アイソコア、等容(変化)
isofemale line : 単雌系統
isofemale offspring : 単雌の子孫
isoform : アイソフォーム、イソ型
isogamy : 同形配偶(同形配偶子の接合)
isolated chromophore : 孤立発色団、分離発色団
isolated habitat : 隔離的な生息地
isolated population : 孤立化した個体群
isolated species : 孤立種
isolating mechanism : 隔離機構
isolation : 隔離、分離
isolation by distance : 距離による隔離(IBD)
isolation of distribution : 分布の断絶
isoleucine-to-methionine substitution : イソロイシンからメチオニンへの置換
isomeric change : 異性体変化
isomerization : 異性化
isoprene production : イソプレン産出
isoprene-emitting plant : イソプレン放出性植物
Isoptera : シロアリ目、等翅目
isotope : アイソトープ、同位体
isotropic light field : 等方性光条件
isoxanthopterin : イソキサントプテリン
issue on larval foodplant : 幼虫の食草問題
ITCZ : 熱帯の赤道収斂域、熱帯収束帯(InterTropical Convergence Zone)
itero- : 繰り返す -〔ラテン語〕
iteroparity : 多回繁殖
iteroparous : 多数回繁殖
iteroparous colonizer : 多回繁殖性の移住種
iteroparous species : 複数回繁殖の種、多回繁殖性種
ithomiine butterfly : トンボマダラ科の蝶
ithomiine community : トンボマダラ科の蝶の群集
ITS : 内部転写スペーサー(internal transcribed spacer)(リボソームDNAを構成する転写単位の1つ)
IUCN : 国際自然保護連合(自然及び天然資源の保全に関する国際同盟)(International Union for Conservation of Nature and Natural Resources)

j

JA : ジャスモン酸(Jasmonic Acid)(信号伝達物質の一種)
JA-dependent wound-induced response : ジャスモン酸依存性傷誘導性反応(Jasmonic Acid)
jacamar : キリハシ科の鳥
jack-of-all-trades : なんでも屋、よろず屋
jagged curve : ぎざぎざの曲線
jagged shape : のこぎり歯のような形
Janus Green B : ヤヌスグリーンB(ミトコンドリアの染色に用いられる色素)

Japan and surrounding East Asian Countries : 日本とその周辺の東アジア諸国
Japan mainland : 日本本土
Japanese : 日本(産)の
Japanese butterfly : 日本産蝶
Japanese Islands : 日本列島
Japanese name : 和名
Japanese Society of Applied Entomology and Zoology : 日本応用動物昆虫学会(JSAEZ)
Japanese type : 日本型
jasmonic acid : ジャスモン酸(JA)(信号伝達物質の一種)
jaw : 大腮(たいさい、おおあご)、大顎(だいがく、おおあご)、口部
jeopardize : 危険にさらす
jerky flight : ぎくしゃくと飛ぶ
jet-black : 漆黒色、まっ黒の
JH : 幼若ホルモン(ようじゃくほるもん)、幼虫ホルモン(Juvenile Hormone)
JH analog methoprene : 幼若ホルモン類似体メトプレン(Juvenile Hormone)
JH production : 幼若ホルモンの産出、JHの生産(Juvenile Hormone)
JH titer : 幼若ホルモンの力価(Juvenile Hormone)
Johnston's organ : ジョンストン器官
joining behavior : 結合行動
JPP-NET : 植物防疫情報総合ネットワーク(Japan Plant Protection general information NETwork system)
JSAEZ : 日本応用動物昆虫学会(Japanese Society of Applied Entomology and Zoology)
JTT substitution : JTT モデルの(アミノ酸)置換(Jones-Taylor-Thornton)
judge : 査読者
jugal lobe : 翅垂
jugum : 翅垂
Jukes-Cantor model : ジュークス・カンターモデル、JC モデル
julia butterfly : チャイロドクチョウ
jumping bean : セバスチアナの種(メキシコ産トウダイグサ科植物の種子内にいるガの幼虫の動きに伴って種子が踊り動く)
junction : 接合部、結合部
junior homonym : 新参同名
junior synonym : 下位同物異名、新参異名
junk DNA : がらくた DNA
justified emendation : 正当な修正名
juvenile : 幼体、若い
juvenile development : 若齢期の発育、幼生期の発育、幼生の生育
juvenile hormone : 幼若ホルモン(ようじゃくほるもん)、幼虫ホルモン
juvenile stage : 幼虫(体)段階
juvenoid : ジュベノイド
juxta (jx) : ユクスタ、挿入器腹板〔ラテン語〕

k

K locus : K 遺伝子座
k-mer length : k マーの長さ(塩基配列の中から任意の長さ k の配列を切り出したもの)(「k」には数字が入る;「mer」はモノマー単位、単量体単位で、「マー」と読み、「monomeric unit」または「monomer unit」の略語である)
K-pest : K 害虫(K 戦略をとる害虫)
K-selection : K- 淘汰、K 選択
K-species : K 種(K 戦略をとる種)
Ka/Ks : 非同義置換と同義置換の比(Non-synonymous (Ka) and synonymous (Ks) nucleotide substitution ratio)
Kailash : カイラス山(チベット)
kairomone : カイロモン
Kaplan Meier method : カプランマイヤー法(生存分析法)

Karakoram areas trip : カラコルム山地旅行
Karner blue butterfly : メリッサミヤマシジミ
karyotype : 核型
Kashimir : カシミール
katepimeron : 下後側板（胸部）
katepisternum : 下前腹板（胸部）
katydid : キリギリス
kb : キロ塩基、キロベース（塩基数の単位）
kDa : キロダルトン（分子量の単位）（Dalton）
ketone : ケトン
key : 検索表
key enzyme : 鍵酵素、鍵となる酵素
key factor : 重要要因、変動主要因
key factor analysis : 変動主要因分析
key pest : 対象とする害虫、キーペスト、重要害虫
key species : 鍵種
key stage : キーステージ
key stimulus : 鍵刺激
key stone species : キーストーン種、中枢種
key word : キーワード
Khabarovsk : ハバロフスク
Khw : 白色抑制遺伝子、キヌレニンヒドロキシラーゼ - ホワイト（Kynurenine hydroxylase-white）
kilobase : キロ塩基、キロベース（塩基数の単位）
kin : 血縁
kin group : 血縁群
kin recognition : 血縁認識
kin selection : 血縁選択
kinase : キナーゼ
kinase activation : キナーゼ活性化
kinesis : 無定位運動性、キネシス、動性
kinetochore : 動原体
king of summer forest : 夏の雑木林の帝王
kingdom : 界（分類階級の「かい」）
Kingdom of Tonga : トンガ王国
Kinmen Island : 金門島
kinship : 血縁関係、血縁、近親関係
kitchen garden : 家庭菜園
knock out : ノックアウトする、だめにする、欠損する
knockdown experiment : ノックダウン実験
knockout-knockin : ノックアウト・ノックイン解析
known as : ～として知られている
known landmark : 既知の陸標、既知の目印
known target : 既知の目標
krill : オキアミ
kynurenine : キヌレニン
Kyrgyz : キルギス、キルギス共和国

l

L : 明期（Light）
L opsin gene sequence : L（長波長型）オプシン遺伝子配列（Long-wavelength）
L photopigment : L 視物質、長波長感受性視物質
L photopigment lineage : L 視物質系統
L-sensitive photopigment : 長波長感受性視物質
labial : 下唇の
labial palp : 下唇鬚（かしんしゅ）〔英語〕
labial palpus : 下唇鬚（かしんしゅ）〔ラテン語〕
labial segment : 下唇節
labial suture : 下唇縫合線
labile defensive lipid : 不安定な防御脂質
labium : 下唇（かしん）、下脣（かしん）〔ラテン語〕
laboratory cage : 実験ケージ
laboratory condition : 実験室条件
laboratory hybridization experiment : 実験室での交雑実験
laboratory stock : 研究室の貯蔵品、研究室ストック
laboratory study : 基礎研究、実験室研究
laboratory-raised female : 実験室で飼育された雌

labral : 上唇の
labrum : (複 . -ra)、上唇(じょうしん)〔ラテン語〕
lacewing : クサカゲロウ
lacinia : (複 . -ae)、内葉、肢内葉〔ラテン語〕
lack of predictability : 予測可能性の欠如、予見可能性の欠如
lactic acetic orcein : 乳酸酢酸オルセイン
lacuna : ラクナ、(骨や組織中の)小腔
lacy : レース状、レース編みの
lady beetle : テントウムシ
ladybug : テントウムシ
lamella : (複 . -lae)、層板、薄片〔ラテン語〕
lamella antevaginalis : 前膣ラメラ、前膣片
lamella postvaginalis : 後膣ラメラ、後膣片
lamellate antenna : (複 . -nae)、ひだ状の触角、(蛾類)
lamina : 視葉板、板部、薄膜、葉片〔ラテン語〕
lamina subgenitalis : 下蓋板(かがいばん)、生殖下板(せいしょくかばん)
laminate : 葉片状(ようへんじょう)の(触角の形状)、板部の
lanceolate : 槍状(やりじょう)の、槍の穂先形の(蛾類の前翅)
land cover : 土地被覆
land management : 土地管理
land-use pattern : 土地利用パターン
landform : 地形
landing : 着地、着陸
landmark : 陸標、目じるし
landmark information : 陸標情報
landmark point : 目標地点、ランドマーク点
landscape : 景観、地形
landscape disturbance : 景観の攪乱
landscape dynamics : 景観動態
landscape ecology : 景観生態学
landscape fragmentation : 景観の分断化、景観断片化
landscape scale : 景観規模
landscape structure : 景観構造
landscape-scale conservation : 景観規模の保全
Langfang city : 廊坊(ランファン)市(中国河北省)
Langkawi : ランカウイ島
Lao P. D. R. : ラオス、ラオス人民民主共和国(Lao People's Democratic Republic)
lapsus calami : 書きまちがい〔ラテン語〕
large : 大型
large continuous population : 広く分布し連続域に生息する個体群
large copper butterfly : オオベニシジミ
large distance : 長距離
large geographic scale : 大規模な地理的規模
large green-banded blue butterfly : タスキシジミ
large posterior eyespot : 後方の大きな眼状紋
large skipper butterfly : コキマダラセセリ
large white butterfly : オオモンシロチョウ
large wood nymph butterfly : オオモンヒカゲ
large-bodied animal : 大型動物
large-scale rearrangement : 大規模な再編成
large-scale study : 大規模な研究
largescale movement pattern : 大規模な移動パターン
larva : (複 . -ae)、幼虫(ようちゅう)、若虫(わかむし)、幼生〔ラテン語〕
larva harboring baculovirus : バキュロウイルスが宿っている幼虫
larval aggregation : 幼虫の集合性
larval body : 幼虫体
larval carnivore : 幼虫の肉食性
larval development : 幼虫発達、幼虫生育
larval diapause : 幼虫休眠
larval diapause induction : 幼虫の休眠誘起
larval feeding efficiency : 幼虫の摂食効率
larval food plant : 幼虫の食草、幼虫の食餌植物
larval foodplant : 幼虫の食草
larval imaginal disc : 幼虫期の成虫原基

larval imaginal wing disc : 幼虫期の翅成虫原基
larval infection : 幼虫の感染
larval instar : 幼虫の齢期
larval integument : 幼虫皮膚
larval life : 幼虫期、幼虫期間
larval midgut : 幼虫の中腸
larval molt : 幼虫脱皮、幼虫期の脱皮
larval mortality : 幼虫死亡数、幼虫死亡率
larval myrmecoxeny : 幼虫の客棲性
larval overwintering : 幼虫越冬
larval parasitoid : 幼虫寄生蜂(寄主の幼虫に産卵する捕食寄生者)
larval performance : 幼虫の発育、幼虫生存力、幼虫の成育
larval pigmentation : 幼虫の色素沈着
larval protection : 幼虫の防御、幼虫の保護
larval rearing temperature : 幼虫の飼育温度
larval stage : 幼虫期
larval survival : 幼虫の生存率
larval time : 幼虫時間、幼虫の時期
larval tissue : 幼虫組織
larval transfer : 移行齢期
larval web count : 幼虫の巣単位での頭数
larval wing disc : 幼虫期の翅原基
lashed eye : まつ毛がある眼
last author : 最後の著者、統括著者
last glacial maximum : 最終氷期最盛期(LGM)
last instar : 終齢
last instar larva : 終齢幼虫
last-larval stadium : 終齢幼虫期、幼虫の終齢期、終齢幼虫
late autumn : 晩秋
late evening : 深夜、遅い夜、夜分
late fifth instar : 五齢後期
late male-killing : 後期型雄殺し
late pupa : 蛹後期
late pupal stage : 蛹後期、蛹期後期
late season growing condition : シーズン後期の成育条件
late-embryogenesis abundant protein : LEAタンパク質、後期胚発生蓄積タンパク質
latency : 潜時、潜伏期間
latency of response : 反応潜時、応答遅延
later analysis : 事後分析、事後解析
later consequence : より後で現れる影響
lateral : 正中線により近い外側(がいそく)、側の、側方の
lateral abdominal gill : 側腹鰓
lateral flange : 側稜、側突出縁
lateral gill : 側鰓
lateral inhibition : 隣接抑制、側抑制、側方抑制
lateral lobe : 側片
lateral neuron : 側方ニューロン
lateral ocellus : 側単眼
lateral oviduct : 側輪卵管(そくゆらんかん)
lateral portion : 側方部、側方部分、側方部位、側部
lateral posterior neuron : 側後方ニューロン
lateral postnotum : 側後盾板、側後背板
lateral spine : 側棘(そくきょく)、側刺(そくし)
lateral spine plate : 側域刺毛板
lateral-basking : 傾斜日光浴、閉翅日光浴、側面日光浴
laterotergite : 側背板
latex : ラテックス、乳液
latex flow : 乳液の流れ、ラテックスの流出
latex quality : 乳液の質、ラテックスの質
latex-bearing leaf : ラテックスを分泌する葉
Latin : ラテン語
Latin name : 学名
latinize : ラテン語化する
latitude : 緯度
latitudinal cline : 緯度クライン、緯度の連続変異
latitudinal gradient : 緯度勾配

latrunculin : ラトランクリン(単量体アクチンに結合して重合を阻害する魚毒)
lauraceous plant : クスノキ科植物
laurel-montane oak forest and relictual type : 照葉 - 山地カシ林遺存型
Law for the Conservation of Endangered Species of Wild Fauna and Flora : 絶滅のおそれのある野生動植物の種の保存に関する法律
lay egg : 産卵する
LD : 明暗周期(Light-Dark)
lead compound : 鉛化合物
leading edge : 先端、先端部、前縁部
leaf : 葉
leaf disk : 葉片円盤、葉円板、円板状の葉の切片
leaf eater : 葉食者
leaf litter : 落葉層、葉リター、葉積層
leaf mimic : 葉に擬態
leaf photosynthetic rate : 葉の光合成率
leaf powder : 葉粉末
leaf roller : ハマキムシ
leaf shelter : 葉のシェルター
leaf surface : 葉の表面
leaf tissue : 葉組織
leaf toughness : 葉の硬さ
leaf vein : 葉脈
leafout : 葉を出す
leafwing : 葉翼、翼葉
Leafwings : フタオチョウ亜科
learning : 学習
least square method : 最小二乗法
least-squares fitting : 最小二乗法
lectin : レクチン
lectotype : レクトタイプ、選定基準標本
leg : 脚(きゃく)、足(あし)、肢(し)
leg armature : 脚の護身器官、脚の防護器官(刺毛など)
leg disc : 肢原基
leg segment : 脚節、脚部分
legally protected species : 法的保護種
legitimate interest : 合法的な関心
legitimate purpose : 正当な目的
lek : レック、集団求愛場、求愛集団
length of flight bout : 飛翔期間(一続きの行動期間)の長さ
length of microtrich : 微毛長、微毛の長さ
lengthwise : 縦方向に
lengthy circular or spiral flight : 長円状またはらせん状の飛翔
Lepidoptera : チョウ目、鱗翅目
lepidopteran abundance : 鱗翅類の個体数
lepidopteran community : 鱗翅類群集
lepidopteran genome : 鱗翅類のゲノム
lepidopteran host cell DNA : 鱗翅類の宿主細胞 DNA
lepidopteran insect : 鱗翅目の昆虫
lepidopteran insect physiology : 鱗翅目昆虫の生理学
lepidopteran lineage : 鱗翅類系統
lepidopteran species : 鱗翅目の種、鱗翅類昆虫
lepidopteran-specific sequence : 鱗翅類に特異的な塩基配列
lepidopterist : 蝶(蛾)の研究者
Lepidopterological Society of Japan : 日本鱗翅学会(LSJ)
lepidopterology : 鱗翅類学
lepidopteron : (複 . -ran)、鱗翅類、チョウ・ガ類〔ラテン語〕
lethal : 致死的、致死
lethal equivalent : 致死相当量
lethal level : 致死(限界)レベル
lethal limit : 致死限界
leucine : ロイシン
leucopterin : ロイコプテリン
levana form : レヴァナ型(アカマダラ)
level of confidence, 0.05 : 信頼水準５%

level of enforcement : 施行水準
level of enrichment : 濃縮水準、富化水準
Levy flight : レヴィ飛行
LGM : 最終氷期最盛期
(Last Glacial Maximum)
library : ライブラリー
lichen : 地衣類
lie beneath : 真下に置かれる
life cycle : 生活環、ライフサイクル、生活史
life cycle characteristic : 生活環特性
life cycle pattern : 生活環の型、生活史の型
life form : 生活形
life habit : 生活様式
life history : 生活史
life history strategy : 生活史戦略
life pattern : 生活型
life phenomenon : 生活現象
life resource : 生活資源
life span : 生涯、寿命
life table : 生命表
life type : 生活型
life zone : 生息帯、生活圏、生態区域、生息域
life-cycle polymorphism : 生活環多型、生活史多型
life-history characteristic : 生活史特性
life-history strategy : 生活史戦略
life-history trait : 生活史形質、生活史特性
ligament : 靭帯
ligand : 配位子、リガンド
ligate : 結紮(けっさつ)する、糸でくくる
ligation experiment : 結紮実験
light : 明るさ
light condition : 光条件
light environment : 光環境、視環境
light gap : 明るい割れ目
light intensity : 照度
light interruption response : 光中断反応
light irradiation : 光照射
light micrograph : 光学顕微鏡写真
light microscope : 光学顕微鏡
light period : 明期
light phase : 明期、明相
light pulse : 光パルス
light time : 明期
light trap : ライトトラップ
light wavelength region : 光の波長領域
light-dark cycle : 明暗周期、明暗サイクル、LD サイクル
light-receptor substance : 光受容体物質
light-sensitive chromophore : 光感受性発色団
lightness constancy : 明るさ恒常性
lightning : 火花、照明
ligula : 唇舌
likelihood : 尤度、可能性、見込み
likelihood approach : 尤度アプローチ
likelihood function : 尤度関数
likelihood ratio test : 尤度比検定
likewise : 同様に
lima bean : ライマメ、ライマビーン
limestone grassland : 石灰岩の草地
limestone plateau : 石灰岩台地
limited geographical coverage : 限定された地理的範囲
limited observation : 限定された観察
LINE : 長鎖散在反復配列、長散在型反復配列(Long INterspersed 〈nuclear〉 Element)
line of weakness : 脆弱線
line transect : 線状観察路、ライントランセクト、線状の調査経路
lineage : 系列、系統、系譜
lineage ancestral : 系統の祖先
lineage sorting : 系統ソーティング
lineage specific loss of duplicate : 系統特異的な重複消失
lineage's exemplar : 系統の典型種
lineage-specific copy : 系統特異的な複製

lineage-specific expansion of chemosensory gene : 化学感覚遺伝子の系統別の拡張
linear : 線形の、細長い(蛾類の前翅)、線形
linear decline : 直線的な衰亡
linear flight : 直線飛翔
linear hierarchy : 線形階層
linear mixed effect model : 線形混合効果モデル
linear morphometrics : 線形的形態測定学
linear ramp : リニアランプ、線形ランプ
linear regression : 線形回帰
linear regression analysis : 線形回帰分析
linkage : 連鎖
linkage between cue and preference : 刺激と選好の間の連鎖
linkage disequilibrium : 連鎖不平衡
linkage group : 連鎖群
linkage group assignation : 連鎖群の割り当て
linkage group association : 連鎖群関係
linkage mapping : 連鎖地図、リンケージマッピング
linkage mapping study : リンケージマッピング研究、連鎖分析研究
linkage order : 連鎖の順序
linked locus : 連鎖する遺伝子座
Linnaean tautonymy : リンネ式同語反復
Linnaeus : リンネウス(スウェーデンの自然科学者)、リンネ
Linne, Carl von : リンネ、カール・フォン・リンネ
Linnean system : リンネ式動植物分類(命名)法
linoleic acid : リノール酸
linolenic acid : リノレン酸
lipid : 脂質、リピッド
lipid accumulation : 脂質蓄積
lipocalin : リポカリン
lipophilic substance : 親油性物質
lipophorin : リポフォリン
lipoxygenase : リポキシゲナーゼ
liquid : 液体
liquid air : 液体空気
liquid diet : 流動食、液状食料
liquid nitrogen : 液体窒素
liquid oxygen : 液体酸素
list of all species collected : 全採集種リスト
List of Available Names in Zoology : 動物学における適格名リスト
list of Japanese butterfly : 日本産蝶類リスト
list of Lycaenidae : シジミチョウ科のリスト
list of works : 著作リスト
live specimen : 生きた標本、生きている標本
livestock : 家畜
livestock grazing : 家畜の放牧
living ancestor : 生きている祖先
living picked flower : 生きたまま摘まれた花
lizard : トカゲ
LL : 全照明、恒明(continuous Light あるいは Light-Light)
Lloyd and Ghelardi J Index : ロイド - ゲラルディの J 指標
LLT : 最低致死温度、低温致死温度(Lower Lethal Temperature)
LMC : 局所的配偶競争(Local Mate Competition)
LN : 側方ニューロン(Lateral Neuron)
lobe : でっぱり、短い突起、突起、葉片、裂片
lobed : 葉状の
loblolly pine : テーダマツ
lobulus : (視)小葉
lobulus varginalis : 腟小葉
local : 局所的な
local adaptation : 局所適応、局部順応
local butterfly lover : 地方の蝶愛好家
local extinction : 局所的な絶滅
local gene duplication : 遺伝子の局所的な複製

local irradiation : 局所照射
local irradiation experiment : 局所照射実験
local irradiation of light : 局所的光照射
local mate competition : 局所的配偶競争
local mimicry polymorphism : 局所擬態多型、地方擬態多型、地域擬態多型
local population : 局所個体群、地域個体群、地域集団
local population dynamics : 地域個体群動態
local population size : 局所個体群サイズ
local population viability : 局所個体群の生存力
local selection : 局所選択、局部選択、その土地による選択
local selective pressure : 局所的選択圧
local specialist : 局所スペシャリスト
local standard time : 現地標準時間
local strain : 地方系統、地域系統
local variation : 地方変異、地域変動、地域多様性
local variety : 地方種、在来種
locally adapted trait : 局所適応した形質
located midway : 中ほどにある
location of sampling point : 採取地点の場所
lock together : しっかりと組み合わせる
locomotor mimicry : 動作擬態
locus : (複 . -ci)、遺伝子座、座位、座、場所、位置〔ラテン語〕
locus coding : 遺伝子座コーディング
locust : バッタ、イナゴ、トビバッタ
LOD : 対数オッズ、ロッド(Logarithm of odds)
lodgepole pine : コントルタマツ
log collated index : 対数形式の照合指標
log male body size : 対数形式の雄の体サイズ
log-odds ratio : 対数オッズ比、LOR
log-rank analysis : 対数順位解析、対数順位検定
log-rank test : ログランク検定、対数順位検定
log-transformed distance : 対数変換された距離
logarithm of odds : 対数オッズ
logged forest : 伐採された森
logged habitat : 伐採された生息地
logging : 伐採
logistic regression : ロジスティック回帰
logistic regression model : ロジスティック回帰モデル
logotype : 後模式標本
long axis : 長軸
long day length : 長い日長
long distance : 長距離
long distance migration : 長距離移動
long distance pheromone : 長距離のフェロモン
long interspersed nuclear element : 長鎖散在(型)反復配列(LINE)
long noncoding RNA : 長鎖非コード RNA、長鎖ノンコーディング RNA
long photoperiod : 長日日長
long region : 長い領域
long term : 長期間、長期
long term dynamics : 長期動態
long terminal repeat : 長鎖末端反復配列(LTR)
long-day exposure : 長日にさらすこと、長日処理
long-day photoperiod : 長日日長
long-day plant : 長日植物
long-day regimen : 長日養生
long-day treatment : 長日処理
long-day type : 長日型
long-day type response : 長日型反応
long-horn type : 長角型
long-horned beetle : カミキリムシ
long-range cue : 広範囲の刺激、長距離の刺激

long-range signal : 長距離シグナル
long-standing controversy : 長年にわたる論争
long-standing question : 長年の疑問
long-tailed blue butterfly : ウラナミシジミ
long-tailed skipper butterfly : アオネオナガセセリ
long-term control : 長期防除
long-term decline : 長期間にわたる衰亡
long-term diapause : 長期休眠
long-term evolution : 長期的進化、長期間の進化
long-term movement : 長期間の移動
long-term range expansion : 長期間にわたる生息範囲の拡大
long-term success : 長期間の成功
long-wavelength-sensitive cone : 長波長感受性錐体
long-wavelength-sensitive photopigment : 長波長感受性視物質
long-winged individual : 長翅型個体
longevity : 寿命、生涯、長生き、長命
longicorn beetle : カミキリムシ
longitudinal : 縦(じゅう)、縦の
longitudinal section : 縦断面
longitudinal vein : 縦脈、縦走脈
longitudinally-oriented distance : 緯度本位の距離
longleaf pine : ダイオウショウ、ダイオウマツ
Longwings : ドクチョウ亜科
loose linkage : 疎性連鎖
loss of *Distal-less* expression : 末端部のない遺伝子発現の消失
loss of eyespot : 眼状紋の消失
loss of function : 機能欠損、機能喪失、機能退化
loss of scale : 鱗粉の脱落
loss of species diversity : 種多様性の喪失
low altitude : 低地
low molecular weight metabolite : 低分子代謝物、低分子代謝産物
low temperature : 低温
low temperature and short day : 低温短日
low temperature phase : 低温期
low temperature shock : 低温ショック
low temperature treatment : 低温処理
low temperature-sensitive stage : 低温の感受期
low wind : 低風、軽風
low-frequency sound : 低周波数音、低周波音
low-growing : 背丈の低い
low-nitrogen diet : 低窒素試料
low-temperature drying period : 低温乾燥期
low-temperature period : 低温期間
low-temperature stimulus : 低温刺激
low-temperature tolerance : 冷(低)温耐性
lower frequency sound : 低周波数音
lower lamina : 翅裏面
lower lethal temperature : 最低致死温度、低温致死温度(LLT)
lower-bound estimate : 下限推定、下限推定値、下限値の推定
lower-quality diet : より低品質な食料、低品質飼料
lower-taxon--higher-taxon ratio : 低次分類数 - 高次分類数の比
lowest AIC : AIC 最小(Akaike's Information Criterion)【統計学】
lowest intensity : 最低強度
lowest temperature : 最低気温
lowland : 低地
lowland population : 平地個体群
lowland species : 平地性種
lowland tropical rainforest : 低地熱帯雨林
LPN : 側後方ニューロン(Lateral Posterior Neuron)
LSJ : 日本鱗翅学会 (Lepidopterological Society of Japan)

LTR : 長鎖末端反復配列(Long Terminal Repeat)
lubentina group : ルベンチーナグループ(アカホシイナズマ〈*Euthalia lubentina*〉が属するイナズマチョウ属〈*Euthalia*〉)
Ludlow's Bhutan Glory butterfly : ブータンシボリアゲハ
Luehdorfia japonica : ギフチョウ
Luehdorfia line : ルードルフィア線
Luehdorfia puziloi : ヒメギフチョウ
luminosity : 照度
lumped species : ひとかたまりの種
lumper : 併合派分類学者(生物の類似点を重視して大きく分類しようとする分類学者)
luna moth : オオミズアオ
lunar compass : 月コンパス
lunule : 半月紋、三日月形
lush green vegetation : 緑豊かな植生、青々とした植生
lusting of ecdysone : エクダイソン欠除
lutein : ルテイン
lutexin : ルテキシン
luxuriant green foliage : 繁茂している緑色の葉
luxury gene : ラグジュアリー遺伝子、ラクシャリー遺伝子(細胞が構成する組織の機能を実現するために、その細胞に特異的に発現している遺伝子)
lycaenid larva : シジミチョウ科の幼虫
lycaenid-ant association : シジミチョウとアリとの関係
Lyme disease : ライム病

m

m-species-list curve : m 種リスト曲線
m-species-list method : m 種リスト法
MacClade : マッククレイド(最大節約法に基づいて系統推定を行うソフトウェアの名称)
macerate : 液体に浸して柔らかくする
maceration : 浸軟(しんなん)、軟化処理
macro lens : 接写レンズ、マクロレンズ
macroevolution : 大進化
macroevolutionary consequence : 大進化的結果、大進化の結果
macroinvertebrate : 大型無脊椎動物
macrolepidoptera : 大型チョウ目、大型鱗翅類
macropter : 長翅型
macropterism : 長翅
macropterous form : 長翅型
macroscopic phenotype : 巨視的な表現型
macular : 黄斑の
Maculinea teleius : ゴマシジミ
Madagascan sunset moth : ニシキオオツバメガ
Magadan region of Russia : ロシア・マガダン地域
magenta : 深紅色、赤紫色、マゼンタ
magnetic bead : 磁性ビーズ
magnetic compass : 磁気コンパス
magnetoreception : 磁気受容、磁気感知
magnifying glass : 虫眼鏡
magnitude and direction : 大きさと方向
magnitude of winter severeness : 冬期の厳しさの度合い
main route : 主要経路
Mainland China : 中国大陸
mainland community : 本土の群集
mainland of Japan : 日本本土
maintenance of diversity : 多様性の維持
maintenance of genetic variation : 遺伝的変異の維持
maintenance of race : 種族維持、系統維持
maize : トウモロコシ
major barrier : 大きな障壁
major ecosystem : 大生態系
major factor : 主要因、主因

major gene : 主遺伝子、主働遺伝子
major histocompatibility complex : 主要組織適合遺伝子複合体（MHC）
major transition : 主要な移行
majority : 大部分、大多数
majority-rule consensus tree : 多数決原理総意樹、多数決合意樹、多数合意樹、多数派支配型コンセンサス樹
malachite butterfly : ミドリタテハ
maladaptation : 不適応
malar space : マーラースペース、磨縁部
Malayan type : マレー型
male : 雄（おす）、♂、オス
male clasper : 雄の把握器
male color preference : 雄の色彩選好
male courtship preference : 雄の求愛選好
male ejaculate : 雄の射精液
male embryo : 雄（の）胚
male fitness : 雄の適応度
male genitalia : 雄性生殖器、雄性交尾器
male genotype : 雄の遺伝子型
male mate choice : 雄の配偶者選択、雄の配偶者選好性
male mate-locating behavior : 雄の雌探し行動
male mating behavior : 雄の配偶行動
male model : 雄モデル
male phenotype : 雄の表現型
male scan : 雄探索、雄走査
male sexual differentiation : 雄の性分化
male type larva : 雄型幼虫
male wing odour : 雄の翅から発生する匂い
male-biased : 雄に偏った、雄偏重
male-biased sexual size dimorphism : 雄の方が雌よりも体長が大きい性的体長二型
male-derived compound : 雄由来化合物、雄由来混合物
male-heterogametic sex chromosome constitution : 雄ヘテロ型性染色体構成
male-informative : 雄 - 有用な情報付き
male-killing : 雄殺し
male-like : 雄のような
male-like and female-like phenotype : 雄と雌の特徴を合わせもつ表現型
male-like genital trait : 雄の特徴を持つ交尾器表現形質
male-like reproductive organ : 雄の特徴を持つ生殖器官
male-like wing trait : 雄の特徴を持つ翅表現形質
male-male competition : 雄間競争
male-specific *dsx* isoform : 雄特異的な *dsx* 遺伝子のイソ型
male-specific embryonic mortality : 雄特有の胚死亡率
male-specific molecular mechanism : 雄に特有の分子機構
male-specific organ : 雄特有の器官
male-transferred anti-aphrodisiac : 雄から引き渡された抗催淫物質
malfunction : 機能不全
mallow : ゼニアオイ
Malpighian tube : マルピーギ管
Malpighian tubule : マルピーギ管
Malpighian tubule cell : マルピーギ管細胞
mammal : 哺乳類
mammalian genome : 哺乳類のゲノム
mammalian herbivore : 哺乳類の草食動物
man-made structure : 人工構造物
managed area : 管理区域
managed bee : 管理されたミツバチ
managed relocation : 人為的な生息地移動
management unit : （個体群）管理単位（MU）
mandatory change : 強制変更
mandible : 大腮（たいさい、おおあご）、大顎（だいがく、おおあご）
mandibular : 大腮の、大顎の
mandibular gland : 大顎腺（おおあごせん）

Manduca sexta : タバコスズメガ(スズメガ科の一種)
mangrove : マングローブ
manifested : 明白な、現れる
manipulation : 操作
Mann-Whitney U-test : マン-ホイットニーの U 検定
manoeuvrability : 操作能力、操縦性
Mantel test : マンテル検定
Mantel-Cox test : マンテル - コックス検定
mantid : カマキリ
Mantodea : カマキリ目、蟷螂目
Mantophasmatodea : マントファスマ目、カカトアルキ目、踵行目
manual adjustment : 手動調整
manufacture : 生産する、製造する
manure : 有機質肥料(特に動物の排出物)
manuscript : 草稿
MAPK : 分裂促進因子活性化タンパク質キナーゼ、MAP キナーゼ(Mitogen-Activated Protein Kinase)
maple tree : カエデの木、モミジの木
mapping : マッピング
mapping procedure : マッピング手順
marbled white butterfly : ヨーロッパシロジャノメ
marginal : 外縁部、外縁部帯
marginal band : 外縁の(縞状)バンド
marginal eyespot : 外縁部の眼状紋
marginal facies : 縁辺相
marginal habitat : 分布辺縁生息地、周辺生息地、辺縁の生息地
marginal value theorem : 限界値定理
marginopleural suture : 外縁側板(溝状)線
marine turtle : ウミガメ
Mariner like transposable element : マリナー様転移因子(MLE)
marjoram : ハナハッカ、マヨラナ(シソ科の低木)
mark and recapture statistics : 標識再捕獲統計、標識再捕獲統計学
mark and recapture trial : 標識再捕実験
mark and release research : リリースするマーキング調査
mark-recapture : 標識再捕獲法、標識再捕法
mark-recapture method : 標識再捕獲法、標識再捕法
mark-release-recapture study : 標識 - 放出 - 再捕獲の研究
marked difference : 著しい差、顕著な相違
markedly lower rate of survival : 著しく低い生存率
marker : マーカー、標識物質
marker gene : マーカー遺伝子
marker locus : マーカー遺伝子座
marking : 斑紋、模様
marking pattern of wing : 翅斑紋パターン
marking pre-pattern : 斑紋プレパターン
marking variation : 斑紋変異
Markov process : マルコフ過程
marsh : 沼地、湿地
marsh fritillary butterfly : チョウセンヒョウモンモドキ
masquerade : みせかけ、仮装、なりすまし、扮装擬態
mass capture : 大量捕獲
mass extinction : 大量絶滅
mass spectrograph : 質量分析計
master control gene : マスター制御遺伝子
master gene : マスター遺伝子
mate acquisition : 交尾相手の獲得
mate assortatively : 同類的に交配する
mate availability : 交尾可能性
mate choice : 配偶者選択
mate choice experiment : 配偶者選択実験
mate competition : 配偶競争、配偶者競争、配偶者争奪戦
mate location : 交尾場所

mate preference cue : 配偶者選好の刺激
mate randomly in nature : 自然ではランダムに交尾する、自然無作為交配
mate recognition : 配偶者認知、配偶者認識
mate recognition signal : 配偶者認知シグナル
mate refusal behavior : 交尾拒否行動
mate search : 配偶者探索
mate seeking territory : 探雌のためのナワバリ
mate-locating behavior : 雌探し行動、配偶者位置探索行動
mate-locating strategy : 配偶者位置探索戦略
mate-refusal posture : 交尾拒否姿勢
mate-searching behavior : 探雌のための行動
mate-seeking response : 探雌反応
mate-selection : 配偶者選択
mated female : 既交尾雌
material and method : 材料と方法
maternal age : 母年齢
maternal effect : 親世代の影響、母性効果
maternal species : 母親種
maternal-form effect : 母体型の影響
maternally expressed gene : 母(親)性発現遺伝子
mathematical expression : 数学的表現
mating : 交尾(こうび)、交配
mating area : 交尾地域、交配場所
mating asymmetry : 交尾の非対称性
mating attractiveness : 交尾誘引性
mating behavior : 配偶行動、交尾行動
mating between the two species : 種間交配(種)
mating discrimination : 交尾識別
mating pair : 交配ペア、交尾対
mating plug : 交尾栓
mating probability : 交尾確率
mating propensity : 交尾傾向
mating strategy : 交尾戦略
mating success : 交配成功率
mating system : 配偶システム、配偶様式
mating tactics : 交尾策略、配偶行動戦術
mating territory : 交尾テリトリー、雌を見つけるためのナワバリ、交尾ナワバリ
mating tube : 交尾管
mating within species : 種内交配(種)
matrix habitat : マトリックス生息地
maturation : 成熟
maturation period : 成熟期
mature egg : 成熟卵
mature grassland : 十分に成長した草地
mature oocyte : 成熟卵母細胞
mature ovary : 成熟した卵巣
mature ovum : 成熟卵
maxilla : (複 . -ae)、小腮(しょうさい、こあご)、小顎(しょうがく、こあご)〔ラテン語〕
maxillary : 小腮の、小顎の
maxillary galea : 小腮外葉
maxillary gland : 小顎腺(こあごせん)
maxillary lobe : 小腮粒状体
maxillary palp : 小腮鬚(しょうさいしゅ)、小顎鬚(しょうがくしゅ、こあごひげ)〔英語〕
maxillary palpus : 小腮鬚、小顎鬚〔ラテン語〕
maxillary segment : 小腮節
maximally dominant : 最大限に優勢な
maximum and minimum thermometer : 最高最低温度計
maximum intraspecific divergence : 最大種内分化
maximum likelihood : 最尤(度)
maximum likelihood method : 最尤法、ML法
maximum likelihood tree : 最尤法による系統樹、ML 法による系統樹
maximum log-likelihood : 最大対数尤度
maximum parsimony : 最大節約法
maximum parsimony method : 最節約法、最大節約法、MP 法
maximum phylogenetic informativeness : 最大系統的情報性、最大系統発生学的情報量

maximum-likelihood ancestral state reconstruction : 最尤法による祖先状態再構成
maximum-likelihood phylogeny : 最尤推定系統樹
maximum-parsimony analysis : 最節約解析
maximum-parsimony ancestral state reconstruction : 最節約法による祖先状態再構成
McDonald-Kreitman test : マクドナルド - クレイトマンテスト
MCMC : マルコフ連鎖モンテカルロ法 (Markov Chain Monte Carlo method)
meadow : 牧草地、採草地
meadow brown butterfly : マキバジャノメ
meal : 食事
mean : 平均
mean crowding : 平均混み合い度
mean depth : 平均深度
mean dispersal propensity : 平均分散傾向
mean expression level : 平均発現水準、平均発現レベル
mean generation time : 平均世代時間
mean minimum temperature : 平均最低気温
mean number of accumulated individual : 平均累積個体数
mean number of stem per quadrat : 方形区当りの平均樹幹数
mean richness value : 平均種数値
mean supercooling point : 平均過冷却点
mean weight : 平均体重
MEAs : 多国間環境協定 (Multilateral Environmental Agreements)
measure of diversity : 多様性指数
measure of patchiness : パッチネス指数
measure of sampling intensity : サンプリング強度の指数
measuring amplifier : 測定増幅器、計測用増幅器
mechanical control : 機械的防除
mechanical damage : 機械的傷害、機械的損害
mechanical isolation : 機械的隔離
mechanical mean : 機械的手段
mechanical sense : 機械(的)感覚
mechanism : 機械論、機構、機作
mechanism of emergence : 出現機構、羽化機構
mechanism of flight orientation : 飛行方向の機構
mechanism of speciation : 種分化機構
mechanism of voltinism change : 化性変化の機構
mechanism of wing folding : 翅の折りたたみ機構
mechanism-specific : 機構特異的
mechanistic basis : 機構的基盤
mechanistic underpinning : 機械論的な土台、機械論的基礎
mechanoreceptive property : 機械感覚特性
mechanoreceptor : 機械感覚器、機械受容器
mechanosensory neuron : 機械刺激ニューロン、機械刺激神経細胞
meconium : 蛹便
Mecoptera : シリアゲムシ目、長翅目
Medea factor : MEDEA 因子(Maternal-effect dominant embryonic arrest)、母性効果優勢胚発育停止因子、メディア因子(利己的な遺伝因子の一種)
media : 中脈、M 脈
medial : 内側(ないそく)、正中線により近い(相対位置)、中央帯、中央の
medial ablation : 内側部位の切除
medial band : 内側の(縞状)バンド
medial crossvein : 中横脈、m 脈
medial microcautery : 内側部位の微細焼灼(しょうしゃく)
medial-cubital crossvein : 中肘横脈、m-cu 脈

medial-lateral axis : 内外軸、ML 軸
median : 中央帯、中央の、正中(せいちゅう)、中央値、メディアン【統計学】
median area : 中央部(蛾類)
median cleft : 中央欠刻、中央裂目
median dash : 中縦線(蛾類)
median gill : 中央鰓
median line : 正中線、中横線(蛾類)
median lobe : 中片、中央片
median notal wing process : 中背板翅突起、中背翅突
median plane : 正中断面
median section : 正中断面
median surface : 正中断面
median threshold : 中央閾(いき)値、メディアン閾値
median value : メディアン値、中央値
median vein : 中脈(ちゅうみゃく)、M 脈
mediate : 媒介する、調節する、媒体となる
mediated contact : 介在する接触
medical freezer : 医療用冷凍庫
medio- : 中央 -、内側 -〔ラテン語〕
medio-cubital crossvein : 中肘横脈、m-cu 脈
medio-lateral axis : 内外軸、ML 軸
mediobasal : 内側基部
mediodiscal : 内側中央帯
mediolateral axis : 内外軸、ML 軸
MeDIP : メチル化 DNA 免疫沈降法(Methylated DNA ImmunoPrecipitation)
Mediterranean plant : 地中海地域の植物、地中海植物
Mediterranean region : 地中海地域
medius : 中脈
medulla : (複 . -ae)、視髄、髄層
mega pixel : メガピクセル
megablast analysis : メガ BLAST 解析(megablast は、BLASTプログラムの 一機能で、大量のクエリシークエンスで高速検索を行うプログラム機能である。)
megascopic : 肉眼で観察した、巨視的な
megaspore : 大胞子、胚のう細胞
meiosis : 減数分裂
meiotic drive : マイオティックドライブ、減数分裂分離ひずみ
melanic : 黒化の
melanic form : 黒化型、黒色型、暗色型
melanic mimicry : 黒化擬態
melanic morph : 黒化型
melanic pigment : メラニン色素
melanic scale : 黒化鱗粉
melanin : メラニン
melanin precursor : 黒化前駆物質、メラニン前駆体
melanism : 黒色化、黒化
melanized warm-season phenotype : 暖候期の黒化表現型
melatonin : メラトニン
melting curve : 融解曲線
membrane : 膜
membrane protein : 膜タンパク質
membranous cuticle : 膜性表皮、膜質表皮層
membranous structure : 膜状の構造、膜構造
memorandum : メモ、備忘録
memorial address : 追悼、弔辞
memorial statement : 追悼文
memorized landmark : 記憶した陸標
memory mechanism : 記憶機構
Mendel, Gregor J. : メンデル、グレゴール・ヨハン・メンデル
Mendelian factor : メンデル因子
Mendelian locus : メンデル遺伝子座
mentum : 下唇基節
meral suture : 頭基節縫合線
mercury vapour lamp : 水銀灯
mes- : 中 -〔ギリシャ語〕
mesepimeron : 中胸後側板
mesepisternal groove : 中胸側板溝
mesepisternum : 中胸前側板、中胸前腹板

mesh : 網
mesial : 中央の、正中の、正中面
meso- : 中 -〔ギリシャ語〕
mesofemur : 中腿節
mesoleg : 中脚
mesonotum : 中胸背板
mesopleuron : 中胸側板
mesopleurosternum : 中胸側腹板
mesoscutellum : 中胸小盾板(ちゅうきょうしょうじゅんばん)
mesoscutum : 中胸盾板
mesoseries : 半環状(蛾幼虫の鉤爪の形態)
mesosoma : 中体節
mesosomal : 中体節の
mesosternum : 中胸腹板
mesotarsus : 中ふ節
mesothoracic wing base : 中胸翅基部
mesothorax : 中胸(ちゅうきょう)
mesotibia : 中脛節
messenger RNA : メッセンジャー RNA (mRNA)
Met Office online database : 英国の Met Office(英国気象庁；UKMO)で運用しているオンラインデータベース
met- : 後 -〔ギリシャ語〕
meta- : 後 -〔ギリシャ語〕
meta-analysis : メタ解析、メタ分析
meta-species : メタ種
metabolic : 代謝(上)の
metabolic activity : 代謝活動、代謝活性
metabolic change : 代謝変化、代謝変動
metabolic cost : 代謝費用、代謝コスト
metabolic cycle : 代謝回路
metabolic detoxification : 代謝(性)解毒
metabolic origin : 代謝の起点、代謝源
metabolic pathway : 代謝経路(細胞の中で起きる連鎖的な化学反応のこと)
metabolic rate : 代謝速度
metabolism : 代謝、物質代謝
metabolite level : 代謝産物レベル、代謝レベル
metabolomics : メタボロミクス、代謝学
metafemur : 後腿節
metagenome : メタゲノム
metagenomics : メタゲノム学、メタゲノム解析
metaleg : 後脚
metallic : 金属色の、金属光沢の
Metalmark : シジミタテハ科の蝶の総称
metamerism : 体分節、環節体制
metamorphic molt : 変態脱皮
metamorphic rock : 変成岩
metamorphic stage : 変態期
metamorphose : 変態する
metamorphosis : 変態
metanotum : 後胸背板
metapleuron : 後胸側板
metapopulation : メタ個体群
metapopulation biology : メタ個体群生態学
metapopulation dynamics : メタ個体群動態
metapopulation dynamics analysis : メタ個体群動態解析
metapopulation dynamics model : メタ個体群動態モデル
metapopulation effect : メタ個体群効果
metapopulation viability : メタ個体群の生存能力、メタ個体群生存率、メタ個体群の存続性
metasoma : 後体節
metasomal : 後体節の
metasternum : 後胸腹板
metatarsus : 後ふ節、基ふ節
metathetely : メタセテリー、翅芽遅滅、後成現象(体の一部が全体より遅れた発生段階にある現象)
metathorax : 後胸(こうきょう)
metatibia : 後脛節
Metazoa : 後生動物

metepimeron : 後胸後側板
metepisternum : 後胸前側板
methanol-acetic acid : メタノール酢酸
methionine : メチオニン
methionine derived compound : メチオニン由来化合物
methionine derived glucosinolate : メチオニン由来グルコシノレート
methionine-rich storage protein : メチオニン型貯蔵タンパク質
Method for Inferring Sequence History In terms of Multiple Alignment : MISHIMA 法
method of comparative morphology : 比較形態学的方法
methodical study : 整然とした調査、整然とした研究、入念な調査、組織的な調査
methodological advance : 方法論的進歩
methyl jasmonate : ジャスモン酸メチル
methyl salicylate : サルチル酸メチル
methylalkylpyrazine : メチルアルキルピラジン
Methylated DNA immunoprecipitation : メチル化 DNA 免疫沈降法（MeDIP）
methylation : メチル化
methyltransferase : メチルトランスフェラーゼ
mevalonate : メバロナート、メバロン酸
MFO : 混合機能酸化酵素（Mixed Function Oxydase）
MHC : 主要組織適合遺伝子複合体（Major Histocompatibility Complex）
Miami blue butterfly : マイアミブルーシジミ
miconium : 蛹便、ミコニウム
micro structure : 微細構造
micro-RNA : マイクロ RNA（miRNA）
microarray : マイクロアレイ
microbe : 病原菌、微生物
microbial assemblage : 微生物集団、微生物群集
microbial community : 微生物群集
microbial control : 微生物的防除
microbial infection : 微生物感染
microbial insecticide : 微生物的殺虫剤
microcautery : 微細焼灼(しょうしゃく)
microcentrifuge tube : 微量遠心チューブ、マイクロ遠心チューブ、微量遠心管
microclimate : 微気象
microclimatic constrain : 微気象制約条件
microcosm : ミクロコスモス、小宇宙、微小生態系
microevolution : 小進化
microgastrinae subfamily : サムライコマユバチ亜科
microhabitat : 微生息場所、マイクロハビタット
microinjection : 微小注射、マイクロインジェクション
microlepidoptera : 小型鱗翅目、小型鱗翅類、小蛾類
micronodule : 微小結節
microorganism : 微生物
microphone : マイクロホン
Microplitis demolitor : コマユバチ科の一種
micropterous form : 短翅型
micropylar : 精孔の、卵門(らんもん)の
micropyle : 精孔、卵門
microRNA : マイクロ RNA（miRNA）
microsatellite : マイクロサテライト
microsatellite marker : マイクロサテライトマーカー
microsatellite polymorphism : マイクロサテライト多型
microscope slide : 顕微鏡用スライドグラス、顕微鏡用スライド
microscopic endosymbiont : 微視的な細胞内共生生物
microscopic observation : 顕微鏡観察
microscopic specimen : 微視的な標本、顕微鏡標本

microsite : 微高地、微環境、微地形、マイクロサイト
microspore : 小胞子
microsporidiosis : 微胞子虫病
microsporidium : (複 . -ia)、微胞子虫
microtiter plate : マイクロタイタープレート、微量定量プレート(多数のくぼみ〈穴またはウェル〉のついた平板からなる実験・検査器具)
microtrich : 微毛
microvillus : (複 . -li)、微絨毛
mid leg : 中脚(ちゅうきゃく)、中肢(ちゅうし)
mid lib : 中軸
mid- : 中 - 〔英語〕
mid-autumn : 秋の半ば
mid-day : 正午
mid-final instar larva : 中終齢幼虫
mid-gut : 中腸(ちゅうちょう)
mid-pupal stage : 蛹中期、蛹期中期
Mid-Tertiary : 第三紀中葉、第三紀中期
middle : 中央の
middle branch : 中枝
middle larval stage : 幼虫期の中頃
middle leg : 中脚(ちゅうきゃく)、中肢(ちゅうし)
middle-wavelength-sensitive cone pigment : 中波長感受性錐体視物質
middle-wavelength-sensitive photopigment : 中波長感受性視物質
middorsal : 背部中央
midgut : 中腸
midgut microbiota composition : 中腸の腸内細菌叢の組成、中腸の腸管微生物叢の組成
midleg : 中脚
midline : 中線、正中線
midline of symmetry system : 相称系の正中線
midnight : 真夜中
midnight sun : 白夜の太陽、真夜中の太陽
midrib : 葉の中央脈、主脈、中肋
midstory hardwood : 中層の広葉樹
midstory pine : 中層のマツ
midsummer : 真夏
midventral : 腹中線、中腹側
midventral suture : 中腹板溝状線
migrant : 移動性種
migrant skipper butterfly : イチモンジセセリ
migrant species : 移動性種
migrate : 渡りをする
migrating bird : 渡り鳥
migration : 移動、渡り、回遊、移住、分散
migration factor : 移動指数
migration group : 移動群
migration rate : 移動率
migration route : 移動経路、移動ルート
migration syndrome : 移動形質群
migration trajectory : 移動軌跡
migration-colonization syndrome : 移動 - 定着行動様式
migratory : 移動性の
migratory butterfly : 移動性の蝶
migratory habit : 移動性、移動習性
migratory pattern : 移住パターン
migratory species : 移動性種
Milbert’s tortoiseshell butterfly : ヤンキーコヒオドシ、アメリカコヒオドシ
milkweed : トウワタ
Milkweed Butterflies : マダラチョウ亜科、マダラチョウ科
milkweed butterfly : オオカバマダラ
milky latex : 乳濁液、乳状液
million : 100 万
mimeographing : 謄写版印刷
mimesis : 隠蔽擬態、隠蔽的擬態(いんぺいてきぎたい)、擬態、ミメシス
mimetic allele : 擬態対立遺伝子
mimetic association : 擬態関係
mimetic chromosome : 擬態型(の)染色体

mimetic coloration pattern : 擬態型色彩パターン
mimetic convergence : 擬態の収斂
mimetic group : 擬態群、擬態グループ
mimetic insect : 擬態昆虫
mimetic lineage : 擬態系統
mimetic morph : 擬態型
mimetic pattern : 擬態パターン
mimetic phenotype : 擬態の表現型
mimetic polymorphism : 擬態多型
mimetic pre-pattern formation : 擬態のプレパターン形成
mimetic pressure : 擬態圧
mimetic-form female : 擬態型雌
mimetic-type sequence : 擬態型配列
mimic : 擬態しているもの、擬態する、擬態種、ミミック
Mimic-Whites : トンボシロチョウ亜科
mimicry : 擬態、標識擬態、標識的擬態、ミミクリー
mimicry association : 擬態関係
mimicry evolution : 擬態進化
mimicry locus : 擬態遺伝子座
mimicry pattern : 擬態のパターン
mimicry polymorphism : 擬態多型
mimicry ring : 擬態リング
mimicry variation : 擬態変異
mimicry wing pattern : 擬態の翅紋様、翅の擬態紋様
mimicry-related locus : 擬態関連遺伝子座
mimicry-related sex-limited polymorphism : 擬態関連の限性多型
mimivirus : ミミウイルス
mineral : ミネラル、無機物
minimal interference : 最小限の干渉
minimum AIC mixed model : AIC 最小混合モデル(Akaike's Information Criterion)
minimum deviation method : 最小偏差法
minimum frequency of variant : 変異の最小頻度
minimum number : 最小個体数
minimum temperature : 最低気温
minimum threshold : 最小閾(いき)値
Ministry of Agriculture, Forestry and Fisheries : 農林水産省(日本)
Ministry of the Environment : 環境省(日本)
minute wasp : 微小な蜂
Miocene series : 中新統、中新世
miriamide : ミリアミド
miRNA : マイクロRNA(microRNA、micro-RNA)
mirrored tapetum : 鏡型反射層板
misapply : 誤適用する
miscellaneous note : 雑記
MISHIMA : MISHIMA 法(Method for Inferring Sequence History In terms of Multiple Alignment)
misidentify : 誤同定する
misleading result : 誤った結果
mismatch : ミスマッチ、不適当な組み合わせ
missense mutation : ミスセンス変異
missing : 欠失、欠損、欠如、欠落
missing haplotype : 欠損ハプロタイプ、欠失単模式種
missing treatment : 欠けた処理、未処理、無処理
Mission blue butterfly : ミッションヒメシジミ
mist : 霧
mistletoe : ヤドリギ
misunderstanding : 誤認
mite : ダニ
mitigation : 影響緩和、環境緩和
mitigation measure : 影響緩和措置
mitochondrial DNA : ミトコンドリア DNA
mitochondrial gene : ミトコンドリア遺伝子
mitochondrial gene encoding : ミトコンドリア遺伝子の符号化

mitochondrial gene sequence : ミトコンドリア遺伝子配列
mitochondrial genetic diversity : ミトコンドリアの遺伝的多様性
mitochondrial genome : ミトコンドリアゲノム
mitogen-activated protein kinase : 分裂促進因子活性化タンパク質キナーゼ、MAP キナーゼ(MAPK)
mitotic frequency : 細胞分裂頻度
mixed breed : 雑種、混交一種
mixed function oxydase : 混合機能酸化酵素
mixed secondary forest : 混合二次林
mixed strategy : 混合戦略
mixture : 混在、混合液
MK test : MKテスト、マクドナルド-クレイトマンテスト(McDonald-Kreitman test)
ML analysis : ML 法の解析(Maximum Likelihood)
ML axis : 内外軸、ML軸(Medial-Lateral axis)
ML tree : ML 法による系統樹、最尤法による系統樹(Maximum Likelihood)
MLE : マリナー様転移因子(Mariner Like transposable Element)
mM : ミリモル濃度、ミリモーラー(ミリモーラーは濃度を表す単位；大文字の「M」はモーラー〈molar〉と読む)
MNPV : 複数のヌクレオキャプシドを含む核多角体病ウイルス、複数のヌクレオキャプシドを含むNPV(Multiple NucleoPolyhedroVirus)
mobile : 移動性の
mobile element : 可動性因子、可動因子
mobile organism : 移動性生物
mobile species : 移動性種
mobility : 移動性、可動性
mode of flight : 飛翔様式
model : モデル
model coefficient : モデル係数
model of bovine rhodopsin : ウシロドプシンモデル
model of effectively neutral mutation : 有効中立突然変異モデル
model selection : モデル選択
model species : モデル種
model system : モデル系
model taxon : (複 . -xa)、モデル分類群
model-predicted hostplant preference : モデルで予想した寄主植物選好性
modeling clay : 模型用粘土
modern human : 現代人、現生人
modern synthesis : 現代の総合、現代的統合
modification-rescue model : 修飾 - 救済モデル
modified sperm : 修飾された精子、精子修飾
modifier : 変更遺伝子、修飾因子、修飾遺伝子
modulate : 調節する
modulating effect : 調節効果
module : モジュール、測定の標準(単位)
MOI : 感染多重度(Multiplicity Of Infection)
moist and warm condition : 湿暖条件
moist paper towel : 湿ったペーパータオル
moisture : 湿気
mole : モグラ
molecular clock : 分子時計
molecular component : 分子成分
molecular counter adaptation : 分子的対抗適応
molecular datum : (複 . -ta)、分子データ
molecular evidence : 分子的証拠、分子生物学的証拠、分子系統学的証拠
molecular evolution : 分子(的)進化
molecular evolutionary analysis : 分子進化(学)的解析
molecular evolutionary approach : 分子進化(学)的研究法
Molecular Evolutionary Genetics Analysis : MEGA(塩基配列解析ソフトウェア)

molecular genetics : 分子遺伝学
molecular level : 分子レベル
molecular marker : 分子マーカー
molecular mechanism : 分子機構
molecular phylogenetic analysis : 分子系統学的解析
molecular phylogenetical analysis : 分子系統学的解析
molecular phylogenetical point of view : 分子系統学的観点
molecular phylogenetical study : 分子系統学的研究
molecular phylogenetical tree : 分子系統樹
molecular phylogenetics : 分子系統(学)
molecular phylogeny : 分子系統(学)
molecular signature : 分子署名、分子シグネチャー、分子指標、分子サイン(遺伝子発現パターンのこと)
molecular systematics : 分子系統学、分子系統分類学
molecular technique : 分子技法
molecule : 分子
mollusk : 軟体動物
molt : 脱皮(する)
molt instar stage : 脱皮齢期
molting : 眠、脱皮(だっぴ)
molting fluid : 脱皮液
molting gel : 脱皮ゲル
molting hormone : 脱皮ホルモン
moltinism : 眠性
momentum : (複 . -ta)、契機、運動量、はずみ
monandrous female : 単数回交尾の雌
monandry : 単数回交尾
monarch butterfly : オオカバマダラ
Mongolia : モンゴル
moniliform antenna : (複 . -nae)、数珠玉状(じゅずだまじょう)の触角
monitoring : モニタリング
monitoring number and species : 個体種数調査
monitoring scheme : 監視計画、モニタリングスキーム
monkey flower : ミゾホオズキ
mono- : 1-、単-〔ギリシャ語〕
monobasic taxa : 一つしか含まれない分類
monocarpic perennial plant : 一回結実型多年生植物
monocarpy : 一回繁殖性
monocotyledon : 単子葉植物
monogamy : 一夫一妻、一夫一婦、単婚
monogenesis : 単為生殖(たんいせいしょく)
monograph : 研究論文、モノグラフ(特定の単一小分野をテーマとする論文)
monogyny : 単婚、単女王性
monolayer sheet : 単分子膜細胞層
monomorphic : 単一型、単型
monomorphic male : 単型の雄
monooxygenase : 一原子酸素添加酵素
monophagous : 単食性の、単食性、単食
monophagy : 単食性
monophyletic : 単系統、単系統の
monophyletic group : 単系統群
monophyletic sister group : 単系統姉妹群
monophyletic species : 単系統種
monophyletic species concept : 単系統種概念
monophyly : 単系統、単系統性
monoplacophora : 単板綱(たんばんこう)
monostemma-dominated neuron : 単一単眼優性ニューロン
monoterpene : モノテルペン
monotrysia : 単門亜目
monotypic : 単型の、単型
monotypic genus : 一属一種の属、単一種属
monotypy : 単型
monsoon : モンスーン
monsoon forest : 雨緑林
monsoon season : モンスーンの季節、雨季

montane : 山地性の
montane oak forest : 山地カシ林
montane population : 山地(性)個体群
monthly average precipitation : 月平均降水量
monthly total precipitation : 月降水量
morph : モルフ、異形態、形態
morph specific pattern of expression : 型特異的な発現パターン
morpho blue butterfly : ブルーモルフォ
morpho butterfly : モルフォチョウ
morphogen : モルフォゲン(位置情報物質)
morphogen gradient : モルフォゲン勾配、形原勾配
morphogen model : 形原モデル
morphogenesis : 形態形成
morphogenetic : 形態形成の、形態形成的、形態発生の
morphogenetic hormone : 形態形成ホルモン
morphogenetic origin : 形態形成起源
morphological adaptation : 形態的適応
morphological analysis : 形態学的解析
morphological and physiological characteristic : 形態的・生理的特性、形態的・生理的特徴
morphological change : 形態変化
morphological character : 形態(的)形質、形態的特徴
morphological characteristic : 形態(的)形質
morphological classification : 形態(学)的分類
morphological datum : (複 . -ta)、形態(学的)データ
morphological difference : 形態的差異
morphological discontinuity : 形態的不連続性
morphological distinction : 形態的区別
morphological divergence : 形態的分化
morphological diversification : 形態的多様性
morphological diversity : 形態的多様性
morphological evolution : 形態的進化
morphological feature : 形態(学)的特徴
morphological measure : 形態的尺度
morphological pattern : 形態的パターン
morphological peculiarity : 形態的特(有)性
morphological phylogenetics : 形態系統学
morphological reconstruction : 形態的再構成
morphological resemblance : 形態(学)的な類似
morphological separation : 形態的分離
morphological similarity : 形態的類似性
morphological speciation : 形態的分化
morphological species concept : 形態学的種概念
morphological study : 形態(学)的研究
morphological trait : 形態(的)形質
morphological tree : 形態系統樹
morphological variation : 形態的変異
morphological, physiological and behavioral study : 形態学的・生理学的・行動学的研究
morphologically : 形態的に
morphologically abnormal adult insect : 形態的に異常な成虫
morphologically similar : 形態的に類似な
morphology : 形態学、形態
morphology examination : 形態学的試験、形態学的検査
morphology measurement : 形態測定
Morphos : モルフォチョウ亜科
morphospecies : 形態(学的)種
mortality : 死亡率
mortality curve : 死亡率曲線
mosaic analysis : モザイク解析
mosaic of genetic diversity : 遺伝的多様性のモザイク
mosaic pattern : モザイク状
most abundant : 最も豊富な

most closely related : 最も近縁な
most notable decrease : 最大の著しい減少
most order : 最上位の遺伝子配列順序
most parsimonious model : 最節約モデル、MP モデル
moth : 蛾
moth-butterflies, the (superfamily Hedyloidea) : シャクガモドキ上科
moth-pollinated flower : 蛾媒花
mother cell : 母細胞
mother plant : 母植物、親株、母株
motif : モチーフ
motile : 自分で動くことができる、自動性の
motion capacity : 運動能力
motion vision : 動態覚、動体視覚、動体視
motmot : ハチクイモドキ(科の鳥)
motor activity : 運動活動、運動活性
motorway : 高速道路
moulting : 脱皮〔英語〕
moultinism : 眠性
mountain : 山地
mountain ecotype : 山地生態型
mountain population : 山地性個体群
mountainous area : 山岳地帯
mountainous species : 山岳性種
mounted body : 固定した胴体
mounted cell : 封入された細胞
mounting : 展翅(てんし)
mounting board : 展翅板
mounting fresh specimen : 生展翅
mounting relaxed specimen : 軟化展翅
mourning cloak butterfly : キベリタテハ
mouse : (複 . mice)、ネズミ、マウス
mouth : 口、口器
mouth parts : 口器(こうき)
mouth-appendage : 口肢(こうし)、口部付属肢、口脚
mouthparts : 口器
movable genetic element : 可動因子
movable hook : 可動鉤(かどうこう)
movable hook seta : 可動鉤上の刺毛
movement : 移動、運動
movement angle and velocity : 移動角度と速度
movement behavior : 移動行動
movement cost : 移動費用、移動コスト
movement decision : 移動に関する意思決定
movement distance : 移動距離
movement ecology : 移動生態学、行動生態学
movement of basalare : 基翅節片の運動
movement pattern : 行動パターン
moving axis : 駆動軸
moving system : 駆動機構
moving trace : 行動トレース
MrBayes : ベイズ法に基づいて系統樹探索を行うソフトウェアの名称
mRNA : メッセンジャー RNA(messenger RNA)
Ms : 質量分析計(Mass spectrograph)
mtDNA : ミトコンドリア DNA
MU : 個体群管理単位(Management Unit)
mucilage sac : 粘液のう
mud : 泥
mud puddle : 泥の水溜まり
mud-puddling behavior : 泥水の吸水行動
muddy ground : ぬかるみ
Mullerian mimicry : ミュラー型擬態、ミュラー擬態
multi- : 多 -〔ラテン語〕
multi-brooded : 多化性の
multi-year timeframe : 複数年期間
multicloning site : マルチクローニングサイト
multidimensional scaling analysis : 多次元尺度分析、多次元尺度解析
multifurcating : 多分岐
Multilateral Environmental Agreements : 多国間環境協定

multilocus color pattern architecture : 多遺伝子座で構成されるカラーパターン構成
multilocus genotypic difference : 多遺伝子座で構成される遺伝子型の相違
multiple alignment : 多重整列
multiple authorship : 多数著者
multiple brood : 多化
multiple generation : 多世代、複数世代
multiple invasion : 多重侵入
multiple mating : 多回交尾
multiple nucleopolyhedrovirus : 複数のヌクレオキャプシドを含む核多角体病ウイルス、複数のヌクレオキャプシドを含む NPV（MNPV）
multiple original spelling : 複数原綴(つづ)り
multiple regression : 重回帰
multiple regression analysis : 重回帰分析
multiple regression equation : 重回帰式
multiple rich resource area : 複数の豊富な資源のある地域
multiple season : 複数（の）季節
multiple-site nester : 複数箇所営巣性種
multiplex PCR : マルチプレックス PCR
multiplicity of infection : 感染多重度（MOI）
multiply : 繁殖する
multispecies community : 多種群集
multistemmata-dominated neuron : 複数単眼優性ニューロン
multitrophic interaction : 多栄養段階（生物間）相互作用
multivariate point pattern analysis : 多変量点のパターン解析（複数の属性からなる点のパターンを解析をする手法）
multivoltine : 多化性
multivoltine group : 多化性集団
multivoltine population : 多化性集団
municipality : 地方自治体、市（町）当局
murexide reaction : ムレキシド反応
muscle receptor : 筋受容器
muscle-filled thorax : 筋肉で満たされた胸部
musculature : 筋肉組織、筋組織
museum specimen : 博物館の標本
mustard : マスタード、カラシ
mustard oil : カラシ油、マスタード油
mustard oil glycoside : マスタード油グリコシド、マスタード油配糖体
mutagenize : 突然変異を起こさせる
mutant : 突然変異種、突然変異体
mutant eyespot phenotype : 突然変異型眼状紋表現型
mutant phenotype : 突然変異体表現型
mutation : 突然変異、変異
mutation rate : 変異速度、変異率、突然変異率
mutation theory : 突然変異説
mute : 音を出さない
mute butterfly : 音を出さない蝶
mutual pursuit : 相互の追跡
mutualism : 相利共生
mutualist ant : 相利共生者のアリ
mutualistic : 相利共生的、相利型
mutualistic interaction : 相利共生的相互作用
mutually exclusive : 相互（に）排他的
mya : 100 万年前（million years ago）
Mycobacterium leprae : らい菌、ライ菌
Mycobacterium tuberculosis : 結核菌
mycorrhizal fungus : （複 . -gi）、菌根菌
Myr : 100 万年（Million years）
myrmecophilous insect : 好蟻性昆虫
myrmecophilous organ : 好蟻性器官
myrmecophily : 好蟻性、アリとの共生
myrmecophyte : アリ植物（アリに食物や巣を提供している植物）
myrmecoxenous : 客棲性（「蟻の客として迎え入れられて蟻と一緒に蟻の巣に棲む性質」のこと）〔ギリシャ語〕
myrmecoxenous relationship : 客棲性的共生関係

mysterious genus : 謎の種類
mystery : 謎
mystery of the life cycle : 生活史の謎
myxomatosis : 粘液腫病

n

N : 窒素（Nitrogen）
N-containing compound : 窒素含有物（Nitrogen）
N-terminal region : N 末端領域
N-terminus : N 末端〔ラテン語〕
N-β-alanyl dopamine synthase : *N*-β- アラニルドーパミンシンターゼ、*N*-β- アラニルドーパミン合成酵素（「*N*-」は置換基が結合している元素を表現し、窒素位〈Nitrogen position〉である）
N-β-alanyldopamine : *N*-β-アラニルドーパミン
nAChR : ニコチン性アセチルコリン受容体（nicotinic Acetylcholine Receptor）
NADH : ニコチンアミドアデニンジヌクレオチドの還元型（Nicotinamide Adenine Dinucleotide Hydrate）
NADH dehydrogenase subunit 5 : NADH デヒドロゲナーゼサブユニット 5
NADPH oxidase complex : NADPH酸化酵素複合体、NADPH オキシダーゼ複合体（Nicotinamide Adenine Dinucleotide PHosphate ; ニコチンアミドアデニンジヌクレオチドリン酸）
name : 名称
name of a family : 科名
name of a genus : 属名
name of a species : 種名
name of a subfamily : 亜科名
name of a subgenus : 亜属名
name of a subspecies : 亜種名
name of a superfamily : 上科名
name of a tribe : 族名
name-bearing type : 担名タイプ
named specimen : 標本の命名
namely : すなわち、つまり
naming system : 命名法
Nanda-Hamner protocol : ナンダ-ハムナープロトコール、ナンダ-ハムナー実験
narrow contact zone : 狭接触帯、狭い接触域
narrow distribution : 狭分布
narrow geographic area : 地理的に狭い地域
narrow zone : 狭い領域
narrow-distributed species : 狭分布種
narrowly distributed species : 狭分布種、狭域分布種
natal dispersal : 出生地からの分散、生まれた場所からの分散
natal island : 出生島
natality : 出生率
national annual index of abundance : 国レベルの豊富さの年次指数
national butterfly : 国蝶
national butterfly of Japan : 日本の国蝶
National Center for Biotechnology Information : 国立生物工学情報センター、国立バイオテクノロジーセンター（NCBI）（米国）
national park : 国立公園
Native American : アメリカ先住民、ネイティブアメリカン
native baculovirus : 天然バキュロウイルス、自生バキュロウイルス
native element : 自生的な要素
native garden plant : 在来の園芸植物
native habitat : 本来の生息地
native plant : 土着植物
native species : 在来種、土着種、固有種、自生種
native wildlife : 在来野生生物
native woodland : 自然のままの森林地帯
natural : 自然の
natural barrier : 自然の障壁
natural character : 本来的性格

natural community : 自然群集
natural competitor : 自然の競争者
natural condition : 自然条件
natural daylength : 自然日長
natural death : 老衰死、自然死
natural disaster : 自然災害
natural distribution of species : 種の自然分布
Natural Earth : 世界地図(ラスターおよびベクター)データを提供するサイトの名称
natural enemy : 天敵
natural environment : 自然環境
natural fragmentation : 自然な断片化
natural granitic outcrop : 自然な花崗岩の露出部
natural group : 自然群
natural habitat : 自然(の)生息地
natural history : 自然史
natural hybrid zone : 自然交雑帯
natural incidence : 自然発生、自然発生率
natural item : 自然のもの、自然の代物
natural light : 自然光
natural monument : 天然記念物
natural pathogen : 自然界の微生物、天然の病原体
natural period : 固有周期
natural predator : 自然の捕食者
natural range : 自然分布域
natural resting position : 自然な形の静止姿勢
natural seasonal variation : 自然界の季節変異
natural selection : 自然選択、自然淘汰
natural selection theory : 自然選択説
natural vegetation : 自然植生
natural vegetation fragment : 自然植生の分断化
naturalization : 帰化
naturalized : 帰化した
naturalized organism : 帰化生物
naturalized species : 帰化種
naturally polymorphic race : 生来の多型系統
naturally-occurring population : 自然発生個体群
nature : 自然(界)
nature conservation : 自然保護、自然環境保全
nature of predation, the : 捕食の本性
nature reserve : 自然保護区
navigation capacity : 航行能力、ナビゲーション能力
navigational skill : ナビゲーション技能
NCBI : 国立生物工学情報センター、国立バイオテクノロジーセンター(米国)(National Center for Biotechnology Information)
ncRNA : 非コードRNA(non-coding RNA)
ND5 : NADH デヒドロゲナーゼサブユニット 5(NADH Dehydrogenase subunit 5)
Neanderthal : ネアンデルタール人
near-identical color-pattern mosaic : ほぼ同一なカラーパターンモザイク
near-isogenic line : 近似同質遺伝子系統
near-perfect local mimetic convergence : ほぼ完全な擬態の局所的な収束
nearby natural habitat : 近くの自然生息地
Nearctic region : 新北区(しんほっく)
nearest relative : 最寄りの近縁
nearly alike : 大同小異
nearly all : ほとんどすべての
nebulous vein : 模糊脈
neck : 頸部、首
necrosis : 壊死(えし)
necrotic tissue : 壊死組織
nectar : 蜜、花蜜〔ラテン語〕
nectar corridor : 花蜜回廊
nectar form flower : 花蜜様態の花
nectar guide : 蜜標
nectar plant : 吸蜜植物

nectar resource plant : 吸蜜(源)植物
nectar source : 吸蜜源
nectar source abundance estimate : 吸蜜の資源量推定値
nectar sucking : 吸蜜
nectar-feeding insect : 蜜を摂取する昆虫
nectaring : 吸蜜
nectaring source : 吸蜜源
nectarivore : 蜜食動物
nectary : 蜜腺
needle-like shape : 針状の形状
negative association : 負の関連、負の相関
negative binomial : 負の二項(分布)
negative control : 負の対照(区)、ネガティブコントロール
negative correlation : 負の相関
negative effect : 負の効果
negative feedback : 負のフィードバック
negative selection : 負の自然選択、負の自然淘汰、負の淘汰
negative strand RNA virus : 負鎖 RNA ウイルス、ネガティブ鎖 RNA ウイルス
negatively affect : 負の影響を与える、悪影響を与える
Nei's D : ネイの遺伝(学)的距離
neighbor : 近隣
neighbor-joining : 近隣結合法、NJ 法
neighbor-joining analysis : 近隣結合解析
neighbor-joining method : 近隣結合法、NJ 法
neighbor-joining tree : 近隣結合系統樹
neighboring competitive plant : 隣接する競合植物
neighboring conspecific plant : 隣接する同種植物
neighboring gene : 隣接する遺伝子、隣接遺伝子
neighboring population : 隣接する個体群
nematode : 線虫、ネマトーダ
nematode genome : 線虫のゲノム
Neo-Darwinism : ネオダーウィニズム、新ダーウィン説
neo-functionalization : 新機能獲得
neo-Hopkins principle : ネオ-ホプキンス則、新ホプキンス則
neofunctionalization : 新機能獲得
neolignoid feeding deterrent : ネオリグノイド摂食抑制剤
neonate : 新生幼虫
neonicotinoid : ネオニコチノイド
neonicotinoid insecticide : ネオニコチノイド(系)殺虫剤
neonicotinoid pollution : ネオニコチノイド汚染
neonicotinoid usage : ネオニコチノイドの利用、ネオニコチノイドの使用
Neoptera : 新翅目、新翅群
neopterobilin : ネオプテロビリン
neotenic reproductive : ネオテニック生殖虫、幼形成熟の生殖虫
neoteny : 幼形成熟、ネオテニー
Neotropic region : 新熱帯区
neotropical butterfly : 新熱帯産の蝶
Neotropical region : 新熱帯区
neotropics : 新熱帯種、新熱帯
neotype : ネオタイプ、新基準標本
nerve branch : 神経枝
nerve cord : 神経索
nerve ending : 神経終末
nerve ganglion : 神経球
nerve root : 神経根
nerve stain : 神経用染料、神経用染色
nervulus : 前区脈、肘臂横脈
nest : 巣
nested structure : 入れ子構造
nesting site : 営巣場所
nestmate recognition : 巣仲間認識
net : 捕虫網

net competitive effect : 正味の競争効果
net diversification rate : 多様化の正味速度、純多様化速度
net rate of cladogenesis : 純分岐進化速度
net rate of diversification : 純多様化速度
nettle-feeding larva : イラクサ摂食幼虫
network : ネットワーク
network tree : ネットワーク樹
neural activity : 神経活動
neural recruitment : 神経(的)漸増
neural response : 神経応答、神経反応
neural signal : 神経信号、神経シグナル
neural spike : 神経スパイク
neural stimulus : 神経刺激
neurobiological difference : 神経生物学的差異
neuroendocrine control : 神経内分泌調節
neuroendocrine mechanism : 神経内分泌機構
neuroendocrine regulation : 神経内分泌制御
neuroendocrine system : 神経内分泌系
neuroethology : 神経行動学
neurohormonally : 神経ホルモン的に
neuromuscular activity : 神経筋活動
neuronal connection : 神経結合
neuropeptide hormone : 神経ペプチドホルモン
neurophysiological recording : 神経生理的記録法
Neuroptera : アミメカゲロウ目、脈翅目
neurosecretory cell : 神経分泌細胞
neutral evolution : 中立(的)進化
neutral evolution theory : 中立進化論
neutral gene : 中立遺伝子
neutral insect : ただの虫
neutral isoallele : 中立対立遺伝子
neutral landscape : 中立的景観
neutral mutation : 中立突然変異
neutral site : 中立(的)部位
neutral theory : 中立説
neutrality : 中立性
neutralizer : 中和物、中和剤
new and revised edition : 新訂版
new and unrecorded butterfly : 新規の未記載蝶
New Caledonia : ニューカレドニア
new combination : 新組合せ
new generation sequencing : 新世代シークエンシング、新世代シークエンサー
New Guinea : ニューギニア(島)
new infestation : 新たな蔓延
new knowledge : 新知見
new record : 新記録
new replacement name : 新置換名
new scientific name : 新学名
new species : 新種
new subspecies : 新亜種
new sympatric habitat : 新混棲地
new taxon : (複 . -xa)、新種、新しい分類群
New World tropics : 新世界熱帯
new-born female : 生まれたばかりの雌
newly discovered food-plant : 新たに発見された食草
newly eclosed virgin female : 新しく羽化した未交尾の雌
newly emerged adult : 新たに発生した成虫、新しく羽化した成虫
newly evolved population : 新しく進化した個体群
newly hatched female : 孵化したての雌
newly hatched larva : 新しく孵化した幼虫
newly recorded : 新たに記録された
newly recorded hostplant : 新食草
newly-eclosed female : 羽化したばかりの雌
newly-hatched larva : 新しく孵化した幼虫
next-less-complex model : 一段階下げた複合モデル

next-more-complex model : 一段階上げた複合モデル
ng : ナノグラム(nanogram)
niche : ニッチ、生態的地位
niche shift : ニッチシフト、ニッチ分化
niche space : ニッチ空間、生息空間
niche theory : ニッチ説
niche-apportionment model : ニッチ分配モデル
niche-assembly model : ニッチ集合モデル
Nicotiana tabacum : タバコ
nicotinic acetylcholine receptor : ニコチン性アセチルコリン受容体
night interruption : 暗期の光中断
night-flying moth : 夜行性の蛾
night-length : 夜長時間
NIL : 近似同質遺伝子系統(Near-Isogenic Line)
nitrate ion : 硝酸イオン
nitrile-forming activity : ニトリル形成能
nitrile-specifier protein : ニトリルスペシファイアータンパク質、ニトリル指定子タンパク質(NSP)
nitrogen : 窒素(N)
nitrogen requirement : 窒素要求量
nitrogen-containing inorganic ion : 窒素含有無機イオン
nitrogen-deficient diet : 窒素欠乏食料
nitrogen-fixing food plant : 窒素固定食草、窒素固定食餌植物
nitrogenous compound : 窒素化合物
nitrogenous substance : 窒素物質
NJ : 近隣結合法(Neighbor-Joining method)
NMR : 核磁気共鳴
(Nuclear Magnetic Resonance)
no conflicting linkage relationship : 無矛盾の連鎖関係
no courtship behavior : 非求愛行動
no significant correlation : 無有意相関
no significant difference : 無有意差
NOAA : アメリカ海洋大気庁
(National Oceanic and Atmospheric Administration)
noctuid moth : ヤガ科の蛾
noctuid moth species : ヤガ科の種
Noctuidae : ヤガ科
nocturnal : 夜行性の、夜間の
nocturnal behavior : 夜行活動
nocturnal lifestyle : 夜行性の生活様式
nocturnal moth-like ancestor : 夜行性蛾のような祖先
nocturnal prosimian : 夜行性原猿
nocturnality : 夜行性
nocturnally active : 夜に活動的な、夜間活動する、夜行性の
node : 結節
nodule : こぶ、瘤
nodus : 中央分節、結節
noise threshold : 雑音閾(いき)値、ノイズ閾値
nom. : 学名命名(nomen)、名前
nom. nov. : 新名(nomen novus)
nomadic movement : 遊牧(的)移動
nomadic pattern : 放牧パターン
nomadism : 遊牧
nomen dubium : (複 . nomina dubia)、疑問名〔ラテン語〕
nomen novum : (複 . nomina nova)、新名〔ラテン語〕
nomen nudum : (複 . nomina nuda)、裸名〔ラテン語〕
nomen oblitum : (複 . nomina oblita)、遺失名〔ラテン語〕
nomen protectum : 擁護名〔ラテン語〕
nomenclator binominalis : 二名法、二名式命名法
nomenclator trinominalis : 三名法、三名式命名法
nomenclatural : 命名法的

nomenclatural act : 命名法的行為
nomenclatural status : 命名法的地位
nomenclature : 命名法
nominal family-group taxon : 名義科階級群タクソン
nominal genus : 名義属
nominal taxa : 名義タクソン名、名義タクソン
nominate : 名義タイプの、承名の
nominate subspecies : 名義タイプ亜種
nominotypical : 名義タイプの
nominotypical taxon : 名義タイプタクソン
non autonomous : 非自律
non-calculated migration : 無目標移動
non-coding RNA : 非コードRNA(ncRNA)
non-coding sequence : 非コード配列、ノンコーディング配列
non-consumptive effort : 非消費効果
non-crop plant : 非作物植物
non-diapause condition : 不休眠条件
non-diapause destined larval period : 非休眠になる場合の幼虫発育期間、非休眠性幼虫発育期間
non-diapause egg : 非休眠卵、不休眠卵
non-diapause larva : 非休眠幼虫、不休眠幼虫
non-diapause selection : 不休眠選択
non-diapausing adult : 不休眠成虫
non-diapausing generation : 不休眠世代
non-diapausing individual : 非休眠個体
non-diapausing pupa : 非休眠蛹、不休眠蛹
non-*Drosophila* system : 非ショウジョウバエ様式
non-emitting transgenic tobacco plant : 非放出性遺伝形質転換タバコ植物
non-fertile sperm : 受精能力をもたない精子
non-flier form : 非移動型
non-focal ablation : フォーカス以外の部位切除
non-freezing cold injury : 凍傷以外の低温障害
non-functional gene : 非機能性遺伝子
non-host species : 宿主ではない種、非寄主の種
non-infected cell : 非感染細胞
non-informative site : 情報を持たないサイト
non-innervated microtrich : 非神経支配の微毛
non-invasive : 侵略性ではない、非侵略性
non-linearity : 非線形性
non-migrant : 非移動性種
non-migratory : 非移動性の
non-migratory butterfly species : 非移動性蝶種
non-mimetic : 非擬態の
non-native species : 非在来種
non-oriented mechanism : 無定位機構、無指向機構
non-overlapping generation of adult : 成虫の離散世代
non-periodic form : 無周期型
non-poisonous species : 無毒種
non-preference : 非選好性
non-random dispersal : 非ランダム分散、非任意分散
non-seasonal environment : 季節がない環境
non-social insect : 非社会性昆虫
non-socketed hair : ソケットなしの毛、ソケットを付けない毛
non-sperm protein : 無精子タンパク質
non-synonymous change : 非同義置換
non-target invertebrate : 非標的無脊椎動物、対象外の無脊椎動物
non-target species : 非標的種、対象外種
non-tree network : 非系統樹ネットワーク
non-trophic interaction : 食う食われる以外の関係
non-uniform diapause : 不斉休眠
non-uniformity : 不斉一化、不斉一性

non-viral pathogen : 非ウイルス性病原体
non-volatile compound : 不揮発性化合物
non-waxy leaf : 光沢のない葉
nonadaptive : 非適応的
nonadaptive recombinant : 非適応的組み換え種
nonasymptotic richness : 非漸近的種数
noncoding region : 非翻訳領域、非コード領域
nondescript dull brown : 目立たないくすんだ茶色
nondispersive movement : 非分散移動
nonindigenous species : 非在来生物種
nonirradiated : 未照射の
nonlinear : 非線形の
nonmimetic intermediate : 非擬態中間型種
nonmimetic phenotype : 非擬態的表現型
nonmimetic recombinant : 非擬態型組み換え種
nonmodel insect species : 非モデル昆虫種
nonnative species : 外来種
nonoverlapping function : 重複しない機能
nonparalogous gene : 非パラロガス遺伝子
nonparametric estimator : ノンパラメトリック推定量
nonplastic, dorsal eyespot : 背側の非可塑性の眼状紋
nonpolar compound : 無極性化合物
nonsense mutation : ナンセンス変異
nonsound-producing : 音を発生しない
nonsynonymous substitution : 非同義置換
nontreated insect line : 無処理の昆虫系統
norm in nature : 自然界の規範
normal activity : 平常活動
normal activity range : 平常な行動範囲
normal band : 正常な縞状バンド列
normal brood : 正常な同腹仔
normal emergence rate : 正常な羽化率
normal female reproductive organ : 正常な雌性生殖器官
normal form : 正常型
normal matriline : 正常な母系群
normal ovary : 正常な卵巣
normal pupa : 正常蛹
normal sex ratio : 正常な性比
normal testis : 正常な精巣
normalization : 規格化、正規化
normalized absorbance spectrum : 規格化された吸収スペクトル
normalized counted datum : (複 . -ta)、正規化された計数データ
normalized gene expression : 正規化された遺伝子発現(量)
normalizing selection : 正常化選択
normally distributed variance : 正規分布した分散
normally spread wing : 正常に伸びた翅
North American buckeye butterfly : アメリカタテハモドキ
North India : 北インド
Northeast India : インド北東部
northerly flight : 北方へ飛行
northern blotting : ノーザンブロッティング
northern character : 北方的性格
northern continent : 北方大陸
northern distribution limit in Honshu : 本州北限地域
northern edge of distribution : 分布北限
northern Europe : 北欧
northern hemisphere : 北半球
northern Japanese population : 北日本個体群、北日本産個体群
northern Laos : ラオス北部
northern margin : 北端、北縁部
northern origin : 北方起源、北方系
northern range expansion : 北への分布拡大
northern species : 北方的な種
northernmost area : 北限地域、最北端地域

northernmost population : 最北個体群、最北端の個体群
northward invasion : 北方への侵入
northward spread : 北方への広がり
Northwestern Himalayas : ヒマラヤ北西部
notably : 著しく、特に
notal : 背板の
notal wing process : 背板翅突起、背翅突、背板の翅突起
notaulix : (複 . -lices)、中胸背縦斜溝(「notauli」は誤り)
notch : 刻み、くぼみ、ノッチ
notched : 鉤爪状の
noteworthy : 顕著である
noticeable : 著しい
Noto peninsula : 能登半島(石川県)
notochord : 脊索
notum : 背板、中胸背盾板
notwithstanding : それにもかかわらず、にもかかわらず
noun phrase : 名詞句
nourishing substance : 栄養物質
nourishment : 栄養物、滋養(物)
nov. : 新規の(novus)、新しい
novel trait : 新規形質
novelty of the result : 結果の新規性
NPV : 核多角体病ウイルス(NucleoPolyhedroVirus ; Nuclear Polyhedrosis Virus)
NR database : NR データベース、非冗長なタンパク質データベース(Non-Redundant protein)
NSP : ニトリルスペシファイアータンパク質、ニトリル指定子タンパク質(Nitrile-Specifier Protein)
nuclear gene : 核遺伝子
nuclear genome : 核ゲノム
nuclear magnetic resonance : 核磁気共鳴、NMR 解析
nuclear marker : 核マーカー
nuclear polyhedrosis virus : 核多角体病ウイルス(NPV)
nucleate spermatozoon : 有核精子
nucleocapsid transport : ヌクレオキャプシド伝播
nucleomorph : ヌクレオモルフ
nucleopolyhedrovirus : 核多角体病ウイルス(NPV)
nucleosome : ヌクレオソーム
nucleotide configuration : 塩基配置
nucleotide datum : (複 . -ta)、塩基データ、ヌクレオチドデータ
nucleotide divergence : 塩基分岐、ヌクレオチド分岐
nucleotide diversity : 塩基多様性、ヌクレオチド多様性
nucleotide level : ヌクレオチド水準、ヌクレオチドレベル、塩基レベル
nucleotide mixture solution : ヌクレオチド混合溶液
nucleotide model : ヌクレオチドモデル、塩基モデル
nucleotide position : 塩基位置、ヌクレオチド位置
nucleotide sequence : 塩基配列、ヌクレオチド配列
nucleotide sequence database : 塩基配列データベース
nucleotide site : 塩基部位、ヌクレオチド部位
nucleotide stretch : ヌクレオチド伸長、ヌクレオチドストレッチ
nucleotide substitution : 塩基置換
nucleotide variable position : 塩基の位置番号、ヌクレオチドの位置番号
nucleus : (複 . -clei)、核、細胞核
nucleus of bursa copulatrix cell : 交尾のう細胞の細胞核
nucleus of cell : 細胞核

nucleus of Malpighian tubule cell : マルピーギ管細胞の細胞核
nudivirus : ヌディウイルス
nudum : 裸〔ラテン語〕
null ancestor : 空祖先
null hypothesis of no introgression : 無遺伝子移入の帰無仮説
null model : 帰無仮説モデル、ヌルモデル、独立モデル
number of adult emerging : 羽化した成虫の個数、成虫羽化数
number of base substitution : 塩基置換数
number of frost day : 氷点下の日数
number of genera : 属数
number of generation : 世代数、発生回数
number of incoming individual : 飛来個体数
number of individual : 個体数
number of instar : 齢数
number of investigation : 調査回数
number of investigation individual : 調査個体数
number of larval instar : 幼虫の齢数
number of oviposition : 産卵数
number of plants/area : 植物数／面積
number of rainy day : 降水日数
number of silk girdle : 帯糸数
number of species : 種数
number of stem : 樹幹数、幹数、本数
number of visit : 訪問回数
numerical response : 数の反応、数量的反応
numerical taxonomy : 数量分類学
numerous effort : たくさんの努力
numerous species : たくさんの種
nuptial color : 婚姻色
nuptial gift : 婚姻贈呈物
nuptial hue : 婚姻色
nursery stock : 苗木
nutrient : 栄養になる食物
nutrient budget : 栄養物質の収入分
nutrient cycling : 栄養循環、物質循環
nutrient reserve : 栄養物質貯留
nutrient reservoir activity : 栄養素貯蔵行動
nutrient-rich spermatophore : 栄養豊富な精包
nutrition : 栄養
nutritional condition : 栄養条件
nutritional ecology : 栄養生態学、栄養学的生態
nutritional mechanism : 栄養学的機構
nutritional requirement : 栄養要求、栄養要求性
nutritional status : 栄養状態
nutritional suitability : 栄養的適合性
nutritious accessory gland product : 栄養に富んだ付属腺物質
NWP : 背板翅突起、背翅突(Notal Wing Process)
nylon membrane : ナイロン膜
nymph : 若虫、ニンフ、幼虫
nymphalid branch : タテハチョウ科の分岐

O

oak : オーク(カシ、ナラ、カシワ)類の木
oak tiger butterfly : オビモンドクチョウ
OB : 包埋体(Occlusion Body)
object : 対象物
objective : 客観的
objective synonym : 客観(的)同物異名
obligate association : 絶対共生的関係
obligatory : 内因性
obligatory diapause : 内因性休眠
obligatory migration : 内因性移動
obligatory taxa : 必須の分類群
OBP : 臭物質結合タンパク質(Odorant Binding Protein)
obscure : 覆い隠す、はっきりしない、不明瞭な
observation : 観察

observation time : 観察時間
observation-area curve : 観察領域の曲線
observational datum : (複 . -ta)、観察データ
observed variation : 観察された変動
obstruct : ふさぐ、塞ぐ
obtect : 殻で覆われた
obtect pupa : 被蛹(ひょう)
obtuse : 鈍角
occasional observation : 予備的観察
occasional phenotypic leap : 偶発的表現型の飛躍
occasional recombinant phenotype : 偶発的組み換え表現型
occasionally : 時折、時々
occipital : 後頭の
occipital carina : 後頭隆起線(頭部)
occipital suture : 後頭縫合線(こうとうほうごうせん)(頭部)
occiput : 後頭(頭部)
occlusion body : 包埋体(OB)
occlusion-derived virion : 包埋型ビリオン、包埋体由来のウイルス粒子
occupancy : 占有性
occupant : 占有者
occupied area : 占有地域
occupied patch : 占有パッチ
occupied site : 占有地
occupied space : 占有空間
occupying individual : 占有個体
occur : 発生する
occurrence of sexual mosaic : 性モザイクの発生
Oceanian region : オセアニア区
ocellar : 単眼の
ocellated marking : 眼状紋
ocellatus : 蛇の目模様、眼状紋〔ラテン語〕
ocellus : (複 . -li)、単眼、眼状紋〔ラテン語〕
octopamine : オクトパミン
octopamine receptor : オクトパミン受容体
odds of diapause : 休眠の確率値、休眠のオッズ
Odonata : トンボ目、蜻蛉目
odor source : 芳香源、匂い源
odor substance : 臭物質
odorant binding protein : 臭物質結合タンパク質
odoriferous compound : 芳香(性)化合物
Oecophoridae : マルハキバガ科
oenocyte : エノサイト
oenocytoid : エノシトイド(昆虫血球細胞の一種)
oesophagus : 食道(しょくどう)
offense : 罪、違反、攻撃
Official Correction : 公式訂正書
official text : 正文、公式原文
offprint : 別刷り、抜き刷り
offspring : 子孫、子
offspring quality : 子孫の質
OFT : 最適採餌理論(Optimal Foraging Theory)
Ogasawara Islands : 小笠原諸島
oilseed rape : アブラナ、ナタネ
oily droplet : オイル液滴
OIPV : 産卵誘導性植物揮発性物質(Oviposition-Induced Plant Volatile)
old-growth forest stand : 老齢林分
olfaction : 嗅覚(きゅうかく)
olfactometer : オルファクトメーター、嗅覚計
olfactory : 嗅覚(の)
olfactory cue : 嗅覚刺激
olfactory gene : 嗅覚遺伝子
olfactory receptor : 嗅覚感覚器、嗅覚受容体
olfactory sense : 嗅覚
olfactory stimulus : 嗅(覚)刺激
oligopause : 弱い休眠、仮性休眠
oligophagous : 狭食性、狭食性の、少食
oligophagy : 狭食性
Oliza sativa : イネ
ommatidial unit : 個眼ユニット

ommatidium :（複 . -tidia）、個眼〔英語〕
ommatine D : オマチン D
ommine : オミン
ommochrome : オモクローム
ommochrome pathway gene : オモクローム系の経路遺伝子
ommochrome pigment : オモクローム系色素
omniphagy : 雑食性
omnivore : 雑食（性）動物
on line : 蝶道型
On the Origin of Species : 種の起原
once-abandoned woods : 一度見捨てられた森
Oncorhynchus : サケ科
one brood : 一化
one generation a year : 年一回発生、年一世代
one-day-old : 1 日齢、生後 1 日
one-dimensional movement : 1 次元移動
one-migrant-per-generation rule : 1 世代 1 移入個体の原則
one-sample t-test : 1 サンプルの t 検定、1 標本 t 検定
one-step growth curve analysis : 一段増殖曲線解析
one-step growth curve assay : 一段増殖曲線試験、一段増殖曲線検定
one-step mutation : 一段階突然変異
one-to-one relationship : 1 対 1 の関係
one-way ANOVA : 一元配置分散分析
one-way migration : 片道移動
ongoing area of research : 発展中の研究分野
ongoing debate : 継続中の論争
ongoing decline : 進行している衰亡
Online Mendelian Inheritance in Man : OMIM（ヒトの遺伝性疾患データベース、人間のメンデルの遺伝のオンライン版）
onset : 開始
onset of diapause : 休眠の開始
onset of metamorphosis : 変態の開始
ontogeny : 個体発生（論）
ontology : オントロジー、概念体系、本体論、存在論
onward : 前方に、先へ
oocyte presence : 卵母細胞の存在
oogenesis : 卵形成
oogenesis-flight syndrome : 卵形成 - 飛翔症候群、卵形成 - 飛翔形質群
oomycete : 卵菌
open area : 開けた地域
open blood-vascular system : 開放血管系（かいほうけっかんけい）
open deciduous forest : 開けた落葉樹林
open documentation : 公開文書
open dot : 白丸、丸点
open population : 開放個体群
open question : 未解決な問題
open reading frame : オープンリーディングフレーム、読み取り枠
open short grassland : 開けた狭い草地
open space : 開けた場所、開放的場所
open sunny habitat : 開けた陽のあたる生息地
open vegetation : 疎生植生
open woodland : 疎林地帯
open-line : 開線、二重線
opening : 木のまばらな空地
openness of habitat : 生息地の開放性
operating principle : 作動様式、操作原理
operational criterion : （複 . -ia）、操作基準、操作的判定基準
operational taxonomic unit : 操作上の分類単位（OTU）
operculum : （下）蓋板（かがいばん）、生殖下板（せいしょくかばん）、卵蓋（らんがい）
operon : オペロン
Opinion : 意見書

opportunistic diapause : 日和見休眠
opportunistic life cycle strategy : 日和見的な生活史戦略
opposite of the syndrome, the : 逆症候群
opposite pole : 反対の極
opsin allele : オプシン対立遺伝子
opsin expression pattern : オプシン発現様式
opsin gene family : オプシン遺伝子ファミリー
opsin gene sequence : オプシン遺伝子配列
opsin protein : オプシンタンパク質
optic lobe : 視葉
optic stalk : 眼茎
optical magnification : 光学倍率
optical signal : 光刺激
optimal ambient temperature : 最適周囲温度、最適環境温度
optimal diet model : 最適餌選択モデル
optimal foraging : 最適採餌
optimal foraging theory : 最適採餌理論
optimal foray search : 最適な餌探索
optimal patch use time : 最適餌場滞在時間
optimal population dispersion : 最適個体群分散
optimal search strategy : 最適な探索戦略
optimal solution : 最適解
optimal spatial distribution : 最適な空間的分布
optimality model : 最適(化)モデル
optimization model : 最適化モデル
optimized random search : 最適なランダム探索
optimum management tool : 最適な管理道具
optix : *optix* 転写制御因子遺伝子
Or : 嗅覚受容体(Olfactory receptor)
oral : 口側、口部
oral cavity : 口腔
oral infection : 経口感染
oral secretion : 口腔分泌物
orally infect : 経口で感染する
orange albatross butterfly : ベニシロチョウ
orange sulphur butterfly : アメリカオオモンキチョウ、オオアメリカモンキチョウ
orange tip butterfly : クモマツマキチョウ
orange-barred giant sulphur butterfly : ベニモンオオキチョウ
orangedog butterfly : クレスフォンテスタスキアゲハ、オオタスキアゲハ
orbicular spot : 環状紋(蛾類)
orbit : 眼縁部
order : 目(分類階級の「もく」)
order of magnitude : マグニチュードオーダー(対数スケールの等級・階級・規模)
ordinary scale : 普通鱗(片)
ordinary year : 平年
ordinary zone : 普通地域
ordination : 配列
Oregon silverspot butterfly : オレゴンギンボシヒョウモン
ORF : オープンリーディングフレーム、オープンリィーディングフレーム、読み取り枠、読み枠(Open Reading Frame)(タンパク質をコードする領域)
ORF length : ORF長(Open Reading Frame)
organella : 細胞内器官
organic chemist : 有機化学者
organism : 生物、生体、有機体
organism's actual distribution : 生物の分布実態
organism's perception : 生物の知覚
organism's phenotype : 生物の表現型、生物表現型
organismic mechanism : 生体機構
organization : 機構
Oriental and tropical origin : 東洋熱帯起源
Oriental region : 東洋区
Oriental tropical zone : 東洋熱帯地域
orientation : 定位、オリエンテーション

oriented mechanism : 定位機構、指向機構
oriented movement : 定位移動、指向移動
origin : 原点、起源、原産国
origin of biodiversity : 生物多様性の起源
origin of diversity : 多様性の起源
origin of Eurasian mountains : ユーラシア山岳起源
origin time : 起源年代
original description : 原記載
original designation : 原指定
original native wildlife : 本来の在来野生生物
original publication : 原公表
original spelling : 原綴(つづ)り
ornithologist : 鳥類学者
orogenic movement : 造山運動
orphan receptor : オーファン受容体
orthogonal 3-way design : 直交三元計画
ortholog : 直系遺伝子、真正相同遺伝子、オルソログ
ortholog group : オルソロググループ、相同グループ
orthologous : 順系相同
orthologous gene : オルソログ遺伝子、相同遺伝子
orthologous relationship : 相同関係
orthologue pair : 直系遺伝子対
Orthoptera : バッタ目、直翅目
osmeterial secretion : 臭角分泌物
osmeterium : (複 . -teria)、臭角〔ラテン語〕
ostium bursa : (複 . -ae)、交尾口(蛾類)〔ラテン語〕
ostium phallus : 交尾器口、挿入器口〔ラテン語〕
other family : 他の科
other species : 他の種
OTU : 操作上の分類単位(Operational Taxonomic Unit)
out-compete native organism : 在来生物との競争に勝ち残る
outbreak : 大発生
outbreed laboratory population : 研究室の非近交系個体群
outbreeding : 外交配、非血縁者との交配
outbreeding depression : 外交配弱勢、異系交配弱勢、他殖弱勢
outbreeding species : 外交配種、異系交配種
outcropping : 露出、発生
outdoor insectary : 野外の昆虫飼育場
outdoor rearing : 野外飼育
outer angle : 後角部、肛角部、後縁角、肛角
outer epicuticle : 外表皮外層
outer margin : 外縁
outer membrane : 外膜
outer membrane surface : 外膜表面
outgroup : 外群、アウトグループ
outgroup taxa : 外群の分類群
outside air : 外気
outspread : 翅を開いた
oval : 卵形
oval-shaped : 卵形の、卵円形の
oval-shaped outer membrane : 卵形の外膜
ovarian cell culture : 卵巣細胞培養
ovarian development : 卵巣発育
ovarian diapause : 卵巣休眠
ovarian dormancy : 卵巣休眠
ovarian dynamics : 卵巣の動態、卵巣動態
ovarian maturation : 卵巣成熟
ovariole : 卵巣小管
ovary : 卵巣、子房
ovary-derived cell line : 卵巣由来細胞株
over-represent : 大きな比率を占める
overdispersion : 過分散
overdominance : 超優性
overdominant selection : 超優性淘汰
overemphasize : 過度に強調する
overexpression : 過剰発現

overgrown : 生い茂った
overheat : 過熱
overlap spatially : 空間的に重ね合う
overlapping fragment : 重なった断片、重複断片
overlook : 調査する、観察する
overlooked species : 調査対象の種
overrepresentation : 過剰発現、過剰表現
overriding role : 中心的な役割
overripe fruit : 熟れすぎた果実
overshooting : 行き過ぎ現象
oversummering : 越夏
overt defense : 表立った防衛
overwhelming evidence : 圧倒的な証拠
overwinter : 越冬する
overwintered generation adult : 越冬世代成虫
overwintering : 越冬
overwintering ability : 越冬能力
overwintering aggregation : 越冬集団
overwintering form : 越冬型、越冬形態
overwintering larva : 越冬幼虫
overwintering season : 越冬期
overwintering stage : 越冬期、越冬齢期、越冬態、越冬ステージ
oviduct : 輸卵管
oviparous : 卵生の、卵を生む
oviparous orifice : 産卵孔
oviposit : 産卵する
ovipositing female : 産卵する雌
ovipositing plant : 産卵植物
oviposition : 産卵
oviposition activity : 産卵活性
oviposition behavior : 産卵行動
oviposition bioassay : 産卵のバイオアッセイ
oviposition deterrent : 産卵抑制物質
oviposition plant : 産卵植物
oviposition preference : 産卵選好、産卵選好性
oviposition regulator : 産卵調節物質
oviposition repellent : 産卵忌避物質
oviposition secretion : 産卵分泌物
oviposition stimulant : 産卵刺激物質
oviposition stimulant reception system : 産卵刺激物質受容系
oviposition-deterring pheromone : 産卵抑制フェロモン
oviposition-grass preference : 産卵草選好性
oviposition-induced plant volatile : 産卵誘導性植物揮発性物質（OIPV）
ovipositional behavior : 産卵行動
ovipositor : 産卵管
ovipositor lobe : 産卵管（蛾類）
ovipositor sheath : 産卵管鞘、産卵鞘
ovoid : 卵円形の、長楕円形の
ovoviviparous : 卵胎生の
ovule : 胚珠
ovum : （複 . -va）、卵（らん）、卵細胞、卵子〔ラテン語〕
owl : フクロウ
owl butterfly : フクロウチョウ
owlet moth : ヤガ
oxgen toxicity : 酸素毒性、酸素中毒
oxidation/hydroxylation : 酸化／水酸化
oxidative status : 酸化状態
oxidative stress : 酸化ストレス
oxyanion : オキシアニオン
oxygen absorption : 酸素吸収
oxygen consumption : 酸素消費（量）
ozone-induced cell death response : オゾン誘導性細胞死応答
ozone-induced damage : オゾン誘導性傷害、オゾン誘発性傷害

p

p : p 値（帰無仮説が棄却できる確率値）
P value : P 値（二元配置分散分析）
p-distance : p 距離（遺伝的距離の一種）

p10 promoter : p10プロモーター、プロモーター p10
Pa : パスカル(圧力や応力の単位)(Pascal)
PA [hg.] : ピロリジジンアルカロイド(Pyrrolizidine Alkaloid)
pacemaker : ペースメーカー
Pacific area : 太平洋地域
paddy field : 水田地帯
paedogenesis : 幼生生殖
painted lady butterfly : ヒメアカタテハ
painted-on eyespot : 絵の具で描かれた眼状紋
pair-clinging behavior : 組しがみつき行動
paired t-test : 対になった t 検定
paired-end read : ペアエンドのリード(同じ DNA 分子配列を二回読むこと)
pairwise alignment : ペア整列
pairwise coancestry coefficient : 対比較コアンセストリー係数、対比較共祖係数
pairwise comparison : 対比較、一対比較
pairwise competitive interaction : 一対の競争的相互作用
pairwise difference : 対差、対差異
pairwise genetic distance : 対比較した遺伝距離、対様式遺伝的距離
palae- : 旧 -、古 -〔ギリシャ語〕
Palaearctic region : 旧北区
palaeo- : 旧 -、古 -〔ギリシャ語〕
palaeontological site : 古生物学の場所
palaeotropical zone : 旧熱帯
palaeotropics : 旧熱帯
palatability : 美味性、嗜好性
palatability spectrum : おいしさのスペクトル、嗜好性の範囲
palatable : 美味
Palau Islands : パラオ諸島
pale beige ventral hindwing : 腹側の青白いベージュ色の後翅
pale forewing : 青白い前翅
pale form : 青白い型
pale medial band : 中央部の青白い縞状バンド
pale summer form : 青白い夏型
pale- : 旧 -、古 -〔ギリシャ語〕
pale-yellow : 淡い黄色
Palearctic region : 旧北区
Palearctic species : 旧北区産の種
paleo- : 旧 -、古 -〔ギリシャ語〕
paleobotanical : 古植物学の
Paleogene period : 古第三紀
paleovegetation reconstruction : 古植生の再構築
palmar : 掌側(しょうそく)の、手掌の
palp : 鬚(ひげ、しゅ)〔英語〕
palpal : 鬚(ひげ、しゅ)の〔英語〕
palpal seta : 側刺毛
palpus : (複 . -pi)、鬚(ひげ、しゅ)〔ラテン語〕
palynological : 花粉学の
palynophagy : 花粉食性
PAML : 最大節約法に基づいて祖先型推定を行うソフトウェアの名称
PAML approach : PAML アプローチ(PAML ; 最尤法に基づいて解析を行うソフトウェアの名称)
pan of water : 水を張った皿、水のパン
pan trap : パントラップ、平皿トラップ
Pan-Japan Sea Area : 周日本海地域
Pan-tropical type : 汎熱帯型
Panamanian forest : パナマの森
panmictic : 任意交配
panmictic population : 汎生殖個体群、任意交配集団
panmictic unit : 任意交配単位
panoistic ovariole : 無栄養型卵巣小管(むえいようがたらんそうしょうかん)
paper board : 台紙
paper kite butterfly : オオゴマダラ
paper model : 色紙モデル

papiliochrome : パピリオクローム
Papilionoidea tree : アゲハチョウ科の系統樹
papilla analis : 肛乳頭、肛乳房突起
pappus : 冠毛
pappus length of achene : 痩果の冠毛の長さ
Papua New Guinea : パプアニューギニア
para- : 側 -〔ギリシャ語〕
parabiosis : 併体結合、パラビオーシス
paracoxal suture : 側基節溝状線(胸部)
paradigm shift : パラダイムシフト、範例シフト
paradise of the butterfly : 蝶の楽園
paradoxical : 逆説的な
parafilm : パラフィルム
paraformaldehyde : パラホルムアルデヒド(PFA)
paraglossa : (複 . -ae)、側舌〔ギリシャ語〕
paralectotype : パラレクトタイプ、副後基準標本
parallel amino acid change : アミノ酸の平行的変化
parallel change analysis : 平行的変化解析
parallel cladogenesis : 平行分岐進化
parallel evolution : 平行進化、並行進化
parallel phenotype : 平行表現型、同位表現型
parallel phenotypic evolution : 平行的表現型進化、表現型の平行進化
parallel radiation : 平行発散
parallelism : 平行進化
paralog : 側系遺伝子、パラログ
paralogous : 傍系相同
paramere : 把握器、交尾鉤、側片
parameter estimate : パラメータ推定値
paranodal plate : 前胸板
paranotum : 側背板
parapatagium : 側頸板
parapatric : 側所的
parapatric distribution : 側所的分布
parapatric race : 側所的亜種、側所的系統
paraphyletic : 側系統の
paraphyletic group : 側系統群
paraphyletic species : 側系統種
paraproct : 肛側板(こうそくばん)、肛側片
parasite : 寄生、寄生虫、寄生者
parasite datum : (複 . -ta)、寄生データ
parasite-host interaction : 寄生者 - 宿主の相互作用
parasitic : 寄生的、寄生型
parasitic interaction : 寄生相互作用
parasitic thrips : 寄生性のアザミウマ
parasitic wasp : 寄生蜂(きせいほう)
parasitism : 寄生関係、寄生
parasitism success : 寄生成功(率)
parasitization : 寄生
parasitize : 寄生する
parasitized caterpillar : 寄生された毛虫
parasitoid : 捕食寄生者、捕食寄生虫、捕食寄生(の)
parasitoid wasp : 捕食寄生蜂(ほしょくきせいほう、ほしょくきせいばち)
parasitoid-attracting cue : 捕食寄生者を誘引する刺激
parasitoid-attracting plant cue : 捕食寄生者を誘引する植物刺激
parasitoid-attracting volatile : 捕食寄生者を誘引する揮発性物質
paratype : パラタイプ、副模式標本、従基準標本
parcel : 区画
parent term : 親用語、上位概念用語(GO用語〈Gene Ontology term〉で、その用語には親子関係がある)
parent-offspring conflict : 親子間対立
parental population : 親の個体群、親種
paris peacock butterfly : ルリモンアゲハ
Parnassians : ウスバアゲハ亜科
Parnassians and Swallowtails : アゲハチョウ科

Parnassius charltonius : カルトニウスウスバシロチョウ
Parnassius citrinarius : ウスバシロチョウ
pars intercerebralis : 脳間部
parsimonious explanation : 控えめな説明
parsimonious tree : 節約な系統樹
parsimony : 節減、倹約
parsimony informative : 節約に関する情報
Parsonsia : ホウライカガミ属
Parsonsieae : ホウライカガミ族
part : 器官、部分
Part of the List of Available Names in Zoology : 動物学における適格名リストの分冊
part of wing membrane : 翅の膜部分
parthenogenesis : 単為生殖(たんいせいしょく)
parthenogenetic induction : 単為生殖誘導
partial bleaching of rhabdom : 感桿の部分退色、感桿の部分脱色
partial degradation of DNA : DNA の部分劣化
partial L opsin gene sequence : L オプシン遺伝子の部分塩基配列
partial pupal period : 蛹期の一部期間
partial solution : 部分的な解決(策)
partial voltine : 部分化性
partially protected species : 部分的保護種
partially purified prothoracicotropic hormone : 部分精製された前胸腺刺激ホルモン
particle : 粒子
particle production : 粒子産出、粒子生成
particular nectar concentration : 特定の花蜜への集中(化)
particular, in : 特に
partition : 分割
partition-specific substitution model : 分割特異的な置換モデル
partivoltine : 部分化性
partly unknown : 部分的に未知
pass winter : 越冬する
passing female : 通過雌
passion-vine butterfly : ヒョウモンドクチョウ
passive diffusion : 受動放散、受動発散
passive loss : 受動的消失
passive movement : 消極的移動、他動的移動、受身的移動
past report : 従来の報告
Pasteur, Louis : パスツール、ルイ・パスツール
pasture : 牧草地
patagium : (複 . -ia)、頸板〔ラテン語〕
patch : 大斑点、パッチ、生息場所
patch colonization : パッチへの移住
patch network : パッチネットワーク
patch occupation rate : パッチ占有率
patchiness : パッチネス、不調和、むらがある
patchiness parameter : 不均一パラメータ
patchy distribution : パッチ状分布
patchy structure : パッチ(状)構造
paternal chromosome : 父系染色体
paternal investment : 雄側の投資
paternal PA : 父方の PA、父方のピロリジジンアルカロイド(Pyrrolizidine alkaloid)
paternally expressed gene : 父親性発現遺伝子(PEG)
paternity : 父性
paternity analysis : 父系分析、父性分析
path curvature : 経路曲率
path integration : 経路統合、推測航法
pathogen : 病原体、微生物
pathogen genetic resource : 病原体の遺伝子資源
pathogen infection : 病原体感染、微生物感染
pathogen recognition : 病原体認識、病原菌認識
pathogenesis-related gene : 感染特異的遺伝子、*PR* 遺伝子(防御遺伝子の一種)
pathogenic virus : 病原性ウイルス
pathway : 経路
pathway's involvement : 経路関与

patrol : 占有する
patrolling : 巡回型、巡回
patrolling behavior : 巡回行動
patrolling flight : 巡回飛翔
patrolling type : 探索型
pattern : 様相、斑紋
pattern formation : 紋様形成
pattern formation mechanism : パターン形成機構
pattern of seasonal change : 季節変化パターン
pattern oriented modeling : パターン指向モデリング
pattern recognition receptor : パターン認識受容体
pattern transformation rule : 紋様の変換則
pattern variation : パターン変化
patterning gene : パターン化遺伝子
Patterson's D-statistic : パターソンの D 統計量
paved road : 舗装道路
pb gene : プロボースィペディア遺伝子(*proboscipedia* ; 口器足)、吻足遺伝子
PBAN : フェロモン生合成活性化神経ペプチド(Pheromone Biosynthesis Activating Neuropeptide)
PBS : リン酸緩衝食塩水、リン酸緩衝生理食塩水(Phosphate Buffered Saline)
PBS washing : PBS 洗浄、リン酸緩衝生理食塩水での洗浄
PBS-BSA : リン酸緩衝生理食塩水-ウシ血清アルブミン(Phosphate Buffered Saline-Bovine Serum Albumin)
PBS-treated control plant : リン酸緩衝生理食塩水で処理された対照植物
PCA : 主成分分析(Principal Component Analysis)
PCMH : 蛹クチクラ黒色化ホルモン(Pupal-Cuticle Melanizing Hormone)
PCR : ポリメラーゼ連鎖反応、PCR 法(Polymerase Chain Reaction)
PCR amplicon : PCR アンプリコン
PCR fragment : PCR 断片
PCR mixture : PCR 混合物
PCR product : PCR 産物
PCR temperature profile : PCR の温度プロファイル
PCR-based marker : PCRに基づくマーカー
PCR-derived probe : PCR 由来プローブ
PD axis : 近位-遠位軸、PD軸、基部-末端軸、基部-外縁軸(Proximal-Distal axis)
pdb : タンパク質構造データバンク(Protein Data Bank)
PDF : 色素拡散因子(Pigment Dispersing Factor)
PDV lectin : PDV レクチン、ポリドナウイルス由来レクチン(PolyDnaVirus)
pea weevil : エンドウゾウムシ
peacock butterfly : クジャクチョウ
peak absorbance : ピーク吸収
peak direction : ピーク方向
peak number : 最高点の個体数、ピーク個体数
peak sensitivity : ピーク感度
pear-shaped : セイヨウナシ形の
pear-shaped eyespot : セイヨウナシ状の眼状紋
Pearson correlation coefficient : ピアソンの相関係数
peat : ピート、泥炭
pecky rice : 斑点米
pectin : 櫛状板(しつじょうばん)、櫛状(しつじょう)突起、櫛状器官
pectinate : 櫛歯状(くしばじょう、くしはじょう)の(蛾類の触角)
peculiar : 特異的な、特殊な
peculiarity : 特異性、特性
pedestrian recorder : 徒歩記録者
pedicel : 梗節(こうせつ)

pedicellus : 梗節(こうせつ)〔ラテン語〕
peer review : ピアレビュー、同等者による査読、同輩審査
PEG : 父親性発現遺伝子（Paternally Expressed Gene）
pellet : 小球を投げつける、ペリット、糞球、小球形にする、ペレット化する
penalty : 罰則
penetrance : 浸透率
penetrated light : 透過光
penicillin : ペニシリン
peninsula : 半島
penis : 陰茎〔ラテン語〕
Pennsylvanian period : ペンシルベニア紀（地質時代の古生代／石炭紀 (Carboniferous period) の一時代）
penultimate instar : 亜終齢
pepper plant : トウガラシ
peppered moth : オオシモフリエダシャク
peptide : ペプチド
peptidoglycan : ペプチドグリカン（PG）
peptidoglycan recognition protein : ペプチドグリカン認識タンパク質
PER : 口吻伸展反射（反応）（Proboscis Extension Reflex 〈Response〉）
per [hg.] : パー遺伝子（遺伝子名：*period*）（時間遺伝子の１つ）
per capita coefficient : １個体当りの係数
per capita interaction : １個体当りの相互作用
per copy of the gene : 遺伝子１コピーあたり
per se : 本質的に、それ自体が〔ラテン語〕
per- : 通して -、非常に -、完全に -〔ラテン語〕
perceptual cue : 知覚刺激
perceptual information : 知覚情報
perch site : 静止場所
perching : 待ち伏せ型、みはり、見張り
perching behavior : 静止行動
perching male : 静止雄
perching time : 静止時間
perennial grass : 多年生の草
performance : 成育達成度、遂行
performance of estimator : 推定量の性能
peri- : 周辺 -、周囲 -〔ギリシャ語〕
pericardial cavity : 囲心腔(いしんこう)
pericardial cell : 囲心細胞(いしんさいぼう)
pericardial gland : 囲心腺(いしんせん)
pericardium : 囲心のう(いしんのう)、心のう、囲心腔(いしんこう)
peril of normalizing richness : 種数を正規化することの危険
perilous : 危険に満ちた
period : 紀（地質時代の「き」）
period [hg.] : ピリオド遺伝子（遺伝子記号：*per*）（時間遺伝子の１つ）
period of irradiation : 照射期間
periodic : 周期的な
periodic form : 周期型
periodic rediscovery : 周期的な再発見
periodic response : 周期反応
periodicity : 周期性
peripatric speciation : 周辺（的）種分化
peripheral nerve : 末梢神経
peripheral projection pattern : 周辺投射パターン、周辺投影様式
periphery of cell : 細胞周辺（部）
perish : 消滅する、消え去る
peritracheal gland : 気管周腺
peritrophic membrane : 囲食膜(いしょくまく)
permanent diapause : 永久休眠
permanent quadrat : 永久コドラート、永久方形区
permeability : 透過性、浸透性
permeable : 透過性のある、浸透できる
permeate : 充満させる、しみ込む
Permian period : ペルム紀（地質時代の古生代の一時代）
persistence : 回復力、存続性、持続性
persistence ability : 存続能力、回復力

persistence of population：個体群の永続性、個体群の持続性
persistent hybridization：存続する交雑、持続する交雑
persisting metapopulation：存続するメタ個体群
personal communication：私信
personality：個性
persuasive evidence：説得力のある証拠
pertaining study：関連調査
perturbation：攪乱(かくらん)、動揺
Peruvian hinterland：ペルー奥地
pervasive：広がる、まん延する
pest：害虫、有害な虫
pest control：害虫防除
pest of rice：米の害虫、イネ害虫
pesticide：殺虫剤、農薬(殺虫剤・除草剤・殺菌剤など)
petiolate：有柄の
petiole：腹柄(ふくへい)、柄部
Petri dish：ペトリ皿、シャーレ
petrolatum：ペトロラタム(石油から採る半固形状のろう)、ワセリン
PFA：パラホルムアルデヒド(リン酸緩衝液の一種)(ParaFormAldehyde)
pFBD-pH vector：pFBD-pH導入ベクター、pFBD-ポリヘドリンプロモーター導入ベクター、pFBD-多角体タンパク質プロモーター導入ベクター(pFastBacDua-polyHedrin promoter)
PG：ペプチドグリカン(PeptidoGlycan)
PGC：始原生殖細胞(Primordial Germ Cell)
PGRP：ペプチドグリカン認識タンパク質(PeptidoGlycan Recognition Protein)
PH promoter：ポリヘドリンプロモーター、多角体タンパク質プロモーター(PolyHedrin)
phagocytosis：食細胞活動、食作用
phagostimulant：摂食刺激物質
phalloidin：ファロイジン
phallus (ph)：挿入器、ファルス
Phanerozoic：顕生代(けんせいだい)(の)
pharate adult：潜成虫、ファレート成虫(「pharate」はギリシャ語で「隠された」の意味)
pharate condition：ファレート状態
pharate larva：ファレート幼虫、潜幼虫
pharate pupa：ファレート蛹、潜蛹
pharmacophagy：薬物食性、薬物摂食
pharynx：咽頭(いんとう)〔ラテン語〕
phase：相、位相
phase adjustment：位相調節
phase gregaria：群生相
phase polymorphism：相多型、相変異
phase related polymorphism：相関連多型、相関連変異
phase response curve：位相反応曲線
phase shift：位相変位
phase solitaria：孤独相
phase transien：転移相
phase variation：相変異
phasic/tonic nature：一過性／持続性の性質
Phasmatodea：ナナフシ目、竹節虫目
phellamurin：フェラムリン
phenanthroindolizidine alkaloid：フェナントリジンアルカロイド
phenetics：表形学
pheno-：表現-〔ギリシャ語〕
phenogram：表現図、樹状図
phenol red colorant：フェノールレッド染料、フェノール赤色染料
phenol-chloroform：フェノール-クロロホルム
phenol-chloroform extraction：フェノール-クロロホルム抽出法
phenol-chloroform extraction procedure：フェノール-クロロホルム抽出法
phenology：季節的消長、生物気象学、生物季節学、フェノロジー

phenomenon : 現象
phenotype : 表現型、表現形
phenotype mechanism : 表現機構
phenotype variation : 表現型変異
phenotypic appearance : 表現型発現
phenotypic change : 表現型の転換
phenotypic character : 表現形質
phenotypic differentiation : 表現型分化
phenotypic divergence : 表現型分岐、表現型発散
phenotypic diversity : 表現型多様性
phenotypic effect : 表現型効果、表現型の効果
phenotypic evolution : 表現型進化
phenotypic intermediate : 中間表現型
phenotypic plasticity : 表現可塑性、表現型可塑性、表現型可変性
phenotypic resemblance : 表現型酷似
phenotypic variation : 表現型変異
phenotypically differentiated population : 表現型的に異なる個体群
phenylalanine : フェニルアラニン
pheromone : フェロモン
pheromone biosynthesis activating neuropeptide : フェロモン生合成活性化神経ペプチド
pheromone precursor : フェロモン前駆体、フェロモン前駆物質
pheromone transport : フェロモンの運搬、フェロモンの輸送
phleophagy : 樹皮食性
phloem-feeding herbivory : 篩管部摂食性植食、維管束の篩管部分から吸汁する植食性
phorcabilin : フォルカビリン
phosphate buffered saline : リン酸緩衝食塩水、リン酸緩衝生理食塩水(PBS)
phospholipid : リン脂質
photo- : 光 -〔ギリシャ語〕
photo-receptor : 光感覚器、光受容器、光受容体
photograph : 写真を撮る、撮影する
photographic plate : 写真図版、写真乾板
photography : 写真撮影
photoisomerize : 光異性化する
photoisomerizing flash : 光異性化閃光
photomicrography : 顕微鏡写真
photoperiod : 光周期、日長
photoperiod treatment : 光周処理
photoperiod-sensitive stage for diapause induction : 休眠誘起の日長感受期
photoperiodic clock : 光周時計
photoperiodic control : 光周調節、日長調節
photoperiodic counter : 光周(期)カウンター、光周計数機構
photoperiodic reaction : 光周反応
photoperiodic response : 光周反応
photoperiodic response curve : 光周反応曲線
photoperiodic signal : 光周信号
photoperiodic time measurement : 光周測時
photoperiodic time-measurement system : 光周測時機構
photoperiodically controlled diapause : 光周性で調節された休眠
photoperiodism : 光周性、日長効果
photophase : 明期
photophil : 親明相
photopigment : 視物質、感光色素、光色素
photopigment sensitivity : 視物質感度
photopigment-containing microvillar membrane : 光色素含有微絨毛膜
photoreceptive interneuron : 光感受性介在神経
photoreceptor : 光感覚器、光受容器、光受容体
photoreceptor cell : 光受容細胞
photoreceptor nucleus : 光受容細胞核
photosensitive part : 感光部位
photosensitive portion : 感光部分

photosynthesis : 光合成
photosynthetic phase : 光合成期
phototaxis : 走光性(そうこうせい)
photothermograph : 光温図表
phragma : 分隔甲(胸部)
phyllophagy : 葉食性
phylogenetic analysis : 系統解析
Phylogenetic Analysis Using Parsimony : PAUP(節約系統解析ソフトウェア)
phylogenetic and coalescent theory : 系統的合祖理論(現在の集団から得られる遺伝情報から過去の集団動態を推測する、集団遺伝学におけるモデルおよびその手法)
phylogenetic biogeography : 系統生物地理学
phylogenetic biogeography method : 系統生物地理学的方法
phylogenetic branch : 系統分岐
phylogenetic community ecology : 系統的群集生態学
phylogenetic estimate : 系統推定
phylogenetic inference : 系統(樹)推定
phylogenetic information : 系統的情報、系統学的情報
phylogenetic informativeness : 系統的情報性、系統発生学的情報量
phylogenetic method in biogeography : 系統生物地理学的方法
phylogenetic network : 系統ネットワーク
phylogenetic order : 系統的進行順序
phylogenetic placement : 系統的位置
phylogenetic position : 系統学的位置
phylogenetic practice : 系統学的慣行
phylogenetic reconstruction : 系統再構築、系統再構成、系統復元
phylogenetic relationship : 系統(的)相互関係、系統関係
phylogenetic scale : 系統学的な尺度
phylogenetic signal : 系統的シグナル、系統発生シグナル、系統的信号
phylogenetic significant unit : 系統的有意単位、系統上の重要な単位
phylogenetic study : 系統分類に関する研究、系統学的研究
phylogenetic tree : 系統樹
phylogenetic uncertainty : 系統的不確実性、系統的な不確実性
phylogenetic utility : 系統学的効用
phylogenetic view point : 系統的視点
phylogenetically important species : 系統上の重要な種
phylogenetically localize : 系統的に局所化する
phylogenetically related species : 系統的に近縁な種、系統的に類縁の種
phylogenic evolution : 系統進化
phylogenomics : ゲノム系統学
phylogeny : 系統発生(進化)、系統分類、系統学、系統
phylogeny reconstruction : 系統発生の再構築、系統発生の再構成
phylogeographic history : 系統地理学的歴史
phylogeographic interpretation : 系統地理学的解釈
phylogeography : 系統地理学、系統地理
phylogram : 系統図、系統樹
phylum : (複 . -la)、門(分類階級の「もん」)
PHYML : 最尤法(maximum likelihood)のソフトウェアの名称
physical characteristic : 物理的性格、物理的特性
physical control : 物理的防除
physical difference : 物理的相違
physical factor : 物理的要因
physical linkage : 物理的連鎖
physical-to-map distance : 物理的距離
physiological adaptation : 生理(学)的適応
physiological and behavioral experiment : 生理学的・行動学的実験

physiological approach : 生理学的アプローチ
physiological background : 生理的背景
physiological change : 生理的変化
physiological characteristic : 生理的特性、生理特性
physiological condition : 生理(的)条件
physiological cost : 生理的コスト
physiological datum : (複 . -ta)、生理データ
physiological difference : 生理学的差異、生理的差異
physiological factor : 生理学的要因、生理学的因子
physiological mechanism : 生理機構
physiological mediation : 生理的媒介
physiological post-mating change : 交尾後の生理的な変化
physiological recording : 生理(学)的記録
physiological regulation : 生理的調節
physiological response : 生理的反応
physiological resurgence : 生理的誘導多発生、生理的リサージェンス
physiological saline : 生理食塩水
physiological salt solution : 生理食塩水
physiological state : 生理状態
physiological study : 生理学的研究
physiological time : 生理(学)的時間
physiological trait : 生理的形質、生理的特性
physiological variable : 生理的変数、生理学的変数
physiologically regulated step by step : 生理的に調整された一連の行動
physiology : 生理学、生理
phyto- : 植物 -、植物の - 〔ギリシャ語〕
phytoalexin : ファイトアレキシン
phytochemical concentration : 植物化学物質濃度
phytoecdysone : 植物エクジソン
phytohormone : 植物ホルモン
phytophagous insect : 植食性昆虫(しょくしょくせいこんちゅう)
phytophagous myrmecophile : 植食性の好蟻性昆虫
phytophagy : 植食性
PI distribution : PI 分布 (Phylogenetic Informativeness)
PI profile : 系統的情報性プロファイル (Phylogenetic Informativeness)
pictorial book : 図鑑、絵本
picture of the flying position : 飛翔写真
piercing-sucking herbivore : 突き刺して吸汁する植食者、吸汁性植食者
pierid butterfly : シロチョウ科の蝶
Pierinae caterpillar : シロチョウ亜科の幼虫
Pierinae exemplar species : シロチョウ亜科の典型種
Pieris : アセビ属
pierisin : ピエリシン
pig : ブタ
pigment : 色素
pigment content : 含有色素
pigment deposition : 色素沈着
pigment dispersing factor : 色素拡散因子、色素拡散ホルモン様ペプチド(PDF)
pigment fraction : 色素分画
pigment layer : 色素層
pigment pathway : 色素(生合成)経路
pigment polymorphism : 体色多型、色素多型
pigmentation : 色素沈着、色素形成
pigmentation gene *ddc* : 色素沈着遺伝子 *ddc*
pigmentation pattern : 着色パターン、着色紋様
pigmentation process : 着色過程、染色過程、色素沈着過程
pile : 山積み
pile up : 山積みにする、作り上げる
pilifer : 有毛の、軟毛のある 〔ラテン語〕

pilot study : 試験的研究
pilotage : 水先案内
pinaculum : (複 . -la)、硬皮板、有棘毛瘤(ゆうきょくもうりゅう)〔ラテン語〕
pine beauty moth : マツキリガ
pine community : マツ群落
pine forest : マツ林
pine white butterfly : マツノキシロチョウ
pine-dominated forest : マツが優占した林
pinewood : 松林
pioneer : 先駆種、開拓者
pioneer individual : パイオニア個体、先駆個体
pioneer plant : 先駆樹種、先駆植物
pipevine : ウマノスズクサ
pipevine swallowtail butterfly : アオジャコウアゲハ
pitfall : ピットフォール、落し穴、注意点
pivotal experimental datum : (複 . -ta)、極めて重要な実験データ
pivotal genus : 中枢の属、極めて重要な属
pl. : 図版、写真図版(plate)
place of occurrence : 発生地
place of origin : 発祥地
placement of sample : サンプルの配置
placental mammal : 有胎盤哺乳類
placoid sensillum : (複 . -la)、板状感覚子、板状感覚器〔英語〕
Plan Position Indicator : 平面位置表示器(PPI)
planetary biodiversity : 地球上の生物多様性
plant : 植物
plant and butterfly lineage : 植物と蝶の系譜
plant as oviposition site : 産卵植物
plant biomass : 植物のバイオマス、植物生物量、植物量
plant bug : カスミカメムシ、カメムシ
plant cell : 植物細胞
plant community : 植物群集、植物群落
plant defense trait : 植物の防御形質
plant density : 植物密度
plant diversity : 植物(の)多様性
plant ecologist : 植物生態学者
plant exudate : 植物の分泌物
plant genome : 植物のゲノム
plant hormone : 植物ホルモン
plant hosting the insect : 昆虫が宿主とする植物
plant material : 植物材料
plant nursery : 園芸店
plant odor : 植物香気、植物の香り
plant outlier : 植物の異常値
plant pigment : 植物色素
plant population : 植物個体群
plant protection : 植物保護
plant quarantine : 植物検疫
plant science : 植物科学
plant specialized metabolite : 植物特化代謝物、植物二次代謝産物
plant species : 植物種
plant species richness : 植物の種数、植物の種多様性
plant survival : 植物の生存
plant treatment : 植物処理
plant trichome : 植物の毛状突起
plant volatile : 植物揮発性物質
plant worm : 冬虫夏草
plant-herbivore community : 植物-草食(性)動物の共同体
plant-insect interaction : 植物と昆虫間の相互作用
plant-plant signaling : 植物間のシグナル伝達
plantar lobe : ふ端(ふ節の末端)中片、ふ節端葉片、ふ節端突起、蹠片(「plantar」は「蹠〈あしうら〉」という意味)
planter : 底側(ていそく)
planthopper : ウンカ(の仲間)
planting : 植林

plasma transmembrane potential : 形質の膜内外電位差、形質膜の内外に生じる電位差
plasmatocyte : プラズマ細胞(血球細胞の一種)
plasmid : プラスミド
plasmid preparation : プラスミド調製、プラスミド精製
plasmid purification method : プラスミド精製法
plastic bag : プラスチック袋
plastic bottle : プラスチック瓶、プラスチックボトル
plastic cup : プラスチックカップ容器
plastic phenotypic response : 表現型の可塑的反応
plasticity : 可塑性
plate : 平板、板、図版、写真図版
platform : 土台、基盤
plausible explanation : もっともらしい説明
playback : 録音の再生
Plecoptera : カワゲラ目、積(せき)翅目
pledge : 誓約する
pleiotropic effect : 多面発現効果
pleiotropy : 多面発現
Pleistocene climate change : 更新世の気候変動
Pleistocene refugium : 更新(洪積)世での隔離
plenary power : 強権
plesiomorph : 原始的初原形質、原始的旧形質
plesiomorphy : 原始形質、祖先形質
pleural : 側板の
pleural groove : 側板溝
pleural lobe : 側板片、側板葉、側板の外葉
pleural sulcus : 側板溝
pleural suture : 側板縫合線、側板溝状線
pleural wing process : 側板翅突起、側翅突、側板の翅突起
pleurite : 側板
pleuron : 側板(そくばん)
pleurosternum : 側腹板
ploidy : 倍数性、倍数関係
ploidy level : 倍数性レベル
plume : 冠毛
plumose antenna : (複 . -nae)、羽毛状の触角(蛾類)
pluvial : 多雨の
pmol : ピコモル(物質量の単位)(picomole)
PMRF : 蛹黒色化抑制因子(Pupal Melanization Reducing Factor)
PNP : 前胸板(ParaNodal Plate)
poikilotherm : 変温動物
poikilothermic vertebrate : 変温性脊椎動物
point of adult emergence : 成虫羽化の時点
pointed apex in forewing : 前翅頂のとがり
pointed tip : とがった先端
poison : 毒
poison-antidote model : 毒性 - 解毒モデル、毒 - 毒消しモデル
poisonous : 有毒な
poisonous snake : 毒蛇
Poisson distribution : ポアソン分布
Poisson process : ポアソン過程
Poisson Tree Process : ポアソン系統樹過程、ポアソン過程に従って起きると仮定した系統樹(PTP)
polar coordinate model : 極座標モデル
polar night : 極夜
polarisation-dependent color vision : 偏波依存色覚
polarization compass : 偏向コンパス
polarized : 分極化された、極性を示す
polarized light : 偏光
poleward shift : 極方向への移動
policing : 警察行動、警備
pollen : 花粉〔ラテン語〕

pollen dispersal : 花粉飛散、花粉分散
pollen feeding : 花粉摂食
pollen flow : 花粉流動
pollen resource : 花粉資源
pollex : 指状突起、距（蛾類）
pollination : 受粉、花粉媒介
pollination ecology : 受粉生態学、送粉生態学
pollination syndrome : 送粉シンドローム
pollinator : 花粉媒介者、花粉媒介昆虫、ポリネーター、送粉者
pollinator-friendly species : 花粉媒介者が好む種
pollinophagy : 花粉食性
pollution : 汚染
poly- : 多 -〔ギリシャ語〕
poly-DNA virus : 多 DNA ウイルス
polyadenylation : ポリアデニル化
polyandry : 一雌多雄、一妻多夫
polydnavirus : ポリドナウイルス（PDV）
polyembryony : 多胚形成
polygamous animal : 一夫多妻動物
polygene : 多遺伝子、量的遺伝子、ポリジーン
polygene control : ポリジーン支配
polygenic : 多遺伝子（性）の、ポリジーンの
polygenic trait : 多遺伝子性形質
polygyny : 一夫多妻、ポリガミー、複女王性
polyhedrin gene : ポリヘドリン遺伝子、多角体タンパク質遺伝子
polyhedrosis : 多角体病
polymer : 多量体、重合体、ポリマー
polymerase chain reaction : ポリメラーゼ連鎖反応（PCR）
polymorphic : 多形型、多型、多型の、多型的
polymorphic Batesian mimic : 多型のベイツ型擬態（種）
polymorphic difference : 多型的差異、多型的な違い
polymorphic locus : （複 . -ci）、多型的遺伝子座
polymorphic mimicry locus : 多型擬態遺伝子座
polymorphic population : 多型個体群
polymorphic race : 多型系統
polymorphic site : 多型部位
polymorphism : 遺伝的多型、多型、多型現象、多型性
polypeptide factor : ポリペプチド因子
polyphagous : 広食性の、広食性、多食性
polyphagy : 広食性
polyphenic development : 表現型多型の発生
polyphenism : 表現型多型、環境的多型現象
polyphenism phenomenon : 多型現象
polyphenol oxidase : フェノールオキシダーゼ
polyphyletic : 多系統的、多系統
polyphyletic group : 多系統群
polyploidization : 倍数体化
polypropylene tube : ポリプロピレン管、ポリプロピレンチューブ
polyspermy : 多精
polytes : ポリ（ュ）テス型（シロオビアゲハの擬態型雌）
polytypic : 多型の、多型的
polyvoltine : 多化性
pomona form : 銀紋型（ウスキシロチョウ）、ポモナ型
pond : 池
pool : 水たまり
pooled individual : プールされた個体
pooled quadrat : プールされた方形区
poor competitor : 弱い競争相手
poor nutrition : 栄養不足、粗末な栄養
poor predictor : 不十分な予測因子
poor summer weather : 悪天候の夏
poplar : ポプラ
popular butterfly species : 話題の蝶種
populate : 住む

population : 個体群【生態学】、集団、個体数、母集団【統計学】
population biology : 個体群生物学
population bottleneck : 集団ボトルネック、集団遺伝学におけるボトルネック効果
population change : 個体数の変化、個体群変動
population density : 個体群密度
population dynamics : 個体群動態
population dynamics model : 個体群動態モデル
population ecology : 個体群生態学、集団生態学
population genetic analysis : 個体群遺伝解析
population genetic approach : 個体群の遺伝学的研究法
population genetic structure : 個体群の遺伝的構造、集団の遺伝的構造
population growth : 個体群成長
population growth rate : 個体群成長率、個体数増加率
population isolation : 個体群の孤立化
population management : 個体群管理
population regulation : 個体群調整、個体数調整、個体群制御
population size : 個体群サイズ、集団の大きさ
population stability : 個体群の安定性
population structure : 個体群構造
population survival : 個体群の生存
population-genetic approach : 個体群遺伝的手法
population-level change : 個体群レベルの変動、個体数レベルの変化
populations of individual : 個体の母集団
Populus trichocarpa : コットンウッド(ポプラの一種)
pore canal : 孔管(こうかん)
pore cupola organ : 有孔の鐘状感覚器

porrect : 前に突き出た
port : 心門
positional cloning of locus : 遺伝子座の位置クローニング、遺伝子座の位置的単離
positional genetic homology : 遺伝子の位置相同性
positional information : 位置情報
positional value : 位置価(分子的番地表示)
positive and negative electrode : 両極電極、正負電極
positive association : 正の関連、正の相関
positive cell : 陽性細胞、ポジティブ細胞
positive clone : 陽性クローン
positive control : 正の対照(区)、ポジティブコントロール
positive environmental factor : 関与する環境要因、正の環境要因
positive feedback : 正のフィードバック
positive function : 正の機能
positive selection : 正の自然選択
positive selection inference : 正の自然選択の継承
positive selective pressure : 正の淘汰圧、正の選択圧
positive slope : 右上がり斜線、上り傾斜、正の傾斜
possess : 持つ、所有する
possession : 所有
possibility of character displacement : 形質置換の可能性
possibility of copulation at the emergence site of female : 雌の羽化場所での交尾可能性
possible overwintering area : 越冬可能地域
post hoc comparison : 事後比較
post teneral flight : テネラル(羽化)以後の移動飛翔
post teneral period : 羽化以後期
post- : 後 -、外 -〔ラテン語〕

post-anthesis : 開花後
post-border : 国境通過後
post-diapause larva : 休眠後幼虫
post-eclosion : 羽化後
post-eclosion female : 羽化後の雌
post-mating isolation : 交配後隔離
postbasal : 外基部帯
postdiapause longevity : 後休眠の寿命、休眠終了後の寿命
postdiapause period : 後休眠期
postdiscal : 外中央帯、外縁寄り中央帯
postdiscal band : 外中央帯
postdiscal line : 外横線、外中央線
posterior : 後、後部の、後方の〔ラテン語〕
posterior apophyses : 後側甲
posterior basalare : 後方の基翅節片
posterior branch : 後枝
posterior crossvein : 後横脈
posterior cubitus vein : 中脈分岐、CuP 脈
posterior edge : 後縁、後端
posterior eyespot : 後方の眼状紋
posterior focus : 後方のフォーカス
posterior margin : 後縁(蛾類)
posterior portion : 後方部、後方部分、後方部位、後部
posterior side : 後側
posteriorly : 後方に
postgena : 後頬
postgenal : 後頬の
postlabium : 後下唇
postman butterfly : ベニモンドクチョウ、メルポメーネベニモンドクチョウ
postman form : ポストマン型
postmating isolation : 交配後隔離
postmedial : 外中央帯、翅外中央帯
postmedial line : 外横線(蛾類)
postmedian : 外中央部
postmedian notal wing process : 中後背板翅突起、(後中背板翅突起)
postmentum : 下唇後基節
postnatal seta : 出生後の刺毛
postnotum : 後背板(胸部)
postpone mating : 配偶行動を延期する
postreceptoral : 後受容の
postulate : 仮定する
postural change : 姿勢変更
posture : 姿勢
postzygotic isolation : 接合後隔離
pot : 鉢、ポット
potassium hydroxide : 苛性カリ、水酸化カリウム
potent source : 有力な源泉
potential : 潜在力、可能性
potential candidate : 潜在的な候補(遺伝子)
potential cryptic species : 潜在的隠蔽(ぺい)種
potential escalation : 潜在的増大、潜在的拡大
potential hearing organ : 潜在的な聴覚器官
potential high-risk species : 潜在的にリスクの高い種
potential hybrid : 潜在的雑種
potential insect pest : 潜在(的)害虫
potential of hydrogen : pH、ペーハー
potential reproductive success : 潜在繁殖成功度
potential target : 潜在的な目標
potential threat : 潜在的脅威
potentially incompatible lineage : 潜在的に不和合な系統
potentially invasive species : 潜在的侵入種
potentially valid name : 潜在的有効名
poverty : 貧困
power of ANCOVA : 共分散分析の検定力、共分散分析の検出力
powerful flier : 力強く飛ぶ
ppb : パーツパービリオン(主に濃度を表す単位で、10 億分の 1、十億分率のこと)(parts per billion)

PPI : 平面位置表示器（Plan Position Indicator）
ppm : 100 万分の 1、百万分率のこと（parts per million）
PR gene : *PR* 遺伝子、感染特異的遺伝子（*Pathogenesis-Related*）
praeopercular organ : 下蓋板前器（かがいばんぜんき）、蓋片前器（がいへんぜんき）
prairie-inhabiting species : 大草原に生息する種
PRC : 位相反応曲線（Phase Response Curve）
pre- : 前 -〔ラテン語〕
pre-adult stage : 幼若成虫期、前成虫期
pre-border : 国境通過前
pre-chilling incubation period : 前低温馴化期間
pre-epimeron : 前方後側板
pre-existing information : 先在情報
pre-freezing : 予備凍結
pre-requisite : 前提条件、必須条件、必要条件
pre-zygotic isolation : 接合前隔離
preadaptation : 前適応
preadaptation to host plant : 寄主植物に対する前適応
preadaptation to survive winter in the temperate climate : 越冬を可能にする気候前適応
precautionary approach : 予防的アプローチ、予防的手法
precedence : 優先権
precipice : 崖
precipitation : 降水（量）、沈殿、沈降
precocious pupa : 早熟蛹
precostal vein : 肩脈
precoxal suture : 前基節溝状線
precursor : 前駆物質、先駆体、プリカーサー
precursor dopamine : 前駆物質ドーパミン、前駆体ドーパミン
precursor reservoir : 前駆物質貯蔵体、前駆体貯蔵庫
predate : 先行する
predation : 捕食
predation pressure : 捕食圧
predation rate : 捕食率
predator : 捕食（性）動物、肉食（性）動物、捕食者、捕食虫
predator avoidance : 捕食者逃避
predator avoidance behavior : 対捕食（者）回避行動
predator escape : 捕食者逃避
predator evasion : 捕食者回避
predator-avoidance : 捕食者回避
predator-avoidance response : 捕食者回避反応
predatory bird : 捕食（性）鳥
predatory mirid bug : 捕食性のカスミカメムシ
predatory mite : 捕食性のカブリダニ
prediapause copulation : 前休眠期の交尾、休眠前の交尾
prediapause period : 前休眠期
predictable : 予測的、予測可能な
predictable difference : 予測可能な差異
predictable direction : 予測可能な方向
predictable environment : 予測可能な環境
predictable seasonal variation : 予測可能な季節変動
predictable way : 予測可能な方法
predicted amino acid sequence : 推定アミノ酸配列
predicted position : 想定された位置、予測された位置
prediction : 予測
predictor : 予測因子、予測変数
predictor variable : 予測変数
predominant factor : 有力な要因、支配的な要因
predominant signal : 支配的なシグナル、支配的信号

predominantly : 支配的で、有力で
predominate : 圧倒的に多い
preepisternal suture : 前前腹板線
preepisternum : 前前腹板、前前側板(胸部)
preference : 選り好み、選好
preference cue : 選好刺激
preference experiment : 選好性実験
preference index : 選好性指標、選好指標、選好指数、選択指数
preference intermediate to the parental species : 親種の中間種を好む
preference locus : 選好に関する遺伝子座
preferentially : 優先的に、選択的に
prefix : 接頭語
preimaginal epidermal organ : 幼虫期の表皮器官
preimaginal stage : 幼生期
preincubate : 前培養をする、プレインキュベートする
prelabium : 前下唇
preliminary field investigation : 事前フィールド調査、予備的フィールド調査
preliminary study : 事前研究、予備的研究
premating isolation : 交配前隔離
premature cessation : 時期尚早の停止
prementum : 下唇前基節
prementum seta : 鰓刺毛
preocular carina : 単眼内縁隆起線
preovipositional period : 産卵前期
preparation : (実験・解剖用動物の)標本、プレパラート、調製、準備行動
prepattern : 予備パターン、プレパターン
prepectus : 前下胸板、前側板(中胸側板の前方にある切片)(胸部)
preprint : 前刷り
prepupa : (複 . -ae)、前蛹(ぜんよう)〔ラテン語〕
prepupal : 前蛹の、前蛹状態
prepupal burst of JH : 前蛹期での幼若ホルモンの噴出
prescribed fire : 所定の野焼き、計画的野焼き
prescutum : 前盾板(胸部)
presence : 存在
presence or absence of a Vogel's organ : フォーゲル器官の存否
preserve : 所蔵する、保存する
preset : プリセット、あらかじめ設定されている
pressure delivery system : 圧力運搬手段
presternum : 前腹板(胸部)
presumably evolve : どうも進化しているらしい(主観的な基準を反映した表現)
presumably non-territorial species : たぶんテリトリーを持たない種
presumptive locus : 原基遺伝子座
pretarsus : 前ふ節(ぜんふせつ)
prevailing method : 広く行われている方法
prevailing usage : 慣用法
prevalent : 流行している
prevention : 予防
previous information : 以前の情報、過去の情報
previous paper : 前報
previously learned : 前もって学習した、予め学習した
previously-threatened butterfly : 以前に絶滅の恐れがあった蝶
prey : えじき、獲物、被食者、捕食する
preytaxis : 獲物走性、走獲物性
prezygotic isolation : 接合前隔離
primary attractant : 一次誘引物質
primary consumer : 一次消費者
primary culture : 初代培養
primary factor : 主要因、主因
primary genitalia : 原交尾器
primary homonym : 一次(異物)同名
primary infection : 一次感染
primary metabolite : 一次代謝産物、一次生産物

primary production : 一次生産
primary reason : 一次的な理由、主要な理由
primary sensory neuron : 一次感覚神経、一次感覚ニューロン
primary seta : 一次刺毛
primary structure : 一次構造
primary study : 初歩的研究
primary type : 第一次基準標本、原模式標本(初めてある種を記載する時に用いた標本;その種の真の基準となる標本)
primary wing nerve branch : 翅の一次神経枝
primate : 霊長類
primate color vision system : 霊長類の色覚系
primate lineage : 霊長類の系統
primate pigment : 霊長類の色素
prime purpose : 主要な目的
primer : プライマー
primer combination : プライマーの組み合わせ
primer design : プライマー設計
primer pair : プライマー対
primer pheromone : 起動フェロモン、誘導フェロモン、プライマーフェロモン
primitive : 原始的な
primitive form : 原始的な形態
primitive lepidoptera : 原始的な鱗翅目
primitive species : 原始的な種、原始種
primordial germ cell : 始原生殖細胞
principal component analysis : 主成分分析
principal factor : 主要因
principle : 原理
Principle of Binominal Nomenclature : 二名法の原理
principle of competitive exclusion : 競争排除則
principle of coordination : 同位の原理
principle of homonymy : 異物同名関係の原理
principle of priority : 先取権の法則、先取権の原理
principle of the first reviser : 第一校訂者の原理
principle of typification : タイプ化の原理
printer's error : 印刷者の誤り、プリンターの間違い
printing on paper : 紙への印刷
prior expectation : 事前の期待、事前期待
prior permission : 事前の許可
priority : プライオリティ、先取権
priority of a name : 学名の先取権
priority of a nomenclatural act : 命名法的行為の先取権
priority, rule of : 先取権の原則
private allele : プライベートアリル、固有の対立遺伝子(座)(同じ種でも集団によって対立遺伝子が特殊であったり固有であること)
pro- : 前 -〔ギリシャ語〕
probabilistic model : 確率的モデル、確率論的モデル
probable model : 可能性のあるモデル
probe : プローブ(特定部位の検出指標となる物質)
problem of cause and effect : 原因と結果の問題
proboscis : 口吻(こうふん)〔ラテン語〕
proboscis extension reflex response : 口吻伸展反射反応、口吻を伸ばしての食物探索
proboscis recoil : 口吻収縮
process : プロセス、過程、突起、隆起
processes of migration, extinction, and recolonization : 移動・絶滅・再定着プロセス
proctodaeal valve : 幽門弁(ゆうもんべん)
procurement : 獲得
procuticle : 原表皮(げんひょうひ)

producer : 生産者
production of all-male brood : すべて雄の同腹仔の産出
production of dimorphism : 二型の産出、性的二型の産出
productivity gradient : 生産性勾配
profemur : 前腿節
profound scholar : 碩学(造詣が深い博学者)
profoundly : おおいに、深遠に
progeny : 子世代、子孫、後代〔ラテン語〕
prognathous type : 前口式(ぜんこうしき)
programmable : プログラム(化)できる
programmed cell dead : プログラムされた細胞死
progress : 進行する
progression of diapause : 休眠の進行
project laterally : 側方に投影する
projection : つの、突出(突起)部
prokaryote : 原核生物
proleg : (幼虫の)腹脚(ふくきゃく)、前脚(ぜんきゃく)
proleucocyte : 原白血球(昆虫血球細胞の一種)
proliferation pattern : 増殖パターン
proline-rich : 高プロリン含有
prolongation : 延長、延長部分
prolongation of the corner of the wing : 翅の辺縁部の延長部
prolonged copulation : 長期間の交尾
prolonged exposure : 長期間暴露
prolonged preovipositional period : 長期の産卵前期間
prolonged selection : 延長選択
prominent : 浮き出ている、突起した、卓越した
prominent dome-shaped inner membrane : 突起したドーム形の内膜
prominent eyespot : 目立つ眼状紋
prominent projection : 突起部
prominent veining : 隆起した翅脈
prominent wing vein : 盛り上がった翅脈
promiscuity : 乱婚
promiscuous exchange : 無差別な交換
promise : 期待、有望
promising avenue : 前途有望な研究分野
promoter : プロモーター、促進因子
prompt : 促す
prone to extinction : 絶滅する傾向
pronotal : 前胸背板の
pronotal collar : 前胸背襟部、襟首
pronotal flange : 前胸凸縁
pronotal lobe : 前胸背側片
pronotal suture : 前胸背板溝状線
pronotum : 前胸背板(ぜんきょうはいばん)
pronounced : 明白な、著しい
pronounced change : 顕著な変化
pronounced color preference : 著しい色彩選好、明確な色彩選好
pronymph : 前若虫
propagate outward : 外側方向に大きくなる、外側方向に増殖する
propagule : 繁殖体、散布体、無性芽
propensity : 傾向
prophenoloxidase-activating factor : プロフェノールオキシダーゼ活性化因子、フェノール酸化酵素前駆体活性化因子
propleuron : (複 . -ra)、前胸側板
propodeal : 前伸腹節の
propodeal carina : 前伸腹節隆起線
propodeum : 前伸腹節
proportion of empty patch : 空きパッチの割合
proportional weight loss : 比率的体重消失、比率的体重喪失
proposal : 提案書、提唱
prorsa form : プロルサ型(アカマダラ)
prosperity : 繁栄、好況、繁盛
prosperity stage : 繁栄の段階

prospero : プロスペロ遺伝子
prosternum : 前胸腹板(ぜんきょうふくばん)
protandry : 雄性先熟
protarsus : 前ふ節
protected area : 保護地域、保護区
protected species : 保護種
protection against baculovirus infection : バキュロウイルス感染に対する防御
protection toward baculovirus : バキュロウイルスに対する防御
protective color polymorphism : 保護色の色彩多型
protective coloration : 保護色
protective coloration effect : 保護色効果
protective effect : 防御的効果
protective envelope : かたい殻、保護外被
protective function : 防御機能
protective role : 防御的役割
protective scale : 保護鱗粉
protective wire fence : 保護用鉄線柵
protein : タンパク質
protein binding : タンパク質結合
protein biosynthesis : タンパク質生合成
protein crosslinker : タンパク質架橋
protein non-coding region : タンパク質非コード領域
protein ortholog group : タンパク質相同グループ
protein sequence : タンパク質のアミノ酸配列
protein sequence database : タンパク質配列データベース
protein synthesis : タンパク質合成
protein-coding gene : タンパク質コード遺伝子
protein-rich food plant : タンパク質の豊富な食餌植物
proteome : プロテオーム
proteomics : プロテオミクス
prothetely : プロセテリー、先成現象(体の一部が他より進んだ発生段階にある現象)、早熟変態
prothoracic gland : 前胸腺(ぜんきょうせん)
prothoracic gland hormone : 前胸腺ホルモン(ぜんきょうせんほるもん)、脱皮ホルモン(だっぴほるもん)、エクジソン
prothoracicotropic hormone : 前胸腺刺激ホルモン(PTTH)
prothorax : 前胸(ぜんきょう)
protibia : 前脛節
protist : 原生生物
protocol : プロトコール、実験プロトコル(実験の手順及び条件等について記述したもの)、実験の手順・観察(などの)記録
protogyny : 雌性先熟(しせいせんじゅく)(雌が雄より先行して羽化すること)
protogyny phenomenon : 雌性先熟現象
protozoon : (複 . -zoa)、原生動物
protrude : 突き出る
proventriculus : 前胃(ぜんい)
proviral form : プロウイルス型
proviral integrated circle sequence : プロウイルスが組み込まれたサークル(環)配列
provision : 条項、規定、支給
provocative : 刺激的な
provoke : 引き起こす、挑発する、誘起する
proximal : 近位、基部に近い、中央に近い、基部
proximal ablation : 基部側の切除
proximal band : 基部側の縞状バンド(内横線)
proximal cause : 近因、最も近い原因
proximal region : 基部領域、近位領域
proximal-distal axis : 近位 - 遠位軸、PD 軸、基部 - 末端軸、基部 - 外縁軸
proximate basis : 至近的基盤
proximate cause : 近因、至近的原因

proximate factor : 至近要因
proximate mechanism : 至近機構、近接的機構、近接的機序、至近要因(生理学的観点を対象)
proximate stimulus : (複. -li)、直接の刺激、主刺激、直前の刺激
proximity : 近接性
proximodistal axis : 近位 - 遠位軸、PD 軸、基部 - 末端軸、基部 - 外縁軸
proximodistal direction : 基部 - 外縁方向
proximodistally periodic pleat : 基部 - 外縁方向の蛇腹形式に折り畳んだひだ
proxy : 代用品、代理
pseudo- : 偽 -、仮 -、擬 - 〔ギリシャ語〕
pseudo-random number : 疑似乱数
pseudogene : 偽遺伝子
pseudogenization : 偽遺伝子化
pseudopupil : 偽瞳孔
Psocodea : カジリムシ目、咀顎目(従来はチャタテムシ目、シラミ目、ハジラミ目に細分されていた。)
psoralen : ソラレン、ソラーレン
pteridine pigment : プテリジン系色素
pteridine ring : プテリジン環
pterin : プテリン系色素
ptero- : 翅 - 〔ギリシャ語〕
pterostigma : 縁紋(えんもん)
pterothoracic ganglion : 翅胸部神経球、翅胸部神経節
pterothoracic pleuron : 有翅胸節側板、翅胸部側板
PTP : ポアソン系統樹過程、ポアソン過程に従って起きると仮定した系統樹(Poisson Tree Process)
PTSH : 前胸腺抑制ホルモン(ProThoracicoStatic Hormone)
PTTH : 前胸腺刺激ホルモン(ProThoracicoTropic Hormone)
pubescence : 軟毛、細短軟毛、微毛
public awareness : 普及啓発
public-domain software : パブリックドメインソフトウェア
publication : 公表
publish : 公表する、出版する
published work : 公表された著作物
puddle : 水たまり
puddle butterfly : 水たまりの蝶
puddler : 吸水動物
puddling : 吸水活動
puddling behavior : 吸水行動
pulvillus : (複 . -li)、褥盤(じょくばん)
pumping : ポンピング
punctate : 点刻、小斑点
pupa : (複 . -ae)、蛹(さなぎ)〔ラテン語〕
pupa collected in the field : 野外から採集された蛹
pupa period : 蛹期
pupa phase : 蛹期
pupa stage : 蛹期
pupa-diapausing species : 蛹休眠種
pupal coloration : 蛹の色彩化
pupal diapause : 蛹休眠
pupal duration : 蛹期間
pupal ecdysis : 蛹の脱皮
pupal epidermal sheet : 蛹の表皮層、蛹の表皮シート
pupal hindwing : 蛹の後翅、蛹期の後翅
pupal mating : 蛹態配偶行動
pupal melanization reducing factor : 蛹黒色化抑制因子
pupal mortality : 蛹死亡率
pupal overwintering : 蛹越冬
pupal period : 蛹期間
pupal weight : 蛹重、蛹の重さ
pupal wing : 蛹翅、蛹期の翅
pupal-cuticle melanizing hormone : 蛹クチクラ黒色化ホルモン
pupal-mate : 蛹態配偶行動

puparium : 蛹殻(ようかく)、囲蛹殻
pupate : 蛹になる、蛹化する
pupated individual : 蛹化した個体
pupation : 蛹化(ようか)
pupation of larva : 幼虫の蛹化
pupation rate : 蛹化率
pupation site : 蛹化場所
pupation timing : 蛹化時期
pure brood : 純粋同腹仔
pure cross : 純粋交配種
pure female : 純粋な雌、非雑種の雌
pure form : 純粋型
pure-species : 純粋種
purge : パージ、除去、処分、空にする
purified protein : 精製タンパク質
purify : 浄化する、精製する
purifying selection : 純化選択、純化淘汰、浄化選択
purine derivative : プリン誘導体
purine ring : プリン環
purple emperor butterfly : チョウセンコムラサキ
pursue : 追い求める
pursuit : 追跡
push-pull strategy : プッシュ・プル法
putative binding protein : 推定結合タンパク質
putative breakpoint : 推定ブレークポイント
putative diverse role : 推定される多様な役割
putative exon : 推定上のエクソン、推定されるエクソン
putative function : 推定機能
putative genome-wide level : 推定上のゲノム規模水準、推定上のゲノムワイドな水準
putative ORF : 推定 ORF、推定上の ORF、推定オープンリーディングフレーム(Open Reading Frame)
putative photoreceptor organelle : 光受容器の可能性をもつ小器官
putative species : 推定種
putative transcription factor : 推定転写制御因子
putatively distinct : 離れているらしい
putatively relate : 関係しているらしい
puzzle : 頭を悩ます、当惑する、困らせる
PWP : 側板翅突起、側翅突(Pleural Wing Process)
pyloric valve : 幽門弁
pyralid moth : メイガ科の蛾
pyramiding : ピラミッド形のもの、角錐
pyrosequencing : パイロシークエンシング、ピロシークエンシング(DNA ポリメラーゼによる伸長反応を基本にしたシークエンシング手法)
pyrrolizidine alkaloid : ピロリジジンアルカロイド

q

qPCR : 定量的 PCR(quantitative PCR)
qRT-PCR : 定量逆転写 PCR(quantitative reverse transcription PCR)
QTL : 量的形質遺伝子座、量的形質座位(Quantitative Trait Locus)
quadrat : コドラート、方形区、方形枠
quadrat method : コドラート法、区画法、方形区法
quadrat-based richness : 方形区数に基づく種数
quadrate wing : 角張った翅
quadratic term : 二次項
quadrifid : 四分裂の
quadripectinate : 4つに枝分かれした櫛歯状の、四裂片の櫛歯状の(蛾類の触角)
quadruped : 四足類、四肢類
qualitative measure of resource availability : 資源利用可能性の定性的尺度

qualitative substance : 質的物質
qualitatively different input information : 定性的に異なる入力情報
qualitatively different method : 定性的に異なる手法
qualitatively similar : 質的に同じ
quality change : 質的変化
quality criterion : (複 . -ria)、質的基準
quality filtering : クオリティーフィルタリング
quantal response : 計数的反応
quantifiable measure : 定量化可能な尺度、定量化可能な指標
quantification : 定量化
quantification of gene expression level : 遺伝子発現レベルの定量化
quantify : 定量化する
quantitative clock model : 量的測時モデル
quantitative method : 定量的手法
quantitative PCR : 定量的 PCR 法、定量 PCR
quantitative PCR analysis : 定量的 PCR 解析
quantitative prediction : 定量的予測
quantitative reverse transcription PCR : 定量的逆転写 PCR、定量 RT-PCR
quantitative RT-PCR Analysis : 定量的 RT-PCR 解析
quantitative substance : 量的物質
quantitative trait locus : 量的形質遺伝子座、量的形質座位
quantitatively inherited : 定量的に継承された
quantity : 多寡、多量、量、定量
quantity change : 量的変化
Quantum GIS : QGIS(地理情報システムのソフトウェアの一種)
quarantine : 検疫
quarantine measure : 検疫措置
quarantine regulation : 検疫規則
quarantine zone : 隔離地域
quarry : 採石場、石切場、源泉、種本(たねほん)
quartet : 四つ組、カルテット、四次の
quasi-national park : 国定公園
quaternary structure : 四次構造
Queen Alexandra's birdwing butterfly : アレキサンドラトリバネアゲハ
queen ant : 女王アリ、女王蟻
queen mandibular gland pheromone : ミツバチ女王蜂の大顎腺フェロモン
Queen of Spain fritillary butterfly : スペインヒョウモン
queen production : 女王(蜂)の産出、女王(蜂)の生産
query : クエリー、問い合わせ(る)
query sequence : 問い合わせ配列
quest : 探求、追求
quiescence : 静止、休止、活動停止
quinone : キノン系色素
quotient : 比、比率、割合、指数

r

R : 径脈(Radial vein)
r-selection : r- 淘汰、r 選択
r-species : r 種(r 戦略をとる種)
r-strategy species : r 戦略種
Rab geranylgeranyl transferase gene : Rab ゲラニルゲラニル転移酵素遺伝子
rabble : 群れ
race : 系統、レース、品種、人種、亜種
race formation : レース形成、系統形成
racial diversification : 系統の多様性
racial evolution : 系統を生む進化
RAD : 制限酵素認識部位に関わる DNA (Restriction-site Associated DNA)
RAD analysis : RAD 解析
RAD resequencing : RAD 再配列決定
radar : レーダー(装置)、電波探知機
radial : 径脈の、放射状の

radial crossvein : 径横脈、r 脈
radial sector : 径分脈、Rs 脈
radial vein : 径脈(けいみゃく)、R 脈
radial-medial crossvein : 径中横脈、r-m 脈、前横脈
radiant heat : 輻射熱
radiate : 照射する、放散する
radiation : 放射線、放散
radiation of metazoan : 後生動物の放散
radicle : 小根、幼根
radioimmunoassay : ラジオイムノアッセイ、放射(線)免疫測定法(RIA)
radius : 径分脈、径脈、Rs 脈〔ラテン語〕
radius increase : 半径の増分
rainfall : 降雨、降雨量
rainforest : 雨林、降雨林
rainy : 雨の多い
rainy season : 雨期
raise : 飼育する、育てる
raising awareness : 意識啓発
Rajah Brooke's birdwing butterfly : アカエリトリバネアゲハ
ramet : ラメート、ラミート(同じ個体から無性的に繁殖させた個体群の一員)
ramped response : ランプ応答
Ramsar Convention : ラムサール条約
random amplified polymorphic DNA : ランダム増幅多型DNA法、RAPD法(RAPD)
random direction : ランダム方向
random drift : ランダム浮動、機会的浮動
random genetic drift : ランダムな遺伝的浮動
random hexamer primer : ランダムヘキサマープライマー
random mating : ランダム交配、任意交配
random order : ランダム順、無作為順序
random ordering : ランダム順、順不同
random placement curve : ランダム配置曲線
random slope model : ランダム傾きを加味したモデル
random walk : ランダム歩行、ランダムウォーク
randomized sample accumulation curve : ランダムなサンプルの累積曲線
randomly captured individual : ランダムに捕獲した個体
randomly generated landscape : ランダムに生成された景観
range : 生息域、範囲
range contraction : 生息域の縮小
range expansion : 生息域の拡大
range fragmentation : 範囲の断片化、範囲の分断化
range margin : 範囲境界、範囲の端部
range of latency : 潜時帯
range of variation : 変異の幅、変異幅
range residency : 生息圏内の定住性
range resident : 生息地内の定住者
rank : (分類)階級、順位
RAPD : ランダム増幅多型DNA法、RAPD法(Random Amplified Polymorphic DNA)
rape seed : セイヨウアブラナの種子
rapid amplification of cDNA ends : RACE法
rapid cold hardening : 急速低温耐性、急速低温馴化(RCH)
rapid desiccation : 急速乾燥
rapid flight : 素早く飛び回る、急速飛翔
rapid membrane vibration : 膜の高速振動
rapid-assessment survey : 迅速評価調査
rapidly radiating clade : 急速に放散している単系統
rare : 希少
rare butterfly : 珍蝶
rare example : 希少な例、稀な例
rare hybrid : 希少雑種、希少交配種
rare species : 稀種、希少種
rare taxa : 希少な分類群
rare, local, and little known butterfly from Japan : 日本の秘蝶

rarefaction : 希薄化
rarefaction curve : 希薄化曲線、希薄曲線
rarefaction formula : 希薄化の計算式
rarefaction method : 希薄化法
rarefied : 希薄な
rarity : 希少性
rash : 発疹、かぶれること
rate heterogeneity : 置換速度の不均質性
rate of development : 発育の比率、発育速度
rate of food consumption : 食料消費率
rate of JH degradation : 幼若ホルモン(JH)の分解速度
rate of match/mismatch : 一致／不一致率
rate of migration : 移動比、移動率
rate of single-nucleotide variation : 一塩基変異率
rate of the mismatch : 不一致率
ratio of richness to area : 種数対面積の比
ratio of species/individual : 種数／個体数の比
ratio of successful emergence : 羽化に成功した割合、羽化の成功割合
rattling noise : カラカラ音
raw datum : (複 . -ta)、生データ
rayed form : 放射状型
RBSDV : イネ黒すじ萎縮ウイルス、イネ黒すじ萎縮病(Rice Black Streaked Dwarf Virus)
RCH : 急速低温耐性、急速低温馴化(Rapid Cold Hardening)
rDNA : リボソーム DNA(ribosomal DNA)
re- : 再 -〔ラテン語〕
re-analysis : 再解析、再分析
re-colonise : 再移住する、再コロニー化する
re-enforcement : 再増強
re-establishment : 再定着
re-introduction : 再導入
re-sampling : 再サンプリング、リサンプリング、非復元抽出
re-scale : リスケールする、大きさを変更する
re-scaling : 再スケール化
re-sight : 再度見つける
re-survey : 再調査
reaction norm : 反応基準、反応規準
reaction-diffusion model of morphogenesis : 紋様形成の反応拡散モデル
reactive oxygen species : 活性酸素種(ROS)
read : リード(DNA 分子配列を読むこと)
Reakirt's blue butterfly : メスキートヒメシジミ
real mating difference : 交尾の実際の相違
real-time fluorescent quantitative PCR technique : リアルタイム蛍光定量的 PCR 法
real-time tracking : 実時間追跡
realized heritability : 実現遺伝率
realized niche : 実現ニッチ
realized tree : 実現系統樹
reappraisal : 再評価
rear : 飼育する
reared male : 飼育された雄
rearing : 飼育
rearing environment : 飼育環境
rearing experiment : 飼育実験
rearing temperature : 飼育温度
reasonable explanation : 妥当な説明、理にかなった説明
received : 受け取った
receiver plant : 受容植物
recent advance : 最近の進歩
recent advancement : 最近の進歩
recent finding : 最近の発見
recent species : 現生種、現棲種
recent survey : 最近の調査
recently described subspecies : 最近記載された亜種
receptaculum seminis : 貯精のう、受精のう

receptive : 感覚の、受容の
receptive female : 交尾できる状態の雌
receptivity period : 受容期間
receptor gene : 受容体遺伝子
receptor population : 受入側個体群、受容体個体群
receptor site : 受入側生息地
recessive : 劣性の、劣勢形質(潜性形質)、不顕性
recessive detrimental allele : 劣性有害対立遺伝子
recessive detrimental gene : 劣性有害遺伝子
recessive lethal allele : 劣性致死対立遺伝子
recessive lethal gene : 劣性致死遺伝子
recessive mutant : 劣性突然変異体
reciprocal cross : 正逆交配、相反交雑、相互交雑
reciprocal F1 progeny : 正逆交配の F1 子孫
reciprocal F1 type : 正逆交配の F1 型
reciprocal gene loss : 相互遺伝子消失、相互遺伝子喪失、相互遺伝子欠損
reciprocal influence : 相互作用
reciprocal pattern : 相互パターン、正逆パターン
reciprocal test : 相互試験、相互検定
reciprocal transplant experiment : 相互移植実験
reclivous : 外斜
recognizable adult morphogenesis : 認識可能な成虫形態形成
recognized sub-species : 認められた亜種
recollection : 想起、回想
recolonization : 再移入、再定着
recolonize : 再移入する
recombinant baculovirus : 組み換えバキュロウイルス
recombinant phenotype : 組み換え表現型
recombinant protein : 組み換えタンパク質
recombinant virus : 組み換えウイルス
recombination : 組み換え
recombination between color and preference : 色と選好の間の組み換え
recombination breakpoint : 組み換え(の)切断点
recombination distance : 組み換え距離
recombination mapping : 組み換えマッピング
recombination process : 組み換え過程
recombination rate : 組換え率
Recommendation : 勧告
reconstruct : 再構成する、再構築する
reconstructed amino acid : 再構成されたアミノ酸
reconstructed flight path : 再構成された飛翔経路
reconstruction of nucleotide sequence : 塩基配列の再構成
reconstruction of organ : 器官の再構成
record : 記録、採集記録
recorded neural activity : 記録された神経活動
recorder home range : 記録者の行動圏
recording electrode : 記録(用)電極
recording intensity : 記録強度
recreational purpose : レクリエーションの目的、保健休養の目的
recruit : 動員する
recruitment : 動員、漸増(ぜんぞう)、加入、リクルートメント、補充
rectangle : 長方形
rectum : 直腸
recur : 繰り返す、立ち戻る
recurrence : 再出現、再発
recurrent pitfall : 頻発する落し穴
red admiral butterfly : ヨーロッパアカタテハ
red data book : レッドデータブック(レッドリストに載せた生物を紹介した本)
red dot : 赤い丸、赤い点

red lacewing butterfly : ビブリスハレギチョウ
red light : 赤色光
red list : レッドリスト(絶滅危惧種の目録)
red list category : レッドリスト区分、レッドリスト部門(レッドリストの段階分け)
red list criterium : レッドリスト基準(レッドリストの段階分けの基準)
red patch : 赤い大斑点
red pigment-concentrating hormone : 赤色色素濃縮ホルモン
red proboscis : 赤色の口吻
red queen hypothesis : 赤の女王仮説
red ray : 赤い放射状線
red spotted pattern : 赤いまだら模様
red-green color vision : 赤緑色覚
Red-spot Baron butterfly : ベニボシイナズマ
reddish brown pupa : 赤褐色の蛹
redeploy : 配置換えをする、別の目的で用いる
redirect : 向け直す
rediscovery : 再発見
redrawn : 描き直す
reduced carbon compound : 希釈炭素化合物
reduced effort scientific method : 軽減効果のある科学的手法
reduced expression : 低下した発現
reduced fecundity : 低下した生殖能力
reduced interspecific gene flow : 種間の遺伝子浸透を減少させる
reduced major axis regression : RMA 回帰、幾何平均回帰(最小二乗法を使わない簡単な直線回帰)
reduced nicotinamide adenine dinucleotide : ニコチンアミドアデニンジヌクレオチド還元型
reduced preference : 弱められた選好性
reduction : 縮小(図)、還元
reduction of posterior eyespot : 後方(の)眼状紋の縮小
redundant contig : 冗長コンティグ
redundant copy : 冗長複製
reduplicated produced leg : 過剰肢
reedit : 再編集する
reexamine : 再検討する、再調査する
referee : 査読者、審査員、レフェリー
refereeing : 審査
reference : 引用文献、参照
reference gene : リファレンス遺伝子、対照遺伝子
reference genome : 基準ゲノム、参照ゲノム、標準ゲノム
reference pressure : 基準圧、基準圧力
reference sequence : 参照配列、基準配列、リファレンスとなるべき配列データベース
reference spectrum : 基準スペクトル
reference ultrametric phylogenetic tree : 参照超計量系統樹、基準超計量系統樹
refining : 精製
reflectance basking : 開翅日光浴
reflectance spectrum of eyeshine : 暗視眼の反射スペクトル
reflected light : 反射光
refractory period : 無反応期、不応期
refrigerate : 冷却する、冷やす
refuge area : 保護区、避難区域
refuge population : 保護地の個体群
refuge site : 保護地、避難場所、生息地
refugia : レフュジア、待避地、待避場所
refute : 間違いを証明する、否定する、論破する
regain : 回復する
regal fritillary butterfly : イダリアギンボシヒョウモン
regenerating limb : 再生肢(さいせいし)
regeneration : 再生
region of reduced recombination : 組み換えを減少させる領域
region sequence : 領域シークエンス

regional biodiversity survey : 地域の生物多様性調査
regional biogeography : 区系生物地理学
regional bond : 地縁的
regional characteristic : 地縁性
regional diversity : 地域多様性
regional fauna : 地域動物相
regional form : 地理型
regionally extinct : 地域的に絶滅した
regionally rare alternative host plant : 地域的にまれな代替寄主植物
register : 登録する、記録する、記載する
regression : 退化、退行、回帰式
regression analysis : 回帰分析
regression coefficient : 回帰係数
regression equation : 回帰式、回帰方程式
regression line : 回帰直線
regression model : 回帰モデル
regular host : 正規の宿主
regular interval : 規則的な時間間隔
regularity : 規則性
regulation : 調節
regulation of innate immunity : 自然免疫調節
regulatory biosynthesis gene : 生合成制御遺伝子
regulatory element expansion : 調節要素の拡大
regulatory gene : 調節遺伝子
regulatory mechanism : 調節機構
regulatory muscle : 調節筋、制御筋
regulatory region : 調節領域
regulatory sequence : 調節配列、制御配列
regulatory signal : 調節信号
regulatory switch gene: 調節スイッチ遺伝子
regurgitation : 吐き戻し
rehydrate : 再水和する
reinforce : 強化する、補強する
reinforcement : 強化
reinsemination : 再受精
reinstate : 復権
reintroduce : 再導入する
reintroduction : 再導入
reject : 拒否する、却下、リジェクト
rejected courtship : 求愛拒否
rejected name : 拒否名
rejected work : 拒否された著作物
rejection behavior : 拒絶行動
related compound : 類縁化合物
related genera : 近縁属
related species : 類縁種、近縁種
relatedness : 近縁性、血縁度
relatedness asymmetry : 血縁度非対称性
relatedness of individuals : 個体間の近縁度、個体間の血縁度
relational : 類縁関係的
relative : 相対的に
relative abundance : 相対的存在量、相対的な豊富さ
relative abundance distribution : 相対的個体数分布
relative amount of niche space : ニッチ空間の相対量
relative amount of resource : 資源の相対量
relative biomass : 相対的バイオマス、相対的生物量、相対資源量
relative direction : 相対方向
relative efficacy : 相対的有効性
relative growth rate : 相対成長速度、相対的成長率(RGR)
relative humidity : 相対湿度(RH)(その温度における飽和水蒸気量を100%として表した空気中の水分量のこと)
relative inactivity : 相対的不活性
relative investment : 相対的投資量
relative light intensity : 相対照度
relative mating probability : 相対的交尾確率
relative niche space control : 相対的ニッチ空間の制御

relative position : 相対位置
relative probability : 相対確率
relative quantification : 相対的定量
relative rate test : 相対速度テスト
relative resource abundance : 相対的な資源量、資源の相対量
relative scale : 相対尺度
relative sensitivity : 相対的感受性
relative species abundance : 相対種個体数(個体数と種数の関係)
relative term : 相対的な用語
relatively few : 相対的に少ない、比較的少ない
relatively low temperature : 比較的低い温度
relatively puny insect : 比較的弱々しい昆虫
relay of short-range signal : 短距離シグナルの中継
release : 放逐(ほうちく)、リリース、放飼
release experiment : リリース実験、放逐実験
release of ecdysteroid hormone : エクジステロイドホルモンの放出
release of latent heat concomitant with freezing : 冷凍で付随して発生する潜熱の放出
releaser : 解発因、リリーサー、レリーサー、解発因子
releaser pheromone : 解発フェロモン、リリーサーフェロモン
releasing area : リリースした地区、放逐区域
releasing butterfly : 放蝶
releasing point : リリース地点、放逐地点
relevance : 関連性、妥当性
reliability of inferred tree : 推定された系統樹の信頼性
reliable cue : 信頼できる刺激、信頼できる手がかり
reliable identification : 信頼性のある同定、信頼できる同定
relic : 名残り、遺存
relic distribution : 遺存分布
relic endemism : 遺存的固有
relict : 遺存種、残存種
relictual butterfly : 遺存種の蝶
relief pattern : レリーフパターン
relieve population pressure : 個体群圧の緩和
remaining tree : 残りの系統樹
remarkable butterfly : 興味深い蝶
remarkable butterfly species : 注目すべき蝶種
remarkable clinal variation : 注目すべきクライン(的)変異
remigium : 主域(「vannus〈扇域〉」を参照)
reminiscence : 追憶、回想
remnant population : 残存個体群
remnants : 残存物、残(留)物、残余
remote island : 離島
remote location : 遠隔地
remote sensing technique : リモートセンシング技法
removable gauze cloth : 取り外し可能なガーゼ布
removal of brain : 除脳、脳除去
removal of identical haplotype : 同一のハプロタイプの除去
remove : 取り除く、除去する
rendezvous site : ランデブーサイト、面会場所
reniform spot : 腎状紋(じんじょうもん)(蛾類の前翅にある紋様)
Rensch's rule : レンシュの法則、レンシュ則(雄と雌の体サイズで、雄の方が大きい時は、そのサイズ差は広がり、雌の方が大きい時は、そのサイズ差は縮まる)
repeated hibernation : 繰り返される越冬
repeated involvement : 反復関与
repeated re-sampling : 反復リサンプリング
repeated recruitment : 反復補充
repeated recruitment of allele : 対立遺伝子の反復補充
repel : 追い払う、寄せ付けない、撃退する

repellent volatile : 忌避剤の揮発成分
repetitive element : 反復要素
repetitive sequence : 反復配列
replacement name : 置換名
replacement reproductive : 置換生殖虫
replacement substitution : 変化置換、置換／代用
replanting : 移植
replenish : 補給する、補充する
replenishment of protein : タンパク質の補給
replicate : 反復実験、レプリケート、反復、再現実験
replicated plant : 重複植物
replicated sample : 重複されたサンプル
replication : 複製
repopulate : 再び居住させる
report on the butterflies from Himalaya : ヒマラヤ蝶紀行、ヒマラヤの蝶類レポート
representative component of biodiversity : 生物多様性の代表的な構成要素
representative image : 代表画像、代表的な画像
representative locus : 代表的な遺伝子座、典型的な遺伝子座
representative specimen : 典型的な標本
representative trace : 代表的なトレース、典型的な痕跡(こんせき)
representative tree topology : 典型的な系統樹形
repress : 抑制する、抑える
repression of recombination : 組換え抑制
reprint : 別刷り
reproduce : 繁殖する、再生する、再現する
reproduce vegetatively : 栄養生殖を行う
reproduction : 繁殖、生殖、増殖
reproduction period : 繁殖期間
reproductive aberration : 生殖異常
reproductive ability : 繁殖能力
reproductive activity : 生殖活動
reproductive behavior : 生殖行動
reproductive capacity : 繁殖力
reproductive capacity of adult : 成虫の繁殖能力、成虫の生殖能力
reproductive caste : 生殖カースト
reproductive control : 生殖制御による防除
reproductive diapause : 生殖休眠
reproductive dormancy : 生殖休眠
reproductive efficiency : 増殖効率
reproductive effort : 繁殖努力
reproductive fairness : 繁殖の公平さ
reproductive fitness : 繁殖適応度
reproductive history of adult : 成虫の繁殖史
reproductive interference : 繁殖干渉
reproductive investment : 繁殖投資、生殖投資
reproductive isolation : 生殖(的)隔離、繁殖隔離
reproductive manipulation : 生殖操作
reproductive organ : 生殖器官
reproductive phenomenon : 生殖現象
reproductive phenotype : 生殖表現型
reproductive potential : 生殖能、繁殖潜在力
reproductive rate : 増殖率
reproductive skew : 繁殖の偏り
reproductive strategy : 繁殖戦略
reproductive success : 繁殖成功度
reproductive system : 生殖系
reproductive tract : 生殖器官
reproductive tract complex : 生殖器官複合体
reproductive tract development : 生殖器官の発育
reproductive tract tissue : 生殖器官組織
reproductive value : 繁殖価
reptile : 爬(は)虫類
Republic of Guyana : ガイアナ(共和国)
repugnatorial gland : 防御腺(ぼうぎょせん)、忌避腺
repulsion : 撃退

repulsive activity : 反発行動
reputable dealer : 信頼できる業者
required heat unit : 有効積算温量
rescue effect : 救済効果、レスキュー効果
rescue translocation : 危険から救う移動
rescuing function : 救済機能
research : 調査
research journey : 調査旅行
research trip : 探訪旅行
resemble : 似ている
resequencing : リシークエンシング
reside : 住む、在住する
residence : 定住(性)
resident : 先住者、居住者
resident dry season form individual : 居住している乾季型の個体
resident male : 先住者の雄、留蝶の雄、居住の雄
resident range : 定住圏
resident species : 定住種、定棲性種
resident strategy : 定住戦略
residential area : 居住地域、居住区
residual : 残差
residual DNA : 残存 DNA、残留 DNA、残差 DNA
residual reproductive value : 残存繁殖価、残存生殖価
residual term : 残余項、残差項
residual value : 残差値
residue : 残留物
resilin : レシリン
resistance : 耐性、抵抗性
resistance against baculovirus : バキュロウイルスに対する抵抗性、バキュロウイルスに対する耐性
resistance mechanism : 抵抗(性)機構
resistance to parasite : 寄生者に対する抵抗性
resistant : 抵抗する、抵抗力がある
resistant host : 抵抗力のある宿主、耐性のある宿主
resolution of mapping : マッピング分解能
resonance effect : 共鳴効果
resource : 資源、原料
resource abundance : 資源量
resource availability : 資源の有用性、資源の利用可能性
resource budget : 原料の収入分
resource defense polygyny : 資源防衛型の一夫多妻
resource density : 資源密度
resource depletion : 資源減少、資源枯渇
resource distribution : 資源分布
resource dynamics : 資源動態
resource environment : 資源環境
resource gradient : 資源勾配
resource landscape : 資源景観
resource patch : 資源パッチ
resource planning : 資源計画
resource selection : 資源選択
resource similarity : 資源の類似性、資源の類似度
resource-holding power : 資源保持力
respective co-mimics : 各々の相互擬態種
respiration : 呼吸、呼吸作用
respiratory system : 呼吸系
respiratory volume : 呼吸量
respond physiologically : 生理的に反応する
response of thermoperiod : 温度周期反応
response pattern : 反応様式
response property : 応答特性、反応特性
response rate : 反応率
response system : 反応系
response threshold : 応答閾(いき)、応答閾値
response to current adversity : 逆境に対する反応
response to pathogen : 病原体に対する反応
response to selection : 選択に対する反応

response variable : 応答変数、反応変数
responsive : 反応する、応答的な
rest : 休む
resting point : 静止位置
resting posture : 休止姿勢
resting style : 静止姿勢
restoration : 回復、復帰
restrain : 抑制する
restraint need : 抑制的な欲求
restrict : 限定する
restriction enzyme : 制限酵素
restriction fragment length polymorphism : 制限酵素断片長多型、制限断片長多型(RFLP)
restriction site : 制限(酵素切断)部位
restriction-site associated DNA : 制限酵素認識部位に関わるDNA、RAD法(制限酵素が認識するサイトの近傍にある塩基配列のみを対象する解析)
restriction-site associated DNA sequencing : RAD-seq法(制限部位に関連したDNA塩基配列法)
restrictive climatic condition : 限定された気象条件
result : 結果
resultant intense competition : 結果として生じる激しい競争
resultant measurement : 計測結果、測定結果
resulted tree : 結果としての系統樹
resurgence : 誘導多発生、リサージェンス
retain : 保持する
retard : 妨げる、遅らせる
retardation : 遅延
retention of duplicate : 重複保持
reticulate : 網(目)状、網状の斑点
reticulation : 網目(構造)、網目組織
retina : 網膜〔ラテン語〕
retinaculum : 抱鉤(ほうこう)、保体(蛾の飛行中に前後翅を結びつける鉤)
retinal photoisomerase : 網膜の光異性化酵素、レチナール光異性化酵素
retinal photoreceptor : 網膜光受容器
retinal regionalization : 網膜の領域化、網膜部域性、レチナール領域化
retractile tentacular organ : 伸縮自在の触手状器官
retro- : 逆-、退-〔ラテン語〕
retrograde fill : 逆行性充填
retrogressive molt : 退行脱皮
retroposon : レトロポゾン
retrotransposon-associated coding region : レトロトランスポゾン関連コーディング領域
retrovirus receptor gene : レトロウイルス受容体遺伝子、レトロウイルスレセプター遺伝子
return : 帰還する、戻る
return flight : 帰還飛翔、帰還移動
return-migration : 帰還移動、帰還移住、戻り移動
reveal : 明らかにする
reverse complement : 逆相補(体)
reverse gene flow : 逆転遺伝子流動
reverse genetics : 逆遺伝学
reverse orientation : 逆向きの方向、逆方向
reverse question : 逆質問
reverse transcriptase : 逆転写酵素
reverse transcription : 逆転写、逆転写反応
reverse transcription polymerase chain reaction : RT-PCR法、逆転写ポリメラーゼ連鎖反応
reverse-transcribed : 逆転写された
reversible : 可逆的
review : 査読、総説
reviewer : 査読者
revise : 訂正
revised edition : 改訂版
revision : 改訂(版)、再検討

revisional note : 補足、改正された知見
revisit : 再訪
revolutionize : 大変革を起こす、一大進歩をもたらす
rewarding plant : 有益な植物
RFLP : 制限酵素断片長多型、制限断片長多型（Restriction Fragment Length Polymorphism）
rG : 遺伝相関（Genetic correlation）（雑種集団の表現型の間の相関のうちで、遺伝子型に基づいて起こるものをいう）
RGR : 相対成長速度、相対成長率（Relative Growth Rate）
RH : 相対湿度（Relative Humidity）
rhabdom : 感桿（かんかん）（個眼内の光を通す円柱構造）
rhabdom waveguide : 感桿の光導波路
rhabdome : 感桿
rhabdomeric opsin : 感桿型オプシン
rhabdomeric photoreceptor cell : 感桿型光受容細胞
rhabdomeric-type : 感桿型
Rhamnaceae : クロウメモドキ科
rhinoceros beetle : サイカブト
rhizome : 根茎
rhodommatin : ロードマチン
rhodopsin type : ロドプシンタイプ
rhodopsin type receptor : ロドプシンタイプ受容体
rhodopterin : ロドプテリン
Rhopalocera : 蝶（チョウ）のこと（棍棒状の触角）
RI : 繁殖干渉（Reproductive Interference）
RIA : ラジオイムノアッセイ、放射（線）免疫測定法（RadioImmunoAssay）
ribosomal gene : リボソーム遺伝子
ribosomal protein : リボソームタンパク質
ribosomal RNA : リボソーム RNA（rRNA）
ribosome : リボソーム
Rice black streaked dwarf virus : イネ黒すじ萎縮ウイルス、イネ黒すじ萎縮病
rice brown planthopper : トビイロウンカ
rice field : 水田、米畑
rice paddy field : 米の水田、水田、稲田
rice paper butterfly : オオゴマダラ
rice plant : イネ、稲
rice skipper butterfly : イチモンジセセリ
Rice stripe virus : イネ縞葉枯ウイルス、イネ縞葉枯病
rice volatile : イネ揮発性物質
rice-stem borer : ニカメイチュウ（「ニカメイガ」の幼虫時代の名称）（ニカメイガ）
rich biodiversity : 豊富な生物多様性
richness comparison : 種数比較
richness estimation : 種数推定
richness estimator : 種数推定量
richness measure : 種数指数
richness per quadrat : 方形区当りの種数
Rickettsia : リケッチア
ridge : 隆起、縦隆起、突起、稜線
ridge of morphogen : モルフォゲンの隆起線
right tympanal membrane, the : 右鼓膜
ring : 環
ring gland : 環状腺
ring pattern : 環パターン
ringed : リング状の
ringed spot : 縁どりのある斑紋
rinse : すすぎ落とす
Rio Declaration : リオ宣言
riparian forest : 渓畔林、河畔林
ripe fruit : 熟した果実
ripeness : 成熟
ripple pattern : さざ波状パターン
rise/fall : 立ち上がり／立ち下がり
rising winter temperature : 冬期の気温の上昇
risk analysis : リスク分析
risk assessment : リスク評価

risk extinction : 絶滅の危機にさらす
risk management : リスク管理、危機管理
rival : 競争相手、ライバル
river bank : 川の土手
river basin : 河川流域
riverine forest type : 川辺林型
riverside : 流域
RMA regression : RMA 回帰、幾何平均回帰(Reduced Major Axis)
RMA slope : RMA傾き(Reduced Major Axis)
RNA interference phenotype : RNA 干渉の表現型、RNAi 表現型
RNA polymerase : RNA ポリメラーゼ
RNA processing : RNA プロセシング
RNA-seq : RNA-seq 解析、RNA 塩基配列解読(RNA-sequencing)
RNAi : RNA 干渉法、RNA 干渉(RNA interference)
RNAzol reagent : RNAzol 試薬(RNA を単離・抽出する試薬)
road kill : ロードキル、道路で車にひき殺された動物の死体
road side plantation : 路傍の植込み
roadside : 路傍
roadside butterfly : 道ばたの蝶
robust : がっしりした、頑丈な
robust correlation : 強固な相関
robust phylogenetic framework : 強固な系統構造、強固な系統発生学的フレームワーク
robust phylogenetic tree : 堅固な系統樹
robust pupal exoskeleton : 強固な蛹の外骨格
robust-bodied butterfly : 頑丈な体を持った蝶
rocky area : 岩場
rolled leaf : 丸まった葉
room condition : 室内条件
roost : とまり木
root : 根
rooted tree : 有根系統樹
ROS : 活性酸素種(Reactive Oxygen Species)
rostal : 吻側
rotten fruit : 腐った果実
rotting fruit : 腐った果物
rough surface : 粗粒面
rough-surfaced endoplasmic reticulum : 粗面小胞体
round flight : 往復飛翔
round Sea of Japan area : 周日本海地域
rounded : 丸みをおびた
rRNA : リボソーム RNA(ribosomal RNA)
RSV : イネ縞葉枯ウイルス、イネ縞葉枯病(Rice Stripe Virus)
RT : 逆転写(Reverse Transcription)
RT-PCR : 逆転写ポリメラーゼ連鎖反応(Reverse Transcription Polymerase Chain Reaction)
RT-PCR-amplified allele : RT-PCR 法により増殖させた対立遺伝子
rudiment : 痕跡的(なもの)、原基
rudiment of an eyespot : 目玉模様の痕跡
rule : 条項
rule, as a : 決まって、いつも
ruling by the Commission : 審議会による裁定
run anteriorly : 前方に走行する
run transversely : 横断的に延びている
run trial : 試行する
runaway process : ランナウェイ過程
runaway selection : ランナウェイ淘汰
rural area : 田園地域
Rutaceae : ミカン科
rutaceous plant : ミカン科植物
rybozyme : リボザイム、リボ酵素
Ryukyu Islands : 琉球列島

S

s. l. : 広義の、広い意味で(sensu lato)、「例

えば、『属名 s. l.』（広義の XXX 属）の形式で表記」、「[例] *Callophrys* s.l.（広義のコツバメ属）」〔ラテン語〕
s. s.：狭義の、厳密な意味で（sensu stricto）、「例えば、『属名 s. s.』（狭義の XXX 属）の形式で表記」、「[例] *Papilio* s.s.（狭義のアゲハチョウ属）」〔ラテン語〕
s. str.：狭義の、厳密な意味で（sensu stricto）、「例えば、『属名 s. str.』（狭義の XXX 属）の形式で表記」、「[例] *Favonius* s. str.（狭義のオオミドリシジミ属）」〔ラテン語〕
SA：サリチル酸（Salicylic Acid）（信号伝達物質の一種）
SA-dependent egg-induced response：SA 依存性卵誘導（性）応答、サリチル酸依存（性）卵誘導性反応（Salicylic Acid）
SA-dependent pathway：SA 依存性生合成経路、SA 依存性生産経路、サリチル酸依存性生合成経路（Salicylic Acid）
SA-responsive gene：SA 応答性遺伝子、サリチル酸応答性遺伝子（Salicylic Acid）
saccharide：糖類
Saccharomyces cerevisiae：パン酵母、出芽酵母
sacculus：小のう、小胞〔ラテン語〕
saccus：（交尾器の）胞のう、中胞〔ラテン語〕
saddle marking：鞍状の模様
safe roost：安全な止まり木、安全なねぐら
safety band：安全帯
sagittal plane：矢状断面（しじょうだんめん）、矢状面
sagittal section：矢状断面
sagittal surface：矢状断面
SAI：性非対称な近親交配（Sex-Asymmetric Inbreeding）
Saipan：サイパン
Sakhalin：サハリン
Sakhalin Island：サハリン島
sal：スパルト遺伝子（*spalt*）
salicylic (SA) pathway-marker：サリチル酸（SA）の生合成経路マーカー
salicylic acid：サリチル酸（SA）（信号伝達物質の一種）
saline：食塩水、塩類
saline intake：ナトリウム塩の摂取、塩分摂取
salivary gland：唾（液）腺（だえきせん）
same orientation：同一方向、同方向
same place：同一地方、同一場所
same relative position：同一相対位置
same species and distinct subspecies：同種異亜種
Samoan Archipelago：サモア諸島
sample heterogeneity：サンプルの不均質性、サンプルの異種混交性
sample site：標本採集地、サンプル採集地、試料採集地
sample-based curve：サンプル数に基づく曲線
sample-based dataset：サンプル数に基づくデータセット
sample-based protocol：サンプル数に基づくプロトコール
sample-size dependence：サンプルサイズ依存性
sampled site：サンプル採集地
samples：サンプル数
sampling curve：サンプリング曲線
sampling device：サンプリング方法
sampling effort：サンプリング努力、サンプリングの手数、サンプリング調査、採取効果
sampling issue：サンプリング問題点
sampling period：採集期間、採取期間
sampling point：採集地点、採取地点
sampling protocol：サンプリングプロトコール
sampling rate：標本抽出率
sampling time：採集時間、採取時間

sampling unit : サンプリング単位
sandwich : 差し込む、間に入れる
sandy area : 砂漠地帯
sap : 樹液
sapling : 若木
sapling dataset : 若木データセット
sapling species richness : 若木の種数
saprophagous insect : 腐食性昆虫
saprophagy : 腐食性
sarpedobilin : サルペドビリン、サーペドビリン
SAS : 統計解析用ソフトウェアの名称（Statistical Analysis System）
satin : サテン地
satoyama : 里山
saturation of local community : 局所群集の飽和
saturation test : 飽和度検査
Saturn butterfly : ルリオビトガリバワモンチョウ
Satyr butterfly : ジャノメチョウ（亜）科の蝶
satyrine butterfly : ジャノメチョウ（亜）科の蝶
satyrine tribe : ジャノメチョウ族
Satyrs and Wood-Nymphs : ジャノメチョウ亜科
savanna : サバンナ
savanna climate : サバンナ気候
saw-tooth cline : 鋸歯状のクライン
sawfly : ハバチ
Sc : 亜前縁脈（Sub coastal vein）
scaffold : 足場、骨組
scaffold position : 足場位置
scale : 鱗粉、鱗毛、鱗片
scale arrangement : 鱗粉の配列、鱗紛列
scale bar : 目盛り棒、スケールバー
scale cell : 鱗粉細胞
scale morphology : 鱗粉の形態
scale mother cell : 鱗粉の母細胞
scale of interest : 関心度、関心度尺度
scale of space and time : 時空間スケール、時空間尺度
scale row : 鱗粉の列
scale row ring : 鱗粉列リング
scale-forming cell : 鱗粉形成細胞
scalloped edge : 凸凹のある縁
scalloping : 凸凹のある
scalloping flight : 扇形飛行
scan sampling method : スキャンサンプリング法、走査サンプリング法（複数の観察対象個体を、一定間隔ですばやく見渡し、各個体のその瞬間の行動を記録する方法）
scanning electron microscope : 走査（型）電子顕微鏡
scanning electron microscopy : 走査（型）電子顕微鏡検査、走査（型）電子顕微鏡法
scanning harmonic radar : 走査型高調波レーダー
scanning laser vibrometry : 走査型レーザー振動計
scanning-electronmicrograph : 走査型電子顕微鏡像
scape : 柄節（へいせつ）
scaphium (sca) : スカヒュウム、竜骨
scar : 跡、傷跡
scarab beetle : スカラベ、タマオシコガネ（コガネムシの仲間）
scarce : 珍しい、まれな
scarce element : 希少成分
scarce species : 希少種
scarcity : まれなこと、希少
scare : 驚かす
scarlet : 緋色（ひいろ）、深紅色
scat : 糞
scattered region : 散在領域、点在領域
scattered woodland : 点在している森林地帯
scavenger : 腐肉食（性）動物、清掃動物
scent gland : 発香腺

scent horn : 臭角
scent marking behavior : 匂い付け行動
scent organ : 発香器官
scent patch : 性紋、性標
scent sack : 香のう
scent scale : 香鱗、発香鱗
schematic diagram : 体制模式図
schematic illustration : 模式図、図解図、概略図、体制模式図
schematic of ommatidium : 個眼の模式図
schematic representation : 模式図
schematic view : 概略図、模式図
scheme of society : 社会機構
Schiff base : シッフ塩基
Schizosaccharomyces pombe : 分裂酵母
scientific evidence : 科学的証拠、科学的根拠
scientific name : 学名
sclerite : 硬皮、節片、骨片、硬片
sclerophyll forest : 硬葉樹林
sclerotization : 硬化
sclerotized : 硬化した、節片化
sclerotized plate : 硬化板
SCN : 視交叉上核（SupraChiasmatic Nucleus）
scolopale cap : 有桿帽鞘、有桿体帽鞘
scolopale cell : 有桿体細胞、有桿細胞
scolopidia : 弦状感覚子、弦音器官
scolopidium : 弦音器官
scolopophore : 弦音器官、（昆虫の）受音波突起
scolopophorous organ : 弦音器官
scolus : （複 . -li）、棘瘤（きょくりゅう、とげこぶ）、枝刺〔ラテン語〕
scopa : （複 . -ae）、刷毛（はけ）〔ラテン語〕
scotophase : 暗期
scotophil : 親暗相
SCP : 過冷却点（SuperCooling Point）
scraping : こする音、削る音
screen house : 網室
screen out : スクリーニング、ふるい分け
screened lid : スクリーン付き蓋
scrobal : 凹溝の
scrobe : 凹溝
scrub : 潅木、雑木林、こする
scrub clearance : 雑木林の撤去
scrubby forest : 雑木林
scrubland : 低木林地帯
sculpture : 印刻（する）
scutal : 盾板の
scutellae : 小盾板の
scutellum : （複 . -la）、小盾板（胸部）〔ラテン語〕
scutum : （複 . -ta）、盾板（じゅんばん）（胸部）〔ラテン語〕
SDI : 性的二型指標（Sexual Dimorphism Index）
SDW : 滅菌蒸留水（Sterile Distilled Water）
SE : 標準誤差（Standard Error）
sea level : 標高、海水面
Sea of Japan Rim Area : 周日本海地域
seabird : 海鳥、ウミドリ
search image : 探索像
search strategy : 探索戦略
searching behavior : 探索行動
searching for early stage : 幼生期調査、初期探索
searching image : 探索像
season : 時期、季節
seasonal abundance pattern : 季節的個体数パターン
seasonal adaptation : 季節適応
seasonal arrangement : 季節配置
seasonal change : 季節的変化、季節（的）変動
seasonal change of host plant : 寄主植物の季節的転換
seasonal cycle : 季節周期
seasonal decline : 季節的低下、季節的減少
seasonal difference in fitness : 適応性の季節的相違

seasonal dimorphism : 季節的二型
seasonal diphenism : 季節性二型
seasonal environment : 季節(的)環境
seasonal fluctuation : 季節(的)変動
seasonal form : 季節型
seasonal form response : 季節型反応
seasonal habitat variability : 生息地の季節的変動性
seasonal index : 季節指標
seasonal information : 季節情報
seasonal migration : 季節的移住、季節移動性
seasonal morph : 季節型
seasonal morph response : 季節型反応
seasonal periodicity : 季節周期
seasonal polymorphic migrant : 季節多型性移動
seasonal polymorphism : 季節(的)多型
seasonal polyphenism : 季節多形、季節的表現多型、季節的多型
seasonal rhythm : 季節的リズム
seasonal variability : 季節的変異性
seasonal variation : 季節(的)変異
seasonal wing-morph : 季節的翅型
seasonal-form determination : 季節型決定
seasonal-form expression : 季節型発現
seasonal-form-determining hormone : 季節型決定ホルモン
seasonal-morph determination : 季節型決定
seasonal-morph expression : 季節型発現様式
seasonal-morph variation : 季節型変異
seasonal-specific differential expression : 季節特異的な異なる発現
seasonality : 季節性
seasonally changing environmental parameter : 季節的に変化する環境因子
seasonally changing resource abundance : 季節的に変化する資源量
second brood : 第二化
second generation : 第二世代
second hypothesis : 第二仮説
second instar : 二齢
second most abundant : 二番目に多い個数
second peak : 二番目の最高点、第二のピーク
second species : 第二の種
second stainless steel electrode : ステンレススチール製二次電極
second submarginal cell : 第二亜外縁室、第二後外縁室
second-growth forest : セカンドグロース林、二次成長林
secondary attractant : 二次誘引物質
secondary compound : 二次代謝産物
secondary contact : 二次的接触
secondary forest : 二次林、二次森林
secondary homonym : 二次同名、二次異物同名
secondary literature : 二次文献
secondary metabolite : 二次代謝産物
secondary plant substance : 二次植物成分
secondary seta : 二次刺毛
secondary sexual trait of male : 雄の第二次性徴(形質)
secondary structure : 二次構造
secondary type : 第二次基準標本、副模式標本
secrete : 分泌する
secreted protein : 分泌タンパク質
secretion : 分泌(物)、分泌状態
secretion of virgin female : 未交尾雌の分泌物
secretory granule : 分泌顆粒
secretory protein : 分泌タンパク質
section : 節、薄片を作る
section longitudinally : 縦方向に切断する
section transversely : 横方向に切断する
sectoral crossvein : 径分横脈、s 脈

sedentary animal : 定住動物、定住性の動物
sedentary range : 定住圏、定住域
sediment : 堆積物
sedimentary rock : 堆積岩
seed : 種子
seed beetle : マメゾウムシ
seed dressing : 種子粉衣(しゅしふんい)(種子消毒の一つで、農薬の粉剤や水和剤を種子にまぶす方法で、種子の状態により乾粉衣と湿粉衣がある)
seed pod : 種子のさや、種鞘、シードポッド
seed set : 結実率、種子形成
seed-dressing neonicotinoid : 種子粉衣用ネオニコチノイド
seed-feeding beetle : 種子食性甲虫、種子摂食甲虫
seed-treated crop : 種子が(農薬)処理された作物
seedling : 実生(みしょう)の苗木(種子から育てた植物のこと)、若木
seemingly complex array : 見かけ上複雑な行列
seemingly similar situation : 一見すると似たような状況
segment : 節、体節、セグメント、分節、切片
segmentum medianum : 中節(ちゅうせつ)、第一腹部背板
segregate : 分離する、隔離する
segregation : 分離
selected site : 選択された部位
selection : 選択、淘汰、選抜
selection coefficient : 選択係数
selection differential : 選択差
selection favoring single-locus control : 一遺伝子座の支配を有利にする選択
selection pressure : 選択圧、淘汰圧
selective abortion : 選択的中絶
selective advantage : 淘汰利益、淘汰の有利性、選択有利性、選択優位性、選択的優勢
selective benefit : 選択的利益、選択的恩恵
selective disadvantage : 淘汰不利益、淘汰の不利性、選択不利性
selective factor : 選択要因
selective felling : 選択(的)伐採
selective force : 選択圧、淘汰圧
selective harassment : 選択的ハラスメント、選択的干渉
selective logging : 選択(的)伐採、択伐林業、択伐
selective pressure : 選択圧、淘汰圧
selective sweep : 選択的一掃
selector gene : 選択遺伝子
self replication : 自己複製
self-assembly : 自己組織化
self-fertilization : 自家受精
self-incompatibility : 自家不和合性
self-matching : 自分と釣り合う(相手との交尾)
self-organization : 自己組織化
self-organizing map : 自己組織化マップ(SOM)
self-pollination : 自家受粉
self-splicing : 自己つぎはぎする
self-sustainability : 自律振動性、自立持続可能性
self-sustaining oscillation : 自律振動
selfing-avoidance : 自己回避
selfish element : 利己的因子
selfish endosymbiont : 利己的な細胞内共生微生物
selfish genetic element : 利己的遺伝因子
semel- : 一度 -、一回 -〔ラテン語〕
semelparity : 一回繁殖
semelparous : 一回繁殖
semelparous animal : 一回繁殖の動物
semelparous colonizer : 一回結実性の移住種、一回繁殖性の移住種
semi arid region : 半乾燥地、半乾燥地域

semi- : 半 - 〔ラテン語〕
semi-deciduous mesophytic forest : 半落葉性の中生植物の森、半落葉性中生林
semi-dominant mutation : 半優性突然変異
semi-natural grassland : 半自然草地
semi-natural habitat : 半自然な生息地
seminal fluid protein : 精液タンパク質(SFP)
seminal fluid substance : 精液物質
seminal idea : 発展性のある考え、独創性に富んだ考え
seminal infusion : 精液注入
seminal protein : 精液タンパク質
seminal vesicle : 貯精のう
seminatural habitat : 半自然の生息地
semiochemical : 信号物質、信号化学物質
semisocial route : 半社会性ルート
semispecies : 半種(形態的な差異は認められるが、生殖隔離がなされていない)
semivoltine : 半化性(生活史が二年がかり)
Senckenberg Museum, The : ゼンケンベルク博物館
senior author : シニアオーサー、論文指導者、首席著者、主席著者
senior homonym : 古参同名、上位異物同名
senior synonym : 古参異名、上位同物異名
sense organ : 感覚器
sense strand : センス鎖、有意鎖
sensilla : (複 . sensila)、感覚子、感覚器〔ラテン語〕
sensilla basiconicum : (複 . -ca)、円錐状感覚子(えんすいじょうかんかくし)
sensillum : (複 . -la)、感覚子、感覚器〔ラテン語〕
sensillum styloconicum : (複 . sensilla styloconica)、歯状感覚子
sensitive period : 敏感期
sensitivity : 感受性
sensory : 感覚
sensory cell : 感覚細胞
sensory cell body : 感覚細胞体
sensory cue : 感覚刺激
sensory hair : 感覚毛
sensory innervation : 感覚神経支配
sensory mechanism : 感覚機構
sensory process : 感覚過程
sensory stimulus : (複 . -li)、感覚刺激
sensory structure : 感覚器官の構造、感覚構造
sensu : の意味で〔ラテン語〕
sensu lato : 広義の、広い意味で、「例えば、『属名 (sensu lato)』(広義の XXX 属)の形式で表記」、「[例] *Vanessa* (sensu lato)(広義のアカタテハ属)」〔ラテン語〕
sensu strictiore : より厳密に言えば〔ラテン語〕
sensu strictissimo : 最も厳密にいえば〔ラテン語〕
sensu stricto : 狭義の、厳密な意味で、「例えば、『属名 (sensu stricto)』(狭義の XXX 属)の形式で表記」、「[例] *Pieris* (sensu stricto)(狭義のモンシロチョウ属)」〔ラテン語〕
sentinel : センチネル、標識、しるし
separate : 抜き刷り
separate species : 独立種
separation point : 分離点
separation zone : 分離帯
sepiapterin : セピアプテリン
septum : 隔膜
sequence : 配列、シークエンス
sequence alignment : 配列アラインメント、配列アライメント
sequence assembly : シークエンスアセンブリー、配列アセンブリー(短い DNA の断片から元の長い塩基配列を再構築すること)
sequence characterization : 塩基配列の特性
sequence chromatogram : 塩基配列クロマトグラム
sequence comparison : 配列比較

sequence conservation : 配列保存
sequence divergence : 配列多様性、塩基配列間の相違、配列間の差異
sequence identity : 配列同一性、配列の一致度
sequence of mitochondrial DNA : ミトコンドリア DNA の塩基配列
sequence statistics : 配列統計、塩基配列統計
sequence the genome : ゲノムの塩基配列を解読する
sequence trace file : 配列追跡ファイル
sequence variation : 配列変異
sequence-based marker : 塩基配列基盤マーカー
sequenced species : 遺伝子配列された種
sequenced taxa : 遺伝子配列された分類群
sequencing technology : 塩基配列決定技術、シークエンシング技術
sequential strategy : 順次戦略
sequestration : 蓄積、滞留
sequestration of flavonoid : フラボノイドの蓄積
Serengeti Plains : セレンゲティ国立公園
series of ocelli : (単 . -lus)、眼状紋列
serine : セリン
serine-to-alanine substitution : セリンからアラニンへの置換
serious decline : 深刻な衰亡
serosa : 漿膜(しょうまく)
serotonin : セロトニン、5-ヒドロキシトリプタミン、5-HT(5-HydroxytrypTamine)
serpentine soil environment : 蛇紋岩地の土壌環境
serrate antenna : (複 . -nae)、鋸歯状の触角(蛾類)
serration : 鋸歯
sesquiterpene : セスキテルペン
sessile : 無柄
seta : (複 . -ae)、刺毛(しもう)、棘毛(きょくもう)、剛毛〔ラテン語〕
setaceous antenna : (複 . -nae)、刺毛(しもう)状の触角
setal : 剛毛の、刺毛の
settle : 定住する
settling contest : 決着済みの競争
seven transmembrance protein : 7 回膜貫通型タンパク質
sever : 切断する、断絶する
severe competition : 激しい競争
severe long-term decline : 厳しい長期間にわたる衰亡
severe repression : 厳格な抑制
severity : 重症度、厳格(度)
sex brand : 性標(斑)
sex chromatin body : 性染色質体
sex chromosome : 性染色体
sex determination : 性決定
sex determination and differentiation : 性決定と性分化
sex difference : 性差
sex lethal gene : 性致死遺伝子
sex mosaic : 雌雄モザイク、性モザイク
sex patch : 性標
sex pheromone : 性フェロモン
sex pheromone formation : 性フェロモン形成
sex pheromone unit : 性フェロモン単位
sex ratio : 性比
sex ratio selection : 性比淘汰
sex-asymmetric inbreeding : 性非対称な近親交配
sex-biased gene expression : 性差のある遺伝子発現、雄か雌かのどちらかに偏った遺伝子発現
sex-dependent selection : 性依存選択
sex-determination master regulator : 性決定のマスター調節因子、性決定の支配制御因子

sex-determining gene : 性決定遺伝子
sex-differentiating hormone : 性分化ホルモン
sex-lethal gene : 性致死的遺伝子(*Sxl*)
sex-limited inheritance : 限性遺伝
sex-limited melanism : 限性黒化
sex-limited mimicry : 限性擬態
sex-linkage : 伴性、性連鎖、伴性遺伝
sex-linked : 伴性(の)、伴性遺伝の
sex-linked diapause response : 伴性の休眠反応
sex-linked gene : 伴性遺伝子
sex-linked inheritance : 伴性遺伝
sex-related gene : 性関連遺伝子
sex-specific melanin synthesis : 性特異的なメラニン合成
sex-specific mRNA splicing : 性特異的なmRNA スプライシング
sex-specific plasticity : 性特異的な可塑性
sex-specific reaction : 性特異的反応
sexual communication : 性的通信
sexual conflict : 性的対立
sexual development time dimorphism : 発育時間の性的二型、性的発育時間二型(→雄性先熟、雌性先熟)
sexual difference : 性差、性的差異
sexual dimorphism : 性的二型(性)
sexual dimorphism index : 性的二型指標(SDI)
sexual hair : 性毛
sexual harassment : 性的干渉、セクシャルハラスメント、性的いやがらせ
sexual isolation : 性的隔離、生殖(的)隔離
sexual marking : 性標
sexual mosaicism : 性的モザイク現象
sexual partner : 性的パートナー
sexual reproduction : 有性生殖
sexual selection : 性選択、性淘汰
sexual size dimorphism : 体長の性的二型、性的体長二型
sexually dimorphic wing morphology : 性的二型の翅形態
sexually intermediate genitalia : 性的に中間の交尾器
sexually intermediate phenotype : 中間的な性表現型
sexually intermediate trait : 性的に中間型の形質、間性形質
sexually mosaic individual : 性モザイク個体
sexually receptive : 性的に交尾可能な状態、性的受容可能な状態
sexually receptive partner : 性的に交尾可能なパートナー、性的受容可能なパートナー
Sf9 : ツマジロクサヨトウ(*Spodoptera frugiperda*)の卵巣細胞由来の樹立培養細胞株
SFP : 精液タンパク質(Seminal Fluid Protein)
shade vegetation : 日陰の植生
shaded habitat : 日陰になった生息地、日陰生息地
shadow hindwing bar : 後翅線条の透視画像
shake : 振盪(しんとう)する
sham operation : 偽手術
Shannon-Wiener Index : シャノン-ウィナーの多様度指数(生物群集内の多様性を示す指数)
shape : 形、形状、形作る
shape parameter : 形状パラメータ
shaped pulse : 整形パルス、形状パルス
shared barcode : 共用バーコード
shared developmental pathway : 共通の発育経路
shared genetic architecture : 共通の遺伝的構造
shared sequence variation : 共有配列変異、共通配列変異
sharp taste : 鋭い味覚
sharply pointed : 鋭くとがった

shear : 切断する
sheath : 鞘(さや)
shed : 与える
shed light on : 光を当てる
sheep : ヒツジ、羊
shell : 殻
shelter : 隠れ家、シェルター
shelter-dwelling caterpillar : 隠れ家で暮らしている幼虫
shift in diversification rate : 多様化速度の変化
shift of voltinism : 化性転換
shifting balance : 平衡遷移、平衡推移、バランス変動、推移平衡
shikimate pathway : シキミ酸経路(植物において芳香族アミノ酸の生合成経路)
shivering : シバリング、粉砕、はく裂
short day length : 短い日長
short DNA sequence : DNA の短配列
short flight bout : 短期の飛翔バウト(一続きの行動期間)
short hair : 短毛
short insertion : 短い挿入
short interfering RNA : 低分子干渉 RNA (siRNA)
short interspersed element : 短鎖散在反復配列(SINE)
short interspersed nuclear element : 短鎖散在(型)核内反復配列(SINE)
short latency : 短潜時
short period of time : 短時間、短期間
short photoperiod : 短日日長
short report : 短報
short return time : 短い再発期間、短い再帰期間
short tandem repeat : 短鎖縦列反復配列、縦列(型)反復配列(STR)
short turf : 丈の短い芝
short-day effect : 短日効果
short-day frequency : 短日サイクル数
short-day photoperiod : 短日日長
short-day treatment : 短日処理
short-day type : 短日型
short-day type response : 短日型反応
short-focus telescope : 短焦点の望遠鏡
short-horn type : 短角型(たんかくがた)
short-lived insect : 短命な昆虫
short-scale movement : 短期の移動
short-tailed form in both sex : 雌雄短尾型
short-wavelength-sensitive cone pigment : 短波長感受性錐体視物質
shortleaf pine : キマツ(米国南部産の大型のマツ)
shotgun sequencing : ショットガンシークエンス法
shotgun sequencing method : ショットガンシークエンス法
shrink : 小さくなる、縮む
shrub : 低木、灌木(かんぼく)
shrubbery : 低木、生け垣
Shune-Dalgarno sequence : シャイン・ダルガノ配列、SD 配列
SI : 群知能(Swarm Intelligence)
sib mating : 同胞交配、近縁交配、兄弟交配
Siberian distributed type : シベリア分布系統
Siberian type : シベリア型
sibling group : 近縁グループ、兄弟グループ
sibling species : 同胞種、近縁種
sickle cell : 鎌状赤血球
side effect : 副作用
side-by-side comparison : 並列比較(二つの物を並べて比べる)
sieving : ふるい分け、ふるい操作
sight : 視覚
sighting : 目撃(例)、目撃情報、見聞
sigma factor : シグマ因子、σ 因子
signal molecule : シグナル分子
signal peptide : シグナルペプチド

signal preference : シグナル選好
signal production : シグナル生産
signal reception : シグナル受容
signal transduction : シグナル伝達、シグナル伝達系
signal transduction pathway : シグナル伝達経路
signaling factor : シグナル因子
signalling : シグナル伝達、信号伝達
signalling pathway : シグナル伝達経路、シグナル伝達系
signalling-pathway gene : シグナル伝達系遺伝子
signature of selection : 自然淘汰のサイン、自然淘汰のしるし
significance threshold : 有意な閾(いき)値、有意性閾値
significant base composition heterogeneity : 塩基組成の有意な不均一性
significant deviation : 有意な偏差
significant difference : 有意な相違
significant number of regulatory element, a : 相当数の調節要素
significantly less : 有意により少ない
significantly positive D-statistic : 有意な正のD統計量
significantly prefer : 有意に好む
significantly worse : 有意に悪い、大幅に劣る
silencer : サイレンサー、消音装置
silent mutation : サイレント変異
silent substitution : サイレント置換
silicon grease : シリコーングリース、シリコングリス
silk : 絹糸
silk girdle : 帯糸(たいし)
silk gland : 絹糸腺
silken girdle : 絹の支持帯
silkworm : カイコガの幼虫、カイコ
silvaniform clade : シルヴァーニ型の分岐群
silver hairstreak butterfly : シラメスアカミドリシジミ
silver-spotted skipper butterfly : ホクベイオオギンモンセセリ、ホソバセセリ
silvicultural system : 造林システム、造林施業システム
simian immunodeficiency virus : サル免疫不全ウイルス(SIV)
similar compound : 類似化合物
similar species : 近似種
similarity : 類似、類似性、類似度
simple antenna : (複 . -nae)、単純形状の触角(糸状の触角などの総称)(蛾類)
simple eye : 単眼
simple family : 単純家族
simple sequence length polymorphism : 単純反復配列多型、単純配列長多型(SSLP)
simple sequence repeat : 単純反復配列(SSR)
simulation model : シミュレーションモデル
simultaneous attack : 同時攻撃
Simultaneous Partitioning Method : 同時分割法
Simultaneous Sequence Joining : SSJ法(多数の近縁な塩基配列データから迅速に系統樹を生成する方法)
SINE : 短鎖散在反復配列(Short INterspersed 〈nuclear〉 Element)
sine wave : サイン波、正弦波
single autosomal locus : 単一の常染色体遺伝子座
single biochemical mechanism : 単一生化学的機構
single butterfly genus : 蝶の単一属
single characteristic estimate : 単一形質の推定値
single clade : 単系統
single clone : 単一クローン

single diapause stage : 一回限りの休眠期、単発休眠期
single gene mutant : 単一遺伝子の突然変異体
single gene mutation : 単一遺伝子の突然変異
single genotype : 単一(の)遺伝子型
single lens reflex camera : 一眼レフカメラ
single locality : 単一局所性
single nucleopolyhedrovirus : 単一のヌクレオキャプシドを含む核多角体病ウイルス、単一のヌクレオキャプシドを含む NPV(SNPV)
single nucleotide polymorphism : 一塩基多型、単一塩基多型(SNP)
single nucleotide site : 一塩基部位、一ヌクレオチド部位
single nucleotide transition difference : 一塩基の転位差
single optical-sensing structure : 単光センシング構造
single origin : 単一起源
single peak : 単一ピーク
single plant : 単独植物
single positive clone : 単一陽性クローン、単一ポジティブクローン
single recorder : 一人の記録者、単独記録者
single round only : 一回限り
single S opsin variant : 単一 S(短波長型)オプシン変異体
single sensillum recording : 単一感覚子記録法
single spectral class of photoreceptor : 光受容器の単一スペクトル型
single switch gene : 単独スイッチ遺伝子
single-brooded : 一化の
single-copy microsatellite : 単コピーマイクロサテライト
single-copy nuclear gene : 単コピー核遺伝子
single-copy nuclear locus : 単コピー核遺伝子座
single-locus inheritance : 単一遺伝子座の遺伝
single-nucleotide variation : 一塩基変異
single-strain-infected normal brood : 単系統に感染した正常な同腹仔
single-strand conformation polymorphism : 一本鎖高次構造多型、SSCP 法(SSCP)
single-stranded sequencing reaction : 一本鎖シークエンス反応
singleton : 1 個体だけ現れた種、単一個体(種)
singly infected female : 単感染の雌
singular point : 特異点
sinigrin : シニグリン
sink habitat : シンク生息地(繁殖はできるが個体数が減少してしまう場所)
Sino-Japanese region : 日中区、日中区系
Sino-Japanese type : 日中系統、日中型
sinuate : 深波状(しんはじょう)の(深く波のような湾入形)
Siphonaptera : ノミ目、隠翅目
sire : (種馬として)子供を作る
siRNA : 低分子干渉 RNA(short interfering RNA ; small interfering RNA)
sister genus : 姉妹属
sister group : 姉妹群
sister haplotype : 姉妹ハプロタイプ
sister species : 姉妹種
sister taxon : 姉妹分類群、姉妹群
SIT : 不妊虫放飼法(Sterile Insect Technique)
site fidelity : 場所固執性
site-branch analysis : サイト枝解析
site-specific recombination : 部位特異的組み換え
site-specific scale : 現地に特有の規模
sitting behavior : 静止行動
sitting time : 静止時間

situate : 置く、位置を定める
SIV : サル免疫不全ウイルス (Simian Immunodeficiency Virus)
size : 大きさ、寸法
size of a potential inversion : 潜在的な逆位のサイズ
size of sample : サンプルの大きさ、サンプルサイズ
size variation : サイズ変異
size-dependent survival : 体格依存の生存(率)
size-related trait : サイズ関連形質
skeletal muscle : 骨格筋
skeletal soil : 粗骨土壌
skeleton : 骨格、枠、スケルトン
skeleton photoperiod : スケルトン光周期、枠光周期
skew : 偏り、ゆがみ、曲がり
skill of searching literature : 文献探索技能
skin : 表皮、クチクラ
skipper butterfly : セセリチョウ科の蝶
skippers, the (superfamily Hesperioidea) : セセリチョウ上科
skull : 頭蓋骨、頭骨
sky compass : 天空コンパス
skylight compass : 天空光コンパス
sleep : 眠り
slender : 細い
slender hair : 細い毛
slice preparation : 薄切標本
slide glass : スライドグラス
sliding-window phylogenetic analysis : スライディングウィンド手法を用いた系統解析
slight local variation : 微妙な局所的変異
slightly deleterious mutation theory : 弱有害突然変異説
slightly distal : わずかに末端へ、わずかに末梢へ
slightly elliptical eyespot : 少し楕円形状の眼状紋
sling psychrometer : 振り回し式乾湿計、振り回し湿度計
slope : 坂、斜面、傾き
slowly walking : ゆっくり歩行
slug-like : ナメクジのような、ナメクジ形の
small cabbage white : モンシロチョウ
small copper butterfly : ベニシジミ
small nuclear RNA : 核内低分子 RNA (snRNA)
small nucleolar RNA : 核小体低分子 RNA (snoRNA)
small postman butterfly : アカスジドクチョウ
small skipper butterfly : スモールスキッパー、スモールセセリ
small tortoiseshell butterfly : コヒオドシ
small white butterfly : モンシロチョウ
small-bodied insect: 小型昆虫
small-scale map : 小縮尺の地図
small-scale rearrangement : 小規模再配置
smaller tea tortrix : チャノコカクモンハマキ
smell : 嗅覚、におい
smooth cline : 滑らかなクライン
smooth curve : 滑らかな曲線
smooth surface : 平滑な面
smoothed rarefaction curve : 滑らかな希薄化曲線
snap-shot camera : コンパクトカメラ
sneaker : スニーカー
snoRNA : 核小体低分子 RNA (small nucleolar RNA)
snout : 鼻
snout butterfly : テングチョウ
Snouts : テングチョウ亜科
SNP : 一塩基多型、単一塩基多型(Single Nucleotide Polymorphism)
SNPV : 単一のヌクレオキャプシドを含む核多角体病ウイルス、単一のヌクレオ

キャプシドを含むNPV (Single NucleoPolyhedroVirus)
snRNA : 核内低分子RNA (small nuclear RNA)
SNV : 一塩基変異 (Single-Nucleotide Variation)
SNV determination : SNV決定、一塩基変異の決定 (Single-Nucleotide Variation)
social colony : 社会コロニー
social communication : 社会的コミュニケーション
social ecology : 社会生態学
social entrainment : 社会的同調
social factor : 社会的要因
social information : 社会的情報
social parasitism : 社会寄生
sociality : 社会性
socio-economic : 社会経済的な
socius : (複. socii)、尾突、側突起、ソキウス (鱗翅目昆虫の雄性外生殖器で、ウンクスから生じる一対の遊離突起)
socket : ソケット
socket-forming cell : ソケット形成細胞
sodium : ナトリウム
sodium carbonate : 炭酸ソーダ、炭酸ナトリウム
sodium-supplemented mud puddle : ナトリウムが補給された泥水
soft leaf : 柔らかい葉
soil : 地面、土壌
soil formation : 土壌形成
soil ingestion : 土壌摂取
soil surface : 地表面、土壌表面
soil water : 土壌水、土壌水分
solar collector : 太陽熱集熱器
solar compass : 太陽コンパス
solar heat : 太陽熱
solar radiation : 太陽放射、日射
solid bar : 黒色の棒、固い棒
solid dot : 黒丸、黒点
solid-line : 実線
solidified agarose : 凝固したアガロース
solitary : 単独性、単生性の、群居しない、群生しない、孤独性の
solitary bee : 単独性バチ、孤独性バチ (社会生活をしないハチの総称)
Solomon archipelagos : ソロモン諸島
Solomon Islands : ソロモン諸島
solubility : 溶解度
solution containing sucrose : ショ糖含有溶液
solvent : 溶媒
solvent ethanol : エタノール溶媒
solvent extraction : 溶媒抽出
solvent-treated control plant : 溶媒処理済みの対照植物
SOM : 自己組織化マップ (Self-Organizing Map) (生物の脳、特に大脳皮質の様々な領野が異なる感覚様相に従って組織化されていることに着想を得て作られた人工ニューラルネットワークの一種)
somatic component : 体成分
somatic growth : 身体の成長、体の成長
somatic maintenance : 身体の維持
somewhat : いくぶん、多少、やや
songbird species : 鳴禽類(めいきんるい)の種
sorbitol : ソルビトール
soret band : ソーレー帯
sound level meter : 騒音計、音圧レベル計
sound production : 音発生
sound production behavior : 発音行動
sound pulse : パルス音、音(波)パルス
sound stimulus : 音刺激
sound-producing : 音を発生する
source : ソース、(供給)源、供給側
source country : 原産国
source habitat : ソース生息地、ソースハビタット (繁殖に好適で子孫が増える場所)

source of noise : 雑音源
source of selection : 選択の起源
source population : 供給(元)個体群、供給側の個体群
source-sink model : ソース - シンクモデル、本土 - 島モデル
South East Asia : 東南アジア
South Vietnam : 南ベトナム
South-East Asia : 東南アジア
southerly flight : 南方へ飛翔
southern dogface butterfly : ミナミイヌモンキチョウ
southern marginal area : 南縁部
Southern rice black streaked dwarf virus : イネ南方黒すじ萎縮ウイルス、イネ南方黒すじ萎縮病
southern species : 南方性の種、南方的な種
southernmost colony : 南限のコロニー
southernmost habitat : 分布南限地
southwest direction : 南西方向
southwest lowland of Japan : 南西日本の低地、日本の南西地方の低地
southwestern island population : 南西諸島の個体群
soyasaponin : 大豆サポニン
sp : スパルト遺伝子(*spalt* gene)
sp. : 種(species)、「『種名が同定できなかったり、未記載種の可能性があったりして、学名が確定できなかった』ことを表し、『属名 sp.』の形式で表記」(「species」は「単複同形」で、「単数」は「sp.」と略す;「sp.」は「ローマン体」で表記)
sp. aff. : 関係種、近似種、「種が確定していない時にとりあえずある種に類似していることを表す(species with affinity)」
sp. nov. : 新種名(species novus)
space : 翅室
space and time : 時空、時空間
space out : 間置き
spalt : スパルト遺伝子
span : 指寸法で計る、架かる
sparse : 点在する、まばらに植えた、希薄な
sparse vegetation : 点在する植生
spatial aggregation of species : 種の空間的集団
spatial anatagonism : 空間拮抗性
spatial and temporal population dynamics : 時空間の個体群動態
spatial arrangement : 空間配列
spatial autocorrelation : 空間的自己相関
spatial configuration : 空間構成
spatial configuration of suitable habitat : 好適な生息地の空間配置構造
spatial distribution : 空間的分布
spatial distribution of habitat : 生息地の空間分布
spatial distribution of individual : 個体の空間分布
spatial ecology : 空間生態学
spatial genetic pattern : 遺伝的空間パターン
spatial heterogeneity of resources : 資源の空間的不均一性、資源の空間的不均質性
spatial information : 空間情報
spatial location : 空間(的)位置
spatial memory mechanism : 空間(的)記憶機構
spatial pattern : 空間分布パターン
spatial pattern of decline : 衰亡の空間的パターン、衰亡の空間分布パターン
spatial pattern of species richness : 種の豊富さの空間的パターン
spatial position : 空間位置
spatial positioning : 空間位置決め
spatial relationship of location : 空間的位置関係
spatial resolution : 空間分解能

spatial scale : 空間スケール、空間尺度
spatial structure : 空間構造
spatial unit : 空間単位
spatial variability : 空間変動性
spatial variation : 空間的変異、空間変異
spatially uncorrelated stochasticity : 空間的に相関関係のない確率性
spatio-temporal resource dynamics : 時空間の資源動態、資源の時空間動態
spatiotemporal variation : 時空間変動、時空間変化【景観生態学】
spawn : 産む、生産する
speaker : スピーカー、語り手
Spearman correlation : スピアマンの相関
special protection zone : 特別保護地区
specialist : スペシャリスト、専食者、専門家
specialist herbivore : スペシャリストである植食者、スペシャリスト植食者
specialist species : スペシャリスト種
specialization : 特化、分化
specialized : 分化した
speciate : 種分化する
speciation : 種分化、分化
speciation date : 種分化の時期
speciation history : 分化の歴史
speciation part : 種分化した部分
speciation rate : 種分化率、種分化速度(SR)
species : (単複同形)、種(分類階級の「しゅ」)〔ラテン語、英語〕
species accumulation curve : 種数累積曲線
species boundary : 種の境界
species complex : 種複合体、複合種
species composition : 種構成、種組成
species conversion : 種の転化
species decline : 種の減少、種の滅亡、種の衰亡
species delimitation : 種の境界設定
species density : 種密度、種の多様性
species distribution : 種の分布
species diversification : 種多様化
species diversity : 種多様性
species endemic to Japan : 日本特産種
species extinction : 種の絶滅
species group : 種階級群
species hibernating as egg : 卵休眠種
species hibernating as pupa : 蛹休眠種
species hot spot : 種のホットスポット
species identification : 種同定
species incidence : 種の出現(率)
species inquirenda : (複 . -ae)、未確定種
species level : 種レベル
species loss : 種の喪失
species name : 種名
species name list of Japanese butterflies : 日本産蝶種名一覧
species per genus : 属あたりの種数
species recognition : 種認識
species replacement : 種交替、種の置換
species richness : 種数、種の豊富さ
species selection : 種選択
species specific : 種特異的
species total : 種の総数
species tree : 種系統樹
species' ancestor : 種の祖先
species' autecology : 種生態学、種の個生態学
species' extinction : 種の絶滅
species--family ratio : 種数 - 科数の比
species-abundance model : 種数 - 個体数モデル
species-group name : 種階級群名
species-level estimate : 種水準の推定値
species-per-individual ratio : 個体数あたりの種数の比
species-poor : 種が少ない、種が乏しい
species-poor community : 種が少ない群集
species-rich : 種が多い、種が豊富な
species-rich clade : 種類が多い単系統

species-rich community : 種が多い群集
species-rich genus : 種数の多い属
species-rich location : 種が豊富な場所
species-specific color preference : 種特異的な色選好性
species-specificity : 種特異性
species-to-genus ratio : 種数対属数の比
species-to-individual ratio : 種数対個体数の比
species/individual ratio : 種数／個体数の比
specific critical weight : 特定の臨界体重、特定の閾(いき)値体重
specific cue : 固有の刺激
specific effector : 特異的なエフェクター
specific elicitor : 特異的エリシター、特異的誘発物質
specific habitat : 特異的な生息地、特定の生息地
specific mating area : 特定の交尾場所、特異的な交尾場所
specific mixture of secretion : 分泌物の特定混合物、分泌物の特異的混合物
specific name : 種小名
specific nature of the association : 結合の特異的な性質
specific organism : 特定の(微)生物、特異的な(微)生物
specific presence-absence correlation : 特異的な有無相関
specific primer : 特異的プライマー
specific site : 特定部位、特殊部位
specific sound frequency : 特定音周波数
specific tissue : 特異的組織
specific to female : 雌に特異的な
specification of focus : フォーカスの仕様、焦点の内訳
specify : 指定する、規定する、特定する
specimen : 標本
specimen voucher : 標本の証拠書
speciose : 種数が多いこと
speck change : 斑紋変化
speckled wood butterfly : キマダラジャノメ
spectacular array : 壮観な配列
spectacular difference : 目を見張る違い
spectral difference : スペクトル差
spectral diversification : スペクトル多様化
spectral phenotype : スペクトル表現型
spectral receptor type for color vision : 色覚用のスペクトル受容体の型
spectral reflectance : スペクトル反射率、分光反射率
spectral representation : スペクトル表現
spectral sensitivity : 分光感度、スペクトル感受性、スペクトル感度、分光応答度
spectral tuning : スペクトル同調、分光学的微調整
spectral tuning effect : スペクトル調整効果
spectral tuning site : スペクトル調節部位
spectral variation : スペクトル変化、スペクトル変動
spectrally distinct class of photoreceptor : 光受容器のスペクトル的に異なる型
spectrally shifted lineage : スペクトル的にシフトされた系統、スペクトル偏移系統
spectrophotometer : 分光光度計
spectrophotometrical characteristic : 分光学的特徴
spectrophotometrically : 分光学的に
speculate : 熟考する、推測する、見当をつける
speculation : 推測、憶測
speculative : 理論的な、推論にすぎない、思いつき程度の
speculum : 検鏡〔ラテン語〕
speed : (飛翔)速度
speed of fluttering : 飛翔速度
SPEEDI : 緊急時迅速放射能影響予測ネットワークシステム(System for Prediction

of Environmental Emergency Dose Information)
spelling : 綴(つづ)り
sperm : 精子(せいし)
sperm competition : 精子(間)競争
sperm count : 精子数
sperm precedence : 精子優先度、精子の優先性
sperm production : 精子の生産
sperma- : 精子 -〔ギリシャ語〕
spermatheca : 受精のう
spermatogenesis : 精子形成
spermatophore : 精包(せいほう)、精球
spermatophore protein : 精包タンパク質
spherical densitometer : 球面密度計、球面濃度計
spheroidal wing surface : 回転楕円体状の翅面
spherulocyte : 小球細胞(昆虫血球細胞の一種)
Sphingidae : スズメガ科
sphingophilous flower : 虫媒花
sphinx moth : スズメガ科の蛾
sphragis : スフラギス、受胎のう、交尾のう
spicular : 針状
spider : クモ
spider mite : ハダニ
spin : 糸を吐く
spinasternum : 棘腹板(きょくふくばん)、刺腹板(しふくばん)(胸部)
spindle fiber : 紡錘糸
spine : トゲ、刺、棘(とげ)
spinneret : 吐糸管
spinning flight : 回転飛翔
spiracle : 気門(きもん)
spiracular : 気門の
spiral flight : らせん飛翔、卍(まんじ)ともえ飛翔
spiraling : らせん飛翔
Spiroplasma : スピロプラズマ
SPL : 音圧レベル(Sound Pressure Level)
splendid insect : 光り輝く虫
splice form : スプライス型、スプライシング型
splice signal : スプライス信号、切り貼り信号
splicing : スプライシング
split : 分岐する、分割
split frequency : 分割頻度
splitter : 細分類学者(類似点よりも相違点を重視する分類学者)
Spodoptera exigua : シロイチモジヨトウ(ヤガ科の一種)
Spodoptera frugiperda : ツマジロクサヨトウ
Spodoptera litura : ハスモンヨトウ
spontaneous feeding response : 自発的な摂食反応、自然発生的な摂食反応
spontaneous flight : 自発飛翔
spontaneous mutant : 自然誘発突然変異体、自然に起こる突然変異体
spontaneous single-gene mutant : 自然(に起こる)単一遺伝子突然変異体
sporadic : 散発的な
sporadically : 散発的に
sporophyte : 胞子体
spot : 斑点
Spotty : 斑点のある
spp. :「『その属の生物を複数種確認したが、種名は同定できなかった』ことを表し、その属の複数の種を指す表記法」、「『属名 spp.』の形式で表記」(「species」は「単複同形」で、「複数」は「spp.」と略す;「spp.」は「ローマン体」で表記)
spp. nov. : 複数個の新種名(species novus)
spra-anal plate : 肛上板(こうじょうばん)
spray : (殺虫剤などの)噴射
spread : 拡大、広がる

spread northward : 北方へと広がる
Spread-wing Skippers : チャマダラセセリ亜科
spreading : 展翅(てんし)
spreading board : 展翅板
spreading fresh specimen : 生展翅
spreading relaxed specimen : 軟化展翅
spring ephemeral : 早春季生物
spring form : 春型
spring generation : 春世代
spring imago : 春型成虫
spring morph : 春型
spring remigration : 春季の再移動
spring-form female : 春型雌
spring-form induction : 春型誘導
SPU : 性フェロモン単位(Sex Pheromone Unit)
spur : 距(きょ)、蹴爪(けづめ)、端刺(たんし)
Spurr's epoxy resin : スパーのエポキシ樹脂
sputter-coated : スパッタリングコートした、スパッタにて被覆した
square bracket : 角括弧
squarish : 角張った
squeezed out of, be : 追い出される、閉め出される
squid photoreceptor cell : イカ光受容体細胞
squid retinochrome : イカ網膜クロム、イカロドプシン
SRBSDV : イネ南方黒すじ萎縮ウイルス、イネ南方黒すじ萎縮病(Southern Rice Black Streaked Dwarf Virus)
Sri Lanka : スリランカ
Ss : 歯状感覚子(Sensillum styloconica)
SSCP : 一本鎖高次構造多型、SSCP 法(Single-Strand Conformation Polymorphism)
SSD : 体格の性的二型、性的体格二型(Sexual Size Dimorphism)
SSH : SSH 法、抑制サブトラクティブ・ハイブリダイゼーション 法(Suppression Subtractive Hybridization)
SSH approach : SSH 法アプローチ(Suppression Subtractive Hybridization)
SSJ : SSJ 法(多数の近縁な塩基配列データから迅速に系統樹を生成する方法)(Simultaneous Sequence Joining)
SSLP : 単純反復配列多型、単純配列長多型(Simple Sequence Length Polymorphism)
ssp. : 亜種(subspecies)、「『属名 種名 ssp. 亜種名』の形式で表記」(「subspecies」は「単複同形」で、「単数」は「ssp.」と略す；「ssp.」は「ローマン体」で表記)
ssp. nov. : 新亜種名(subspecies novus)
sspp. : 亜種(subspecies)の複数形の略語、「その属(種)の複数の亜種を指す表記法で、『属名 種名 sspp.』の形式で表記」(「subspecies」は「単複同形」で、「複数」は「sspp.」と略す；「sspp.」は「ローマン体」で表記)
SSR : 単純反復配列(Simple Sequence Repeat)
ST-GMYC analysis : ST-GMYC 解析、単閾(いき)値的 GMYC 解析(Single Threshold-Generalized Mixed Yule Coalescent)
stability effect : 安定効果
stability mechanism : 安定機構
stabilization of national distribution : 国レベルの分布の安定化
stabilizing selection : 安定化選択、安定性選択、安定化淘汰
stable asymptote : 安定的漸近
stable environment : 安定した環境
stable isotope experiment : 安定アイソトープ実験、安定同位体実験
stable landscape heterogeneity : 安定した景観の異質性
stable polymorphism : 安定多型

stable reference gene : 安定した参照遺伝子、安定した対照遺伝子
stadium : 齢、期
stage : ステージ、期、段階
stain : 着色する、染色する
stainless steel hook electrode : ステンレス鋼製フック電極、ステンレススチール製フック電極
stalk-eyed fly : 有柄眼のハエ
stand : 林分
standard currency : 標準的に通用している
standard curve : 標準曲線
standard deviation : 標準偏差
standard Lepidoptera primer : 標準の鱗翅類のプライマー、標準のチョウ目のプライマー
standard marker : 基準マーカー、標準マーカー
Standard of Butterflies in Japan, The : 日本産蝶類標準図鑑
standard primer : 標準プライマー
standard procedure : 標準手順書
standard protocol : 標準プロトコル、標準的な実験手順書
standard rearing temperature : 標準的な飼育温度
standard sample : スタンダードサンプル、標準的標本
standardized comparison : 標準的比較、標準化された比較
standardized regression coefficient : 標準化された回帰係数
standardized survey : 標準化された調査
standpoint of phylogeny : 発生的な立場、系統的な立場
star compass : 星コンパス
star symbol : 星形記号
star-like tree : 星状樹
starch : デンプン
startle : 威嚇(いかく)する
startling eyespot : 脅かしの目玉模様
starvation : 絶食、飢餓
starvation endurance : 餓死耐久性
starve : 餓死する
starve to death : 餓死する
state of origin : 原産国
state variable : 状態変数
stationary molt : 定常脱皮
statistical analysis : 統計解析
statistical comparison : 統計的比較
statistical expectation : 統計的期待値
statistical significance : 統計的有意性
statistical study : 統計学的研究
statistical universe : 統計母集団、統計集団
statistically significant : 統計的に有意な
statistically significant relationship : 統計的に有意な関係
statistics : 統計(学)
status alteration : 地位変更
steam distillation : 水蒸気蒸留法
steep decline : 急激な衰亡
stem (of a name) : 語幹(学名の)
stem : 枝、茎、幹、止める
stem borer : ズイムシ(髄虫)、メイチュウ、メイガの幼虫、ニカメイチュウ
stem borer moth : メイガ成虫
stem density : 樹幹密度、幹数密度、本数密度
stem mother : 幹母
stem of leaf : 葉柄
stemma : (複 . -ata)、点眼(完変態の幼虫の頭側部にある個々の眼)、単眼
steppe environment : ステップ的環境
stepping stone habitat : 足がかりの生息地、踏み石の生息地
stepwise AIC procedure : 逐次 AIC 手順(Akaike's Information Criterion ; 赤池情報量基準)

stepwise analysis of deviance : 逐次的逸脱度分析、逸脱度逐次分析
stepwise clustering method : 段階的探索法
stepwise linear regression : 逐次線形回帰
stepwise multiple regression : 逐次重回帰、段階的重回帰
stereo microscope : 双眼実態顕微鏡、双眼実体顕微鏡
stereo power amp : ステレオパワーアンプ
stereogram : ステレオグラム、立体画像
sterile : 滅菌する、殺菌する
sterile caste : 不妊カースト
sterile distilled water : 滅菌蒸留水（SDW）
sterile insect technique : 不妊虫放飼法
sterility : 不妊性、不稔性
sternaulus : （複 . -li）、胸側溝、中胸側線〔ギリシャ語〕
sternite : 腹板
sternopleural suture : 腹上側板溝状線、腹胸側線
sternum : 腹板
Stibochiona nicea : シロヘリスミナガシ、ルリボシスミナガシ
stiff : 堅い、固い
stigma : （複 . -ata）、性標（斑）、縁紋、柱頭、気門(きもん)〔ラテン語〕
stigmergy : スティグマジー、徴候
still-steep sample-based rarefaction curve : 急傾斜型のサンプル数に基づく希薄化曲線
stimulant : 刺激因子
stimulate : 刺激する、誘導する
stimulus : （複 . -li）、刺激
stimulus duration : 刺激持続期間
stimulus effect : 刺激効果
stimulus intensity : 刺激強度
sting : 刺針(ししん)
stinging hair : 刺毛
stinging nettle : （セイヨウ）イラクサ
stinging of non-host species : 非宿主種を刺す行動、非宿主種への刺入(しにゅう)
stink club : 悪臭を放つ棍棒
stipes : （複 . -pites）、蝶咬節(ちょうこうせつ)〔ラテン語〕
stipital : 蝶咬節の、柄の【植物学】
stipple : 点々を付ける
stochastic rainfall event : 確率（論）的な降雨イベント
stochasticity : 確率性
stolon : 匍匐枝(ほふくし)、匍匐茎(ほふくけい)、走根、芽茎
stomodeal valve : 噴門弁(ふんもんべん)
stop codon : 終止コドン、停止コドン
storage protein : 貯蔵タンパク質
store sperm : 精液を貯蔵する、精子を貯蔵する
STR : 短鎖縦列反復配列、縦列（型）反復配列（Short Tandem Repeat）
straight flight : 直進飛翔
straight white band : 一文字状白帯
straightforward : 分かりやすい、すっきりとした、単刀直入な、ごまかしのない
strain : 系統、株、菌株
strataum : 層
strategy : 戦略
stratified random design : 層化無作為抽出法
stratify : 階層化する、階層に分ける
stray : 道に迷う
stray butterfly : 迷蝶(めいちょう)
stray light : 迷光
stray species : 迷い種
streak : 条
stream : 小川、流れ
stream assemblage : 水生生物群集
stream invertebrate : 小川に生息する無脊椎動物、河川性無脊椎動物
Strepsiptera : ネジレバネ目、撚翅目(でんしもく)
streptomycin : ストレプトマイシン

stress factor : ストレス因子、ストレス要因
stress-induced volatile : ストレス誘導性揮発性物質
stress-induced volatile response : ストレス誘導性揮発物質応答
stressful climatic condition : ストレスが多い気象条件
stretch : 広げる
stretch out : 広げる
stretch receptor : 伸張受容器
strict clock : 精密な時計
stridulation : 摩擦発音
striking : 派手な、著しい、際立った
striking discovery : 際立った発見
striking eyespot : 派手な眼状紋、目立つ眼状紋
striking variation : 著しい変動
strikingly similar : 極めて類似な
stringent condition : 過酷な条件、厳しい条件
stripe : 縞
striped policeman butterfly : シロオビアフリカアオバセセリ
striping : 縦縞模様
strong daylight : 昼間の強い自然光
strong developmental constraint : 発生上の強い制約、成育上の強い制約
strong dominance : 強力な優性
strong magnetic field : 強磁場
stronghold : 生息地
strongly relate : 強く関係する
strongly suggest : 強く示唆する
structural : 構造的な
structural color : 構造色
structural difference : 構造的差異、構造的差
structural elaboration : 構造的同化
structural protein : 構造タンパク質
structure determination : 構造の決定
structure of gene : 遺伝子の構造
structure-activity relationship : 構造活性相関
study area : 調査地域、研究対象地域
study organism : 研究対象(微)生物
study region : 調査地域
study species : 研究対象の種
stylate : 披針(ひしん)形の、針状の、尖筆状(せんぴつじょう)の(触角の形状)
stylet : 口針
Styrofoam plank : スタイロフォーム厚板(発砲スチロールの厚板)
sub coastal vein : 亜前縁脈、Sc 脈
sub- : 亜 -、下 -〔ラテン語〕
sub-cloned : サブクローン化した(DNA から特定の部位を切り出してベクターにつなぎ単離すること)
subadult : 終齢、終令
subalare : 亜翅節片、後翅基骨(胸部側板の節片)、後側板関節節片
subalpine zone : 亜高山帯
subapical : (翅)亜頂室、(翅)亜端室、亜頂端の、亜翅頂部、亜翅頂部帯、亜翅端部
subapical cell : 亜頂室、亜端室
subarctic species : 亜北極圏種
subarctic zone : 亜寒帯
subclass : 亜綱(あこう)
subcosta : 亜前縁
subcostal vein : 亜前縁脈(あぜんえんみゃく)、Sc 脈
subdiscal line : 内横線
subdiscipline : 副(次)専門分野
subdorsal : 亜背域
subequal : ほぼ同じ大きさの、ほぼ等長の
subesophageal ganglion : 食道下神経節、食道下神経球
subfamily : 亜科
subfamily name : 亜科名
subfunctionalization : 潜在的機能分化
subg. : 亜属(subgenus)の略語
subgeneric name : 亜属名

subgeneric speciation : 亜属的分化
subgenital plate : 生殖下板(せいしょくかばん)、下蓋板(かがいばん)、亜生殖板(腹部)
subgenual organ : 膝下(しっか)器官
subgenus : 亜属
subgenus name : 亜属名
subjective : 主観的
subjective synonym : 主観同物異名、主観異名
sublateral : 下外側
sublethal dose : ほとんど致死量に近い用量、亜致死投与量
sublethal effect : 致死量寸前の効果、致死量に近い効果
submandibular gland : 顎下腺
submarginal : 亜外縁部
submarginal band : 亜外縁の縞状バンド
submedial : 亜中央帯
submentum : 下唇亜基節、亜基節(あきせつ)、亜基板
submetapleural carina : 後胸側板下縁隆起線(後胸側板の腹方縁にある隆起線)
submitted : 投稿中
subocular groove : 眼下線
suboesophageal ganglion : 食道下神経節、食道下神経球
suborder : 亜目
subordinate taxon : 従属的タクソン
subpopulation : 部分個体群【生態学】、サブ個体群、分集団、部分母集団【統計学】
subreniform spot : 亜腎状紋、亜腎臓紋(蛾類)
subsample : 副(次)標本(を取る)
subsequent analysis : その後の解析
subsequent designation : 後指定、その後の指示
subsequent monotypy : 後世の単型
subsequent spelling : 後綴(つづ)り
subsidence : 沈降、沈下、陥没
subsist : 生存する、存続する
subsocial : 亜社会性
subsocial route : 亜社会性ルート
subsp. : 亜種(subspecies)の略語(「ssp.〈亜種〉」を参照)
subspecies : 亜種
subspecies name : 亜種名
subspecific name : 亜種小名
subspp. : 亜種(subspecies)の複数形の略語(「sspp.〈亜種〉」を参照)
substantial change : 実質的な変化
substantial difference : 実質的な差異、相当な差異
substantial fitness cost : 相当な適応度費用
substantial saturation : 十分な飽和
substantial time : 実質的な時間
substantial variation : 実質的な変異、有意な変異
substitute name : 代用名
substitution : 置換
substitution model : 置換モデル
substitution pattern : 置換パターン
substitution rate : 置換速度、置換率
substitutional saturation : 置換の飽和
substrate : 基質、底質(例えば、湖の中の基質)
substrate-borne chemical cue : 基質由来化学的刺激、基質を介した化学的刺激
subtaxa : 細分類
subtaxon--taxon ratio : 細分類数 - 分類数の比
subterminal : 亜末端の、亜端の
subterminal area : 亜外縁部(蛾類)
subterminal line : 亜外縁線(蛾類)
subtle climatic : 敏感な気候
subtracted library : 差引きライブラリー、差分ライブラリー
subtraction efficiency : 差引き効率
subtractive hybridization : サブトラクショ

ン法
subtribe : 亜族
subtribe name : 亜族名
subtropical : 亜熱帯
subtropical area : 亜熱帯地域
subtropical district : 亜熱帯地域
subtropical group : 亜熱帯グループ
subtropical population : 亜熱帯の個体群、亜熱帯性個体群
subtropical semideciduous forest : 亜熱帯性半落葉樹林
subtropical species : 亜熱帯性の種
subunit : サブユニット
suburbs of Tokyo : 東京近郊
subzero-temperature : 零下の気温、氷点下の気温
successful amplification : 成功した増殖
successful copulation : 交尾成立
successfully emerged adult insect : 首尾よく羽化した成虫
succession : 遷移、継承、サクセッション
succession of butterfly fauna : 蝶相の変遷
succession of host plant : 寄主植物の遷移
successive collection : 歴代のコレクション、継続的なコレクション
successive evolutionary process : 連続的進化過程
successive instar : 連続的な齢期
succulent plant : 多肉植物
suck nectar : 吸蜜する
sucker : ひこばえ、吸盤、吸器
sucrose : ショ糖、スクロース
sucrose intake : ショ糖摂取
sucrose solution : ショ糖溶液、スクロース溶液
sudden dash : 急激な突進
sufficient datum : (複 . -ta)、十分なデータ
suffix : 接尾語
sugar solution : 砂糖溶液
sugar transporter gene : シュガートランスポーター遺伝子、ショ糖トランスポーター遺伝子
sugar-alcohol : 糖アルコール
suggest : 示唆する、提案する
suitability : 適切性、適合性
suitable breeding habitat : 好適な繁殖地
sulcus : (複 . -ci)、溝〔ラテン語〕
Sulphurs : モンキチョウ亜科
summary : 摘要、要約
summative : 加算的
summer diapause : 夏眠、夏休眠
summer form : 夏型
summer form adult : 夏型成虫
summer form induction : 夏型誘導
summer generation : 夏世代
summer morph : 夏型
summer temperature : 夏の気温、夏場の気温
summer-form female : 夏型雌
summit of a mountain : 山頂
sun bathing : 日光浴
sun compass : 太陽コンパス
sun's azimuth : 太陽方位
sun-worshipper butterfly : 日光浴好きな蝶
sunflower : ヒマワリ
sunlight : 日光、太陽光
sunlight intensity : 太陽光の強度、太陽光の強さ
sunny area : 日当たりのよい地域
sunny condition : 日当たり条件
sunset : 日没
sunset azimuth : 日没の方位、日没の方位角
Sunset moth : ニシキオオツバメガ
sunspot : 日溜り、太陽黒点、日光浴に出かける場所、日光に恵まれた地点
super gene : 超遺伝子
super- : 上 -〔ラテン語〕
supercool : 過冷却
supercooling point : 過冷却点

superfamily : 上科(じょうか)、超科
superfamily name : 上科名
superficial : 浅(せん)、表面(上)の、みかけの
superficial taxonomic decision : 表面的な分類学的意思決定、表面上の分類学的(意思)決定
supergene : 超遺伝子、スーパー遺伝子
supergene architecture : 超遺伝子構成
supergene family : 超遺伝子族
superior : 上部の、上(方)の、上(じょう)〔ラテン語〕
supernumary eyespot : 過剰に形成された眼状紋
supernumerary instar : 過剰な齢、過剰脱皮齢
supernumerary juvenile instar : 過剰脱皮の若齢
supernumerary limb : 過剰肢
superorder : 上目
superoxide dismutase : 超酸化物不均化酵素、超酸化物イオン分解酵素、スーパーオキシドディスムターゼ
supplementary feeding : 補助的摂食
supplementary note : 補注
supplementary queen : 補充女王
supplementation : 補強、強化
supplier's instructions : 供給者(側)の指示書
supply : 供給量
support value : 支持値、支持度の値、サポートバリュー、支持度
supporting structure : 支持構造
supposed territory : 想定上の縄張り
suppressed infection density : 抑制された感染密度
suppressed name : 抑制名、抑圧名
suppressed work : 抑制された著作物
suppression : 抑制、抑圧
suppression factor : 抑制要因
suppression of ovarian development : 卵巣の発達停止
suppression subtractive hybridization : 抑制サブトラクティブ・ハイブリダイゼーション法、SSH 法、抑制差引ハイブリダイゼーション法(2つの異なるRNAサンプルから、片方にのみ特異的に存在する遺伝子を濃縮する方法)
supra- : 上の -、超えた -〔ラテン語〕
suprachiasmatic nucleus : 視交叉上核
supraclypeal area : 頭盾上区
suprageneric : 属よりも高位の
supralateral : 上外側
suralare : 胸背板の前翅突起(胸部背板の盾板の一部)
surface area : 表面領域、表面積
surface structure : 表面構造
surface topography : 表面形状
surface water : 地表水
surface/mass ratio : 表面積/体重比、表面積/質量比
surgical experiment : 外科的実験
surgical manipulation : 外科的操作
surplus stock : 過剰ストック、余剰ストック
surprisingly common : 驚くほど普通
surrogate of activity : 活動の代行物、活動の代替物
surround : 辺縁、周辺、周囲
surrounding : 周辺、周囲
surrounding area : 周辺域
surrounding vegetation : 周辺植生、周縁植生
survey design : 調査計画
survival : 生き残り、生存
survival analysis : 生存分析、生き残り解析、生存時間解析
survival rate : 生存率
survival value : 生存価、生存上の利益
survivin : サバイビン(アポトーシス抑制タンパク質)
surviving progeny : 生存する子孫
survivor : 生き残り、遺存種

survivorship curve : 生存曲線
susceptibility : 感受性、易感染性
susceptible : 感染しやすい、感受的な
sustain : 支える、維持する
sustainable development : 持続可能な開発
sustainable existence : 持続可能な生存
sustainable woodland management : 持続可能な森林地帯の管理
sustainable yield : 持続収穫量
sustaining mechanism : 維持機構、持続機構
sustenance : 食物源
suture : 縫合線(ほうごうせん)
swallowtail butterfly : 燕尾系アゲハチョウ、尾状突起付きアゲハチョウ
Swallowtails : アゲハチョウ亜科、アゲハチョウ科
swamp : 湿地、沼地
sward : 草地
swarm : 群飛、昆虫の群れ
swarm intelligence : 群知能
swarm robot : 群ロボット
sweep net : 捕虫網、地引網
sweetish substance : 甘露(物質)
swidden : 焼畑
switch off : やめさせる、止める、スイッチを切る
switch supergene : スイッチ超遺伝子
Switching Mechanism At 5'-end of the RNA Transcript : SMART 法
swollen : ふくれた
sword-grass brown butterfly : ベニモンクロヒカゲ
sword-like tail : 剣状の尾状突起
Sxl : 性致死的遺伝子(*Sex-lethal* gene)
symbiont density regulation : 共生微生物密度の調節
symbiont infection : 共生微生物感染
symbiont-depleted insect : 共生微生物が激減した昆虫
symbiont-induced reversal of insect sex : 共生微生物による昆虫の性転換
symbiosis : (複 . -ses)、共生〔ラテン語〕
symbiote : 共生者
symbiotic association : 共生関係
symbiotic bacteria : 共生細菌
symbiotic microorganism : 共生微生物
symbiotic virus : 共生ウイルス
symmetrical topology : 対称的樹形、相称的樹形
symmetrical wing damage : 対称的な翅の損傷
symmetry system of band : 縞状バンドの相称系、相称系の縞状バンド
symmetry system of colored band : 着色された縞状バンドの相称系
sympatric : 同所性の、同所性
sympatric allele : 同所的対立遺伝子
sympatric co-mimetic postman butterfly : 同所的相互擬態型のベニモンドクチョウ
sympatric distribution : 同所的分布
sympatric form : 同所的変異型
sympatric mimetic taxon : 同所的擬態分類単位
sympatric morph : 同所的形態
sympatric race : 同所的系統
sympatric speciation : 同所的種分化
sympatric species : 同所(的)種
sympatrically : 同所的に
sympatry : 同所性
sympatry/allopatry relationship : 同所性/異所性の関係
symplesiomorphy : 共有祖先形質、祖先形質共有
symptom : 病徴、症状
syn- : 共に -、同時に -、類似 -、合成の -〔ギリシャ語〕
syn. : 同物異名(synonym の略語)、シノニム、異名

synapomorphy : 共有派生形質、派生形質共有
synchronic : 共時性、共時的な
synchronicity : 共時性(形態種 : 形態的特徴で区別された種)
synchronized biennial butterfly : 同期化した二年生の蝶
synchronized emergence : 斉一的発生、斉一的羽化
synchronous : 斉一の、同時(性)の
synchrony : 同時発生
syndrome : 症候群、形質群、シンドローム
synephrine : シネフリン
synephrine receptor : シネフリン受容体
synergia : 相乗作用、共力作用〔ラテン語〕
synergist : 協力者
synergistically : 相乗的に
syngamy : 配偶子合体
synomone : シノモン
synonym : 同物異名、シノニム、異名
synonymic note : 同物異名の検討
synonymous : 同義の
synonymous change : 同義置換
synonymous site : 同義サイト
synonymous substitution : 同義置換
synonymy : 異名表、異名関係、異名リスト
synteny : シンテニー(同一の染色体に数個の遺伝子がのっていること)
synthesis : 総合、合成、統合、統一
synthesis of ecdysteroid hormone : エクジステロイドホルモンの合成
synthesize : 合成する
synthetic understanding : 総合的理解
synthorax : 合胸
syntype : シンタイプ、等価基準標本
system biology : システム生物学
system of nature : 自然の体系
systematic coverage : 計画的被覆
systematic difference : 系統的相違、系統的差異
systematic investigation : 系統的な調査
systematic study : 系統学的研究、系統分類学的研究
systematics : 分類学、体系学、系統学、系統分類学
systemic : 全身の、組織(体系)的な

t

t-test : t 検定
tactics : 戦術
tactile sense : 触覚
tactile seta : 触知性刺毛、触刺毛
tactile signal : 触覚刺激
tagging : タギング、タグ付け
tagma : (複 . -ata)、合体節、連続環節体
taiga : タイガ、北方針葉樹林
tail : 尾状突起、尾
tail photoreceptor : 尾端光受容器
tailed female form : 雌有尾型、有尾型雌
tailless : 尾状突起を欠く
tailless female form : 無尾型雌
tailwind drift : 追い風ドリフト、追い風偏流
Talbotia nagana : カルミモンシロチョウ
tall grassland : 背の高い草地
tally : 集計する、一致する、符合する
tandem birth : 縦列出現、縦列発生
tandem duplication : 縦列重複
tannin content : タンニン含量
tanning : 硬化・着色(黒く硬化する)
tansy extraction : ヨモギギク抽出
tap water : 水道水
taper off : 次第に減る
tapered : だんだん細くなる
tapetum : タペータム、反射層板、内面層、タペート
target gene : 標的遺伝子
target location : 目標位置
target organ : 標的器官
target PCR product : 目的の PCR 産物

target species : 対象種
tarsal : ふ節の
tarsal claw : 脚ふ節の鉤爪
tarsal comb : ふ節の剛毛列
tarsal contact chemosensory hair : ふ節の接触化学感覚毛
tarsal formula : ふ節式(ふ節を形成する環節の個数で、前脚・中脚・後脚のふ節の数を示したもの)
tarsomere : ふ節亜節、ふ節小節
tarsus : (複. -si)、跗節(ふせつ)、ふ節〔ラテン語〕
taste : 味覚、味見
taste sensillum : 味覚感覚子
taste-enhancing effect : 味覚増強効果
tautonymous name : 反復名
tautonymy : 同語反復
taxa group : 分類群
taxis : 走性
taxon : (複. -xa)、タクソン、分類単位、分類学的単位、分類群〔ラテン語〕
taxon coverage : 分類学的範囲
taxon density : 分類密度
taxon occurrence : 分類出現頻度
taxon oriented survey : 分類本位の調査
taxon richness : 分類数
taxon sampling : 分類サンプリング
taxon sampling curve : 分類サンプリングの曲線
taxon-rich group : 分類群が多いグループ
taxonomic array : 分類配列、分類系列
taxonomic boundary : 分類学的境界
taxonomic composition : 分類組成、系統組成、分類学的な構成
taxonomic designation : 分類学的名称、分類学上の名称
taxonomic distribution : 分類学的分布
taxonomic field : 分類分野
taxonomic group : 分類(学的)群
taxonomic hierarchy : 分類の階層構造
taxonomic information : 分類学(的)情報
taxonomic placement : 分類学的位置
taxonomic position : 分類学的位置
taxonomic rank : 分類学的階級、分類階級、分類順位
taxonomic ratio : 分類比
taxonomic review : 分類学的検証
taxonomic status : 分類学的位置づけ
taxonomic study : 分類学的研究
taxonomic taxon : 分類学的タクソン
taxonomical status : 分類学的位置、分類学上の位置
taxonomy : 分類法、分類学、分類
TE : 転移因子(Transposable Element)
tea leaf : 茶葉
tear : 引き裂く、引きちぎる
tear-shaped : ナスビ形の
technical assistance : 技術支援
technical replicate : 技術的反復実験(実験操作や装置に起因する誤差を小さくするために行う反復実験)
Teg : 翅基片(Tegula)
tegula : (複. -ae)、肩板(けんばん)、翅基片〔ラテン語〕
tegular arm : 肩腕
tegumen (tg) : テグメン、覆片
tegument : 外被、外皮、皮膚
telemetry datum : (複. -ta)、テレメーターデータ、テレメトリデータ、遠隔計測データ
tell apart : 識別する、区別する
telomere : テロメア、末端小粒
telotarsus : 末ふ節
telson : 尾節(びせつ)
temperate area : 温度領域(低温域、常温域、高温域の各温度領域)
temperate climate : 温帯気候
temperate forest : 温帯林
temperate forest and relictual genus of Asian origin : アジア温帯林遺存属

temperate forest and wide-distributed genus : 温帯森林広分布属
temperate forest and wide-distributed genus of American origin : アメリカ温帯林広分布属
temperate forest and wide-distributed genus of Asian origin : アジア温帯林広分布属
temperate genera : 温帯属
temperate genotype : 温帯性遺伝子型、温帯産遺伝子型
temperate group : 温帯グループ
temperate latitude : 温帯緯度、温帯地方
temperate pierid butterfly : 温帯(性)のシロチョウ科の蝶
temperate population : 温帯域個体群
temperate rainforest : 温帯雨林
temperate region : 温帯地域
temperate species : 温帯性の種
temperate zone : 温帯
temperature : 温度
temperature compensation : 温度補償(性)
temperature condition : 温度条件
temperature cycle : 温度周期
temperature dependence : 温度依存性
temperature effect : 温度効果、温度の影響
temperature environment : 温度環境
temperature fall : 気温低下
temperature law : 温度法則
temperature sensitibity : 温度感受性
temperature-induced sex difference : 温度誘導性性差
temperature-size rule : 温度 - 体格則、温度 - サイズ則
template : 鋳型、テンプレート
template absorbance spectrum : 吸収スペクトルテンプレート
temporal and spatial information : 時空情報
temporal and spatial separation : 時空分離
temporal autocorrelation : 時間的自己相関
temporal dependence : 時間依存性
temporal distribution : 一時的分布(発生)、時間的分布
temporal environmental variation : 一時的な環境変動、環境の時間的変化
temporal heterogeneity : 時間的不均一性、時間的異質性、一時的不均質性
temporal isolation : 時間的隔離、一時的隔離
temporal landscape dynamics : 時間的な景観動態、景観の時間的動態
temporal predictability : 時間的予測可能性
temporal prosperity and decline : 時間的盛衰
temporal scale : 時間スケール、時間尺度
temporal variability : 時間的変化、時間的変動
tendency : 傾向
tendon cell : 腱細胞
teneral : テネラル、不整成虫(羽化したてでまだ体がやわらかい成虫のこと)
tentacle : 触角、触手、伸縮突起
tentacle nectary organ : 伸縮突起様蜜腺、伸縮蜜腺
tentacle organ : 伸縮突起器官
tentacular organ : 触手状器官
tentatively : 一応
teratological specimen : 奇形標本
tergal : 背板の
tergal fissure : 背板裂、背板溝、背板の割目
tergite : 背板
tergopleural muscle : 側背板筋、背側板筋
tergum : 背板
termen : 外縁
terminal : 末端
terminal area : 外縁部(蛾類)
terminal branch : 末端分岐、末端の枝
terminal filament : 端糸(たんし)、ターミナルフィラメント
terminal fusion : 末端融合型
terminal line : 外縁線(蛾類)

terminal mechanism : 末端機構
terminal spur : 末端突起、末端の距
terminal stage of larva : 幼虫末期
terminal taxa : 末端分類群
terminalia : 外部生殖器、腹部末端節
terminate diapause : 休眠を終了させる
terminus : (複 . -ni)、末端
termite : シロアリ
terpene : テルペン
terpene synthase : テルペン合成酵素
terpenoid defensive secretion : テルペノイド防御分泌物
terrestrial arthropod : 陸生節足動物、陸域節足動物、陸産節足動物
terrestrial biodiversity : 陸生動物の生物多様性
terrestrial biome : 地球上の生物群系、陸のバイオーム
terrestrial crustacean : 陸域甲殻類、陸産甲殻類、陸上甲殻類
terrestrial ecosystem : 陸域生態系、陸上生態系
terrestrial habitat : 地上の生息地
terrestrial life : 地球生物、陸上生物
territorial activity : 縄張り活動、縄張り行動
territorial behavior : 占有行動、テリトリー行動、ナワバリ行動
territorial defence : テリトリーの防衛
territorial fluctuation : テリトリーの変動
territorial male : テリトリー占有雄
territorial marker : テリトリーマーカー
territorial tenure : テリトリーの保有期間、テリトリーの保有
territorialism : ナワバリ制
territoriality : ナワバリ制
territorium : ナワバリ〔ラテン語〕
territory : テリトリー、縄張り、ナワバリ
territory contact area : テリトリー接触域
Tertiary period : 第三紀(地質時代の新生代の一時代)
tertiary structure : 三次構造
test of historical biogeographical hypothesis : 歴史的生物地理学的仮説の検定
test plant : 検定(対象)植物
test set : 試験セット
testing hypothesis : 仮説を検定する
testis : (複 . -es)、精巣〔ラテン語〕
Tethyan region : テチス区
Tethyan type : テチス型
tetracycline : テトラサイクリン
tetracycline hydrochloride : テトラサイクリンヒドロクロリド、 テトラサイクリン塩酸塩
tetracycline-containing artificial diet : テトラサイクリン入り人工飼料
tetracycline-supplemented diet : テトラサイクリン補給飼料
tetrapyrrolic pigment : テトラピロール系色素
Tg mouse : トランスジェニックマウス(Transgenic)
tga : 肩腕(tegular arm)
TH : チロシンヒドロキシラーゼ、チロシン水酸化酵素(Tyrosine Hydroxylase)
thale-cress genome : シロイヌナズナのゲノム
thaw : 解凍する
thelytoky : 雌性産生単為生殖(しせいさんせいたんいせいしょく)、産雌単為生殖
theoretical argument : 理論的論争
theoretical ground : 理論的基礎、理論的根拠
theoretical investigation : 理論的調査
theoretical model : 理論的モデル
theoretical progress : 理論的進展、理論的展開
theoretical study : 理論的研究
theoretical treatment : 理論的取り扱い
theoretical work : 理論的研究
theoretician : 理論家

theory of island biogeography : 島の生物地理学理論
theory of speciation : 種分化(の)理論
thereafter : その後は
therein : その中に
thermal constant : 有効積算温量定数、有効積算温度定数
thermal cycling profile : 温度サイクルプロファイル
thermal hysteresis protein : 不凍タンパク質
thermal inertia : 熱慣性、温度慣性、温度遅鈍
thermal reaction : 温度反応
thermo sensor : 温度計、温度センサー
thermocouple : 熱電対
Thermocycler : サーモサイクラー(精密な温度制御と迅速な温度変化によってポリメラーゼ連鎖反応〈PCR〉を行う装置)
thermoperiodic response : 温度周期反応
thermophase : 高温期、温度相
thermoreceptor : 温度受容器
thermoregulation : 体温調節、温度調節
thermoregulatory function : 体温調節機能
Theropods : 獣脚類
thiacloprid : チアクロプリド(「ネオニコチノイド系」の殺虫剤)
thiamethoxam : チアメトキサム(「ネオニコチノイド系」の殺虫剤)
thick : 厚い
thick ridge of cuticle : 表皮の厚みのある隆起、クチクラの厚みのある隆起
thin membrane : 薄膜
thinning the canopy : 林冠の間引き
thoracic : 胸部の
thoracic flight muscle : 胸部の飛翔筋
thoracic leg : 胸脚、胸部の脚
thoracic muscle : 胸部筋肉(組織)
thoracic nervous system : 胸部神経系
thoracic structure : 胸部構造
thoracic temperature : 胸部温度
thorax : 胸部(きょうぶ)〔ラテン語〕
thorax garter : 胸部の帯糸
thorax mass : 胸部重量
thorax muscle : 胸部筋、胸部の筋肉
thorax shape : 胸部の形
thorax width : 胸部幅
thorn : とげ
thorough system of quality assurance : 一貫した品質保証体制
threat : 脅威
threaten : 脅かす
threatened species : 絶滅危惧種
threatened wildlife group : 絶滅の恐れのある野生生物群、絶滅危惧の野生生物群
three dimensional structure : 三次元構造
three generations a year : 年三世代
three prime untranslated region : 3’ 非翻訳領域(3’UTR)
three-allele system : 三つの対立遺伝子システム
threshold : 閾(いき)値
threshold intensity : 閾(いき)値強度、閾値の強さ
threshold level : 閾(いき)値レベル
threshold size : 閾(いき)値サイズ
threshold temperature for development : 発育限界温度
thrive : 繁殖する、繁栄する
throughout the year : 年中
thyridium : (複 . -dia)、鏡紋〔ラテン語〕
Thysanoptera : アザミウマ目、総翅目
Thysanura : シミ目、総尾目
tibia : (複 . -ae)、脛節(けいせつ)〔ラテン語〕
tibial : けい(脛)側、脛節の
tibial claw : 脛節の爪
tibial spur : 距刺(きょし)、脛節棘
tibial tuft : 脛節の毛房
tide : 風潮、流れ
Tien-Shan Mountains : 天山山地、天山山脈

tiger moth : ヒトリガ
tiger pierid butterfly : ベニオビコバネシロチョウ
tiger swallowtail butterfly : (メスグロ)トラフアゲハ
tiger-patterned morph : トラ模様の形態
tight genetic association : 緊密な遺伝相関
tightly linked genetically : 遺伝的に強く連鎖している
tightly linked locus : 密に連鎖した遺伝子座
Tillyard notation : ティリャード表記法
Tillyard's notation : ティリャード表記法
Tillyardian notation : ティリャード表記法
timber : 材木、立木
time estimate : 時代推定
time limitation : 時間制約、時間制限
time measuring mechanism : 測時機構
time of adult emergence : 成虫の羽化時期
time of appearance : 活動時間、出現時刻、姿を見せる時刻
time of disappearance : 終了時刻、姿を消す時刻
time of the day : その日の時刻
time scale : 時間スケール、時間尺度
time signal : 時刻信号
time to death : 死亡までの時間
time tree : 時間樹
time-calibrated phylogenomic tree : 時間較正したゲノムベースの系統樹
time-consuming task : 時間を費やす作業、時間がかかる作業
time-measurement function : 測時機能
time-measuring response : 測時反応
time-period : 時期
time-series datum : (複 . -ta)、時系列データ
time-since-colonisation : 定着からの経過時間、移入からの経過時間
Timeline : 年譜、歴史年表
timing of flowering : 開花時期、開花期
timing of pupation : 蛹化時期
timing of the origin, the : 起源時期
tip : 先端、末端、崩す
tip node of tree : 系統樹の先端結節点
tip of abdomen : 尾端、腹部末端
tip of the iceberg : 氷山の一角
tissue : 組織(動植物の細胞の)
tissue damage : 組織損傷
tissue paper : ティッシュペーパー
tissue tropism : 組織向性、組織親和性
tissue-specific expression : 組織特異的な発現
tissue-specific quantitative PCR technique : 組織特有の定量的 PCR 法、組織特異的定量 PCR 法
titer : タイター、力価
title : 表題
titration experiment : 滴定実験
tobacco budworm moth : オオタバコガ
tobacco hornworm moth : タバコスズメガ
token stimulus : 誘導する要因
tolerance : 耐性
tolerance of shady habitat : 日陰生息地の許容性、日陰生息地の耐性
tolerance to cold : 耐寒性
Toluidine blue : トルイジンブルー
tongue : 口吻(こうふん)
tonic : 緊張性の、持続性の
tooth-like : 歯形の
toothed : ぎざぎざの
top-down effect : トップダウン効果
topographic map : 地形図
topographical : 地形的
topographical model : 地形図モデル
topography : 地形
topological distance : 樹形距離
topology : 樹形、トポロジー、位相(幾何学)
topotype : トポタイプ、原地基準標本、原地模式標本

torchlight : たいまつの火、懐中電灯の明り
tormogen : 窩生(かせい)
tormogen cell : ソケット細胞、窩生細胞
tornal : 後角部、肛角部、後縁角、肛角
tornal dash : 後角縦線(蛾類)
tornus : 後角部、肛角部、後縁角、肛角(「円」や「盤」を意味するラテン語)
torpor : 休眠状態、冬眠状態
torrid zone : 熱帯
torsion : ねじれ
torulus : (複 . -li)、ソケット環節、球節、触角挿入孔〔ラテン語〕
total dry mass : 総乾重量、総乾燥重量
total effective temperature : 有効積算温度、有効温量、有効積算温量
total evidence analysis : 全証拠解析
total genome size : 全ゲノムサイズ
total map length : 地図の全長
total mass : 総重量
total number of genes in human genome : ヒトゲノム中の遺伝子総数
total number of nucleotide per human genome : ヒトゲノムの塩基総数
total range : 総行動範囲
totipotency : 全能性
touch : 触覚、触感、触れること
tough leaf : 固い葉
toughening of leaf : 葉の硬化
tourism : 観光、観光事業
toxic : 有毒な、毒性の
toxic breakdown product : 毒素分解産物
toxic compound : 毒性化合物
toxic model : 有毒のモデル
toxicity : 毒性
toxin : 毒素、毒物
trabecula : 柱、小柱、トラベキュラ
trachea : (複 . -ae)、気管(きかん)〔ラテン語〕
trachea system : 気管系
tracheal air sac : 気管の気のう
tracheal branch : 気管分枝、気管分岐
tracheal system : 気管系
tracheal twig : 気管小分枝、気管小分岐
tracheolar layer : 毛細気管層
tracheole : 気管小枝
tractable model : 扱いやすいモデル
trade-off : トレードオフ、拮抗的関係
trading partner : 貿易相手
traditional approach : 伝統的なアプローチ、従来のアプローチ
traditional morphology-based taxonomy : 伝統的な形態分類学
trail marking : 道しるべ
trail pheromone : 道しるべフェロモン
trailing margin : 後縁(蛾類)
trait : 形質
trait complexity : 形質の複雑性
trait specific : 形質特異的、形質固有
trait-mediated indirect effect : 形質の変化を介する間接効果
trajectory of evolution : 進化軌跡、進化軌道
trampling : 踏みつける
trans effect : トランス効果(陰性の配位子〈ligand〉がシス位よりもトランス位にある配位子に大きな不安定性をもたらすこと ;「cis effect〈シス効果〉」を参照)
trans- : 越えて -、他の側へ -、別の状態へ -〔ラテン語〕
trans-element : トランスエレメント(他からやってきてシスエレメントの塩基配列を認識して結合し、遺伝子の転写を調節するタンパク質)
trans-membrane receptor : トランスメンブラン受容体、膜貫通受容体
trans-migration : 通過移動
transaminase : アミノ基転移酵素
transcriptase : トランスクリプターゼ、転写酵素
transcription : 転写

transcription factor : 転写制御因子、転写因子
transcription start site : 転写開始点（TSS）
transcriptional regulator : 転写制御因子
transcriptional repressor : 転写リプレッサー、転写産物抑制因子
transcriptional reprogramming : 転写（の）リプログラミング
transcriptome : トランスクリプトーム（特定の状況下において細胞中に存在するすべての mRNA〈ないしは一次転写産物〉の総体を指す呼称）
transcriptomics : トランスクリプトミクス（トランスクリプトームを扱う学問）、分子遺伝学
transduce : 変換する
transect : 観察路、観察地、トランセクト
transect census : 観察路調査、トランセクト調査
transect method : トランセクト法
transect recorder : トランセクトの記録者
transect survey : 観察路調査、トランセクト調査
transfection : 遺伝子導入、核酸導入、形質移入、トランスフェクション（人為的操作で遺伝物質を導入すること）
transfer : 媒介する、移動、注入する
transfer protein : 運搬タンパク質
transfer RNA : 転移 RNA、トランスファーRNA（tRNA）
transferred gene : 導入された遺伝子
transform : 変態する
transformation : 変態、形質転換
transformation mechanism : 変態機構
transformed distance method : 変換距離法
transformer : 変化させるもの
transgenerational induction of defence : 世代をわたる防御誘導、世代を越える防御誘導
transgenic mouse production : トランスジェニックマウスの作製、トランスジェニックマウスの作出、遺伝子導入マウスの産出
transgenic plant : 遺伝子導入植物、トランスジェニック植物、形質転換植物、遺伝子組換え植物
transgenic pollen : トランスジェニック花粉、遺伝子導入花粉、形質転換花粉
transgression : 違法、違反、犯罪
transient : 移行期
transient rise in temperature : 過渡的な温度上昇
transiently : 過渡的に
transit flight : 通過移動飛翔
transit record : 移動記録、移動履歴
transit time : 移動時間（目的地までの所要時間）
transition : 転移型置換、転位、塩基転位
transition from race to a species : 亜種から種への転移
transitional form : 移行型
transitional zone of food plant : 食草の移行帯
transliteration : 換字
translocated birth : 転座した発生
translocation : 移動、移植、転地、転座
translucent : 半透明の
transmembrance domain : 膜貫通領域、膜貫通ドメイン
transmission : 伝播、伝達、伝染
transmit vertically : 垂直に伝播する
transparent bluish white marking : 青白色の透明斑
transparent patch : 透明な斑紋部分
transparent region : 透過域
transpiration : 蒸散
transplant : 移植する
transponder : 送受信機、トランスポンダー
transport protein : 運搬タンパク質

transporter : トランスポーター、運搬役、膜輸送体
transposable element : 転移因子(TE)
transpose : 転移する
transposition : (遺伝子)転移
transposon : トランスポゾン
transversal : 横(おう)、横(の)
transverse : 横縫合線、横径の、横行の
transverse band : 横行眼状紋
transverse line : 横断横線(蛾類)
transverse plane : 横断面
transverse section : 横断面
transverse surface : 横断面
transversion : 転換型置換、転換
trap : トラップ
trap lining : トラップを敷くこと、トラップ覆工、トラップ用道づくり、わな道
trapezoidal sound pulse : 台形状の音波、台形波のパルス音
trapping : 罠(わな)、トラップを仕掛ける
traumatic insemination : トラウマを被った受精
travel up : 走る(神経が)
traveling time : 探索時間
Tre group : Tre 系統
treacle : 糖蜜
treated and control larva : 処理区の幼虫および対照区(非処理区)の幼虫
treated and untreated larva : 処理済みの幼虫および未(非)処理の幼虫
treated crop : (農薬)処理された作物
treated normal matriline : 処理された正常な母系群
treatise : 論文、学術論文、論説
treatment : 処理、トリートメント、取り扱い
treatment chamber : 処理室、処理チェンバー
treatment group : 処理群、処理グループ
treatment temperature : 処理温度
tree : 系統樹、木
tree canopy : 樹冠
tree diagram : 樹形図
tree nymph butterfly : オオゴマダラ
tree sap : 樹液
tree species : 樹種
tree trunk : 樹幹、木の幹
treefall gap : 倒木ギャップ
treeline : 森林限界、高木限界線
treetop : こずえ、木の頂上
trehalase : トレハラーゼ
trehalose : トレハロース
TRI reagent : トータル RNA 抽出試薬(商品名)(Total RNA Isolation)
tri- : 3-、三-〔ギリシャ語、ラテン語〕
triangle envelope : 三角紙
triangular-shaped : 三角形の
tribal maintenance : 種族維持
tribe : 族(分類階級の「ぞく」)
tribe level : 族水準、族レベル
tribe name : 族名
Tribolium castaneum : コクヌストモドキ(ゴミムシダマシ科の甲虫)
trichogen : 毛生、生毛、発毛、毛母
trichogen cell : 毛母細胞、生毛細胞
Trichogramma egg parasitoid : *Trichogramma* 属の卵寄生蜂(寄主の卵に産卵する寄生蜂)
Trichogramma wasp : *Trichogramma* 属のハチ
trichoid sensillum : 感覚毛、毛状感覚子
trichome : 毛、毛状突起
Trichoptera : トビケラ目、毛翅目
trichromatic : 三色性の
trichromatic color vision : 三色型色覚
tricin : トリシン
trifid : 三裂の
trifurcate : 三枝に分かれる、三叉に分かれる

trim : 切り取る
trimodal distribution : 三峰性分布
trinomen : 三語名〔ラテン語〕
trinomina : 三語名〔ラテン語〕
trinominal name : 三語名
triose phosphate isomerase : トリオースリン酸イソメラーゼ
triple copulation : 三重交尾
triplet : 三つ組、トリプレット、連続する塩基３個の１組
Tris-EDTA buffer : TE バッファー
Triton X-100 : トリトン X-100（タンパク質を可溶化する非イオン性界面活性剤）
tritrophic : 三栄養
trivial movement : 日常飛翔、日常行動
trivoltine zone : 三化性地域
TRIzol reagent : TRIzol試薬、トータルRNA分離用試薬（Total RNA Isolation）
tRNA : 転移 RNA、トランスファー RNA（transfer RNA）
trochantellus : （複 . -li）、第二転節、下転節、小転節〔ラテン語〕
trochanter : 転節（てんせつ）
trochantin : 小転節、副転節
Troidini : ジャコウアゲハ族
trophic level : 栄養段階（食物連鎖上の生産者、第一消費者、第二次消費者といった段階のこと）【生態学】
trophobiotic partner : 栄養共生パートナー
tropical arthropod : 熱帯産節足動物
tropical arthropod dataset : 熱帯産節足動物データセット
tropical beetle : 熱帯産甲虫、熱帯性甲虫
Tropical Brushfoots : カバタテハ亜科
tropical butterfly : 熱帯産の蝶、熱帯性の蝶、熱帯蝶
tropical climate : 熱帯気候
tropical crepuscular butterfly : 熱帯産の薄明活動性蝶
tropical dry : 乾燥熱帯性
tropical dry vegetation : 乾燥熱帯的植生
tropical forest : 熱帯林
tropical invertebrate community : 熱帯産無脊椎動物群集
tropical monsoon : 熱帯モンスーン
tropical monsoon region : 熱帯モンスーン地域
tropical morph : 熱帯型
tropical old-growth forest : 熱帯の老齢林
tropical rainforest : 熱帯雨林、熱帯降雨林
tropical region : 熱帯地方
tropical relic : 熱帯遺存
tropical second-growth forest : 熱帯の二次植生林
tropical seedbank dataset : 熱帯種子銀行データセット
tropical swallowtail moth : オオツバメガ
tropical trees : 熱帯林
tropical zone : 熱帯
tropics : 熱帯地方
true alpine butterfly : 純高山性蝶、真性高山蝶
True Brushfoots : タテハチョウ亜科
true butterflies, the (superfamily Papilionoidea) : アゲハチョウ上科
true diapause : 真正休眠
true leg : 真脚、真足
true phylogeography : 真の系統地理学
true species richness : 真の種数
true tree : 真の系統樹
true tropical species : 純熱帯性の種
truncate : 切断状
truncated version : 切断型
tryptophan : トリプトファン
TSS : 転写開始点（Transcription Start Site）
tube : 管鞘、管
tubercle : 伸縮突起、結節
tubular : 筒状の、管状の

tubular gland : 管状腺
tubular vein : 管状脈
tuft : 房、房状
Tukey least significant difference comparison : チューキー型最小有意差法の比較、チューキー型 LSD 法の比較
Tukey test : チューキー検定、チューキー法
Tukey's honestly significant difference test : チューキーの HSD 検定
Tukey's HSD post-hoc comparison : チューキーの HSD 事後比較法 (Honestly Significant Difference)
tundra : ツンドラ
tundra environment : ツンドラ的環境
turf : 芝生、芝土
Turing model : チューリングモデル(斑紋形成の反応拡散系モデルで、2つの仮想的な化学物質が、ある条件を満たして互いの合成をコントロールし合うとき、その物質の濃度分布は均一にならず、濃い部分と薄い部分が、空間に繰り返しパターン〈反応拡散波〉を作って安定し、シマウマのストライプ模様やキリンの網目模様、豹の斑点模様をも形成できることを示す数学的モデル)
Turkey : トルコ
turn, in : 順番に、順次
turning angle : 旋回角(度)
turnover rate of population : 個体群の入れ替わり率、個体群の回転率
turquoise : 青緑色の
tussock moth : ドクガ科の蛾の総称
twice as likely : たぶん二倍
twig : 小枝
twig mimic : 小枝に擬態
twilight : 日暮
twilight signal : 日暮信号
two brood : 二化
two remarkably distinct forms : 顕著に異なる二型
two strikingly distinct phenotypes : 二つの顕著に区別できる表現型
two-species stable equilibrium : 二種の間での安定平衡
two-tailed Z-test : 両側 Z 検定
two-way absorbance spectrum : 双方向吸収スペクトル
two-way factorial ANOVA : 二元配置分散分析法(ANalysis Of VAriance)
TYLCV : トマト黄化葉巻ウイルス、トマト黄化葉巻病(Tomato Yellow Leaf Curl Virus)
tympanal chamber : 鼓膜室、膜チャンバー
tympanal ear : 鼓膜耳、鼓膜を持つ耳
tympanal membrane : 鼓膜
tympanal organ : 鼓膜器官
tympanal surface : 鼓膜表面
tympanum : (複 . -na)、鼓膜〔ラテン語〕
type : タイプ
type fixation : タイプ固定
type genus : タイプ属
type horizon : タイプ層準
type host : タイプ宿主
type locality : タイプ産地
type of distribution : 分布系統、分布型
type series : タイプシリーズ
type species : タイプ種、模式種
type specimen : タイプ標本、模式標本、基準標本
type-specific input parameter : 型特異的な入力パラメータ
typical : 特有の、典型的な
typical autumn form : 典型的な秋型
typification : タイプ化
tyrosine : チロシン
tyrosine hydroxylase : チロシンヒドロキシラーゼ、チロシン水酸化酵素

Tyrosine recombinase gene : チロシンリコンビナーゼ遺伝子、チロシン組換え酵素遺伝子

u

U3X : *U3X* 遺伝子（３つのエキソンからなる非コード RNA をコードする遺伝子）（*Unknown-3-Exon*）
ubiquitous : 至る所にある、偏在する
ubiquitously expressed transcript : UXT、遍在発現転写体（多くの生物種で保存されている転写調節因子）
Ubx : 超双胸遺伝子（*Ultrabithorax* gene）
UK Butterfly Monitoring Scheme : 英国蝶類モニタリング計画（UKBMS）
UKBMS : 英国蝶類モニタリング計画（UK Butterfly Monitoring Scheme）
ulnar : 尺側の、尺骨（側）の
ultimate factor : 究極要因
ultimate mechanism : 究極機構、究極的機序、究極要因（進化学的観点を対象）
ultimate pattern : 究極パターン
ultimately : 最終的に、最後に、結局
ultra-pure water : 超純水、超高純度の水
ultrabithorax gene : 超双胸遺伝子
ultraconserved element : 極保存配列
ultrametric tree : 超計量的樹形図、ウルトラメトリック系統樹、超計量系統樹
ultrasonic click : 超音波クリック音
ultrasonic echolocation call : 超音波の反響定位音、超音波の反響定位鳴声
ultrasound cry of bat : コウモリの超音波音の鳴き声
ultrasound-sensitive : 超音波音感受性の
ultrasound-sensitive ear : 超音波音感受性の耳
ultraviolet : 紫外線
ultraviolet ray : 紫外線
ultraviolet ray reflection : 紫外線反射
ultraviolet region : 紫外域
Ulysses butterfly : オオルリアゲハ
un- : でない -、反対に -〔英語〕
unaffect : 影響を受けない
unaltered white : 白色のまま、無変更の白色
unaltered yellow : 黄色のまま、無変更の黄色
unambiguously demonstrate : 明快に実証する
unambiguously identified : 明快に同定された、明確に特定された
unavailability : 不適格性
unavailable name : 不適格名
unavailable nomenclatural act : 不適格な命名法的行為
unavailable work : 不適格な著作物
unbiased way : 偏りのない方法
unblemished : きずのない
unburned : 未野焼き、非野焼き、焼かれていない
uncharacteristic : 特質のない、特性のない
uncharacterized : 性質不明の
unclear : 正体が分からない
uncoating : 脱け殻(ぬけがら)、脱殻(だっかく)、脱外被
uncommon species : 珍しい種
uncus (un) : （交尾器の）ウンクス、鉤、交尾器〔ラテン語〕
under-represent : 小さな比率を占める
underbrush : 下草
undergo : 経験する、受ける、耐える
undergrowth : 下草
underline : 強調する、明白にする、際立たせる
underlying adaptive difference : 根本的な適応的差、根本的な適応的相違
underlying gene : 根元となる遺伝子、基本遺伝子
underlying membrane : 基底膜

underlying population trend : 個体数減少傾向
underneath : 下面
underpin : 強化する
underscore : 強調する、力説する
underside : 裏面、下側
understory : 低木層、下層
underway : 進行中の
undeveloped ovariole : 未発育の卵巣小管
undisturbed control : 無攪乱(むかくらん)の対照
undisturbed environment : 攪乱されていない環境、未攪乱環境
undulate : 起伏
UNEP : 国連環境計画(United Nations Environment Programme)
unequal crossing over : 不等交叉
unequal variance : 不等分散
uneven crowding : 不均一な混み具合
uneven sampling : 一様でないサンプリング
unexpected mechanism : 予想されていなかった機構、予期されなかった機構
UnF : 前翅裏面(Underside of Forewing)
unfavorable abiotic condition : 不都合な非生物的条件
unfavorable condition : 不都合な条件、不利な条件
unfavorable habitat : 不良な生息地、好ましくない生息地
unfavorable season : 好ましくない季節
unfertilized egg : 未受精卵
unfertilized haploid egg : 未受精の半数体卵
UnFH : 両翅裏面(Underside of Forewing and Hindwing)
unfit : 適応しない
unguis : (複 . -ues)、爪〔ラテン語〕
ungulates : 有蹄類(ゆうているい)
UnH : 後翅裏面(Underside of Hindwing)
uni- : 1 -、単 -〔ラテン語〕
uniform distribution : 一様分布
uniform dryness : 均一な乾燥
uniform environment : 均一な環境
uniform species density : 一様な種密度
uniformity : 斉一化(せいいつか)、斉一性
unifying framework : 統一的な枠組み
unilateral pursuit : 一方的な追跡、片側追跡
unimproved grassland : 未改良の草地
uninfected female : 非感染雌
uninominal : 一語名の
uninominal name : 一語名
unintentional introduction : 非意図的導入、意図しない移入
uninvestigated : 調査されていない
unique courting organ : ユニークな求愛器官
unique nymphalid species : 特異なタテハチョウ種
unisexual sterility : 単性不妊性、単性不稔性
unit area : 単位地域
unit biomass : 単位生物量、単位バイオマス
unit of molecular clock : 分子時計の単位
unit of replication : 複製単位
unite : 結合する、合体する
United Nations Environment Programme : 国連環境計画
universal : 普遍的
universal characteristic : 普遍的な形質
universal genetic code : 普遍遺伝暗号
universal occurrence : 普遍的な発生
universal pattern : ユニバーサルパターン、普遍的パターン
univoltine : 一化性(いっかせい)
univoltine group : 一化性集団
univoltine population : 一化性個体群
univoltine type of egg overwintering : 卵越冬一化型
unjustified emendation : 不当な修正名
unknown : 未知の
unknown ecological factor : 未知の生態学的要因

unknown-3-exon : *U3X* 遺伝子（3 つのエキソンからなる非コード RNA をコードする遺伝子）
unlinked color pattern locus : 非連鎖カラーパターン遺伝子座
unlinked downstream gene : 未結合の下流遺伝子
unlinked region : 未結合領域
unlogged forest : 伐採されなかった森
unmanipulated control : 無操作の対照
unmated female : 未交尾雌
unobservable : 観察不可能
unoccupied site : 非占有場所
unpalatability : 不快性
unpalatable : まずい、好みに合わない
unpalatable butterfly : まずい味の蝶
unprecedented : 先例のない
unprecedented decrease : 空前の減少
unprecedented resolution : 先例のない解決策、前例のない解決策
unpredictable : 予測不能な
unpredictable environment : 予測不能な環境
unpredictable environmental exigency : 予見不可能な環境急変
unpredictable movement : 予測不可能な移動
unpublished : 未発表、未出版
unpublished datum : （複 . -ta）、未発表データ、非公開データ
unpublished work : 未発表の著作物
unravel : 解明する
unrelated gene : 無関係遺伝子
unrelated species : 無縁種
unreplicated mass sample : 重複なしの大量サンプル
unrestrained collecting : 自制のない採集
unrooted tree : 無根系統樹
unseasonable development : 季節はずれの発育
unsettled summer weather : 不安定な夏の天候
unstable topology of tree : 系統樹の不安定な樹形
untimely development : 時期はずれの発育
untreated feminized matriline : 無処理の雌化した母系
untreated insect : 無処理の昆虫
unusual case : 珍しい例、変わった（事）例
unusual type : 異常型、変わった型
unveil : 明らかにする、ベールを取る
unwary : 油断、不注意
Unweighted Pair-Group Method with Arithmetic mean : 非加重結合法、非加重平均結合法、平均距離法（UPGMA）
up-regulate : 上方制御する
UpF : 前翅表面（Upperside of Forewing）
UpFH : 両翅表面（Upperside of Forewing and Hindwing）
UPGMA : 非加重結合法、非加重平均結合法、平均距離法（Unweighted Pair-Group Method with Arithmetic mean）
UpH : 後翅表面（Upperside of Hindwing）
uphold : 維持する
upland : 高地
upland forest : 高地の森林、高地森林帯
uplift : 隆起
upper lamina : 翅表面
upper leaf side : 葉の表面
upper part of warm-temperate forest : 暖帯林上部
upper surface : 表翅、表面
upper threshold temperature : 上方臨界温度
upperside (ups) : 表面
upregulate : 上方調節する、上方制御する
upregulated and downregulated genes : 上方制御遺伝子および下方制御遺伝子（発現が正／負に制御された遺伝子）
upregulation of antimicrobial peptide : 抗菌性ペプチドの上方制御

upward selection : 上方選択
Urania moth : ニシキオオツバメガ、ツバメガ科オオツバメガ亜科の蛾
urban area : 都市(的)地域
urban cover : 都市型被覆
urban land use : 都市型土地利用、都市的土地利用
urge : 推進する
uric acid : 尿酸
uric acid derivative : 尿酸誘導体
uridine 5'-triphosphate : ウリジン-5-三リン酸
urine : 尿、小便
Urochordata : 尾索動物
urosome : 腹部
used species : 利用種
Ussuri type : ウスリー型
utility : 効用、有用、有益
utopian : 夢物語(的な)、理想郷の
UTR : 非翻訳領域(UnTranslated Region)
UV : 紫外線(UltraViolet)
UV-C-irradiated plant : UV-C(遠紫外線)照射された植物(UV-C ＝ UltraViolet-C)
UXT : UXT、遍在発現転写体(多くの生物種で保存されている転写調節因子)(Ubiquitously eXpressed Transcript)

V

V : 裏面、下面、腹側(Ventral)
V-shaped marking : V 字形の斑紋
vacant patch : 空きパッチ
vacant site : 空白地、空き地
vacant spot : 空白地帯
vacuole : 空胞、液胞、小胞
vacuum : 真空(状態)
valid name : 有効名
valid nomenclatural act : 有効な命名法的行為
validated : 有効にされた
validity : 有効性
valuable guideline : 価値ある指針
valuable tool : 価値のある道具
valva : (複 . -ae)、(交尾器の)バルバ、交尾弁〔ラテン語〕
valve : 弁
valvula : (複 . -lae)、弁、小弁、産卵弁片〔ラテン語〕
van't Hoff's law : ファントホッフの法則
van't Hoff's principle : ファントホッフの法則
vannus : 扇域、扇状部(「remigium〈主域〉」を参照)
vantage point : 有利な点、地の利
var. : 変種(varietas の略語)、「『属名 種名 var. 品種名』の形式で表記」
variability : 変異性、変動性、多様性
variable phenotype : 可変表現型
variance : 分散
variant : 変異体、変動体、多様体、変異集団
variant population : 変異個体群
variant spelling : 変体綴(つづ)り
variation : 変異(多様性)、変動
variation pattern : 変異様式
variation series : 変異系列
variation speed : 変異速度
variety : 変種、品種、系統、多様性
variety of habitat : 生息地の多様性
various fields of research : 研究の様々な分野
vary : 変える、変化する
varying mixtures of herb and shrub : 草本と低木との変化に富んだ混合
vas deferens : 輸精管
vascular : 維管束の、管(導管、脈管、血管など)の
vascular plant : 維管束植物
vascular system : 維管束系、脈管系(統)
vast majority : 大部分、圧倒的多数
vector : 媒介者、ベクター、媒介する
vector clipping : ベクタークリッピング

vector control : 媒介者防除
vector of disease : 病原菌媒介生物、疾病媒介生物
vector-borne disease : 媒介者によって媒介される疾病、ベクター媒介性疾患
vegetation : 植生、植物
vegetation change : 植生転換、植生変化、植生改変、植生変動
vegetation classification : 植生分類
vegetation structure and composition : 植生の構造と(種)組成
vegetation survey : 植生調査
vegetation type : 植生型
vegetation zone : 植生帯
vegetative growth : 栄養成長
vegetative part : 茎葉部、生長力のある部分
vegetative plant tissue : 成長する植物組織、植物の茎葉部組織
vegetative reproduction : 栄養繁殖
vegetative requirement : 栄養要求
vegetative volatile : 生長力のある揮発性物質
vein : 翅脈、葉脈、脈
vein pattern : 翅脈パターン
vein-cutting behavior : 葉脈切断行動
veinal : 翅脈の
velocity and direction : 速度と方向
venation : 脈相、脈系、翅脈
Venn diagram : ベン図(表)
ventilation : 換気、通気性
ventral : 腹部の、腹側の、裏面の、腹側(ふくそく)
ventral abdomen board : 腹部腹板(ふくぶふくばん)
ventral hindwing : 腹側の後翅
ventral light : 腹側からの光
ventral retina : 腹側網膜
ventral side : 腹側
ventral surface : 腹側表面
ventral view : 腹側から見た図
ventral wing marking : 腹側の翅の斑紋
ventral wing surface : 腹側の翅面、翅の裏面
verge of extinction : 絶滅の瀬戸際
verity : 多様性
vermilion : 朱色
Vermont stream : バーモント州の河川
vernacular name : 俗名、通俗名
vernal : 春の
vernalization : 春化(しゅんか)、春化処理
vertebrate : 脊椎動物
vertebrate biology : 脊椎動物の生物学
vertebrate genome : 脊椎動物のゲノム
vertebrate predator : 脊椎動物の捕食者
vertebrate retina : 脊椎動物網膜
vertex : 頭頂部(とうちょうぶ)
vertical dashed line : 垂直破線
vertical distribution : 垂直分布
vertical infection : 垂直感染
vertical line : 垂直線
vertical stratification : 垂直層別
vertical transmission : 垂直伝達、垂直伝播
vesica (ve) : 膀胱、のう、内鞘、内陰茎端節
vestibulum : 腟前室
vestigial mutualistic benefit : 退化した相利共生的利益
vestigial organ : 痕跡器官、退化した器官
vestiture : 被毛、装飾
viability : 生存率、生存度、生存力、存続性
viability selection : 生存力選択
viable : 生存力のある
viable offspring : 生存可能な子孫
viable population : 存続可能個体群
vial : 小瓶
vibration property : 振動特性
vibrational property : 振動特性
vibratory papilla : 振動突起
vicariance : 分断分布
vicariant speciation : 分断種分化、分断分

布を示す種分化（１つの種の地理的分布が２つ以上の大きな独立した個体群に分断された後で、生殖隔離が進化した場合をいう）
vice versa, and : 逆もまた同様、逆に
Viceroy butterfly : チャイロイチモンジ、カバイロイチモンジ
vicinity of Beijing, the : 北京市郊外
vicious circle : 悪いサイクル、悪循環
Victoria era : ビクトリア朝
vinculum (vin) : 繋帯、ヴィンクルム、ビンクルム
vine : ブドウの木、草本性つる植物
vineyard : ぶどう園
viral binding : ウイルス結合
viral circle : ウイルスの環
viral concentration : ウイルス濃度
viral DNA : ウイルス DNA
viral genome : ウイルスゲノム
viral infection : ウイルス感染
viral inoculum : ウイルス接種液
viral invader : ウイルス性侵入者
viral manipulation : ウイルス操作
viral motility : ウイルス運動性
viral particle : ウイルス粒子
viral pathogen : ウイルス（性）病原体
viral product : ウイルス産物
viral replication : ウイルス複製
viral titer : ウイルスの力価、ウイルス価、ウイルス感染力価
virgin female : 未交尾雌
virion : ビリオン（感染性を有するウイルス粒子）
virtual tautonymy : 疑似同語反復
virulence : 病毒性、病原力、毒性
virulence factor : 毒性因子、毒素因子、病原性因子
virulence gene : 毒素遺伝子、毒性遺伝子、病原性遺伝子
virus : （複 . -es）、ウイルス〔ラテン語〕
virus entry to cell : 細胞へのウイルス侵入
virus particle production : ウイルス粒子産出、ウイルス粒子生成
virus preincubation : ウイルス前培養
virus resistance : ウイルス抵抗力、ウイルス抵抗性
virus transcript : ウイルスの転写産物
virus-lectin : ウイルス - レクチン
viscous : 粘性のある
visible light spectrum : 可視光スペクトル
visual : 視覚
visual cell : 視細胞
visual communication : 視覚による通信、視覚通信
visual contact : 視覚的な接触
visual cue : 視覚刺激
visual identification : 視覚識別、目視識別
visual inspection : 目視検査
visual organ : 視覚器官
visual pigment : 視物質
visual sense : 視覚
visual signal : 視覚信号
visual stimulus : （複 . -li）、視覚刺激〔ラテン語〕
visual system : 視覚系
visualize : 視覚化する
vitalism : 生気論
vitamin : ビタミン
vitellin : ビテリン
vitelline membrane : 卵黄膜
vitellogenesis : 卵黄形成
vitellogenin : ビテロジェニン
vivid color pattern : 鮮やかなカラーパターン
viviparous : 胎生の
VOC : 揮発性有機化合物（Volatile Organic Compound）
vocalization of predatory bird : 捕食鳥のさえずり、捕食鳥の鳴き声

Vogel's organ : フォーゲル器官
volatile chemical cue : 揮発性化学的刺激
volatile compound : 揮発性化合物
volatile emission : 揮発性物質の放出
volatile fatty acid derivative : 揮発性脂肪酸誘導体
volatile ketone : 揮発性ケトン
volatile organic compound : 揮発性有機化合物（VOC）
volatile profile : 揮発性物質プロファイル
volatile ratio : 揮発率
volatile terpene : 揮発性テルペン
volcanic : 火山性の
volcanic activity : 火山活動
volcanic island : 火山島
volsella : （複 . -ae）、陰具小片〔ラテン語〕
voltinism : 化性（かせい）
voucher : 証拠物件、証票、証憑
voucher number : 証書番号（vn）
voucher specimen : 証拠標本
vulnerable : 危急
vulnerable body : 傷つきやすい身体、傷つきやすい体
vulnerable site : 傷つきやすい場所
vulnerable speices : 危急種（ききゅうしゅ）（VU ＝ VUlnerable）

W

W-mark : W 字状の斑点（蛾類）
waggle dance of bee : ハチの尻振りダンス
waiting type : 待ち伏せ型
Wald statistic : ワルド統計量
walking : 歩行
walking speed : 歩行速度
wall brown butterfly : メゲラツマジロウラジャノメ
wall pattern : 壁模様
Wallace's line : ウォーレス線
wandering : 歩き回る
wandering behavior : 徘徊行動
wandering movement : 徘徊移動
wandering stage : 遊走期、さまよい期
war of attrition : 消耗戦
warm blooded : 定温の、温血の
warm-season species : 暖地性種
warm-temperate ancestral type : 暖温帯性の祖先型
warm-temperate forest and wide-distributed genus of Asian origin : アジア暖帯林広分布属
warm-temperate species : 暖地性種、暖温帯の種
warmer condition : より温暖な条件
warmer summer temperature : より暖かい夏の気温
warmer tropical rainforest : より暖かい熱帯雨林
warming : 温暖化
warn : 警告する
warning behavior : 威嚇（いかく）、警告行動
warning color : 警戒色
warning color pattern : 警告色パターン
warning coloration : 警告色、警戒色
warning pattern : 警告パターン
warning signalling : 警告信号
warningly colored and mimetic butterfly : 警告色化と擬態化した蝶
warningly colored butterfly : 警告色の蝶
warrant : 保証する
wasabi : ワサビ
wash twice : ２回洗浄する
Washington Convention : ワシントン条約
wasp : 黄蜂、ハチ、スズメバチ
wasp egg : 蜂の卵
wasp gene set : 蜂の遺伝子セット
wasp larva : 蜂の幼虫
wasp ovary : 蜂の卵巣
waste product : 老廃物

watching : 観察
water body : 水域、水体
water course : 水路、水流
water loss : 水分喪失
water purification : 水質浄化
water soluble : 水に溶ける、水溶性の
water source : 水資源
water strider : アメンボ
waterproofing : 防水効果、防水性
wavelength region : 波長領域
wavelength specific behavior : 波長特異的な行動
wavy line : 波状の線
wax layer : ロウ層
WCMC : 世界自然保全モニタリングセンター（World Conservation Monitoring Centre）
weakened state : 弱体化した状態
weakly developed eyespot : 発達の悪い目玉模様
weakly negative relation : 弱い負の関係
weakly supported node : 弱く支持された結節点
weather condition : 気象条件
weather station : 気象観測所、気象測候所
web : クモの巣
Weber's line : ウェーバー線
wedge-shaped : くさび形の
weed : 雑草
weigh : 体重を量る
weight loss : 体重の減少（率）
weight of post-diapause larva : 休眠後の幼虫の体重、後休眠期の幼虫の体重
Welch's t test : ウェルチの t 検定
well : くぼみ、穴、ウェル
well developed prairie vegetation : 良く発育した大草原植生
well studied host plant and herbivore system : 良く研究された寄主植物と草食生物とのシステム
well-burned forest : 十分に野焼きされた森
well-defined sister relationship : 明確に定義された姉妹関係、十分に定義された姉妹関係
well-diverged lineage : よく分岐した系統
well-recorded area : 記録が多い地域
well-understood molecular mechanism : よく理解された分子機構
West-Chinese element : 中国西部系要素
western blotting : ウェスタンブロッティング
western pygmy blue butterfly : アレチコビトシジミ（アレチコシジミ：改訂新名称）
Western Tibet : チベット西部、西部チベット
wet forest : 湿潤森林
wet lowland grass : 低湿地の草、湿地草
wet sand : 湿った砂
wet season : 雨季
wet season form : 雨季型
wet soil : 湿った土、湿った土壌
wet weight : 湿重量、湿潤重量
wet-dry seasonal environment : 乾湿の季節的環境
wetland : 湿地
wetland vegetation : 湿地性植生、湿原植生
wg : 無翅遺伝子（*wingless* gene）
WGD : 全ゲノム重複（Whole Genome Duplication）
WGS : 全ゲノムショットガン（Whole Genome Shotgun）
WGS data base : WGS データベース、全ゲノムショットガンデータベース（Whole Genome Shotgun）
whale : クジラ
wheat field : 小麦畑
whip-like : 鞭状の
white band : 白色の帯
white band and yellow band forms : 白帯型と黄帯型

white forewing and hindwing shutter allele : 白色の前翅と後翅のシャッター対立遺伝子
white forewing portion colored yellow : 前翅の白色部分を黄色に変えた
white form : 白化型
white line : 白い線
white list : ホワイトリスト
white marking : 白色の斑紋
white morph : 白化型
white morph formation : 白化型形成
white morph induction : 白化型誘起
white morph production : 白化型生産
white patch : 白い大斑点
white pigmentation : 白色素沈着
white pupation board : 白色の蛹化台紙
white scale : 白色鱗粉
white streak : 白いすじ
white-blue screening : 白青スクリーニング、青白スクリーニング
white/yellow color shift : 白色／黄色の色彩転換
white/yellow switch : 白色／黄色スイッチ
Whites : シロチョウ亜科、シロチョウ科
Whites and Sulphurs : シロチョウ科
whitish : 白っぽい、やや白い
whitish form : 白化型
whole genome duplication : 全ゲノム重複（WGD）
whole genome shotgun : 全ゲノムショットガン（WGS）
whole genome shotgun contig data bank : 全ゲノムショットガンコンティグデータバンク
whole genome shotgun sequencing : 全ゲノムショットガン配列
whole mount : 全載標本、ホールマウント、全組織標本、伸展標本
whole view : 全体像
whole-changing pattern : 全体が変化した斑紋
whole-genome bacterial artificial chromosomal library : 全ゲノムバクテリア人工染色体ライブラリー
whole-genome sequence : 全ゲノム配列
wide distribution : 広分布、広域分布
wide variation ability : 広い変異能力
wide-distributed species : 広分布種、広域分布種
widely accepted views : 広く受け入れられた見方
widely distributed genera : 広分布の属、広分布属
widely distributed species : 広分布種、広域分布種
widely distributed species of the Northern Hemisphere : 北半球に広く分布する種
widening ride : 乗馬道の拡大
wider countryside species : 広範な里山にいる種
widespread butterflies : 広範な蝶類、広汎な蝶類
widespread coexistence : 広範囲にわたる共存
widespread extinction : 広範囲に及ぶ絶滅
widespread phenomenon : 広範囲に及ぶ現象
widespread species : 広く分布する種
width of membrane : 膜（の）幅
width of outer membrane : 外膜（の）幅
wild adult : 野外成虫
wild bee : 野生のミツバチ
wild crucifer : 野生の十字花科植物、野生のアブラナ科植物
wild *Drosophila* : 野生のショウジョウバエ
wild flower : 野草、野生の花
wild individual : 野生個体
wild plant : 野生植物
wild population : 野生個体群

wild silkworm : 野蚕、クワコやヤママユガ科の蛾
wild type : 野生型
wild, in the : 野外では
wild-caught individual : 野外採集個体
wild-caught male : 野外採集雄、野外で採集した雄
wild-type individuals : 野生型個体群
wild-type larva : 野生型幼虫
wildfire : 自然火災、山火事、森林火災
wildlife sanctuary : 自然保護区
wind : 風
wind speed : 風速
wind-borne movement : 風まかせ移動
wind-tunnel assay : 風洞実験法
window : 窓
window screen : 窓(の)網戸
wing : 翅(はね)
wing area : 翅面積
wing base : 翅基部
wing base nerve branch : 翅基部の神経枝
wing beat : 羽ばたき
wing bud : 翅芽
wing cell : 翅細胞
wing color preference cue : 翅色選好の刺激
wing condition : 翅の損傷度、翅の状態
wing configuration : 翅型
wing dimorphism : 翅二型
wing disc : 翅原基
wing edge : 翅端
wing epidermis : 翅の表皮細胞
wing expansion : 開翅、翅の開張
wing flick : 翅による摩擦音、翅によるフリック音
wing folding system : 翅の折りたたみ系
wing length : 翅長
wing load : 翼(翅)荷重
wing loading : 翅荷重、翅面荷重
wing margin : 翅の外縁部
wing marking variation : 斑紋変異
wing morphogenesis : 翅の形態形成
wing morphology : 翅形態
wing pattern : 翅の斑紋
wing pattern diversification : 翅パターンの多様性
wing pattern element : 翅の紋様要素、翅紋要素
wing pattern formation : 翅の紋様形成
wing pattern polyphenism : 翅パターンの表現型多型
wing pattern variation : 斑紋変異
wing patterning candidate gene : 翅のパターン形成候補遺伝子
wing patterning gene : 翅パターン化遺伝子
wing polymorphism : 翅多型、翅型多型
wing portion : 翅の部位
wing scale : 翅の鱗粉
wing sharp : 翅形
wing sheath : 翅芽
wing span : 翅長、開張
wing subdivision : 亜翅室
wing surface : 翅(表)面
wing tissue : 翅組織
wing toughness : 翅の強度、翅の硬さ
wing upperside : 翅の表面
wing variation : 翅変異
wing vein : 翅脈(しみゃく)
wing vein pattern : 脈相
wing venation : 翅の脈相、翅の脈系
wing-coupling mechanism : 連結器官
wing-dimorphic bush cricket : 翅二型のキリギリス(翅二型は長翅 - 短翅あるいは有翅 - 無翅である)
wing-pigment : 翅の色素
wing-related trait : 翅関連形質
wing-tip : 翅の先端

wingbeat : はばたく
winged form : 有翅型、翅型
wingless form : 無翅型
wingless gene : 無翅遺伝子
wingless species : 無翅種
wingspan : 開張、翅幅
winter annual : 越年草、越冬一年生植物
winter bud : 冬芽
winter cold : 冬期の寒さ
winter day : 冬日
winter diapause : 冬眠、冬休眠
winter survival : 越冬の成功率、冬期の生存率、冬の生存率
winter temperature : 冬期の気温、冬季気温
wintering colony : 越冬集団繁殖地
winterkill : 冬枯
wire fence : 鉄線柵
wire mesh enclosure : 金網で囲まれた入れ物
withdraw : 立ち去る、引き下がる、引っ込む
withdrawal : 回収、撤回
withhold : 保留する、差し控える
within-island effect : 島内効果
within-morph courtship : 同色の翅内の求愛
within-network variation : ネットワーク内変異
within-plant signaling : 植物内のシグナル伝達
without irradiation : 暗条件で、無照射で
withstand : 耐える
WntA : *WntA* スイッチ遺伝子(ドクチョウの模様形成に関与する遺伝子)
Wolbachia : ウォルバキア、ボルバキア
Wolbachia density : ボルバキア密度
Wolbachia-induced feminization : ボルバキアによる雌化
Wolbachia-induced male killing : ボルバキアによる雄殺し
Wolbachia-induced parthenogenesis : ボルバキアによる単為生殖
Wolbachia-induced reproductive manipulation : ボルバキアによる生殖操作
Wolbachia-mediated addiction : ボルバキア媒介による中毒
Wolbachia-mediated addiction hypothesis : ボルバキア媒介による中毒説
Wolbachia-mediated addiction mechanism : ボルバキア媒介による中毒機構
wood nymph butterfly : ペガラオオモンヒカゲ(暫定和名)(ジャノメチョウ(亜)科の一種)
Wood Nymphs : ジャノメチョウ(亜)科
wood pasture : 森林放牧地
wooded strip : 樹木に覆われた細長い土地、樹木が茂った細長い土地
woodfuel : 木質燃料、木材燃料
woodland : 森林地帯、森林地、森林性の、林地
woodland butterfly : 森林性の蝶
woodland environment : 森林的環境
woodland habitat : 森林性生息場所
woodland species : 森林性種
woodlouse : ワラジムシ
woody area : 森林地域
woody plant parasite : 木本寄生
woody understory cover : 木本の林床被覆(率)
woolly bear : 毛虫
work : 著作物
work of an animal : 動物の仕業
worker ant : 働きアリ
worker policing : ワーカーポリシング
working hypothesis : 作業仮説
World Conservation Monitoring Centre : 世界自然保全モニタリングセンター
world's famous butterfly : 世界の名蝶

worn : 擦り減った、擦り切れた、使い古した
worth noting : 指摘しておく価値がある、価値ある指摘事項
worthwhile : やりがいのある、費やした時間に相当する
wound : 傷
wound-induced defence : 傷害で誘導される防御、傷害誘導防御
wounded leaf : 傷ついた葉
wounding : 傷を付けること
wriggling movement : 蠕動運動（くねらせて出てくること）
Wright-Fisher model : ライト-フィッシャーモデル
wsp gene segment : wsp 遺伝子断片（Wolbachia surface protein）

x

xanthopterin : キサントプテリン
xenobiotic interaction : 生体異物の相互作用
xenologous : 異種相同（の）
Xenopus laevis : アフリカツメガエル
Xenopus tropicalis : ネッタイツメガエル
xg :「『x』は読まず、『g』はグラムではなくてジー（重力；gravity）のこと（遠心分離の単位）」、「かけるジー（重力加速度 *g* の何倍に当たるかの力）」
Xinjiang genotype : 新疆（しんきょう）産遺伝子型

y

y-intercept : Y 切片
Y-tube olfactometer : Y 字管オルファクトメーター、Y 字管型嗅覚計
YAC : 酵母人工染色体（Yeast Artificial Chromosome）
year-to-year variation : 年変動
yearly variation of temperature : 温度の年変化、温度の年変動
yeast : 酵母
yeast artificial chromosome : 酵母人工染色体（YAC）
yeast species : 酵母種
yellow band : イエローバンド
yellow dung fly : ヒメフンバエ
yellow pigment : 黄色色素
yellow stripes colored white : 黄色の縞模様を白色に変えた
yellow-banded female model : 黄色翅の雌モデル
yellow-green pigment : 黄緑色素
yellowing syndorome : 黄変症候群
yellowish white ray : 黄色ぽい白色光線
yielded tree : 産出された系統樹
young caterpillar : 若齢幼虫
young fruit : 未熟果
young pupa : 蛹初期
Yucca giant skipper butterfly : イトランセセリ
yucca moth : ユッカガ
Yule process : ユール過程
Yunnan Province : 雲南省（中国）

z

Z chromosome : Z 染色体
(*Z*)-3-hexenol : (*Z*)-3-ヘキセノール、シス-3-ヘキセノール（*cis*-3-Hexenol）（「(*Z*)-」は二重結合の周りの幾何異性を区別する記号で、ドイツ語の「zusammen〈同じ側の〉」に由来している；「(*Z*)-」が「*cis*-」に対応する）
(*Z*)-3-hexenylvicianoside : (*Z*)-3-ヘキセニルビシアノシド
Z-linked single copy nuclear locus : Z 染色体と連鎖した単コピー核遺伝子座

zebra longwing butterfly : キジマドクチョウ
zebra swallowtail butterfly : トラフタイマイ
zebrafish : ゼブラフィッシュ
Zeitgeber : 同調因子、体内時計の周期に影響を与える外的因子〔ドイツ語〕
Zeitgedachtnis : 時間記憶〔ドイツ語〕
zen gene : ゼン遺伝子、ツェアクヌルト(*zerknullt*)遺伝子(キイロショウジョウバエのホメオティック遺伝子)〔ドイツ語〕
Zephyrus : ゼフィルス
zero growth isocline : ゼロ成長の等傾斜線
zigzag pattern : ジグザグ模様
zinc finger protein : ジンクフィンガータンパク質
Zizina emelina : シルビアシジミ
Zizina otis : ヒメシルビアシジミ
zodiac moth : オジロルリツバメガ、フトオビルリツバメガ
zonation : 帯状分布
zone (zn) : ゾーン
zoocarotenoid : 動物性カロチノイド
zoogeographical region : 動物地理区
zoological formula : 動物定型名
zoological name : 動物学的名称
zoological nomenclature : 動物命名法
zoological taxon : 動物学的タクソン
zoologist : 動物学者
Zoraptera : ジュズヒゲムシ目、絶翅目
ZW : ZW 型(雄の性染色体がZ染色体同士の相同染色体の対になり、雌の性染色体がZ染色体とW染色体の組となる性決定様式)
zygote : 接合子
zygote mortality : 接合子の死亡率

付録：日本産蝶類名称の英和／和英編

■ 英和編

A

African Grass Blue : ハマヤマトシジミ
Albocaerulean : サツマシジミ
Alphabetical Hairstreak : ウラミスジシジミ
Alpine Clouded Yellow : ミヤマモンキチョウ
Ambigua Fritillary : コヒョウモンモドキ
Anadyomene Fritillary : クモガタヒョウモン
Angled Castor : カバタテハ
Angled Sunbeam : ウラギンシジミ
Angulated Grass Yellow : ツマグロキチョウ
Apefly : シロモンクロシジミ
Arctic Skipper : タカネキマダラセセリ
Argus Rings : ヒメウラナミジャノメ
Argyrognomon Blue : ミヤマシジミ
Arran Brown : クモマベニヒカゲ
Asahina's Skipper : アサヒナキマダラセセリ
Asama Silver-studded Blue : アサマシジミ
Asama White Admiral : アサマイチモンジ
Asamana Arctic : タカネヒカゲ
Asian Comma : キタテハ
Asian Swallowtail : アゲハ
Autumn Leaf : イワサキコノハ

B

Bamboo Treebrown : シロオビヒカゲ
Banana Skipper : バナナセセリ
Bianor Peacock : カラスアゲハ
Black Cupid : クロツバメシジミ
Black Hairstreak : リンゴシジミ
Black Veined Tiger : スジグロシロマダラ
Black-banded Hairstreak :
ミズイロオナガシジミ
Black-veined White : エゾシロチョウ
Blackburn's Blue : サツマシジミ
Blue Admiral : ルリタテハ
Blue Branded King Crow :
マルバネルリマダラ
Blue Hairstreak : ウラゴマダラシジミ
Blue Pansy : アオタテハモドキ
Blue Quaker : ツシマウラボシシジミ
Blue Spotted Crow : ミダムスルリマダラ
Blue Tiger : ミナミコモンマダラ
Blue-branded King Crow :
マルバネルリマダラ
Blue-spotted Crow : ミダムスルリマダラ
Brilliant Hairstreak : アイノミドリシジミ
Brimstone : ヤマキチョウ
Broad-bordered Grass Yellow :
ホシボシキチョウ
Brown Awl : タイワンアオバセセリ
Brown-banded Hairstreak :
ウスイロオナガシジミ

C

Cabbage White : モンシロチョウ
Camberwell Beauty : キベリタテハ
Cassia Butterfly : ウスキシロチョウ
Ceylon Blue Glassy Tiger :
リュウキュウアサギマダラ
Ceylon Lesser Albatross : ナミエシロチョウ
Checkered Skipper : タカネキマダラセセリ
Chequered Skipper : タカネキマダラセセリ
Chestnut Tiger : アサギマダラ
China Flat : ダイミョウセセリ
Chinese Bushbrown : ヒメジャノメ
Chinese Comma : キタテハ
Chinese Peacock : カラスアゲハ
Chinese Windmill : ジャコウアゲハ
Chinese Yellow Swallowtail : アゲハ
Chocolate Albatross : タイワンシロチョウ
Chocolate Argus : イワサキタテハモドキ
Chocolate Pansy : イワサキタテハモドキ
Citrus Swallowtail : オナシアゲハ
Cognatus Green Hairstreak :
ジョウザンミドリシジミ
Comma : シータテハ
Common Albatross : ナミエシロチョウ
Common Awl : テツイロビロウドセセリ
Common Banded Awl :
オキナワビロウドセセリ
Common Bluebottle : アオスジアゲハ
Common Crow Butterfly : ガランピマダラ
Common Eggfly : リュウキュウムラサキ
Common Evening Brown :
ウスイロコノマチョウ

Common Five-ring : ヒメウラナミジャノメ
Common Grass Yellow :
キチョウ(ミナミキチョウ)
Common Hedge Blue : ヤクシマルリシジミ
Common Indian Crow : ガランピマダラ
Common Jay : ミカドアゲハ
Common Leopard : ウラベニヒョウモン
Common Lineblue : ヒメウラナミシジミ
Common Map : イシガケチョウ
Common Mormon : シロオビアゲハ
Common Palm Dart : ネッタイアカセセリ
Common Rose : ベニモンアゲハ
Common Rose Swallowtail : ベニモンアゲハ
Common Sailer : コミスジ
Common Sergeant : シロミスジ
Common Tiger : スジグロカバマダラ
Common Yellow Swallowtail : キアゲハ
Compton Tortoiseshell : エルタテハ
Constable : スミナガシ
Copper Hairstreak : クロミドリシジミ
Cranberry Blue : カラフトルリシジミ
Crow Eggfly : ヤエヤマムラサキ
Cycad Blue Butterfly :
クロマダラソテツシジミ
Cycad Butterfly *** : ソテツシジミ

D

Daisetsuzana Arctic : ダイセツタカネヒカゲ
Danaid Eggfly : メスアカムラサキ
Daphne Fritillary : ヒョウモンチョウ
Dark Cerulean : ルリウラナミシジミ
Dark Evening Brown : クロコノマチョウ
Dark Grass Blue : ハマヤマトシジミ
Dark Green Fritillary : ギンボシヒョウモン
Diamina Fritillary :
ウスイロヒョウモンモドキ
Diana Treebrown : クロヒカゲ
Dingy Line Blue : マルバネウラナミシジミ
Dingy Lineblue : マルバネウラナミシジミ
Dod-dash Sailer : ホシミスジ
Double-branded Crow : ルリマダラ
Dragon Swallowtail : ホソオチョウ
Dryad : ジャノメチョウ
Dryad Butterfly : ジャノメチョウ

E

East European Sailer : フタスジチョウ
Eastern Large Blue : オオゴマシジミ
Eastern Pale Clouded Yellow : モンキチョウ
Eastern Silverstripe :
ウラギンスジヒョウモン
Eastern Wood White : ヒメシロチョウ
Essex Skipper : カラフトセセリ
European Beak : テングチョウ
European Map Butterfly : アカマダラ
European Purple Emperor : コムラサキ
European Skipper : カラフトセセリ
Evening Brown : ウスイロコノマチョウ
Eversmann's Parnassius :
キイロウスバアゲハ
Eyed Pansy : アオタテハモドキ

F

False Comma : エルタテハ
False Heath Fritillary :
ウスイロヒョウモンモドキ
False Ringlet : ヒメヒカゲ
Flower Swift : オオチャバネセセリ
Forest Pierrot : ゴイシシジミ
Forest Quaker : リュウキュウウラボシシジミ
Forget-me-not :
ムラサキオナガウラナミシジミ
Formosan Swift : ユウレイセセリ
Freija Fritillary : アサヒヒョウモン
Freya's Fritillary : アサヒヒョウモン
Freyer's Purple Emperor : コムラサキ
Fujisan Green Hairstreak : フジミドリシジミ

G

Glacial Apollo : ウスバアゲハ
Glacial Parnassius : ウスバアゲハ
Glassy Tiger : ヒメアサギマダラ
Golden Birdwing : キシタアゲハ
Golden Hairstreak : ウラキンシジミ
Goschkevitshi's Labyrinth :
サトキマダラヒカゲ
Gram Blue : オジロシジミ
Grass Demon : オオシロモンセセリ

Gray-pointed Pierrot : クロシジミ
Gray-veined White : スジグロシロチョウ
Great Eastern Silverstripe :
　オオウラギンスジヒョウモン
Great Eggfly : リュウキュウムラサキ
Great Mormon : ナガサキアゲハ
Great Nawab : フタオチョウ
Great Orange Tip : ツマベニチョウ
Great Purple : オオムラサキ
Great Purple Emperor : オオムラサキ
Green Flash : イワカワシジミ
Green Hairstreak : ミドリシジミ
Green-veined White :
　エゾスジグロシロチョウ
Green-veined White :
　ヤマトスジグロシロチョウ
Grizzled Skipper : ヒメチャマダラセセリ

H

Hayashi Hairstreak : ハヤシミドリシジミ
Heath Fritillary : コヒョウモンモドキ
High Brown Fritillary : ウラギンヒョウモン
Hill Hedge Blue : ルリシジミ
Hisamatsu Green Hairstreak :
　ヒサマツミドリシジミ
Holly Blue : ルリシジミ
Hungarian Glider : フタスジチョウ
Hylas Common Sailer : リュウキュウミスジ

I

Indian Awlking : アオバセセリ
Indian Cabbage White :
　タイワンモンシロチョウ
Indian Cupid : タイワンツバメシジミ
Indian Fritillary : ツマグロヒョウモン
Indian Leaf Butterfly : コノハチョウ
Indian Palm Bob : クロボシセセリ
Indian Red Admiral : アカタテハ
Ino Fritillary : コヒョウモン
Iphigenia Fritillary : カラフトヒョウモン

J

Jacintha Eggfly : リュウキュウムラサキ
Janson's Swift : ミヤマチャバネセセリ
Japanese Argus : ベニヒカゲ
Japanese Circe : ゴマダラチョウ
Japanese Dart : キマダラセセリ
Japanese Emperor : オオムラサキ
Japanese Flash : トラフシジミ
Japanese Labyrinth : ヤマキマダラヒカゲ
Japanese Luehdorfia : ギフチョウ
Japanese Oakblue : ムラサキシジミ
Japanese Rings : ウラナミジャノメ
Japanese Scrub Hopper :
　ホシチャバネセセリ
Japanese Silverlines : キマダラルリツバメ
Japanese Swift : コチャバネセセリ
Jezo Green Hairstreak : エゾミドリシジミ
Jonasi Orange Hairstreak : ムモンアカシジミ

K

Korean Hairstreak : チョウセンアカシジミ

L

Large Banded Swift : トガリチャバネセセリ
Large Brown : オオヒカゲ
Large Cabbage White : オオモンシロチョウ
Large Comma : エルタテハ
Large High Brown :
　オオウラギンヒョウモン
Large Map Butterfly : サカハチチョウ
Large Sailer : オオミスジ
Large Shijimi Blue : オオルリシジミ
Large Skipper : コキマダラセセリ
Large Tortoiseshell : ヒオドシチョウ
Large Tree Nymph : オオゴマダラ
Large White : オオモンシロチョウ
Latifasciatus Green Hairstreak :
　ヒロオビミドリシジミ
Lemon Butterfly : オナシアゲハ
Lemon Emigrant : ウスキシロチョウ
Lemon Migrant : ウスキシロチョウ
Leoninus Skipper : スジグロチャバネセセリ
Lesser Albatross : ナミエシロチョウ
Lesser Brimstone : スジボソヤマキチョウ
Lesser Grass Blue : シルビアシジミ
Lesser Marbled Fritillary : コヒョウモン
Lilacine Bushbrown : コジャノメ

Lime Butterfly : オナシアゲハ
Lime Swallowtail : オナシアゲハ
Long Tail Spangle : オナガアゲハ
Long-streak Sailer : ミスジチョウ
Long-tailed Blue : ウラナミシジミ
Lycormas Blue : カバイロシジミ

M

Maackii Peacock : ミヤマカラスアゲハ
Maculatus Skipper : チャマダラセセリ
Malayan : タイワンクロボシシジミ
Malayan Crow : シロオビマダラ
Malayan Eggfly : ヤエヤマムラサキ
Mandarin Grass Yellow : キタキチョウ
Mangrove Tree Nymph : オオゴマダラ
Map Butterfly : アカマダラ
Marbled Fritillary : ヒョウモンチョウ
Marginalis Treebrown : クロヒカゲモドキ
Masaki's Rings : マサキウラナミジャノメ
Melissa Arctic : ダイセツタカネヒカゲ
Mera Black Hairstreak : ミヤマカラスシジミ
Metallic Cerulean : シロウラナミシジミ
Moore's Cupid : ゴイシツバメシジミ
Moorland Clouded Yellow :
ミヤマモンキチョウ
Mottled Emigrant : ウラナミシロチョウ
Mountain Tortoiseshell : コヒオドシ
Mourning Cloak : キベリタテハ

N

Nettle-tree Butterfly : テングチョウ
Northern Checkered Skipper :
カラフトタカネキマダラセセリ
Northern Orange Hairstreak :
カシワアカシジミ(キタアカシジミ)

O

Ocean Tree-Nymph : オオゴマダラ
Ochracea Skipper : ヒメキマダラセセリ
Ogasawara Hedge Blue : オガサワラシジミ
Ogasawara Swift : オガサワラセセリ
Okinawa Peacock : オキナワカラスアゲハ
Old World Swallowtail : キアゲハ
Orange Emigrant : キシタウスキシロチョウ
Orange Hairstreak : アカシジミ
Orange Oak Leaf : コノハチョウ
Orange Oakleaf : コノハチョウ
Orange Tiger : スジグロカバマダラ
Orange Tip : クモマツマキチョウ
Oriental Black-veined White :
ミヤマシロチョウ
Oriental Blue Tiger : ウスコモンマダラ
Oriental Chequered Darter : アカセセリ
Oriental Hairstreak : オオミドリシジミ
Oriental Palm Bob : クロボシセセリ
Orion Blue : ジョウザンシジミ

P

Painted Lady : ヒメアカタテハ
Pale Clouded Yellow : モンキチョウ
Pale Grass Blue : ヤマトシジミ
Pale Hedge Blue : タッパンルリシジミ
Pale Palmdart : ネッタイアカセセリ
Pallas's Fritillary : ウラギンスジヒョウモン
Palm Bob : クロボシセセリ
Paper-butterfly : オオゴマダラ
Pea Blue : ウラナミシジミ
Peacock : クジャクチョウ
Peacock Pansy : タテハモドキ
Plain Tiger : カバマダラ
Plains Cupid : クロマダラソテツシジミ
Plumbago Blue : カクモンシジミ
Poplar Admiral : オオイチモンジ
Powdered Oakblue : ムラサキツバメ
Pseudo-Labyrinth : キマダラモドキ
Psyche : クロテンシロチョウ
Purple Beak : ムラサキテングチョウ

Q

Quaker : ヒメウラボシシジミ

R

Red Helen : モンキアゲハ
Red-ring Circe : アカボシゴマダラ
Red-spotted Hairstreak :
ベニモンカラスシジミ
Restricted Demon : クロセセリ
Reverdin's Blue : ミヤマシジミ

Rice Paper Butterfly : オオゴマダラ
Rice Swift : ユウレイセセリ
Riukiuana Chinese Bushbrown :
　リュウキュウヒメジャノメ
Riukiuana Rings :
　リュウキュウウラナミジャノメ
Rustic : タイワンキマダラ

S

Sagana Fritillary : メスグロヒョウモン
Sailer : ミスジチョウ
Saphirinus Green Hairstreak :
　ウラジロミドリシジミ
Scarce Heath : シロオビヒメヒカゲ
Scarce Large Blue : ゴマシジミ
Scotosia Fritillary : ヒョウモンモドキ
Sericin Swallow-tail Butterfly :
　ホソオチョウ
Short-tailed Blue : ツバメシジミ
Sicelis Treebrown : ヒカゲチョウ
Silver Forget-me-not :
　ウスアオオナガウラナミシジミ
Silver Hairstreak : ウラクロシジミ
Silver-lined Skipper : ギンイチモンジセセリ
Silver-spotted Skipper : ホソバセセリ
Silver-studded Blue : ヒメシジミ
Silver-washed Fritillary : ミドリヒョウモン
Sivery Hedge Blue : ルリシジミ
Sky Blue : アサマシジミ
Small Branded Swift : チャバネセセリ
Small Cabbage White : モンシロチョウ
Small Copper : ベニシジミ
Small Grass Yellow : ホシボシキチョウ
Small Labyrinth : ヒメキマダラヒカゲ
Small Luehdorfia : ヒメギフチョウ
Small Straight Swift : ヒメイチモンジセセリ
Small Tortoiseshell : コヒオドシ
Small White : モンシロチョウ
Smaragdinus Green Hairstreak :
　メスアカミドリシジミ
Snout Butterfly : テングチョウ
Spangle : クロアゲハ
Spring Flat : ミヤマセセリ
Staff Sergeant : ヤエヤマイチモンジ
Strait Swift : イチモンジセセリ
Striped Blue Crow : ツマムラサキマダラ
Sugitani's Hedge Blue : スギタニルリシジミ
Sulphur : ヤマキチョウ
Swinhoe's Chocolate Tiger :
　タイワンアサギマダラ
Sylvaticus Skipper :
　ヘリグロチャバネセセリ

T

Tailed Cupid : タイワンツバメシジミ
Tailed Jay : コモンタイマイ
Tailless Bushblue : ルーミスシジミ
Tailless Hairstreak : コツバメ
Teleius Large Blue : ゴマシジミ
Thor's Fritillary : ホソバヒョウモン
Thore Fritillary : ホソバヒョウモン
Three Spots Grass Yellow : タイワンキチョウ
Three-spot Grass Yellow : タイワンキチョウ
Tiny Grass Blue : ホリイコシジミ
Transparent 6-Line Blue :
　アマミウラナミシジミ
Transparent Six-line Blue :
　アマミウラナミシジミ
Trebellius Flat : コウトウシロシタセセリ
Tree Nymph Butterfly : オオゴマダラ
Two-brand Crow : ルリマダラ

V

Varied Eggfly : リュウキュウムラサキ

W

Walnut Hairstreak : オナガシジミ
White Admiral : イチモンジチョウ
White Albatross : カワカミシロチョウ
White Banded Awl :
　タイワンビロウドセセリ
White Clouded Parnassius : ヒメウスバアゲハ
White Mountain Arctic :
　ダイセツタカネヒカゲ
White-bordered : キベリタテハ
White-letter Hairstreak : カラスシジミ
White-tipped Woodland Brown :
　ツマジロウラジャノメ

Wonderful Green Hairstreak :
キリシマミドリシジミ
Wood White : エゾヒメシロチョウ
Woodland Brown : ウラジャノメ

Y

Yayeyamana Peacock *** :
ヤエヤマカラスアゲハ
Yayeyamana Rings :
ヤエヤマウラナミジャノメ
Yellow Apollo : キイロウスバアゲハ
Yellow Awl : キバネセセリ
Yellow Tip : ツマキチョウ
Yellow-legged Tortoiseshell :
ヒオドシチョウ

Z

Zebra Blue : カクモンシジミ
Zebra Hairstreak : ウラナミアカシジミ
Zigzag Fritillary : アサヒヒョウモン

■ 和英編

ア

アイノミドリシジミ : Brilliant Hairstreak
アオスジアゲハ : Common Bluebottle
アオタテハモドキ : Blue Pansy
アオタテハモドキ : Eyed Pansy
アオバセセリ : Indian Awlking
アカシジミ : Orange Hairstreak
アカセセリ : Oriental Chequered Darter
アカタテハ : Indian Red Admiral
アカボシゴマダラ : Red-ring Circe
アカマダラ : European Map Butterfly
アカマダラ : Map Butterfly
アゲハ : Asian Swallowtail
アゲハ : Chinese Yellow Swallowtail
アサギマダラ : Chestnut Tiger
アサヒナキマダラセセリ :
Asahina's Skipper
アサヒヒョウモン : Freija Fritillary
アサヒヒョウモン : Freya's Fritillary
アサヒヒョウモン : Zigzag Fritillary
アサマイチモンジ : Asama White Admiral
アサマシジミ : Asama Silver-studded Blue
アサマシジミ : Sky Blue
アマミウラナミシジミ :
Transparent Six-line Blue
アマミウラナミシジミ :
Transparent 6-Line Blue

イ

イシガケチョウ : Common Map
イチモンジセセリ : Strait Swift
イチモンジチョウ : White Admiral
イワカワシジミ : Green Flash
イワサキコノハ : Autumn Leaf
イワサキタテハモドキ : Chocolate Argus
イワサキタテハモドキ : Chocolate Pansy

ウ

ウスアオオナガウラナミシジミ :
Silver Forget-me-not
ウスイロオナガシジミ :
Brown-banded Hairstreak
ウスイロコノマチョウ :
Common Evening Brown
ウスイロコノマチョウ : Evening Brown
ウスイロヒョウモンモドキ :
Diamina Fritillary
ウスイロヒョウモンモドキ :
False Heath Fritillary
ウスキシロチョウ : Cassia Butterfly
ウスキシロチョウ : Lemon Emigrant
ウスキシロチョウ : Lemon Migrant
ウスコモンマダラ : Oriental Blue Tiger
ウスバアゲハ : Glacial Apollo
ウスバアゲハ : Glacial Parnassius
ウラギンシジミ : Angled Sunbeam
ウラキンシジミ : Golden Hairstreak
ウラギンスジヒョウモン :
Eastern Silverstripe
ウラギンスジヒョウモン : Pallas's Fritillary
ウラギンヒョウモン : High Brown Fritillary
ウラクロシジミ : Silver Hairstreak
ウラゴマダラシジミ : Blue Hairstreak
ウラジャノメ : Woodland Brown

ウラジロミドリシジミ :
Saphirinus Green Hairstreak
ウラナミアカシジミ : Zebra Hairstreak
ウラナミシジミ : Long-tailed Blue
ウラナミシジミ : Pea Blue
ウラナミジャノメ : Japanese Rings
ウラナミシロチョウ : Mottled Emigrant
ウラベニヒョウモン : Common Leopard
ウラミスジシジミ : Alphabetical Hairstreak

エ

エゾシロチョウ : Black-veined White
エゾスジグロシロチョウ :
Green-veined White
エゾヒメシロチョウ : Wood White
エゾミドリシジミ : Jezo Green Hairstreak
エルタテハ : Compton Tortoiseshell
エルタテハ : False Comma
エルタテハ : Large Comma

オ

オオイチモンジ : Poplar Admiral
オオウラギンスジヒョウモン :
Great Eastern Silverstripe
オオウラギンヒョウモン : Large High Brown
オオゴマシジミ : Eastern Large Blue
オオゴマダラ : Large Tree Nymph
オオゴマダラ : Mangrove Tree Nymph
オオゴマダラ : Ocean Tree-Nymph
オオゴマダラ : Paper-butterfly
オオゴマダラ : Rice Paper Butterfly
オオゴマダラ : Tree Nymph Butterfly
オオシロモンセセリ : Grass Demon
オオチャバネセセリ : Flower Swift
オオヒカゲ : Large Brown
オオミスジ : Large Sailer
オオミドリシジミ : Oriental Hairstreak
オオムラサキ : Great Purple
オオムラサキ : Great Purple Emperor
オオムラサキ : Japanese Emperor
オオモンシロチョウ : Large Cabbage White
オオモンシロチョウ : Large White
オオルリシジミ : Large Shijimi Blue
オガサワラシジミ : Ogasawara Hedge Blue
オガサワラセセリ : Ogasawara Swift
オキナワカラスアゲハ : Okinawa Peacock
オキナワビロウドセセリ :
Common Banded Awl
オジロシジミ : Gram Blue
オナガアゲハ : Long Tail Spangle
オナガシジミ : Walnut Hairstreak
オナシアゲハ : Citrus Swallowtail
オナシアゲハ : Lemon Butterfly
オナシアゲハ : Lime Butterfly
オナシアゲハ : Lime Swallowtail

カ

カクモンシジミ : Plumbago Blue
カクモンシジミ : Zebra Blue
カシワアカシジミ(キタアカシジミ) :
Northern Orange Hairstreak
カバイロシジミ : Lycormas Blue
カバタテハ : Angled Castor
カバマダラ : Plain Tiger
カラスアゲハ : Bianor Peacock
カラスアゲハ : Chinese Peacock
カラスシジミ : White-letter Hairstreak
カラフトセセリ : Essex Skipper
カラフトセセリ : European Skipper
カラフトタカネキマダラセセリ:
Northern Checkered Skipper
カラフトヒョウモン : Iphigenia Fritillary
カラフトルリシジミ : Cranberry Blue
ガランピマダラ : Common Crow Butterfly
ガランピマダラ : Common Indian Crow
カワカミシロチョウ : White Albatross

キ

キアゲハ : Common Yellow Swallowtail
キアゲハ : Old World Swallowtail
キイロウスバアゲハ :
Eversmann's Parnassius
キイロウスバアゲハ : Yellow Apollo
キシタアゲハ : Golden Birdwing
キシタウスキシロチョウ : Orange Emigrant
キタキチョウ : Mandarin Grass Yellow
キタテハ : Asian Comma
キタテハ : Chinese Comma

キチョウ(ミナミキチョウ) :
Common Grass Yellow
キバネセセリ : Yellow Awl
ギフチョウ : Japanese Luehdorfia
キベリタテハ : Camberwell Beauty
キベリタテハ : Mourning Cloak
キベリタテハ : White-bordered
キマダラセセリ : Japanese Dart
キマダラモドキ : Pseudo-Labyrinth
キマダラルリツバメ : Japanese Silverlines
キリシマミドリシジミ :
Wonderful Green Hairstreak
ギンイチモンジセセリ :
Silver-lined Skipper
ギンボシヒョウモン : Dark Green Fritillary

ク

クジャクチョウ : Peacock
クモガタヒョウモン :
Anadyomene Fritillary
クモマツマキチョウ : Orange Tip
クモマベニヒカゲ : Arran Brown
クロアゲハ : Spangle
クロコノマチョウ : Dark Evening Brown
クロシジミ : Gray-pointed Pierrot
クロセセリ : Restricted Demon
クロツバメシジミ : Black Cupid
クロテンシロチョウ : Psyche
クロヒカゲ : Diana Treebrown
クロヒカゲモドキ : Marginalis Treebrown
クロボシセセリ : Indian Palm Bob
クロボシセセリ : Oriental Palm Bob
クロボシセセリ : Palm Bob
クロマダラソテツシジミ :
Cycad Blue Butterfly
クロマダラソテツシジミ : Plains Cupid
クロミドリシジミ : Copper Hairstreak

コ

ゴイシシジミ : Forest Pierrot
ゴイシツバメシジミ : Moore's Cupid
コウトウシロシタセセリ : Trebellius Flat
コキマダラセセリ : Large Skipper
コジャノメ : Lilacine Bushbrown
コチャバネセセリ : Japanese Swift
コツバメ : Tailless Hairstreak
コノハチョウ : Indian Leaf Butterfly
コノハチョウ : Orange Oak Leaf
コノハチョウ : Orange Oakleaf
コヒオドシ : Mountain Tortoiseshell
コヒオドシ : Small Tortoiseshell
コヒョウモン : Ino Fritillary
コヒョウモン : Lesser Marbled Fritillary
コヒョウモンモドキ : Ambigua Fritillary
コヒョウモンモドキ : Heath Fritillary
ゴマシジミ : Scarce Large Blue
ゴマシジミ : Teleius Large Blue
ゴマダラチョウ : Japanese Circe
コミスジ : Common Sailer
コムラサキ : European Purple Emperor
コムラサキ : Freyer's Purple Emperor
コモンタイマイ : Tailed Jay

サ

サカハチチョウ : Large Map Butterfly
サツマシジミ : Albocaerulean
サツマシジミ : Blackburn's Blue
サトキマダラヒカゲ :
Goschkevitshi's Labyrinth

シ

シータテハ : Comma
ジャコウアゲハ : Chinese Windmill
ジャノメチョウ : Dryad
ジャノメチョウ : Dryad Butterfly
ジョウザンシジミ : Orion Blue
ジョウザンミドリシジミ :
Cognatus Green Hairstreak
シルビアシジミ : Lesser Grass Blue
シロウラナミシジミ : Metallic Cerulean
シロオビアゲハ : Common Mormon
シロオビヒカゲ : Bamboo Treebrown
シロオビヒメヒカゲ : Scarce Heath
シロオビマダラ : Malayan Crow
シロミスジ : Common Sergeant
シロモンクロシジミ : Apefly

ス

スギタニルリシジミ : Sugitani's Hedge Blue
スジグロカバマダラ : Common Tiger
スジグロカバマダラ : Orange Tiger
スジグロシロチョウ : Gray-veined White
スジグロシロマダラ : Black Veined Tiger
スジグロチャバネセセリ : Leoninus Skipper
スジボソヤマキチョウ : Lesser Brimstone
スミナガシ : Constable

ソ

ソテツシジミ : Cycad Butterfly ***

タ

ダイセツタカネヒカゲ : Daisetsuzana Arctic
ダイセツタカネヒカゲ : Melissa Arctic
ダイセツタカネヒカゲ :
White Mountain Arctic
ダイミョウセセリ : China Flat
タイワンアオバセセリ : Brown Awl
タイワンアサギマダラ :
Swinhoe's Chocolate Tiger
タイワンキチョウ : Three Spots Grass Yellow
タイワンキチョウ : Three-spot Grass Yellow
タイワンキマダラ : Rustic
タイワンクロボシシジミ : Malayan
タイワンシロチョウ : Chocolate Albatross
タイワンツバメシジミ : Indian Cupid
タイワンツバメシジミ : Tailed Cupid
タイワンビロウドセセリ :
White Banded Awl
タイワンモンシロチョウ :
Indian Cabbage White
タカネキマダラセセリ : Arctic Skipper
タカネキマダラセセリ : Checkered Skipper
タカネキマダラセセリ : Chequered Skipper
タカネヒカゲ : Asamana Arctic
タッパンルリシジミ : Pale Hedge Blue
タテハモドキ : Peacock Pansy

チ

チャバネセセリ : Small Branded Swift
チャマダラセセリ : Maculatus Skipper
チョウセンアカシジミ : Korean Hairstreak

ツ

ツシマウラボシシジミ : Blue Quaker
ツバメシジミ : Short-tailed Blue
ツマキチョウ : Yellow Tip
ツマグロキチョウ : Angulated Grass Yellow
ツマグロヒョウモン : Indian Fritillary
ツマジロウラジャノメ :
White-tipped Woodland Brown
ツマベニチョウ : Great Orange Tip
ツマムラサキマダラ : Striped Blue Crow

テ

テツイロビロウドセセリ : Common Awl
テングチョウ : European Beak
テングチョウ : Nettle-tree Butterfly
テングチョウ : Snout Butterfly

ト

トガリチャバネセセリ : Large Banded Swift
トラフシジミ : Japanese Flash

ナ

ナガサキアゲハ : Great Mormon
ナミエシロチョウ :
Ceylon Lesser Albatross
ナミエシロチョウ : Common Albatross
ナミエシロチョウ : Lesser Albatross

ネ

ネッタイアカセセリ : Common Palm Dart
ネッタイアカセセリ : Pale Palmdart

ハ

バナナセセリ : Banana Skipper
ハマヤマトシジミ : African Grass Blue
ハマヤマトシジミ : Dark Grass Blue
ハヤシミドリシジミ : Hayashi Hairstreak

ヒ

ヒオドシチョウ : Large Tortoiseshell
ヒオドシチョウ :
Yellow-legged Tortoiseshell

ヒカゲチョウ : Sicelis Treebrown
ヒサマツミドリシジミ :
Hisamatsu Green Hairstreak
ヒメアカタテハ : Painted Lady
ヒメアサギマダラ : Glassy Tiger
ヒメイチモンジセセリ :
Small Straight Swift
ヒメウスバアゲハ :
White Clouded Parnassius
ヒメウラナミシジミ : Common Lineblue
ヒメウラナミジャノメ : Argus Rings
ヒメウラナミジャノメ : Common Five-ring
ヒメウラボシシジミ : Quaker
ヒメギフチョウ : Small Luehdorfia
ヒメキマダラセセリ : Ochracea Skipper
ヒメキマダラヒカゲ : Small Labyrinth
ヒメシジミ : Silver-studded Blue
ヒメジャノメ : Chinese Bushbrown
ヒメシロチョウ : Eastern Wood White
ヒメチャマダラセセリ : Grizzled Skipper
ヒメヒカゲ : False Ringlet
ヒョウモンチョウ : Daphne Fritillary
ヒョウモンチョウ : Marbled Fritillary
ヒョウモンモドキ : Scotosia Fritillary
ヒロオビミドリシジミ :
Latifasciatus Green Hairstreak

フ

フジミドリシジミ : Fujisan Green Hairstreak
フタオチョウ : Great Nawab
フタスジチョウ : East European Sailer
フタスジチョウ : Hungarian Glider

ヘ

ベニシジミ : Small Copper
ベニヒカゲ : Japanese Argus
ベニモンアゲハ : Common Rose
ベニモンアゲハ : Common Rose Swallowtail
ベニモンカラスシジミ :
Red-spotted Hairstreak
ヘリグロチャバネセセリ : Sylvaticus Skipper

ホ

ホシチャバネセセリ : Japanese Scrub Hopper
ホシボシキチョウ :
Broad-bordered Grass Yellow
ホシボシキチョウ : Small Grass Yellow
ホシミスジ : Dod-dash Sailer
ホソオチョウ : Dragon Swallowtail
ホソオチョウ : Sericin Swallow-tail Butterfly
ホソバセセリ : Silver-spotted Skipper
ホソバヒョウモン : Thor's Fritillary
ホソバヒョウモン : Thore Fritillary
ホリイコシジミ : Tiny Grass Blue

マ

マサキウラナミジャノメ : Masaki's Rings
マルバネウラナミシジミ :
Dingy Line Blue
マルバネウラナミシジミ : Dingy Lineblue
マルバネルリマダラ :
Blue Branded King Crow
マルバネルリマダラ :
Blue-branded King Crow

ミ

ミカドアゲハ : Common Jay
ミズイロオナガシジミ :
Black-banded Hairstreak
ミスジチョウ : Long-streak Sailer
ミスジチョウ : Sailer
ミダムスルリマダラ : Blue Spotted Crow
ミダムスルリマダラ : Blue-spotted Crow
ミドリシジミ : Green Hairstreak
ミドリヒョウモン : Silver-washed Fritillary
ミナミコモンマダラ : Blue Tiger
ミヤマカラスアゲハ : Maackii Peacock
ミヤマカラスシジミ : Mera Black Hairstreak
ミヤマシジミ : Argyrognomon Blue
ミヤマシジミ : Reverdin's Blue
ミヤマシロチョウ :
Oriental Black-veined White
ミヤマセセリ : Spring Flat
ミヤマチャバネセセリ : Janson's Swift
ミヤマモンキチョウ :
Alpine Clouded Yellow
ミヤマモンキチョウ :
Moorland Clouded Yellow

ム

ムモンアカシジミ : Jonasi Orange Hairstreak
ムラサキオナガウラナミシジミ :
Forget-me-not
ムラサキシジミ : Japanese Oakblue
ムラサキツバメ : Powdered Oakblue
ムラサキテングチョウ : Purple Beak

メ

メスアカミドリシジミ :
Smaragdinus Green Hairstreak
メスアカムラサキ : Danaid Eggfly
メスグロヒョウモン : Sagana Fritillary

モ

モンキアゲハ : Red Helen
モンキチョウ :
Eastern Pale Clouded Yellow
モンキチョウ : Pale Clouded Yellow
モンシロチョウ : Cabbage White
モンシロチョウ : Small Cabbage White
モンシロチョウ : Small White

ヤ

ヤエヤマイチモンジ : Staff Sergeant
ヤエヤマウラナミジャノメ :
Yayeyamana Rings
ヤエヤマカラスアゲハ :
Yayeyamana Peacock ***
ヤエヤマムラサキ : Crow Eggfly
ヤエヤマムラサキ : Malayan Eggfly
ヤクシマルリシジミ : Common Hedge Blue
ヤマキチョウ : Brimstone
ヤマキチョウ : Sulphur
ヤマキマダラヒカゲ : Japanese Labyrinth
ヤマトシジミ : Pale Grass Blue
ヤマトスジグロシロチョウ :
Green-veined White

ユ

ユウレイセセリ : Formosan Swift
ユウレイセセリ : Rice Swift

リ

リュウキュウアサギマダラ :
Ceylon Blue Glassy Tiger
リュウキュウウラナミジャノメ :
Riukiuana Rings
リュウキュウウラボシシジミ :
Forest Quaker
リュウキュウヒメジャノメ :
Riukiuana Chinese Bushbrown
リュウキュウミスジ : Hylas Common Sailer
リュウキュウムラサキ : Common Eggfly
リュウキュウムラサキ : Great Eggfly
リュウキュウムラサキ : Jacintha Eggfly
リュウキュウムラサキ : Varied Eggfly
リンゴシジミ : Black Hairstreak

ル

ルーミスシジミ : Tailless Bushblue
ルリウラナミシジミ : Dark Cerulean
ルリシジミ : Hill Hedge Blue
ルリシジミ : Holly Blue
ルリシジミ : Sivery Hedge Blue
ルリタテハ : Blue Admiral
ルリマダラ : Double-branded Crow
ルリマダラ : Two-brand Crow

日本語用語索引

1．本書掲載の和訳語を原則としてすべて抽出し、アイウエオ順索引とした。
2．同一頁に同じ訳語が複数掲載されている場合、索引では1回のみ掲載した。本書はアルファベット順配列の英和辞典であるため、同一頁に同義語が掲載される可能性が高く、索引からの用語の検索時には注意を要する。
3．「DNA」や「RNA」などの例のように訳語にアルファベットなどが含まれる場合、日本語の音に読み替え、該当する五十音の位置に掲載した。
4．基本的な用語のうち、音訓両方の読みが考えられるものについては、できる限り両方の読みの該当箇所に掲載することとした。

ア

亜－　209
アーカイブ保存　22
RACE法　178
RAD解析　177
RAD再配列決定　177
RAD-seq法　186
RAD法　186
Rabゲラニルゲラニル転移酵素遺伝子　177
RAPD法　178
*ras*遺伝子を含有する染色体　39
Rs脈　178
RNAi表現型　188
RNA塩基配列解読　188
RNA干渉　188
RNA干渉の表現型　188
RNA干渉法　114, 188
RNA-seq解析　188
RNAzol試薬　188
RNAプロセシング　188
RNAポリメラーゼ　188
RMA回帰　181, 188
RMA傾き　188
r-m脈　178
r種　177
r選択　177
r戦略種　177
RT-PCR法　186
RT-PCR法により増殖させた対立遺伝子　188
r-淘汰　177
*RpL3*遺伝子の発現水準　79
r脈　178
愛好家　73, 114
アイソコア　117
アイソトープ　117
アイソフォーム　117
間に入れる　190
間の－　58
相手を下方に抑え込もうとする飛翔行動　57
相反する　106
曖昧な特性　17
相矛盾する　48
アイランドホッピング　117
アウトグループ　154
青々とした植生　127
青いケシ　30
青色型の幼虫　30
青色側へのシフト　30
青色光　30
青色色素　30
青色シフト系統　30
青色スペクトル範囲へのシフト　30
青色のスペクトル領域へのシフト　30
青色翅の雌モデル　30
青色ビリン色素　30
青色ビリン色素の生成　30
青色ビリンの蓄積　30
青色ビリン量　18
青色スペクトルシフト　30
アオカケス　30
アオジャコウアゲハ　165
青白い型　156
青白い前翅　156
青白い夏型　156
アオスジアゲハ　44
アオネオナガセセリ　126
青葉組織　97
青葉の揮発性物質　97
青葉の植食　97
アオムシサムライコマユバチ　50
あおる　89
亜科　209
－亜科　11
亜外縁線　210
亜外縁の縞状バンド　210
亜外縁部　210
赤池情報量基準　15
赤い大斑点　181
赤い点　180
赤い半月紋　44
赤い放射状線　181
赤いまだら模様　181
赤い丸　180
赤色の口吻　181
アカエリトリバネアゲハ　178
アカスジドクチョウ　200
アカタテハ属の蝶　14
赤の女王仮説　181
アカマダラ　76
亜科名　142, 209
明るい割れ目　123
明るさ　123
明るさ恒常性　123
アガロースゲル　15
アガロースゲル電気泳動　15
亜寒帯　209
秋型　25, 80
秋型成虫　25
秋型誘導条件　111
秋から翌春にかけて　88
秋雨　25

空き生息地 72
亜基節 210
亜基線 26
空き地 228
秋の半ば 135
空きパッチ 228
空きパッチの割合 173
亜基板 210
明らかな 77
明らかな漸近線 41
明らかにする 108, 186, 227
悪影響を与える 144
悪臭を放つ棍棒 208
悪循環 230
悪条件 60
アクセスが可能な 12
アクセスが容易 12
アクセス道路 12
アクセス容易性 12
アクセプト 12
アクチノマイシンD 13
アクチン重合 13
アクチン染色 13
悪天候の夏 167
アクトグラム 13
アクレア 13
アゲハチョウ亜科 213
アゲハチョウ科 157, 213
アゲハチョウ科の系統樹 157
アゲハチョウ上科 223
あご 96
亜綱 209
亜高山帯 209
アゴニスト 15
アザミウマ目 218
鮮やかなカラーパターン 230
鮮やかな翅の色彩 32
足 122
アジア温帯林広分布属 216
アジア温帯林遺存属 215
アジア暖帯林広分布属 231
アジア・太平洋地域 23
アジア大陸とその属島 23
足がかりの生息地 207
味刺激 98
亜翅室 234
亜翅節片 209
亜翅端部 209
亜翅頂部 209
亜翅頂部帯 209
脚の護身器官 122
脚の防護器官 122
味の悪い 62
味の悪い種 62
味の悪いモデル 62
足場 190
足場位置 190
脚発生 58
味見 215
亜社会性 210
亜社会性ルート 210
亜種 177, 206, 210
亜終齢 160
亜種から種への転移 221
亜種間差 115
亜種間の 114
亜種間の交雑帯 115
亜種間の差異 115
亜種間の接触帯 115
亜種小名 210
亜種名 142, 210
亜種よりも低位の 111
亜種よりも低位の学名 111
亜種よりも低位のタクソン 111
アシル化反応 13
亜腎状紋 210
亜腎臓紋 210
小豆色の地色 56
アズキノメイガ 14
預ける 57
アスパラギンからセリンへの置換 23
アスパラギン酸アミノ基転移酵素 11, 23
アスパラギン酸アミノトランスフェラーゼ 11, 23
アスペクト比 23
あずまや 22
亜生殖板 210
アセタミプリド 13
アセビ属 164
亜前縁 209
亜前縁脈 190, 209
アセンブリしたコンティグ 23
アセンブルコンティグ 23
アセンブルされたゲノム上の足場 23
亜属 209, 210
亜族 211
亜属的分化 210
亜属名 142, 209, 210
亜族名 211
値する 57
与える 46, 197
頭を悩ます 176
新しい 149
新しい分類群 145
新しく羽化した成虫 145
新しく羽化した未交尾の雌 145
新しく進化した個体群 145
新しく発生したトランスクリプトームの生成 55
新しく孵化した幼虫 145
亜端室 209
亜端の 210
亜致死投与量 210
亜中央帯 210
亜頂室 209
亜頂端の 209
厚い 218
暑い期間 106
暑い夏 106
悪化 57
扱いやすいモデル 220
悪化させる 78
悪化した食物条件 57
悪化しつつある環境 57
悪化する 56, 57
圧縮した 46
アッセイプレート 23
圧倒的多数 228
圧倒的な証拠 155
圧倒的に多い 171
集まり 15
集まる 47

集めて分析する 43
圧力運搬手段 171
アデノシン三リン酸 14, 24
アデノシン三リン酸合成酵素 24
アデノシンデアミナーゼ関連増殖因子 14
あてはめ値 84
当てはめられた平均値 84
当てはめられたモデル値 84
充てる 58
後 169
跡 190
後指定 210
後綴り 210
穴 232
孔 86
アナグラム 18
アニーリング温度 19
アニオン 19
アニオンチャ(ン)ネル 19
亜熱帯 211
亜熱帯グループ 211
亜熱帯性個体群 211
亜熱帯性の種 211
亜熱帯性半落葉樹林 211
亜熱帯地域 211
亜熱帯の個体群 211
アネリフェル 19
アノナセウスアセトゲニン 19
亜背域 209
アフタヌーンピーク 15
油紙のような 97
アブラナ 151
アブラナ科 32
アブラナ科植物 32, 52
アブラナ目 32
アブラナ目を摂食するモンシロチョウ科の蝶 32
アブラムシ 21
アフリカオナシアゲハ 40
アフリカサバクシジミ 44
アフリカツメガエル 236
アプリコン 18
アホウドリ 15
亜北極圏種 209
アポトーシス 21
アポトーシス様細胞死 21
アポミクシス 21
アポリシス 21
アポロウスバシロチョウ 21
アマゾン川流域 17
アマゾン源流 100
アマゾンの蝶 17
アマチュア研究者 17
アマチュアの自然科学研究者 17
亜末端の 210
網 133
網(目)状 186
アミノ基転移酵素 220
アミノ酸 17
アミノ酸型脂質共役体 18
アミノ酸残基 17
アミノ酸置換 17
アミノ酸置換数の推定 76
アミノ酸置換頻度 17
アミノ酸置換モデル 17
アミノ酸トリプトファン 18
アミノ酸の収束的変化 49
アミノ酸の分泌物 17
アミノ酸の平行的変化 157
アミノ酸配列 18
アミノ酸配列の違い 17
アミノ酸部位 17
アミノ酸変異 17
網目(構造) 186
アミメカゲロウ目 145
網目組織 186
網目模様 52
アムール型 18
雨の多い 178
アメリカオオモンキチョウ 153
アメリカ温帯林広分布属 216
アメリカ海洋大気庁 146
アメリカ起源草原属 97
アメリカコヒオドシ 135
アメリカ先住民 142
アメリカタテハモドキ 33, 148
アメリカタバコガ 50
アメリカテングチョウ 17
アメリカベニシジミ 17
アメリカヤドリギシジミ 97
アメンボ 232
亜目 210
危うくする 46
怪しげな 67
誤った結果 136
誤り 75
誤りの度合 56
誤りを実証する 62
粗い 42
アラインメント 16
アラインメント長 16
予め学習した 171
あらかじめ設定されている 171
争い 62
アラタ体 34, 50
新たな種の出現 23
新たな蔓延 145
新たに 55
新たに記録された 145
新たに発見された食草 145
新たに発生した成虫 145
アラトスタチン 16
アラトスタチンホルモン 16
アラトトロピン 16
現れる 129
アラントイン 16
アラントイン酸 16
アリ 20
蟻 20
あり合わせ製作 32
アリー効果 16
アリが豊富な生息地 20
アリコロニー最適化法 13, 20
アリ植物 141
アリとの共生 141
アリルイソチオシアネート 17

アリルフォリン 23
アリルホリン遺伝子 23
アルカロイド 16
歩き回る 231
アルコール 15
アルコールブアン固定液 15
アルゴリズム 15
アルゴリズム的アプローチ 16
アルゼンチンアリ 22
アルデヒド 15
アルドキシム産物 15
アルドキシム中間体化合物 114
アル配列 17
α-多様性 17
アルファ多様性 17
α-チューブリン 17
アルファチューブリン 17
アレキサンドラトリバネアゲハ 177
アレクサンダー・ケーニヒ博物館 15
荒地 100
アレチコビトシジミ 232
アレリックリッチネス 16
アレル 16
アレロケミカル 16
アレロパシー関連化合物 16
アロザイム 16
アロザイム対立遺伝子頻度 16
アロザイム多様性 16
アロザイム電気泳動 16
アロザイム分化 17
アロタイプ 16
アロメトリー 16
アロメラニン 16
アロモン 16
アロリウム 17, 22
淡い黄色 156
暗黄褐色の眼状紋 30
アンカー 19
アンカー遺伝子座 19
アンカー配列 19
暗化型 54
暗化擬態 54
暗期 54, 191
暗期測定 54
暗期の光中断 146
アンキリンリピート 19
案件 36
暗号 42
暗黒 45
暗黒処理 54
安山岩質の 19
暗視眼 80
暗視眼の反射スペクトル 181
暗順応性の生きた昆虫 54
暗所 54
暗条件で 235
鞍状の模様 189
暗色化遺伝子 67
暗色型 54, 132
暗色斑 30
暗赤褐色 54
安全帯 189
安全な止まり木 189
安全なねぐら 189
暗相 54
暗騒音 25
アンチコドン 21
アンチセンス RNA 21
安定アイソトープ実験 206
安定化選択 206
安定化淘汰 206
安定機構 206
安定効果 206
安定した環境 206
安定した景観の異質性 206
安定した参照遺伝子 207
安定した対照遺伝子 207
安定性選択 206
安定多型 206
安定的漸近 206
安定同位体実験 206
アンテナペディア遺伝子 20, 21
アントキサンチン 20
暗箱 54
アンヒドロビオシス 19
アンプリコン長の変異 18
アンモニア 18
アンモニア摂取 18
アンモニアの吸収 18
アンモニアの摂取 18
アンモニウム 18
アンモニウム塩化物 18

イ

異 - 16
EST 解析 79
EST データベース 75
EST 配列 75
ESU の境界設定 76
EF1α 遺伝子 69
イーオン 74
異域性の 16
E 値のカットオフ値 67
イエローバンド 236
イオンの 116
異化 36, 62
威嚇 231
威嚇する 207
鋳型 216
イカ光受容体細胞 206
イカ網膜クロム 206
イカロスシジミ 44
イカロドプシン 206
易感染性 213
維管束系 228
維管束植物 228
維管束の 228
維管束の篩管部分から吸汁する植食性 162
生き生きと出現した標本 88
行き過ぎ現象 155
生きた標本 124
生きたまま摘まれた花 124
閾値 218
閾値強度 218
閾値サイズ 218
閾値の強さ 218
閾値レベル 218
生きている祖先 124
生きている標本 124
生き残り 212

生き残り解析　212
イクオリン　14
イクオリン遺伝子　14
育仔　36
育種価　32
いくぶん　201
池　167
異形花柱性　102
異系交配弱勢　154
異系交配種　154
異型性　101
異型性の染色体　101
異型接合体　102
異形態　139
異型配偶　19
異型配偶子　101
異型配偶子性の染色体　101
生け垣　100, 197
生け垣管理　100
意見書　152
移行型　221
移行期　221
移行齢期　121
意識啓発　178
維持機構　213
石切場　177
意志決定　55
意思決定　55
維持する　213, 227
異時性　16
異質性　61, 101
異質選択　102
異質な環境　102
異質な種　102
異質な生息地　102
異質倍数性　16
異質倍数体　16
遺失名　146
異時的隔離　16
異時的な　16
イシノミ目　22
異種　59
移住　43, 78, 135
移住して行くこと　72
移住して来ること　109
移住する　72
移住能力　43
移住パターン　135
異種間の水平移動　115
異種間の水平伝播　115
異種交配種　52
異種交配できる　114
異種交配様式　52
異種混交性　101
異種接合性　102
異種接合体遺伝子型　102
異種接合体の　102
異種接合体の個体　102
異種選択　102
異種相同(の)　236
移出　72
移出種　72
移出率　72
異種の雌　102
異種発現されたタンパク質　102
異常　11, 12
異常型　11, 227
異常気候　12
異常行動　12
異常な　24
異常な突起部　117
異常な翅　12
異常な割合　80
移植　184, 221
移植したフォーカス　96
移植実験　96
移植する　221
移植組織　96
異色の翅間での求愛　28
移植片　96
囲食膜　160
異所性　16
異所性の　16
異所的眼状紋　69
異所的種分化　16
異所的な遺伝的種分化　16
囲心腔　160
囲心細胞　160
異人種間差　115
異人種間の交雑帯　115
異人種間の接触帯　115
囲心腺　160
囲心のう　160
異性化　117
異性間選択　115
異性体変化　117
以前に衰亡しつつあった蝶　87
以前に絶滅の恐れがあった蝶　171
以前の情報　171
位相　161, 219
位相調節　161
位相反応曲線　161, 170
位相変位　161
イソ型　117
イソキサントプテリン　117
イソプレン産出　117
イソプレン放出性植物　117
イソロイシンからメチオニンへの置換　117
遺存　183
遺存種　183, 212
遺存種の蝶　183
依存する　48
遺存的固有　183
依存パターン　57
遺存分布　183
板　166
イダリアギンボシヒョウモン　181
いたるところに　77
至る所にある　225
位置　125
1－　138, 226
一遺伝子座の支配を有利にする選択　193
一塩基多型　199, 200
一塩基の転位差　199
一塩基部位　199
一塩基変異　199, 201
一塩基変異の決定　201
一塩基変異率　179
一応　216
位置価　168
位置関係の依存性　48
一眼レフカメラ　199
一元配置分散分析　152
1個体当りの係数　160
1個体当りの相互作用　160

１個体だけ現れた種 199
一語名 226
一語名の 226
１サンプルのｔ検定 152
一次感覚神経 172
一次感覚ニューロン 172
一次感染 171
イチジクコバチ 83
１次元移動 152
一次構造 172
一次刺毛 172
一次消費者 171
一次生産 172
一次生産物 171
一次代謝産物 171
一時的隔離 216
一時的な環境変動 216
一次的な理由 172
一時的にホバリングして 106
一時的不均質性 216
一時的分布（発生） 216
一次（異物）同名 171
一次誘引物質 171
位置情報 168
著しい 149, 173, 209
著しい差 129
著しい色彩選好 173
著しい変動 209
著しく 149
著しく低い生存率 129
１新属１新種 11
１世代１移入個体の原則 152
一属一種の属 138
１対１の関係 152
一大進歩をもたらす 187
一段階上げた複合モデル 146
一段階下げた複合モデル 145
一段階突然変異 152
一段増殖曲線解析 152
一段増殖曲線検定 152
一段増殖曲線試験 152
一度－ 193
一度見捨てられた森 152
一ヌクレオチド部位 199
一年中 16
一年生草本 20
一年生の草 20
１標本ｔ検定 152
１ヘクタール 11
イチモンジセセリ 135, 187
イチモンジチョウ亜科 14
一様でないサンプリング 226
一様な種密度 226
一様分布 226
一齢 83
一連の情況証拠 40
一連の状況証拠 40
位置を定める 200
一化 152
一回－ 193
一回限り 199
一回限りの休眠期 199
一回結実型多年生植物 138
一回結実性の移住種 193
一回交雑する 52
一回繁殖 193
一回繁殖性 138
一回繁殖性の移住種 193
一回繁殖の動物 193
一化性 226
一化性個体群 226
一過性／持続性の性質 161
一化性集団 226
一化の 199
一貫した増幅 47
一貫した品質保証体制 218
一原子酸素添加酵素 138
一見すると似たような状況 193
一妻多夫 167
一雌多雄 167
一緒に 17
逸脱度逐次分析 208
一致した 46
一致したパターン 46
一致しない 110
一致する 43, 214
一致性 47
一致／不一致率 179
一対の競争的相互作用 156
一対比較 156
一定温度条件 48
一定短日日長 48
一定の齢数 84
一定方向を目指して 61
一般化 91
一般化線形モデル 91, 95
一般化融合（混合）Yule-Coalescent モデル 91
一般時間反転可能モデル 91
一般的な機能分類 91
一般的なサンプリング法 91
一般的な表皮細胞 91
一般的なプロトコル 91
一般的に使用されている植物 45
一般的にはあまりない 103
一般的にはあり得ない 103
一般用語 91
一夫一妻 138
一夫一婦 138
一夫多妻 167
一夫多妻動物 167
一方的な追跡 226
一本鎖高次構造多型 199, 206
一本鎖シークエンス反応 199
いつも 188
遺伝 112
遺伝暗号の進化 77
遺伝暗号の方言 58
遺伝暗号表 92
遺伝因子型 29
遺伝学 93
遺伝学的構成 92
遺伝荷重 92
遺伝型 92, 94
遺伝型の違い 94
遺伝機構 92
遺伝共分散 92
遺伝（的）距離 92
遺伝子 90

遺伝子位置 93
遺伝子1コピーあたり 160
遺伝子移入 116
遺伝子オントロジー 91, 96
遺伝子解析 92
遺伝子概念体系 91, 96
遺伝子獲得 90
遺伝子型 94
遺伝子型解析 94
遺伝子型－環境相互作用 94
遺伝子型同定 94
遺伝子型と環境との相互作用 94
遺伝子型の違い 94
遺伝子型の分離対立遺伝子 94
遺伝子型判定 94
遺伝子型を決定する 94
遺伝子間領域 108, 114
遺伝子銀行受入れ番号 90
遺伝子組換え植物 221
遺伝子組み換え生物 93, 96
遺伝子系図 91
遺伝子系図学 91
遺伝子系統学 91
遺伝子系統樹 91
遺伝子欠失 91
遺伝子欠損 91
遺伝子交流 91
遺伝子コピー数変異 42, 50
遺伝子座 125
遺伝子座あたりの平均対立遺伝子数 25
遺伝子座コーディング 125
遺伝子座の位置クローニング 168
遺伝子座の位置的単離 168
遺伝子座の有効数 70
遺伝子産物 91
遺伝子順序 91
遺伝子情報 92
遺伝子浸透 116
遺伝子侵入 116
遺伝子喪失 91
遺伝子族 91
遺伝子対遺伝子共進化 91
遺伝子多様性 91
遺伝子地図作製 92
遺伝子注釈 90
遺伝子重複 91
遺伝子重複の年齢分布 15
遺伝子伝播因子 91, 98
遺伝子淘汰 93
遺伝子導入 221
遺伝子導入花粉 221
遺伝子導入植物 221
遺伝子導入体 91, 98
遺伝子導入マウスの産出 221
遺伝子特異的逆転写用プライマー 91
遺伝子特異的プライマー 91
遺伝子突然変異 91
遺伝子内組換え 115
遺伝子の－ 90, 93
遺伝子の位置相同性 168
遺伝子の局所的な複製 124
遺伝子の系譜 91
遺伝子の構造 209
遺伝子のコーディング領域 91
遺伝子の消失 91
遺伝子の新生 90
遺伝子の水平伝播 102, 105
遺伝子の流れ 91
遺伝子配列 91
遺伝子配列された種 195
遺伝子配列された分類群 195
遺伝子配列順 91
遺伝子破壊 91
遺伝子発現 91
遺伝子発現解析 91
遺伝子発現カスケード 91
遺伝子発現研究 91
遺伝子発現レベルの定量化 177
遺伝子ファミリー 91
遺伝子ファミリー動態 91
遺伝子プール 91
遺伝子複合体 45
遺伝子分化係数 42
遺伝子分析 92
遺伝子変換 90
遺伝子マーカー 92
遺伝子マッピング 92
遺伝情報 92
遺伝子流動 91
遺伝子領域 91
遺伝子レベル 91
遺伝子連鎖地図 92
遺伝性 101
遺伝生化学的手法 92
遺伝性変異 92
遺伝相関 92, 187
遺伝的アルゴリズム 90, 91
遺伝的影響 92
遺伝的応答 92
遺伝的雄 92
遺伝的雄個体群 93
遺伝的雄部位 93
遺伝的確率性 92
遺伝的加重 92
遺伝的関係 92
遺伝的完全性 92
遺伝的起源 92
遺伝的基礎 92
遺伝的基盤 92
遺伝的休眠 92
遺伝的空間パターン 202
遺伝的区別 92
遺伝的結合 92
遺伝的原因 92
遺伝的交雑 92
遺伝的構成 92
遺伝的構造 92
遺伝的交流 91, 92
遺伝的差異 92
遺伝的翅形異常 101
遺伝的種分化 92
遺伝的刷り込み 93
遺伝的制御 92
遺伝的制約 92
遺伝的相違 92

遺伝的相同性 92
遺伝的多型 92, 167
遺伝的多様性 92
遺伝的多様性の衰退 75
遺伝的多様性の衰亡 55
遺伝的多様性の喪失 75
遺伝的多様性の低下 55, 75
遺伝的多様性のモザイク 139
遺伝的特殊性 92
遺伝的特徴 93
遺伝的な役割分業 92
遺伝的に決まっている形質 92
遺伝的に決まっている選好性 92
遺伝的に継承された 93
遺伝的に決定する 92
遺伝的に相関した 92
遺伝的に多様な個体群 93
遺伝的に多様な種 93
遺伝的に強く連鎖している 219
遺伝的に最も離れている 93
遺伝的に類似な個体群 93
遺伝的に類似な集団 93
遺伝的背景 92
遺伝的反応 92
遺伝的不一致 92, 93
遺伝的負荷 92
遺伝的浮動 92
遺伝的不和合性 92, 93
遺伝的分化 92
遺伝的分岐した系統 93
遺伝的変異 92, 112
遺伝的変異性 92
遺伝的変異体 92
遺伝的変異の維持 127
遺伝的変動 92
遺伝的防除 92
遺伝的雌部位 93
遺伝的モザイク現象 92
遺伝的要因 92
遺伝的レベル 92
遺伝的連関 92
遺伝的連鎖地図 92
遺伝統計学 92
遺伝の－ 90, 93
遺伝標識 92
遺伝分散 92
遺伝変異性 92
遺伝マーカー 92
遺伝モデル 92
遺伝率 101
糸 83
緯度 121
移動 61, 135, 140, 221
移動角度と速度 140
移動型 84
移動軌跡 135
移動距離 85, 140
移動記録 221
移動群 135
移動形質群 135
移動経路 85, 135
移動行動 140
移動コスト 140
移動時間 221
移動指数 135
移動習性 135
移動性 135, 137
移動性種 135, 137
移動性生物 137
移動生態学 140
移動性の 135, 137
移動性の蝶 135
移動･絶滅･再定着プロセス 172
異動態的 101
異動態的な発生型 101
移動多型 85
移動－定着行動様式 135
移動に関する意思決定 140
移動の駆動因子 66
移動の契機 27
移動比 179
移動飛翔 84
移動費用 140
移動率 135, 179
移動率の進化 77
移動履歴 221
移動ルート 135
緯度クライン 121
緯度勾配 121
意図しない移入 226
意図的導入 113
糸でくくる 123
緯度の連続変異 121
緯度本位の距離 126
イトランセセリ 236
イトランセセリ亜科 95
糸を吐く 205
イナゴ 97, 125
稲田 187
1日齢 152
移入 43, 109, 116
移入からの経過時間 219
移入交雑 116
移入雑種形成 116
移入されたコロニー 116
移入種 16, 109, 116
移入種の園芸植物 78
移入植物 116
移入率 43
イネ 151, 187
稲 187
イネ害虫 161
イネ揮発性物質 187
イネ黒すじ萎縮ウイルス 179, 187
イネ黒すじ萎縮病 179, 187
イネ縞葉枯ウイルス 187, 188
イネ縞葉枯病 187, 188
イネ南方黒すじ萎縮ウイルス 202, 206
イネ南方黒すじ萎縮病 202, 206
居場所 104
イバラのやぶ 38
違反 151, 221
ε-多様性 75
イプシロン多様性 75
異物同名関係の原理 172
イベリア半島の蝶 108
違法 221
異方性 19
違法導入 108

イマーゴ　109
今話題の蝶種　106
イミダクロプリド　109
異名　213, 214
異名関係　214
異名表　214
異名リスト　214
イモムシ　36
イモ虫型　75
囲蛹　42
囲蛹殻　176
囲蛹殻に包まれている　42
イラクサ　208
イラクサ摂食幼虫　145
イラスト　108
入り江　26
イリドイド　116
イリドイドグルコシド　116
イリドイド配糖体　116
医療用冷凍庫　132
異類交配　61
イルミナ技術　108
入れ子構造　144
色－　39
色鮮やかな花　44
色紙モデル　156
色恒常性　43
色情報をコードする要素的なニューロン　71
色図版　44
色選好性　44
色対比型応答　44
色対立型応答　44
色と選好性の関連　23
色と選好の間の組み換え　180
色留め剤　84
岩の裂け目　41
岩場　188
陰イオン　19
陰イオンチャ(ン)ネル　19
インカ　112
因果関係　36
インカム型ブリーダー　110
インカムブリーダー　110
陰具小片　231
イングループ　112
陰茎　160
陰茎支持片　19
陰茎包隔膜　19
陰茎盲のう　42
咽喉　98
印刻(する)　191
インサイチュー･ハイブリダイゼーション法　109, 117
印刷者の誤り　172
印刷中　109
隠翅目　199
インシュリン　113
インシュリン経路　113
インシリコ　109
インスリン　113
インスリン信号伝達経路　113
インセクト　112
インセット　113
インテグラーゼ　113
インデル　110
咽頭　161
インド・オーストラリア区　111
インドール　111
インドールキノン化合物　111
インドールグルコシノレート　111
インドールメラニン　111
インドシナ半島と近隣地域　111
インド北東部　148
イントロン　116
イントロン－エクソン構造　116
イントロンスプライシング　116
イントロン前生説／後生説　116
インバージョン　116
インビトロ　109
インビボ　109
インフォームド・コンセント　111
インブリーディング　109
インプリンティング　109
インフルエンザウイルス　111
隠蔽　35, 52
隠蔽型擬態　52
隠蔽擬態　135
隠蔽種　52
隠蔽色　52
隠蔽性　52
隠蔽的異物擬態　16
隠蔽的擬態　135
隠蔽的多様性　52
隠蔽的な生物多様性　52
引用文献　181

ウ

ウイルス　230
ウイルス運動性　230
ウイルス価　230
ウイルス感染　230
ウイルス感染力価　230
ウイルス結合　230
ウイルスゲノム　230
ウイルス産物　230
ウイルス性侵入者　230
ウイルス接種液　230
ウイルス前培養　230
ウイルス操作　230
ウイルスDNA　230
ウイルス抵抗性　230
ウイルス抵抗力　230
ウイルス濃度　230
ウイルスの環　230
ウイルスの転写産物　230
ウイルスの力価　230
ウイルス(性)病原体　230
ウイルス複製　230
ウイルス粒子　230
ウイルス粒子産出　230
ウイルス粒子生成　230
ウイルス－レクチン　230
ヴィンクルム　230
ウェーバー線　232
ウェスタンブロッティング　232
上の　212
上の－　212
ウェル　232

ウェルチのｔ検定 232
ウォーレス線 231
ウォルバキア 235
羽化 68, 71
羽化以後期 168
羽化機構 131
羽化後 169
羽化後の雌 169
羽化時間 72
羽化時期 68
羽化した成虫 14
羽化した成虫の個数 150
羽化したばかりの雌 145
羽化した雌 68
羽化消去 68
羽化する 68, 71
羽化成虫 72
羽化に成功した割合 179
羽化の成功割合 179
羽化パターン 72
羽化ホルモン 68
羽化率 68, 72
雨期 178
雨季 138, 232
雨季型 232
浮き出ている 173
受入側個体群 180
受入側生息地 180
受入コード 12
受入番号 12
受け取った 179
受身的移動 158
受ける 225
受けるに足る 57
ウサギコウモリ 95
ウシ胎児血清 81, 82
ウシテンプレート 32
ウシロドプシン 32
ウシロドプシンモデル 137
薄明かりの単色閃光セット 60
薄切標本 200
薄暗い光 60
薄暗がり 67
ウスグロミヤマシロチョウ 21
ウスバアゲハ亜科 157
ウスバシロチョウ 158
薄めたハチミツ 60
ウスリー型 228
ウスルリシジミ 44
疑いなく正しい 37
疑わしい 67
内側－ 132
内側の縞状バンドの発育不全 109
内側の(縞状)バンド 131
内気 51
うっそうと茂った森の景観 100
促す 173
ウマノスズクサ 165
ウマノスズクサ科 22
生まれたばかりの雌 145
生まれた場所からの分散 142
ウミガメ 129
ウミドリ 191
海鳥 191
産む 32, 203
羽毛状の触角 166
羽毛状の斑紋 81
羽毛のようなトゲ 81
裏返しにする 76
裏付ける 46, 50
ウラナミシジミ 126
裏翅 27
ウラミドリシジミ 102
裏面 27, 226, 228
裏面の 229
ウリジン-5-三リン酸 228
雨緑林 138
ウリ類退緑黄化ウイルス 36
ウリ類退緑黄化病 36
雨林 178
ウルトラメトリック系統樹 225
熟れすぎた果実 155
鱗模様状の鱗粉 109
ウンカ(の仲間) 165
ウンクス 225
運動 140
運動活性 140
運動活動 140
運動能力 140
運動量 138
雲南省 236
運搬タンパク質 221
運搬役 222
運命づける 65
運命的に決定された翅 81

エ

鋭角 13
鋭角的な 19
鋭角に屈曲した触角 93
永久休眠 160
永久コドラート 160
永久方形区 160
影響緩和 136
影響緩和措置 136
影響緩和のための指針原則 98
影響を与える 109
影響を受けない 225
影響を受ける在来種 15
英国自然史博物館 31, 33
英国人蝶類採集家 73
英国蝶類モニタリング計画 225
営巣場所 144
HR エリシター 106
HR 発現する植物 106
H 遺伝子座 98
H 遺伝子座の初期スクリーニング 112
HSD 検定 104, 106
HSD 法 104, 106
HSP 遺伝子 100, 106
H-B アイレット 103
HPLC 法 106
h 脈 106
鋭敏な時空調節 79
栄養 150
栄養学的機構 150
栄養学的生態 150
栄養共生パートナー 223
栄養循環 150
栄養条件 150
栄養状態 150

栄養生殖を行う　184
栄養生態学　150
栄養成長　229
栄養素貯蔵行動　150
栄養段階　223
栄養的適合性　150
栄養に富んだ付属腺物質　150
栄養になる食物　150
栄養繁殖　229
栄養不足　167
栄養物　149
栄養物質　149
栄養物質貯留　150
栄養物質の収入分　150
栄養分枝系　41
栄養豊富な精包　150
栄養要求　150, 229
栄養要求性　150
ARG 上清　22
ARG 貯蔵器　22
AIC 最小　126
AIC 最小混合モデル　136
Alu 配列　17
ATP 合成酵素　24
エーテル　76
A ピーク　15
ABGD　25
AP 軸　20, 21
A 脈　18
描く　74
液状食料　124
液状糞　98
エキス　80
腋節片　25
エキソクチクラ　78
液体　85, 124
液体空気　124
液体酸素　124
液体窒素　124
液体に浸して柔らかくする　127
液体分泌　85
益虫　27
液胞　228
エクアドル　69
エクオリン　14
エクオリン遺伝子　14
エクジステロイド　68
エクジステロイドの放出　68
エクジステロイドホルモン　68
エクジステロイドホルモンの合成　214
エクジステロイドホルモンの放出　183
エクジステロイド量　68
エクジステロン　68
エクジソン　68, 174
エクジソン受容体のアイソフォーム　68
エクソン　78
エクソンかきまぜ説　78
エクダイソン　68
エクダイソン欠除　127
エコシステム　69
エコロジカルエンジニアリング　68
エコロジカルニッチ　69
餌の型　59
餌の質　59
餌の転換効率　81
餌場　82
壊死　143
えじき　171
壊死組織　143
壊死の強度　113
壊死の重症度　113
SA 依存性生合成経路　189
SA 依存性生産経路　189
SA 依存性卵誘導(性)応答　189
SA 応答性遺伝子　189
SA(サリチル酸)の信号伝達経路の機能性下流　89
SSH 法　206, 212
SSH 法アプローチ　206
SSCP 法　199, 206
SSJ 法　198, 206
SNV 決定　201
Sc 脈　209
ST-GMYC 解析　206
SD 配列　197
エステル　76
エステル化する酸　76
s 脈　192
似而非近縁種　80
エソグラム　76
エゾスジグロシロチョウ　97
枝　32, 207
エタノール　76
エタノール沈殿させた DNA サンプル　76
エタノール保存されていた筋肉組織　76
エタノール溶媒　201
枝分れしたトゲ　32
エタン酸エチルの蒸気　76
エチオピア区　15, 76
エチレン　76
エチレンジアミン四酢酸　69
エチレン生合成　76
越夏　155
エッジ効果　69
越冬　102, 155
越冬一年生植物　235
越冬型　102, 155
越冬可能地域　168
越冬期　155
越冬形態　102, 155
越冬集団　155
越冬集団繁殖地　235
越冬状態　102
越冬ステージ　155
越冬する　102, 155, 158
越冬生息場所　102
越冬世代成虫　155
越冬態　155
越冬能力　155
越冬の成功率　235
越冬幼虫　102, 155
越冬齢期　155
越冬を可能にする気候前適応　170
越年生の　28
越年草　235
越年卵　102
エデアグス　14

EditSeq プログラム 69
NR データベース 149
NADH デヒドロゲナーゼサブユニット 5 142, 143
NADPH オキシダーゼ複合体 142
NADPH 酸化酵素複合体 142
NMR 解析 149
NJ 法 144
N-β-アラニルドーパミン 142
N-β-アラニルドーパミン合成酵素 142
N-β-アラニルドーパミンシンターゼ 142
N 末端 142
N 末端領域 142
エネルギー需要 73
エネルギー摂取(量) 73
エネルギー的な制約 73
エネルギー等価則 73
柄の 208
絵の具で描かれた眼状紋 156
エノサイト 151
エノシトイド 73, 151
エバゴラスヒスイシジミ 44
エピクチクラ 74
エピ顕微分光測光法 75
エピジェネティクス 75
エピジェネティック 74
エピジェネティック制御 74
エピスタシス 75
エフェクター分子 70
F 値 80
FBLRH プロジェクト 81
FPKM 値 88
エボ - デボ 77
絵本 164
MEDEA 因子 131
MAP キナーゼ 129, 137
ML 軸 132, 137
ML 法 130
ML 法による系統樹 130, 137
ML 法の解析 137
MK テスト 137
m-cu 脈 131, 132
m 種リスト曲線 127
m 種リスト法 127
MP 法 130
MP モデル 140
エムポディウム 72
M 脈 131, 132
獲物 171
獲物走性 171
えり 43
襟首 173
選り好み 171
エリシター 71
エリシターの一種 34
襟状部 43
エリスロプテリン 75
LEA タンパク質 121
L オプシン遺伝子の部分塩基配列 158
L(長波長型)オプシン遺伝子配列 119
L 視物質 119
L 視物質系統 119
LD サイクル 123
エルニーニョ現象 71
エレガンス線虫 34
エレクトロポレーション法 71
エレクトロポレーション法を用いた低分子干渉 RNA の取込み 71
エロンゲーションファクター 71
遠位 62
遠位領域 62
塩化アンモニウム 18
遠隔計測データ 215
遠隔地 183
沿岸の 42
塩基 26
塩基位置 149
塩基組成の有意な不均一性 198
塩基多様性 149
塩基置換 26, 149
塩基置換数 26, 150
塩基長 32
塩基対 26, 32
塩基データ 149
塩基転位 221
塩基の位置番号 149
塩基配置 149
塩基配列 26, 149
塩基配列間の相違 195
塩基配列基盤マーカー 195
塩基配列クロマトグラム 194
塩基配列決定技術 195
塩基配列データベース 149
塩基配列統計 195
塩基配列による遺伝子型解析 94
塩基配列の再構成 180
塩基配列の出現頻度 26
塩基配列の特性 194
塩基頻度 26
塩基部位 149
塩基分岐 149
塩基モデル 149
遠距離場の音の圧力成分 81
塩基レベル 149
エングレイルド遺伝子 72, 73
園芸 90
園芸家 105
園芸店 165
園芸品種 53
エンコーディング C 型レクチン 72
エンザイムエステラーゼ 74
エンザイムチロシンヒドロキシラーゼ 74
遠心分離 37
遠心分離機にかける 37
遠心分離を行う 37
円錐形 46
円錐状感覚子 194
円錐晶体 52
円錐ツイーター 46

遠征報告　78
円柱のニューロン　44
延長　173
延長選択　173
延長部分　173
エンドウゾウムシ　159
エンドクチクラ　72
円盤　61
エンハンサー　73
円板状の葉の切片　122
円盤方程式　61
燕尾系アゲハチョウ　213
塩分摂取　189
エンベロープ　73
縁辺相　129
縁毛　88
縁毛状の　83
縁毛色　88
縁毛帯　88
縁紋　175, 208
塩類　189

オ

尾　214
追い風ドリフト　214
追い風の方向　66
追い風偏流　214
生い茂った　155
おいしさのスペクトル　156
追い出される　206
追い払う　183
追い求める　176
オイル液滴　151
負う　110
横　222
横隔膜　59
凹形　71
横径の　222
凹溝　191
横行眼状紋　222
横行の　222
凹溝の　191
欧州分子生物学研究所　71, 76
黄色色素　236
黄色翅の雌モデル　236
黄色対立遺伝子のホモ接合　104
黄色対立遺伝子頻度　88
横断横線　222
横断的に延びている　188
横断面　222
凹頭　71
応答閾　185
応答閾値　185
応答期間　67
応答遅延　121
応答的な　186
応答特性　185
応答変数　186
黄斑の　127
往復移動の能力　12
往復飛翔　188
黄変症候群　236
横縫合線　222
横脈　52
横脈くぼみ　61
横脈月状紋　61
横脈欠落部　61
横脈紋　61, 67
凹面　46
応用昆虫学　21
黄緑色素　236
ORF 長　153
大顎　118, 128
大腮　118, 128
大顎腺　128
大顎の　128
大腮の　128
大当たり　31
大雨　100
オオアメリカモンキチョウ　153
覆い隠す　150
おおいに　173
覆う　42
大写し　41
大型　120
大型チョウ目　127
大型動物　120
大型無脊椎動物　127
大型鱗翅類　127
オオカバマダラ　135, 138
大きさ　200
大きさと方向　127
大きさを変更する　179
大きな障壁　127
大きな比率を占める　154
オーク（カシ、ナラ、カシワ）類の木　150
オオクロムクドリモドキ　44
オオゴマダラ　156, 187, 222
大ざっぱに言えば　33
オージオグラム　24
オージオグラム実験　24
オオシモフリエダシャク　160
オーストラリア区　24
オオタスキアゲハ　95, 153
オオタバコガ　50, 219
オオタバコガの仲間（*Helicoverpa virescens*）の化学受容体 4　107
オオタバコガの卵巣由来の樹立培養細胞株　108
覆った　51
オオツバメガ　223
オートミクシス　25
大幅に劣る　198
オーファン受容体　154
オープンリィーディングフレーム　153
オープンリーディングフレーム　152, 153
オオベニシジミ　120
オオミズアオ　127
オオモンシロチョウ　120
オオモンヒカゲ　120
オオヤドリギシジミ　94
おおよそ　22
オオルリアゲハ　225
小笠原諸島　151
小川　208
小川に生息する無脊椎動物　208
オキアミ　119
オキシアニオン　155
置く　57, 200

憶測 204
奥地 114
オクトパミン 151
オクトパミン受容体 151
遅らせる 109, 186
遅れた幼虫 56
抑える 184
オジロルリツバメガ 237
♂ 128
オス 128
雄 128
雄型幼虫 128
雄か雌かのどちらかに偏った遺伝子発現 195
雄から引き渡された抗催淫物質 128
雄側の投資 158
雄間競争 128
雄殺し 128
雄走査 128
雄探索 128
雄特異的な *dsx* 遺伝子のイソ型 128
雄特有の器官 128
雄特有の胚死亡率 128
雄と雌 90
雄と雌の特徴を合わせもつ表現型 128
雄に偏った 128
雄に特有の分子機構 128
雄の遺伝子型 128
雄の求愛選好 128
雄の色彩選好 128
雄の射精液 128
雄の性分化 128
雄の第二次性徴(形質) 192
雄の適応度 128
雄の特徴を持つ交尾器表現形質 128
雄の特徴を持つ翅表現形質 128
雄の特徴を持つ生殖器官 128
雄の把握器 128
雄の胚 128
雄の配偶行動 128
雄の配偶者選好性 128
雄の配偶者選択 128
雄の翅から発生する匂い 128
雄の表現型 128
雄の方が雌よりも体長が大きい性的体長二型 128
雄の雌探し行動 128
雄のような 128
雄ヘテロ型 101
雄ヘテロ型性染色体構成 128
雄偏重 128
雄雌 90
雄モデル 128
雄 - 有用な情報付き 128
雄由来化合物 128
雄由来混合物 128
オセアニア区 151
汚染 167
汚染する 48
遅い夜 121
おそらく 22, 46
オゾン誘導性細胞死応答 155
オゾン誘導性傷害 155
オゾン誘発性傷害 155
お互いの - 16
穏やかな気象 35
落ちる 80
落し穴 165
音刺激 13, 201
音刺激の強度 113
音の刺激強度 113
音発生 201
音(波)パルス 201
おとり装置 56
衰える 57
驚かす 190
驚くほど普通 212
音を出さない 141
音を出さない蝶 141
音を発生しない 148
音を発生する 201
オナシアゲハ 40
同じ種類 47
尾の 36
各々の相互擬態種 185
帯 26, 95
オビカレハ幼虫 87
帯状地域 27
帯状分布 237
オビモンドクチョウ 26, 150
脅かし戦術 88
脅かしの目玉模様 207
脅かす 218
オプシン遺伝子配列 153
オプシン遺伝子ファミリー 153
オプシン対立遺伝子 153
オプシンタンパク質 153
オプシン発現様式 153
optix 転写制御因子遺伝子 153
オペロン 152
オマチン D 152
オミン 152
思いつき程度の 204
思いつきの放蝶 36
オモクローム 152
オモクローム系色素 152
オモクローム系の経路遺伝子 152
重さ 97
面白いことに 114
表立った防衛 155
表面 54, 227
親株 140
親子間対立 157
親種 157
親種の中間種を好む 171
親世代の影響 130
親の継続的な抗生物質処理 48
親の個体群 157
親用語 157
およそ 40
及ぼす 78
オリエンテーション 153
折りたたむ 86
オルソログ 154
オルソログ遺伝子 154
オルソロググループ 154
オルファクトメーター 151

オレゴンギンボシヒョウモン 153
オレンジ色の地色に黒斑のあるヒョウモンチョウの翅模様 88
音圧レベル 205
音圧レベル計 201
音響刺激 13
音響の 13
温血の 73, 231
温室 97
温室効果 97
温室効果ガス 97
温帯 216
温帯域個体群 216
温帯緯度 216
温帯雨林 216
温帯気候 215
温帯グループ 216
温帯産遺伝子型 216
温帯森林広分布属 216
温帯性遺伝子型 216
温帯性の種 216
温帯属 216
温帯地域 216
温帯地方 216
温帯(性)のシロチョウ科の蝶 216
温帯林 215
温暖化 231
温度 216
温度依存性 216
温度環境 216
温度感受性 216
温度慣性 218
温度計 218
温度効果 216
温度サイクルプロファイル 218
温度 - サイズ則 216
温度周期 101, 216
温度周期反応 185, 218
温度受容器 218
温度条件 216
温度センサー 218
温度相 218
温度 - 体格則 216
温度遅鈍 218
温度調節 218
温度の影響 216
温度の年変化 236
温度の年変動 236
温度反応 218
温度法則 216
温度補償(性) 216
温度誘導性性差 216
温度領域 215
オントロジー 152

カ

- 科 11
科 81
窩 81, 86
下 111
下 - 111, 209
蛾 140
カースト 36
ガーデニング 90
カーブフィッティング外挿法 53
カール・フォン・リンネ 124
界 119
外 79
外 - 78, 168
概 40
ガイアナ(共和国) 184
外因性 80
外因性休眠 80
外因性 β-グルコシダーゼ 78
外因的 80
外縁 31, 154, 216
外縁黒帯 30
外縁線 216
外縁側板(溝状)線 129
外縁の先端 62
外縁の(縞状)バンド 129
外縁部 129, 216
外縁部帯 129
外縁部の眼状紋 129
外縁寄り中央帯 169
外横線 62, 169
外温生物 69
外温の 69
開花 20
下位概念用語 39
外界の刺激 79
開花期 30, 219
開花季節 30
外殻 116
開花後 169
開花作物 85
開花時期 219
開花フェノロジー 85
開眼 80
外観 21, 80
海岸域 42
外気 154
外気温 17
回帰係数 182
回帰式 182
回帰性 104
外寄生者 69
回帰直線 182
外基部帯 169
回帰分析 182
回帰方程式 182
回帰モデル 182
階級 178
階級群 97
外クチクラ 78
外クチクラ形成 78
外群 154
外群の分類群 154
解決する 61
外見 21, 80
外見的には死んだ蛹 21
外原表皮 78
外原表皮形成 78
カイコ 31, 198
外交配 154
外交配弱勢 154
外交配種 154
カイコガの幼虫 198
外国種 78
外骨格 78
カイコの卵巣由来の樹立培養細胞株 31
カイコ類の塩基配列 31
開墾 53

開墾する 53
介在する接触 132
開始 44, 152
開翅 234
χ二乗検定 39
カイ二乗検定 39
概日仮説 40
概日振動 40
概日性 40
概日時計 40
概日リズム 40
概して 33
開翅日光浴 65, 181
外斜 180
回収 235
海水面 191
外生アンモニア 78
改正された知見 187
開線 152
改善する 17
回想 180, 183
外挿 80
階層化する 208
階層的F統計 102
階層的G検定 102
階層的順序 102
階層的状態空間モデル 102
階層的序列 102
階層的対数尤度比検定 102
階層的優性 102
階層的尤度比検定 102
階層に分ける 208
解像能力を有する眼 108
開拓者 165
開拓地 41
開拓地に適応した個体群 41
害虫 112, 161
外中央線 169
外中央帯 169
外中央部 169
害虫管理 112
懐中電灯の明り 220
害虫防除 161
開張 234, 235
改訂 186
改訂版 186
外敵に満ちた 106
外的符合モデル 79
外転する 76
回転楕円体状の翅面 205
回転飛翔 40, 205
解糖 96
解糖系 96
解凍する 217
下位同物異名 118
概念体系 152
概念的超遺伝子 46
概年リズム 40
～開発 58
海抜 11, 23
解発因 183
解発因子 183
解発する 71
解発フェロモン 183
蓋板 152
外皮 113, 215
外被 116, 215
回避する 27
回避反応 75
外表皮 74
外表皮外層 154
外表皮内層 112
回復 186
回復する 181
回復力 160
外部形質 79
外部形態(学) 79
外部結節 79
外部構造 79
外部刺激 79
外部生殖器 79, 217
外部組織 79
外部の 79
外部の解剖学的構造 79
外分泌 78
外分泌物 78
外分泌物質 78
蓋片前器 170
解剖 19, 62
蓋帽 35
解剖学的に 19
解剖技法 62
開放血管系 152
開放個体群 152
解剖手法 62
解剖する 62
解剖体 62
開放的場所 152
外膜 154
外膜(の)幅 233
外膜表面 154
解明 71
解明する 71, 108, 227
壊滅的な打撃 58
回遊 72, 135
回廊 50
概要 12
外葉 90
概要塩基配列 66
概要ゲノム 66
海洋底生の大量採取品 28
外来種 16, 78, 148
外来種の影響の予防 98
外来性害虫 86
外来の 78
カイラス山(チベット) 118
解離 103
概略図 191
改良草地 109
カイロモン 118
下咽頭 107
カウンター戦術 51
カエデの木 129
カエル 88
変える 228
科階級群 81
科階級群名 81
下外側 210
下蓋板 120, 152, 210
下蓋板前器 170
花外蜜腺 80
化学－ 38
化学感覚 38
化学感覚遺伝子の系統別の拡張 124
化学感覚子 38
化学感覚神経細胞 38
化学感覚タンパク質 38, 52
化学感覚ニューロン 38

化学感覚による摂食行動制御　38
化学感覚毛　38
化学擬態　38
化学群　38
化学構造　38
化学刺激　38
化学受容　38
化学受容器　38
化学受容体　38
化学進化　38
化学生態学　38
化学的感覚　38
化学的感覚遺伝子　38
化学的軍拡競争　38
科学的根拠　191
科学的証拠　191
化学的な暈　38
化学的な刺激　38
化学的媒介による信号　38
化学的媒介による相互作用　38
化学的変化　38
化学的変形　38
化学的防除　38
化学物質　38
化学変化　38
化学防御　38
化学薬品　38
カカトアルキ目　129
踵行目　129
鏡型反射層板　136
架かる　202
科間　114
花冠深度　50
鉤　225
花器形成　85
鍵酵素　119
鍵刺激　119
鍵種　119
鉤状の　99
鉤状の前翅先端　53
鉤爪　52
鉤爪　52, 105
鉤爪状の　105, 149
鉤爪状の尾状突起　105
鍵となる酵素　119
描き直す　181
垣根　100
書きまちがい　120
可逆的　186
格　36
核　149
-学　11
額　88
顎　96
核遺伝子　149
顎下腺　210
核型　119
角括弧　206
核ゲノム　149
拡散　60, 61, 62
拡散過程　60
拡散共進化　60
拡散共進化の相互作用　60
核酸導入　221
拡散と寒冷押し戻し　60
拡散に基づくランダム歩行　60
拡散反応系モデル　60
拡散方程式　60
核磁気共鳴　146, 149
確実な記録　37
確実に正しい　37
革翅目　57
学習　122
学術論文　222
確証する　50
核小体低分子 RNA　200
角状突起　105
角状突起長　105
角状突起の形態　105
確証例　49
確信させてくれる例　49
額伸縮突起　89
革新的な変化　112
隠す　46
角錐　176
覚醒　25
額線　88, 89
拡大　75, 205
拡大した領域　73
拡大図　73
拡大断面図　41
核多角体病ウイルス　149
拡張した気管　73
額頭盾縫合線　89
獲得　172
獲得されたアンモニア　13
核内低分子 RNA　200, 201
核内有糸分裂　73
確認された個体群　46
角張った　19, 206
角張った翅　176
角張った斑点　19
額板　88
核マーカー　149
角膜　50
隔膜　59, 194
角膜レンズ　50
学名　121, 191
学名の先取権　172
学名命名　146
攪乱　63, 161
攪乱強度　63
攪乱された疎林　63
攪乱されていない環境　226
攪乱処理　63
攪乱頻度　63
隔離　117
隔離機構　117
隔離する　193
隔離地域　177
確立　76
確立された手法　75
確立した　73
確立する　75
確率性　208
確率(論)的な降雨イベント　208
確率的モデル　172
確率論的モデル　172
隔離的な生息地　117
隔離分布　61
額瘤　89
額隆起線　88
隠れ家　197
隠れ家で暮らしている幼虫　197
隠れ層　102
隠れた多様性　102

隠れた雌による選り好み 38, 52
隠れ場所 51
顎腕 96
崖 41, 170
掛け合わせ様式 52
花形 85
芽茎 208
欠けた処理 136
蜉蝣目 74
カゲロウ目 74
崖を住居とする 41
下限推定 126
下限推定値 126
下限値の推定 126
過誤 75
籠 34
河口 76
下降－ 55
下口橋 107
下後側板 119
河口底生無脊椎動物群集 28
下口隆起線 107
過酷な環境 74
過酷な条件 209
禾穀類 37
過去の情報 171
籠複合体 34
風下に 66
風下の方向 66
重なった断片 155
重なり 86
飾る 56
火山活動 231
火山性の 231
加算的 211
火山島 231
花糸 83
可視光スペクトル 230
餓死する 207
餓死耐久性 207
果実 28
果実に擬態 89
カシミール 119
果樹園 89
花種選好性 85
花序 111
過剰寄生 107
過剰肢 181, 212
過剰ストック 212
過剰脱皮 79
過剰脱皮の若齢 212
過剰脱皮齢 212
過剰な齢 212
過剰に形成された眼状紋 212
過剰発現 154, 155
過剰表現 155
カジリムシ目 175
下唇 119
下脣 119
下唇亜基節 210
下唇基節 132
下唇後基節 169
下唇鬚 119
下唇節 119
下唇前基節 171
下唇の 119
下唇縫合線 119
課す 73
加水分解 107
科数－目数の比 81
ガスクロマトグラフィー 90
ガスクロマトグラフィー－マススペクトロメトリー 90
ガス交換 90
数の反応 150
カスミカメムシ 165
カスリタテハ属の蝶 51
風 234
化性 231
寄生 220
苛性カリ 169
火成岩 108
仮性休眠 151
寄生細胞 220
化性転換 197
化性変化の機構 131
風がない天候 35
化石 87
化石記録に基づく年代較正 87
仮説上の概念 108
仮説的概念 108
仮説的超遺伝子 46
仮説を検定する 217
風まかせ移動 234
河川性無脊椎動物 208
下前腹板 119
河川流域 188
下層 226
仮装 129
仮想事例 108
仮想的サンプル数に基づく希薄化曲線 108
仮想的分類単位 106, 108
仮想例 108
加速化された変態 12
可塑性 166
可塑性の遺伝的制御 92
型 87
堅い 208
固い 208
かたい殻 174
過体重 110
固い葉 220
硬い葉 100
固い棒 201
片側追跡 226
かたき 86
肩透かし 64
形 87, 196
形作る 196
型特異的な入力パラメータ 224
型特異的な発現パターン 139
片道移動 152
傾き 200
偏り 28, 200
偏りのない方法 225
語り手 203
花壇 85
価値ある指針 228
価値ある指摘事項 236
カチオン 36
家畜 65, 124
家畜の放牧 124
価値のある道具 228

価値のない視点 116
画期的発見 32
褐色がかった 33
褐色がかった壊死組織 33
褐色色素 33
褐色の腐葉(層) 33
褐色斑紋 33
がっしりした 188
渇水 67
活性化刺激 13
活性化する 13
活性酸素種 179, 188
活性脳 13
合体する 226
合体節 214
合体理論 42
合着した系統樹の事前情報 42
合着した系統樹の事前分布 42
合着した部分 42
活動 27
活動位相 13
活動期 13
活動季節 13
活動休止状態の休眠 65
活動時間 219
活動時間配分 13
活動時期 13
活動スケジュール 13
活動相 13
活動中心部 50
活動停止 177
活動的にする 13
活動点 13
活動の代行物 212
活動の代替物 212
活動場所 13
カットオフ水準 53
ガットパージ 98
合併 110
過程 172
家庭菜園 104, 119
仮定する 108, 169
カテゴリ - サブカテゴリの比 36
カテゴリ - サブカテゴリの分類数比 36
下転節 223
かど(角) 50
可動因子 137, 140
~かどうかを議論する 62
可動鉤 140
可動鉤上の刺毛 140
可動性 137
可動性因子 137
過渡的な温度上昇 221
過渡的に 221
過度に強調する 154
過度の - 107
金網で囲まれた入れ物 235
必ず 111
かなり多くの 62
カニアシシジミ亜科 100
カニバリズム 35
加入 180
過熱 155
可能性 123, 169
可能性のあるモデル 172
カバーグラス 51
カバースケール 51
カバースリップ 51
カバー率 51
蛾媒花 140
カバイロイチモンジ 230
カバタテハ亜科 223
河畔林 187
下皮 107
カビ 89
花標 98
過敏感型反応 107
過敏感反応 106, 107
過敏感反応性エリシター 106
過敏感反応に類似する壊死 106
過敏感反応マーカー 106
過敏感反応様ネクローシス 106
過敏感反応を発現する植物 106
過敏感様反応 107
株 208
カフェイン感受性神経細胞 34
カフェイン感受性ニューロン 34
カブトムシの角 27
下部の 111
カプランマイヤー法 118
かぶれること 179
花粉 166
花粉学の 156
過分散 154
花粉資源 167
花粉食性 156, 167
花粉摂食 167
花粉媒介 167
花粉媒介昆虫 167
花粉媒介者 167
花粉媒介者が好む種 167
花粉媒介者の誘引 24
花粉飛散 167
花粉分散 167
花粉流動 167
壁模様 231
可変表現型 228
下方選択 66
加法的な 14
加法的に 14
下方の 111
河北省 100
カマキリ 129
カマキリ目 129
蟷螂目 129
鎌状赤血球 197
鎌状の 80
咬み跡 27
咬み型口器 30
カミキリムシ 125, 126
花蜜 85, 143
花蜜回廊 143
花蜜様態の花 143
紙への印刷 172
夏眠 14, 76, 211
夏眠する 14
咬む 30
カムフラージュ 35
科名 81, 142
カメムシ 165

カメムシ目 100
下面 226, 228
下目 111
殻 113, 197
カラー 43
カラー図解 44
カラーパターンの多様性 43
カラーパターン変化 43
カラーパターン変動 43
カラーパターン領域 44
カラーフィールド 43
カラカラ音 179
がらくたDNA 118
カラコルム山地旅行 119
カラザ 38
カラシ 141
カラシ油 141
カラシ油配糖体 96
ガラスキャピラリー 95
カラスシジミ亜科 99
ガラス製毛細管 95
ガラスナイフ 95
ガラス微小電極 95
空祖先 150
体 31
体の大きさ 31
体の大きさの指標 31
体の器官 31
体の形態 31
体の構造 19
体の成長 201
殻で覆われた 151
空にする 176
カラフトセセリ 75
カラムを構成しているニューロン 44
仮 - 175
カリブー 36
下流経路 66
顆粒細胞 96
顆粒病 97
顆粒病ウイルス 97, 98
カルシウム依存性タンパク質キナーゼ 35
カルシウム依存的発光タンパク質 35
カルシウムイメージング(検定)法 35
カルシウム濃度 34
カルディニウム 36
カルテット 177
カルデノリド 35
カルデノリド含有 35
カルデノリド毒 36
カルデノリド濃縮 35
カルトニウスウスバシロチョウ 158
カルフォルニアイヌモンキチョウ 35
カルボキシルエステラーゼ 35
カルミモンシロチョウ 214
過冷却 211
過冷却点 191, 211
花歴学 85
枯れ葉 55
枯れ葉の切断 53
ガレリア森林 90
ガロアムシ目 98
かろうじて発現する 79
カロチノイド 36
カワゲラ目 166
カワスズメ 40
変わった型 227
変わった(事)例 227
川の土手 188
川辺林型 188
カワラケツメイ属の一種 36
瓦状に重なった鱗粉 109
環 187
管 67, 223
間 - 113
眼縁部 153
灌漑 117
灌漑草地 117
灌漑された草地 117
灌漑地 117
考えられる 46
感覚 194
感覚過程 194
感覚器 194
感覚器官の構造 194
感覚機構 194
感覚構造 194
感覚細胞 194
感覚細胞体 194
感覚子 194
感覚刺激 194
感覚神経支配 194
感覚の 180
感覚毛 194, 222
環化した 20
眼下線 210
乾型 67
感桿 187
感桿型 187
感桿型オプシン 187
感桿型光受容細胞 187
感桿の光導波路 187
感桿の部分退色 158
感桿の部分脱色 158
換気 229
乾季型 67
間期細胞核 115
柑橘類の木 40
完気門式 103
環境 73
環境意識が高い国民 74
環境因子 74
環境影響評価 70, 74
環境(的)感受性 74
環境緩和 136
環境共変量 74
環境経験 74
環境傾度 74
環境決定要因 74
環境ゲノム学 74
環境勾配 74
環境刺激 73, 74
環境支配 74
環境周期 74
環境収容力 36
環境省(日本) 136
環境条件 17, 73
環境操作 74
環境体験 74
環境多様性 74
環境抵抗 74
環境的 80

環境的確率性 74
環境的飼育条件 74
環境的多型現象 167
環境的敏感性 74
環境特異的な 48
環境特異的なやり方 48
環境にやさしい 74
環境の時間的変化 216
環境パラメーター 74
環境への影響が少ない 74
環境変化 73
環境変数 74
環境変数の操作 74
環境変動 73, 74
環境変動の確率性 74
環境保健 74
環境保護 74
環境保護運動 74
環境保全 74
環境保全主義者 47
環境要因 74
環境要因の相互作用 114
環境リスク 74
眼茎 153
関係しているらしい 176
関係種 202
環形成の障害 109
完系統的 103
完系列 40
還元 181
間腔 115
観光 220
寒候期の表現型 43
感光色素 162
観光事業 220
刊行日 55
感光部位 162
感光部分 162
勧告 180
観察 150, 232
観察員 116
観察された変動 151
観察時間 151
観察する 155
観察地 221
観察データ 151
観察不可能 227
観察領域の曲線 151
観察路 221
観察路調査 221
換字 221
監視計画 138
乾湿気候の周年サイクル 19
乾湿の季節的環境 232
感受性 194, 213
感受的な 213
干渉 99
管鞘 223
干渉型競争 114
環状化捕獲法 35
干渉行動 99
干渉する 114
管状腺 224
環状腺 187
緩衝帯 33
頑丈な 188
頑丈な体を持った蝶 188
管状の 223
眼状斑点 80
冠状部 50
管状脈 224
環状紋 153
眼状紋 80, 151
眼状紋関連遺伝子 80
眼状紋形成 80
眼状紋サイズ 80
眼状紋の外見上の融合 21
眼状紋の消失 80, 126
眼状紋の発育不全 109
眼状紋フォーカス 80
眼状紋列 195
関心度 190
関心度尺度 190
幹数 150
幹数密度 207
幹数密度効果 69
間性 115
間性形質 196
間性欠陥 115
間性欠陥説 115
間性現象 115
間性障害 115
間性障害説 115
間性表現型 115
間接効果 110
間接相互作用網 110
環節体制 133
間接的安定化選択 110
間接的証拠 110
間接的選択 110
間接防衛 110
完全－ 103
感染雄 111
感染後(経過)時間 106
完全黒化型 45
感染細胞 111
感染しやすい 213
感染する 111
感染性 111
感染性細菌 111
感染性単為生殖 111
完全総当りの整列 16
感染多重度 137, 141
完全同語反復 12
感染特異的遺伝子 158, 170
完全な開翅 45
完全な成体形成 45
完全な成虫の形態形成 45
完全な漸近線 41
完全な翅の伸長 45
完全に－ 160
完全に成長した幼虫 89
完全２分岐樹探索法 45
完全に保護されている種 89
感染病 111
感染頻度 111
完全変態 45, 103
完全変態する昆虫類 103
完全変態の 103
完全保護 89
完全保護種 89
感染密度 111
乾燥 57, 67
乾燥化 57, 66, 67
乾燥気候 67
乾燥休眠 19, 52
乾燥高地の草 67
乾燥森林 67

乾燥森林地域 67
乾燥森林地帯 67
乾燥地 22
乾燥地帯 22, 67
乾燥熱帯性 22, 223
乾燥熱帯的植生 223
乾燥標本 66
乾燥有刺低木林の生息地 67
寒帯高山性の分類群 31
寒帯種 31
寒帯地域 43
寒暖気候の周年サイクル 19
寒暖と乾湿の組合せ 44
乾地草 67
巻頭写真 51
管の 228
環のある 20
環の組み込み能力 113
環の再組み込み 40
環パターン 187
干ばつ 67
旱魃 67
ガンビエハマダラカ 20
間氷期 114
カンブリア爆発 35
幹母 89, 207
潅木 191
灌木 197
潅木管理 100
陥没 210
環北極種 40
γ線照射 90
ガンマ線照射 90
ガンマ速度のカテゴリ数 90
γ-多様性 90
ガンマ多様性 90
ガンマ分布 90
ガンマ分布比 90
ガンマ率のカテゴリ数 90
乾眠 19
顔面 80
顔面の 80
冠毛 157, 166
完模式標本 103
含有色素 164
慣用法 171
関与している 109
関与する環境要因 168
管理 49
管理区域 128
管理されたミツバチ 128
管理単位 128
寒冷化 49
寒冷順化 43
寒冷ショック 43
寒冷地の長野県 43
寒冷麻痺 43
関連解析 23
関連研究 23
関連性 183
関連調査 161
甘露 104, 213

キ

紀 160
期 207
木 222
擬 - 175
偽 - 175
キーステージ 119
キーストーン種 119
偽遺伝子 175
偽遺伝子化 175
奇異な 30
キーペスト 119
キイロショウジョウバエの味覚受容体 43a 64
黄色ぽい白色光線 236
黄色の縞模様を白色に変えた 236
黄色のまま 225
キーワード 119
起因子 36
基因する 24
偽陰性 80
消え去る 160
記憶機構 132
記憶した陸標 132
気温 15
気温低下 216
帰化 143
飢餓 207
機械(的)感覚 131
機械感覚器 131
機械感覚特性 131
機械刺激神経細胞 131
機械刺激ニューロン 131
機械受容器 131
機会的 80
機械的隔離 131
機械的手段 131
機械的傷害 131
機械的損害 131
機会的単為生殖 80
機会的浮動 178
機械的防除 131
奇怪な 30
機械論 131
機械論的基礎 131
機械論的な土台 131
基角 26
規格化 148
規格化された吸収スペクトル 148
帰化した 143
帰化種 16, 143
帰化生物 143
幾何平均回帰 181, 188
器官 158
基環 26
気管 220
擬眼 80
帰還移住 186
帰還移動 186
気管系 220
気管周腺 160
気管小枝 220
気管小分岐 220
気管小分枝 220
帰還する 186
気管の気のう 220
器官の再構成 180
帰還飛翔 186
気管分岐 220
気管分枝 220
器官を刺激する 112
危機管理 188

危機的な在来の絶滅危惧種　52
聞き取れる音　24
危急　231
危急種　231
キク科　46
ぎくしゃくと飛ぶ　118
奇形の精巣　56
奇形の翅　56
奇形標本　216
起源　154
危険から救う移動　185
起源時期　219
棄権宣言　61
危険にさらす　118
危険に満ちた　160
起源年代　154
起源の中心　37
危険分散　28
危険分散戦略　28
危険分散適応　28
危険分散の効果　28
気候　41
気孔　14, 32
機構　131, 153
気候条件　41
気候帯　41
気候調節　41
気候データ　41
気候適応　41
機構的基盤　131
気候的差異　41
気候的な致死限界　41
機構特異的　131
気候の温暖化　41
気候の冷涼化　41
既交尾雌　130
気候変数　41
気候変動　41
気候変動に関する政府間パネル　114, 116
気候変動の影響　69
機作　131
記載　57
記載項目間の遺伝距離　114
記載する　57, 182
ぎざぎざの　219
ぎざぎざの曲線　117
刻み　149
キサントプテリン　236
擬死　55
基翅甲　26
基翅節片溝　26
基翅節片の運動　140
キシタアゲハ　96
基質　210
基質と酵素の区画化　45
基質由来化学的刺激　210
基質を介した化学的刺激　210
疑似同語反復　230
キジマドクチョウ　237
希釈液　60
希釈系列　60
希釈効果　60
希釈する　60
希釈炭素化合物　181
希釈物　60
希釈溶液　60
寄主　105
稀種　178
基縦線　26
寄主起源の認定　105
寄主系統　105
寄主個体群　105
寄主残留物　105
偽手術　196
寄主植物　105
寄主植物化学物質　106
寄主植物種　105
寄主植物特異性　106
寄主植物に対する前適応　170
寄主植物の季節的転換　191
寄主植物のグルコシノレート含有量　106
寄主植物の遷移　211
寄主植物の被視認性　105
寄主植物の目立ちやすさ　105
寄主植物利用　106
寄主植物量の密度　105
寄主選好性　105
寄主選好性転換　105
技術支援　215
技術的反復実験　215
寄主転換　105
寄主によるシアン形成　105
寄主によるシアン発生　105
寄主の協力　105
寄主範囲　105
寄主被子植物　19
寄主標識　106
寄主品種　105
寄主マーカー　106
寄主マーキングフェロモン　105
基準圧　181
基準圧力　181
基準ゲノム　181
基準シードバンクデータセット　27
基準種子銀行データセット　27
基準スペクトル　181
基準超計量系統樹　181
基準配列　181
基準標本　224
基準マーカー　207
希少　178, 190
気象観測所　232
気象現象　24
希少交配種　178
希少雑種　178
希少種　178, 190
気象条件　232
希少性　179
希少成分　190
気象測候所　232
希少な分類群　178
希少な例　178
疑似乱数　175
傷　30, 236
傷跡　190
帰すことができる　24
傷ついた葉　236
傷つきやすい体　231
傷つきやすい身体　231
傷つきやすい場所　231
傷つける　30
きずのない　225

傷を付けること 236
寄生 157
犠牲が多い 50
寄生型 157
寄生関係 157
寄生された毛虫 157
寄生者 157
寄生者－宿主の相互作用 157
寄生者に対する抵抗性 185
寄生する 157
寄生成功(率) 157
寄生性のアザミウマ 157
寄生相互作用 157
寄生虫 157
寄生データ 157
寄生的 157
寄生蜂 157
基節 51
季節 191
季節移動性 192
季節型 192
季節型決定 192
季節型決定ホルモン 192
季節型発現 79, 192
季節型発現様式 192
季節型反応 192
季節型変異 192
季節がない環境 147
季節(的)環境 192
季節別繁殖動態的 101
季節指標 192
季節周期 191, 192
季節情報 192
季節性 192
季節性二型 192
季節(的)多型 192
季節多形 192
季節多型性移動 192
季節的移住 192
季節適応 191
季節的減少 191
季節的個体数パターン 191
季節的翅型 192
季節的消長 161
季節的多型 192
季節的低下 191
季節的二型 192
季節的に変化する環境因子 192
季節的に変化する資源量 192
季節的表現多型 192
季節的変異性 192
季節的変化 191
季節的リズム 192
季節と温度との間の相互作用 114
季節特異的な異なる発現 192
季節配置 191
季節外れの夏に降りた霜 23
季節はずれの発育 227
季節(的)変異 192
季節変化の効果 69
季節変化パターン 159
季節(的)変動 191, 192
基線 26
帰巣行動 104
寄贈者 65
寄贈する 65
帰巣ナビゲーション 104
帰巣能力 104
規則性 182
規則的な時間間隔 182
基礎研究 119
基礎情報 26
基礎的手法 97
既存の個体群 78
期待 173
擬態 35, 135, 136
擬態圧 136
擬態(の)遺伝学 93
擬態遺伝子座 136
擬態遺伝子の固定 84
擬態型 136
擬態型色彩パターン 136
擬態型(の)染色体 135
擬態型配列 136
擬態型雌 136
擬態関係 135, 136
擬態関連遺伝子座 136
擬態関連の限性多型 136
擬態グループ 136
擬態群 136
擬態系統 136
期待系統樹 78
擬態行動 52
擬態昆虫 136
擬態した翅の腹側 35
擬態しているもの 136
擬態種 136
擬態進化 136
擬態する 61, 136
擬態対立遺伝子 135
擬態多型 136
期待値のカットオフ値 67
擬態の収斂 136
擬態のパターン 136
擬態の翅紋様 136
擬態の表現型 136
擬態のプレパターン形成 136
擬態パターン 136
擬態表現型の運命 81
擬態表現型の最終結果 81
擬態変異 136
擬態リング 136
北インド 148
来た同じ道を高頻度で引き返す 102
北朝鮮 54
北日本個体群 148
北日本産個体群 148
北半球 148
北半球に広く分布する種 233
北への分布拡大 148
既知の目印 119
既知の目標 119
既知の陸標 119
既知の隣者とよそ者との間の識別 61
キチン(質) 39
キチン環 39
キチン結合性タンパク質 39
気づく 61
拮抗作用 20
拮抗的関係 220

拮抗的多面発現　20
キツネノマゴ科　12
規定　174
基底グループ　26
基底細胞　26
規定する　204
基底ノード　26
基底膜　26, 225
基底面積　25
擬頭　80
偽瞳孔　175
起動フェロモン　172
危篤種　51
輝度処理　32
キナーゼ　119
キナーゼ活性化　119
気に入られている　81
絹の支持帯　198
キヌレニン　119
キヌレニンヒドロキシラーゼ - ホワイト　119
キネシス　119
機能解析　89
機能解析研究　89
機能獲得　90
機能形態学　89
機能欠損　126
機能研究　89
機能亢進　90
機能しなくなる　12
機能性雌　89
機能喪失　126
機能退化　126
機能多様性　89
機能注釈　89
機能的遺伝子座　89
機能的機構　89
機能的形態学　89
機能的構成　89
機能的構築　89
機能的差異　89
機能的な雌　89
機能的に近縁な遺伝子　89
機能的パフォーマンス　89
機能的反応　89
機能的部位の収斂進化　49
機能的分岐　89
機能的役割　89
機能の反応　89
機能発散　89
機能不全　128
機能分化　89
機能分担　89
木の頂上　222
木のまばらな空地　152
木の実　28
木の幹　222
キノン系色素　177
希薄化　179
希薄化曲線　179
希薄化の計算式　179
希薄化法　179
希薄曲線　179
希薄な　179, 202
黄蜂　231
偽発見率　80, 81
揮発性化学的刺激　231
揮発性化合物　231
揮発性ケトン　231
揮発性脂肪酸誘導体　231
揮発性テルペン　231
揮発性物質プロファイル　231
揮発性物質の放出　231
揮発性有機化合物　230, 231
揮発率　231
基盤　166
忌避行動　25
忌避剤の揮発成分　184
忌避作用　25
厳しい条件　209
厳しい長期間にわたる衰亡　195
厳しくない経験則　55
忌避性　25
忌避腺　184
機敏　15
機敏性　15
基部　26, 174
基部 -　26
基部 - 外縁軸　159, 174, 175
基部 - 外縁方向　175
基部 - 外縁方向の蛇腹形式に折り畳んだひだ　175
基部側の縞状バンド　174
基部側の切除　174
基部側へ　26
起伏　226
基腹板　27
基腹板 - 側前側板溝状線　27
基部群　26
基部種　26
寄付する　65
基ふ節　27, 133
基部帯　26
ギフチョウ　127
ギフトオーサーシップ　95
基部に近い　174
基部の　26
基部の結節　26
基部の縞状バンド　26
基部 - 末端軸　159, 174, 175
基部領域　174
キベリタテハ　35, 140
基本遺伝子　225
基本骨格　26
基本生息場所　89
基本転写因子　91, 98
基本ニッチ　89
基本プラン　26
キマダラジャノメ　204
キマツ　197
決まって　188
帰無仮説モデル　150
キメラ　39
擬目　80
気門　32, 205, 208
気門の　205
疑問名　146
規約　42
脚　122
逆 -　186
逆位　116
逆位した H 対立遺伝子　116
逆位多型　116
逆遺伝学　186

逆位の切断点　116
逆位のブレイクポイント　116
逆効果　48
逆質問　186
逆症候群　153
逆正弦変換　22
客棲性　141
客棲性的共生関係　141
脚節　122
逆説的な　157
逆相補(体)　186
逆転遺伝子流動　186
逆転写　186, 188
逆転写酵素　186
逆転写された　186
逆転写反応　186
逆転写ポリメラーゼ連鎖反応　186, 188
逆に　230
逆に言えば　49
脚の護身器官　122
脚の防護器官　122
逆比　116
脚ふ節の鉤爪　215
脚部分　122
逆変換　25
逆方向　116, 186
逆向きの方向　116, 186
逆もまた同様　230
却下　182
客観的　150
客観(的)同物異名　150
逆境　14
逆境淘汰　14
逆境に対する反応　185
逆行性充填　186
キャップ　35
ギャップ　90
キャップ形成　35
ギャップ生息地　90
ギャップペナルティ　90
キャピタル型ブリーダー　35
キャピラリー　35
キャベツ　32, 34
キャベツの作物変異(多様性)　34
キャラクターマッピング　38
キャラクターマップ　38
級　40
旧－　156
求愛拒否　182
求愛行動　51, 74
求愛行動を抑制する　53
求愛誇示　51
求愛集団　122
求愛する　51
求愛選好　51
求愛ダンス　51
求愛の行動要素　51
求愛フェロモン受容体　51
嗅覚　151, 200
嗅覚遺伝子　151
嗅覚感覚器　151
嗅覚計　151
嗅覚刺激　151
嗅覚受容体　151, 153
吸管　100
球桿状の触角　35
吸器　211
究極機構　225
究極的機序　225
究極パターン　225
究極要因　225
急傾斜型のサンプル数に基づく希薄化曲線　208
急激な衰亡　207
急激な適応放散　79
急激な突進　211
急降下する　66
休耕地　80
救済機能　185
救済効果　185
休止　37, 177
QGIS　177
嗅(覚)刺激　151
休止姿勢　186
休止状態　65
休止相　65
吸収　12
吸収管　100
吸収極大　12
吸収効率　12
95% 信頼区間　40
吸収スペクトル　12
吸収スペクトルテンプレート　216
吸収スペクトルの極大波長　12
吸収する　12
吸汁性植食者　164
吸収性のティッシュ　12
吸収値　12
吸収波長域　12
吸収ピーク　12
96 穴プレート　11
吸水　66
吸水活動　175
吸水行動　175
吸水する　66
吸水中　66
吸水動物　175
急性の　13
球節　220
急増した　117
急速乾燥　178
急速低温馴化　178, 179
急速低温耐性　178, 179
急速に変化する環境　81
急速に放散している単系統　178
急速飛翔　178
キューティクル　53
旧熱帯　156
吸盤　211
旧北区　156
旧北区産の種　156
吸蜜　144
吸蜜源　144
吸蜜源の量　12
吸蜜植物　143, 144
吸蜜する　211
吸蜜中　81
吸蜜の資源量推定値　144
休眠　58, 65
休眠維持機構　59
休眠因子　59
休眠解除率　59

休眠覚醒　59
休眠型成虫　59
休眠期　58, 59
休眠期間のエネルギー要求量　73
休眠経路　59
休眠後の幼虫の体重　232
休眠後幼虫　169
休眠蛹　59
休眠持続時間時計　59
休眠終了後の寿命　169
休眠消去　59
休眠条件　59
休眠症候群　59
休眠状態　59, 220
休眠深度　59, 113
休眠シンドローム　59
休眠成虫　59
休眠成長　59
休眠性幼虫　59
休眠性幼虫発育期間　59
休眠相　59
休眠阻止効果　59
休眠特性　59
休眠突入　59
休眠になる場合の幼虫発育期間　59
休眠能力　59
休眠のオッズ　151
休眠の開始　152
休眠の確率値　151
休眠の進行　173
休眠関連の生物時計に関わるタンパク質　59
休眠の発生　109
休眠の深さ　59
休眠場所　59
休眠発育　59
休眠発育期　59
休眠反応　59
休眠表現型　59
休眠プログラム　59
休眠ホルモン　58, 59
休眠前の交尾　170
休眠誘起　59
休眠誘起の光周期　59
休眠誘起の日長　59
休眠誘起の日長感受期　162
休眠誘起率　59
休眠幼虫　59
休眠予定蛹　59
休眠予定幼虫　59
休眠卵　59
休眠率　59, 109
休眠齢期　59
休眠を回避した幼虫　115
休眠をして発育した個体　59
休眠をしないで発育した個体　60
休眠を終了させる　217
球面濃度計　205
球面密度計　205
丘陵地帯　103
キュビタスインターラプタス　53
距　167, 206
脅威　218
狭域分布種　142
教育　69
強化　73, 113, 182, 212
境界　31, 56
境界設定　56
境界線　31
境界層　32
境界領域　32
強化されたストレス耐性　73
強化する　73, 182, 226
共感染する　43
狭義の　189, 194
胸脚　218
供給側　201
供給側の個体群　202
供給(元)個体群　202
供給者(側)の指示書　212
供給量　212
強権　166
競合説　45
胸高断面積　25
凝固したアガロース　201
強固な系統構造　188
強固な系統発生学的フレームワーク　188
強固な蛹の外骨格　188
強固な相関　188
供試昆虫　112
共時性　214
供試虫　112
共時的な　214
強磁場　209
凝縮性染色質体　46
共焦点観察　46
共焦点顕微鏡　46
共焦点顕微鏡観察　46
共焦点分析　46
狭食性　151
狭食性の　151
共進化　42
共進化の共同者　42
共進化の相互作用　42
共進化のパートナー　42
強心配糖体　35, 36, 38
強心配糖体に対する非感受性　36
共生　213
偽陽性　80
共生ウイルス　213
共生関係　213
共生細菌　213
共生者　213
擬陽性発見率　102
共生微生物　213
共生微生物が激減した昆虫　213
共生微生物感染　213
共生微生物による昆虫の性転換　213
共生微生物密度の調節　213
強制変更　128
狭接触帯　142
共線性　43
競争　45
競争相手　188
競争相手の競争戦術　48
競争者との関係　45
競争種　45
競争的相互作用　45
競争排除則　172
競争力　45
競争理論　45

胸側溝 208
共存 42
共存系統 42
共存種 42
兄弟グループ 197
兄弟交配 197
強調する 225, 226
共直線性 43
共著者 42
共通 - 42
共通環境 44
共通環境での飼育 45
共通擬態種 42
共通祖先 44
共通祖先種 44
共通祖先の年代決定 55
共通庭園実験 44
共通庭園条件 44
共通の遺伝的構造 196
共通のウイルス病原体 45
共通の現象 44
共通の祖先群 44
共通の発育経路 196
共通の病原体 44
共通の物理的条件 45
共通配列 47
共通配列変異 196
共同 - 42
共同研究者 42
共同行為 13
共同採餌行動 45
共同執筆者 42
共同して行動する 13
強度 - 応答関係 113
胸背板の前翅突起 212
胸部 218
胸部温度 218
胸部の形 218
胸部筋 218
胸部筋肉(組織) 218
胸部構造 218
胸部重量 218
胸部神経系 218
胸部に折り畳まれている前脚 86
胸部の 218
胸部の脚 218
胸部の筋肉 218
胸部の刺毛相 38
胸部の帯糸 218
胸部の飛翔筋 218
胸部幅 218
共分散構造 51
共分散分析 18, 19, 51
共分散分析の検出力 169
共分散分析の検定力 169
狭分布 142
狭分布種 142
共分離 50
共変数 51
共変動する 51
共変量 51
鏡胞 22
興味深い見識 114
興味深い蝶 183
興味を釘付けにする標本 114
興味をそそられる 81
興味をそそる 115
興味をひくように 115
共鳴効果 185
鏡紋 218
共有環境 44
共優性 42
共優性分子マーカー 42
共有祖先形質 213
共有配列変異 196
共有派生形質 214
共用バーコード 196
供与菌 65
供与者個体群 65
供与植物 65
供与体 65
共力作用 214
協力者 214
協力的に働く 13
強力な優性 209
許可 47
極域周辺の 40
極限状態 80
極座標モデル 166
極周辺の 40
局所擬態多型 125
局所群集の飽和 190
局所個体群 125
局所個体群サイズ 125
局所個体群の生存力 125
局所照射 125
局所照射実験 125
局所スペシャリスト 125
局所選択 125
局所適応 124
局所適応した形質 125
局所的選択圧 125
局所的な 124
局所的な絶滅 124
局所的配偶競争 124, 125
局所的光照射 125
極性を示す 166
曲折させる 56
曲線当てはめ型外挿法 53
極相 41
極端な不均一 80
極端な不均等 80
極端な変動 80
極東ロシア 81
棘腹板 205
局部順応 124
局部選択 125
極方向への移動 166
棘毛 195
極夜 166
棘瘤 191
距刺 218
鋸歯 195
鋸歯状のクライン 190
鋸歯状の触角 195
巨視的な 132
巨視的な表現型 127
居住区 185
居住している乾季型の個体 185
居住者 185
居住地域 185
居住の雄 185
拠水林 90
拒絶行動 182
巨大グリア細胞 95
巨大木 72
拒否された著作物 182
拒否する 182

拒否名　182
距離閾値　62
距離行列法　62
距離効果　62
距離尺度　62
距離節約法　62
寄与率　49
距離による隔離　108, 117
距離ワグナー法　62
魚類のゲノム　84
霧　136
キリギリス　97, 119
霧雨　66
ギリシャ語　97
切り捨て水準　53
切り取った翅　62
切り取る　223
キリハシ科の鳥　117
切り花　53
切り貼り信号　205
キルギス　119
キルギス共和国　119
ギルド内捕食　115
切れ込んでいる　71
キロ塩基　119
記録　180
記録が多い地域　232
記録強度　180
記録された神経活動　180
記録者の行動圏　180
記録する　182
記録(用)電極　180
キロダルトン　119
キロベース　119
議論　61
議論の余地がある　49
議論の余地のある話題　49
際立たせる　225
際立った　209
際立った発見　209
極めて重要な実験データ　165
極めて重要な属　165
きわめて詳細　97
極めて多様な分類　107
極めて類似な　209
筋　84
近位　174
近位－遠位軸　159, 174, 175
均一な環境　226
均一な乾燥　226
均一に分布した　104
近位領域　174
近因　174
近縁　41
近縁グループ　197
近縁交配　197
近縁種　16, 17, 41, 182, 197
近縁性　182
近縁属　182
近縁で同所的に生育するペア　41
近縁な分類群　41
近縁のウイルス　41
近縁配列　41
金環　96
緊急時迅速放射能影響予測ネットワークシステム　204
緊急時対応計画　48
近交　109
近交係数　109
近交系幼虫　109
近交弱勢　109
均衡度　77
菌根菌　141
近似種　15, 198, 202
均質　75
均質化　75
均質化された卵　104
均質化処理をされた卵　104
均質化する　104
均質性　104
近似同質遺伝子系統　143, 146
近似の　22
菌株　208
筋受容器　141
近親関係　119
近親交配　109
近親交配弱勢　109
近親交配種　109
近親交配の　109
近接性　175
近接的機構　175
近接的機序　175
金属光沢の　133
金属色の　133
筋組織　141
緊張性の　219
均等度　77
均等に分布した　104
筋肉　84
筋肉組織　141
筋肉で満たされた胸部　141
金パラ　96
金パラジウム　96
緊密な遺伝相関　219
銀紋型　167
金門島　119
近隣　144
近隣結合解析　144
近隣結合系統樹　144
近隣結合法　144, 146
菌類　89
菌類感染　89

ク

区　42
グアム島　98
空間(的)位置　202
空間位置決め　202
空間(的)記憶機構　202
空間拮抗性　202
空間構成　202
空間構造　203
空間尺度　203
空間情報　202
空間スケール　203
空間生態学　202
空間単位　203
空間的位置関係　202
空間的自己相関　202
空間的集団効果　69
空間的に重ね合う　155
空間的に相関関係のない確率性　203
空間的分布　202
空間的変異　203

空間の詳細スケール　83
空間配列　202
空間範囲　22
空間分解能　202
空間分布パターン　202
空間変異　203
空間変動性　203
空気　24
空気伝播コミュニケーション　15
空気力学　14
食う食われる以外の関係　147
空前の減少　227
空地　41
空祖先　150
空中戦　14
空中戦闘　14
空中相互作用　14
空白地　228
空白地帯　228
偶発的組み換え表現型　151
偶発的表現型の飛躍　151
空胞　228
クエリー　177
クオリティーフィルタリング　177
区画　45, 64, 157
区画化　45
区画法　176
区間マッピング法　115
茎　207
区系生物地理学　182
クサカゲロウ　120
草木のない土地　26
草地　97, 213
腐った果実　188
腐った果物　188
くさび形の　232
櫛歯状の　159
クジャクチョウ　159
クジラ　232
崩す　219
クスノキ科植物　122
クソバエ　67
管　67, 223
管の　228
果物食者　89
口　140
口絵　89
クチクラ　53, 200
クチクラ組成　53
クチクラの厚みのある隆起　218
クチクラ遊離　21
朽ちる　55
屈筋　84
クックフォン国立公園　53
駆動機構　140
駆動軸　140
国レベルの分布の安定化　206
国レベルの豊富さの年次指数　142
首　143
区分　64
区別する　61, 215
区別的発音符　58
くぼみ　46, 149, 232
くぼんだ　46, 110
クマリン　50
組合せ確率　44
組合せ操作　44
組合せ(理)論　44
組み換え　180
組み換えウイルス　180
組み換え過程　180
組み換え距離　180
組み換え(の)切断点　180
組み換えタンパク質　180
組み換えバキュロウイルス　180
組み換え表現型　180
組み換えマッピング　180
組換え抑制　184
組換え率　180
組み換えを減少させる領域　181
組み込まれたプロウイルス型　113
組み込み型　113
組しがみつき行動　156
クモ　205
クモの巣　232
クモマツマキチョウ　153
クライ　52
クライン　41
クライン的　41
クライン的変異　41
グラシン(紙)の三角紙　95
暮らす　81
クラスター　41
クラスター構造　41
クラスターを構成する遺伝子　41
クラスタリング　42
クラスパー　40
グラディエント　96
クラドジェネシス　40
クラブ　41
グランヴィルヒョウモンモドキ　95
グランドスケール　97
グリア細胞　95
グリア細胞層　95
クリーバーゼ断片長多型　41
クリーム　51
クリーンな環境　41
繰り返される越冬　183
繰り返す　180
繰り返す－　117
グリコーゲン　96
グリコーゲン含量　96
グリコシド　96
グリコシド結合揮発性物質　96
クリサリス　39
CRISPR-Cas9 技術　51
クリスパーキャスナイン技術　51
グリセリン　96
グリセロールの蓄積　96
クリ帯　38
グリッド四方　97
クリプトクロム　52
クリプトビオシス　52
グループ行動　97
グループ内変異　114
グルコース　95

グルコース -6- リン酸デヒドロゲナーゼ　90, 95
グルコシノレート　96
グルコシノレート化合物　96
グルコシノレート含有植物　96
グルコシノレート生合成　96
グルコシノレートの解毒化　96
グルコシノレート防御システム　96
グルタチオン S- 転移酵素　96
グルタチオン・S・トランスフェラーゼ　96, 98
グレース昆虫培地　96
Grace 培地　96
クレード　40
クレードの支持度　40
グレゴール・ヨハン・メンデル　132
クレスフォンテスタスキアゲハ　95, 153
クロウメモドキ科　187
クローズアップ　41
クローニング効率　41
グローバリゼーション　95
グローバル侵入種登録簿　95, 97
クローン　41
クローン化プライマー　41
クローン生存効率　41
クロガラシ　30
クロカレ型　52
黒く塗りつぶされた棒　83
黒く縁どられた　30
クロス　52
クロスハッチング領域　52
黒ずんだ　112
クロチアニジン　41
黒っぽい翅表　30
クロバエ　30
クロマチン　39
クロマチン構成　39
クロマチン免疫沈降シークエンシング　39
クロマチンモデリング　39
クロマトグラム　39
黒丸　30, 201
黒味をおびた褐色　30
クロラムフェニコール　39
クワコ　31
クワゴ　31
クワコやヤママユガ科の蛾　234
群　97
軍拡競争　22
群居　97
群居しない　201
群居する　97
群居性　97
群居性の　97
群集　23, 29, 45, 52
群集(群落)構成　45
群集構造　45
群集水準の間伐説　23
群集生態学　45
群集生態学的研究　45
群集・生態系遺伝学　45
群集多様性　45
群集モジュール　45
群集レベルの間伐説　23
群集レベルの間引き説　23
燻蒸剤　89
群生　97
群生しない　201
群生する　97
群生性の　97
群生相　97, 161
群選択　97
群知能　197, 213
群淘汰　97
群内種　109
群飛　213
群葉　86
群葉密度　86
群ロボット　213

ケ

毛　99, 222
経緯　103
径横脈　178
警戒色　21, 231
警戒色の　21
警戒色の幼虫　21
計画的野焼き　171
計画的被覆　214
景観　120
景観規模　120
景観規模の保全　120
景観構造　120
景観生態学　120
景観断片化　120
景観動態　120
景観の撹乱　120
景観の時間的動態　216
景観の分断化　120
契機　138
軽減効果のある科学的手法　181
形原勾配　139
経験する　225
経験ベイズ法　72
形原モデル　139
傾向　173, 216
蛍光　85
蛍光インサイチュー・ハイブリダイゼーション法　84, 85
経口感染　153
蛍光染色用封入剤　85
経口で感染する　153
蛍光灯電球　85
蛍光塗料　85
蛍光標識したプライマー　85
警告擬態　21
警告行動　231
警告誇示　21
警告色　21, 231
警告色化と擬態化した蝶　231
警告色の蝶　231
警告色パターン　231
警告信号　231
警告する　21, 231
警告性昆虫　21
警告戦略　21
警告パターン　231

警告率 15
経済的被害許容水準 69, 71
警察行動 166
計算集約的 46
計算性を強化したもの 46
形式的論証 87
経時性 58
形質 38, 220
形質移入 221
形質移入6日後 55
形質群 214
形質固有 220
形質状態 38
形質置換 38
形質置換の可能性 168
形質転換 221
形質転換花粉 221
形質転換植物 221
形質特異的 220
形質の複雑性 220
形質の変化を介する間接効果 220
形質の膜内外電位差 166
形質膜の内外に生じる電位差 166
傾斜 41, 55
傾斜日光浴 121
継承 112, 211
形状 196
形状パラメータ 196
形状パルス 196
計数的反応 177
形成過程 87
形成機構 87
形成史 58
繋節 89
脛節 218
脛節棘 218
繋節の 89
脛節の 218
脛節の爪 218
脛節の毛房 218
頸節片 37
けい(脛)側 218
計測結果 186
継続実験 89
継続中の論争 152
継続的なコレクション 211
継続的な支出 48
継続飛翔 48
計測用増幅器 131
形態 139
繋帯 230
形態学 139
形態学的解析 139
形態学的検査 139
形態学的試験 139
形態学的種概念 139
形態学的・生理学的・行動学的研究 139
形態(的)形質 139
形態形成 139
形態形成起源 139
形態形成的 139
形態形成の 139
形態形成ホルモン 139
形態系統学 139
形態系統樹 139
形態(学的)種 139
形態測定 139
形態(学的)データ 139
形態的かつ遺伝的に分化している 60
形態的区別 139
形態(学)的研究 139
形態的差異 139
形態的再構成 139
形態的尺度 139
形態的進化 139
形態的・生理的特性 139
形態的・生理的特徴 139
形態的多様性 139
形態的適応 139
形態的特(有)性 139
形態的特徴 139
形態(学)的特徴 139
形態(学)的な類似 139
形態的に 139
形態的に異常な成虫 139
形態的に類似な 139
形態的パターン 139
形態的不連続性 139
形態的分化 139
形態的分離 139
形態(学)的分類 139
形態的変異 139
形態的類似性 139
形態発生の 139
形態変化 139
径中横脈 178
系統 91, 123, 163, 177, 208, 228
系統維持 127
系統解析 163
系統学 163, 214
系統学的位置 163
系統学的慣行 163
系統学的研究 163, 214
系統学的効用 163
系統学的情報 163
系統学的な尺度 163
系統関係 163
系統形成 177
系統再構成 163
系統再構築 163
系統樹 91, 163, 222
系統樹の先端結節点 219
系統樹の不安定な樹形 227
系統上の重要な種 163
系統上の重要な単位 163
系統進化 163
系統図 163
系統推定 163
系統(樹)推定 163
系統生物地理学 163
系統生物地理学的方法 163
系統(的)相互関係 163
系統ソーティング 123
系統組成 215
系統地理 163
系統地理学 163
系統地理学的解釈 163
系統地理学的歴史 163
系統的位置 163
系統的群集生態学 163
系統的合祖理論 163
系統的差異 214
系統的シグナル 163
系統的視点 163
系統的情報 163
系統的情報性 163

系統的情報性プロファイル　164
系統的信号　163
系統的進行順序　163
系統的相違　214
系統的な立場　207
系統的な調査　214
系統的な不確実性　163
系統的に局所化する　163
系統的に近縁な種　163
系統的に類縁の種　163
系統的不確実性　163
系統的有意単位　163
系統特異的な重複消失　123
系統特異的な複製　123
系統ネットワーク　163
系統の先験的明確化　11
系統の祖先　123
系統の多様性　177
系統の典型種　123
系統発生(進化)　163
系統発生学的情報量　163
系統発生シグナル　163
系統発生の再構成　163
系統発生の再構築　163
系統復元　163
系統分岐　163
系統分岐の支持度　40
系統分類　163
系統分類学　214
系統分類学的研究　214
系統分類に関する研究　163
系統変化　91
系統を生む進化　177
頸板　158
渓畔林　187
警備　166
系譜　91, 123
頸部　143
軽風　126
頸部シールド　37
径分横脈　192
径分脈　178
警報フェロモン　15
兄妹交配　114
兄妹交尾　33
径脈　177, 178
径脈の　177
茎葉部　229
系列　123
経路　158
経路関与　158
経路曲率　158
経路統合　158
K 遺伝子座　118
K 害虫　118
ケージ　34
K 種　118
ケーススタディ　36
K 選択　118
ゲート　90
K- 淘汰　118
ゲートキーパー　90
k マーの長さ　118
毛がある眼　99
外科的実験　212
外科的操作　212
激化させる　78
撃退　184
撃退する　183
劇的な改変　66
ゲストオーサーシップ　98
削る音　191
毛束　99, 106
毛束状の　81
血 -　101
血液　30, 101
血液細胞　101
血液中のエクジステロイド量　101
血縁　119
血縁関係　119
血縁群　119
血縁選択　119
血縁度　182
血縁度非対称性　182
血縁認識　119
結果　186
結核菌　141
結果として生じる激しい競争　186
結果としての系統樹　186
結果の新規性　149
血球　99, 101
血球細胞　99
結局　225
結合　44, 110, 114
結合行動　118
結合する　226
結合の特異的な性質　204
結合部　118
結合母音　47
結紮実験　123
結紮する　123
血色素　101
欠失　56, 136
欠失単模式種　136
結実率　193
欠翅目　98
欠如　136
結晶形成　52
楔状紋　41
結節　146, 223
欠損　136
欠損する　119
欠損ハプロタイプ　136
血体腔　101
決着済みの競争　195
決定機構　57
決定的なステップ　55
決定的な役割　52
決定的な予測　46
決定要因　49, 57
決定論的モデル　57
蹴爪　206
欠落　136
血リンパ　99, 101
血リンパ - エクジステロイド価　101
血リンパのサンプル　99
結論　46
解毒遺伝子　57
解毒酵素　57
解毒する　58
ケトン　119
ゲニタリア　93
ゲニタリアチューブ　93
毛のない眼　99
ゲノミクス　94
ゲノミクスの手法　94
ゲノム　93

ゲノムアセンブリー 93
ゲノム位置 93
ゲノム遺伝子座 93
ゲノムインプリンティング 93
ゲノムウォーキング 93
ゲノム塩基配列の再解析 94
ゲノム機構 94
ゲノム規模での重複状態 93
ゲノム規模の 93
ゲノム規模の一塩基多型データ 93
ゲノム規模の SNP データ 93
ゲノム系統学 163
ゲノム系統樹 93
ゲノム研究 94
ゲノム細菌人工染色体 93
ゲノムサイズ 93
ゲノムシークエンシング 93
ゲノム資源 94
ゲノム資源の欠乏 55
ゲノム資源の不足 55
ゲノム尺度 93
ゲノム上の足場 93
ゲノム情報ブローカー 93, 95
ゲノム進化学 77
ゲノムスキャフォールド 94
ゲノム刷り込み 93
ゲノム重複 93
ゲノムデータ 93
ゲノムデータの指数的上昇 79
ゲノムデータの指数的増加 79
ゲノム突然変異 94
ゲノム内闘争 115
ゲノムの塩基配列を解読する 195
ゲノムのカバー率 93
ゲノム配列 93
ゲノム配列歩行 93
ゲノムバクテリア人工染色体 93
ゲノム BAC 93
ゲノム不安定化 93
ゲノム不安定性 93
ゲノムブロック 93
ゲノムプロファイリング法 93
ゲノム平均の関係 93
ゲノム骨組 94
ゲノムリード 94
ゲノムリソース 94
ゲノム領域 94
ゲノムワイドでの重複状態 93
ゲノムワイドの遺伝子移入 93
ケミカルエコロジー 38
ケミカルハロー 38
ケムシ 36
毛虫 235
毛虫状の 75
ケモカイン受容体５ 36, 38
ゲラニオール 95
ゲル 90
源 201
原位置標識法 109
原因と結果の問題 172
原因病原体 36
検疫 177
検疫規則 177
検疫措置 177
肩横脈 106
弦音感覚器官 39
弦音器官 39, 191
限界光量 51
限界照度 51
限界値定理 129
厳格(度) 195
幻覚作用 99
原核生物 173
厳格な抑制 195
顕花個体 85
顕花植物 85
原基 61, 188
原基遺伝子座 171
原記載 154
研究基礎資料 26
研究室ストック 119
研究室の貯蔵品 119
研究室の非近交系個体群 154
研究対象(微)生物 209
研究対象地域 209
研究対象の種 209
研究の様々な分野 228
研究論文 138
検鏡 204
健康的な環境 100
減光波長 54
原交尾器 171
原公表 154
堅固な系統樹 188
顕在化 71
顕在化する 71
腱細胞 216
原材料 112
検索表 119
原産国 154, 201, 207
原産の 110
絹糸 198
原始形質 166
原子座標 24
原始種 172
顕示性(度) 21
絹糸腺 198
原指定 154
原始的旧形質 166
原始的初原形質 166
原始的な 172
原始的な形態 172
原始的な種 26, 172
原始的な鱗翅目 172
検出可能な関連 57
検出可能な酵素活性 57
検出された転写産物 57
検出された転写物 57
現象 162
弦状感覚子 191
減少した幼若ホルモン力価 60
減少する 56, 66
減少する存在量 56

剣状の尾状突起　213
剣状紋　26
原色　44
原色蝶類検索図鑑　44
原色日本昆虫図鑑　44, 108
原色日本蝶類図鑑　44
原色版　44
減じる　109
減衰　24
懸垂器　51
減衰させる　24
減衰された雌化　24
減衰された雌化行動　24
減衰する　24
減数分裂　132
減数分裂分離ひずみ　132
顕性　65
現生　79
限性遺伝　196
限性擬態　196
顕性形質　65
限性黒化　196
現棲種　179
現生種　79, 179
現生人　137
原生生物　174
顕生代(の)　161
原生動物　174
源泉　177
健全な草　100
現存　79
現存個体群　79
現存している　79
現存の集団繁殖地　79
現代人　137
現代的統合　137
現代の総合　137
原地基準標本　219
現地に特有の規模　199
現地標準時間　125
原地模式標本　219
原腸のう　90
顕著さ　48
顕著である　149
顕著な色　48
顕著な生物学的差異　48
顕著な相違　129
顕著な変化　173
顕著に異なる二型　224
原綴り　154
限定された観察　123
限定された気象条件　186
限定された地理的範囲　123
検定(対象)植物　217
限定する　186
検定統計量　80
原点　154
見当をつける　204
原白血球　173
肩板　215
顕微鏡観察　134
顕微鏡写真　162
顕微鏡標本　134
顕微鏡用スライド　134
顕微鏡用スライドグラス　134
顕微鏡用のスライドガラス　95
原表皮　172
見聞　197
厳密な意味で　189, 194
肩脈　106, 170
原模式標本　172
倹約　158
肩葉　106
原理　172
原料　185
原料の収入分　185
肩腕　215, 217

コ

子　151
古－　156
小顎　130
小腮　130
小顎腺　130
小顎の　130
小腮の　130
小顎鬚　130
コア生合成経路　50
コアレセンス　42
故意の導入　56
故意の放蝶　56
請う　27
綱　40
孔　86
後－　133, 168
高圧滅菌された　24
恒暗　55
好意　51
広域総合的害虫管理　22, 25
広域的総合的害虫管理　22, 25
広域分布　233
広域分布種　233
合意系統樹　47
高位の分類群　103
降雨　178
抗ウイルス応答　21
抗ウイルス作用　21
抗ウイルス防御　21
降雨量　178
降雨林　178
幸運　31
後縁　66, 112, 169, 220
後縁角　154, 220
後縁帯　66
後横脈　169
恒温　48
高温　102
高音圧　102
高温感受性　102
恒温器　110
高温期　218
高温系列　100
高温湿潤期　106
高温・湿潤な森林　106
恒温条件　48
光温図表　163
高温多湿な森林　106
硬化　191
公開文書　152
効果器　70
肛角　154, 220
光学顕微鏡　123
光学顕微鏡写真　123
後角縦線　18, 220
高拡大図　102
光学倍率　153
後角部　154, 220

肛角部　18, 154, 220
甲殻類レクチン　52
硬化した　191
後下唇　169
硬化・着色　214
効果的な成長期　70
効果的な発育時期　70
硬化板　191
孔管　168
口器　140
後期型雄殺し　121
好蟻性　141
好蟻性器官　141
好蟻性昆虫　141
広義で　33
広義の　188, 194
後期胚発生蓄積タンパク質　121
後脚　103, 133
口脚　140
後休眠期　169
後休眠期の幼虫の体重　232
後休眠の寿命　169
後胸　103, 133
後頬　169
好況　173
合胸　214
工業暗化　111
後胸後側板　134
工業黒化型　111
高偽陽性比率　102
高擬陽性率　102
後胸前側板　134
後胸側板　133
後胸側板下縁隆起線　210
高強度　102
後頬の　169
後胸背板　133
後胸腹板　133
抗菌性　21
抗菌性ペプチドの上方制御　227
高グリシン含有　96
好景気 - 不景気サイクル　31
後継事業　86
後脛節　133
攻撃　24, 151
攻撃的な幼虫　15
抗原　21
荒原　57
高原　103
口腔　153
光合成　163
光合成期　163
航行能力　143
口腔分泌物　153
交互作用　114
交互作用項　114
交互作用の強さ　114
交互に入れ替わる温度環境　17
交互に入れ替わる季節　17
交互に入れ替わる季節環境　17
交互に発現する成虫の表現型　17
交互に発現する表現型　17
交叉　39
抗催淫剤　20
抗催淫性化合物　20
交差慣化　52
耕作　53
耕作化　81
耕作地　53
耕作農業　22
交差項　114
交差耐性　52
考察　61
交雑　52, 107
交雑実験　107
交雑種　52, 107
考察する　74
交雑帯　107
交雑による遺伝子交換　107
交雑による種形成イベント　107
交雑による種分化イベント　107
好サボテン性ショウジョウバエ　34
交差率　52
交叉率　52
抗酸化酵素　21
高山性　17
高山性生物　17
高山性草原　17
高山性の　17
高山帯　17
高山蝶　17, 102
高山ツンドラ　17
高山ツンドラ草地　17
後翅　98, 103
後肢　103
口肢　140
後枝　169
後翅裏面　226
後翅縁　103
後翅芽　103
格子囲い領域　52
高時間分解能　103
公式原文　151
後翅基骨　209
公式訂正書　151
高次元モデル　102
格子縞　38
後翅線条の透視画像　196
高湿度　102
後指定　210
後翅内縁　18
後翅の結合部　103
後翅表面　227
高次分類　103
高次分類間の分岐　63
高次分類数　103
光周(期)カウンター　162
光周期　162
光周期変化　38
光周計数機構　162
光周処理　162
光周信号　162
光周性　162
光周性で調節された休眠　162
光周測時　162
光周測時機構　162
光周調節　162
光周時計　162
高周波数　103
光周反応　162
光周反応曲線　162

耕種的防除　53
後受容の　169
肛上板　75, 205
肛上片　75
抗植食者防御　20
広食性　167
広食性の　167
広食性の蝶　91
後翅を覆っている前翅　87
口針　209
更新(洪積)世での隔離　166
更新世の気候変動　166
肛錐　18
降水(量)　170
洪水緩和　85
降水日数　150
高スコア分節対　102, 106
合成　214
後成学　75
後成現象　133
抗生作用　20
構成種　46
合成する　214
後成的機構　74
後生的進化形質　21
後成的な　74
構成的ヘテロクロマチン　48
構成的類似性　46
後生動物　133
後生動物の放散　178
高精度の航空写真　103
合成の－　213
後世の単型　210
抗生物質　20
抗生物質入りの飼料　20
抗生物質処理　20
構成分子　45
構成要素　45
洪積世　60
肛節　18
梗節　159, 160
酵素　74
合祖　42
構造化　45
構造活性相関　209
構造色　209
構造タンパク質　209
構造的差　209
構造的差異　209
構造的同化　209
構造的な　209
構造の決定　209
高層ビル　103
肛側　18
口側　153
後側　169
高速液体クロマトグラフィー　102
高速液体クロマトグラフィー分析法　106
後側甲　169
高速道路　140
後側板　75
肛側板　157
後側板関節節片　209
後側板線　75
後側板の　75
後続プロジェクト　86
肛側片　157
合祖系統樹の事前情報　42
合祖系統樹の事前分布　42
酵素コード等価物　74
酵素コード同等物　74
合祖時間　42
酵素的活性化　74
酵素分類　68, 74
合祖理論　42
抗体　21
後代　173
後退する　66
後体節　133
後腿節　133
後体節の　133
後退的モデル選択法　25
広大な帯状土　106
後退流跡線解析　25
光沢の　116
光沢のない葉　148
後端　169
後端にある　36
耕地　53
高地　227
高地森林帯　227
高地性種　103
後膣片　120
後膣ラメラ　120
高地の森林　227
甲虫の角　27
後中背板翅突起　169
後肘脈　53
コウチュウ目　43
後腸　103
高調波レーダー　100
好都合な寄主植物　81
好適な生息地の空間配置構造　202
好適な繁殖地　211
後天的新形質　21
高度　17, 71
後頭　151
行動　27
行動学的アプローチ　27
行動学的観察　27
行動学的プロタンドリー　27
行動観察　27
行動結果として生じる　22
行動圏　15, 104
後頭孔　86
行動シンドローム　27
行動生態学　27, 140
行動的応答　27
行動的隔離　27
行動的・神経生理的応答　27
行動的反応　27
行動的不一致　27
行動特性　27
行動トレース　140
行動によるもの　27
後頭の　151
行動の　27
行動の時間配分　13
行動パターン　140
行動範囲　13
行動半径　13
行動ベースモデル　27
行動変化　27
後頭縫合線　151
行動面での雄性先熟　27

行動目録　76
行動様式　99
後頭隆起線　151
行動レベル　27
高度シフト　17
高度的クライン　17
高度的最適　17
高度に改変された生息地　103
高度変化　17
高度変動　71
高入射角　102
肛乳頭　157
肛乳房突起　157
香のう　30, 191
高濃度　102
交配　52, 114, 130
勾配　41, 96
勾配群の融合　89
交配後隔離　169
交配成功率　130
交配世代の　83
交配前隔離　171
交配場所　130
後背板　169
勾配プロファイル　96
勾配プロフィール　96
交配ペア　130
勾配変異　96
高倍率　102
喉板　98
広範囲な調査　79
広範囲に及ぶ現象　233
広範囲に及ぶ絶滅　233
広範囲にわたる観察　79
広範囲にわたる共存　233
広範囲の刺激　125
広汎種　50
広範な里山にいる種　233
広汎な蝶類　233
広範な蝶類　233
広範な地理的サンプリング　46
硬皮　191
交尾　50, 130
交尾相手の獲得　129
交尾確率　130
交尾可能性　129
交尾管　67, 130
交尾器　40, 93, 225
交尾器口　154
交尾拒否行動　130
交尾拒否姿勢　130
交尾傾向　130
交尾口　154
交尾孔　50
交尾鉤　157
交尾行動　130
交尾誘引性　130
交尾個体　50
交尾後の生理的な変化　164
交尾棍　14
交尾策略　130
交尾識別　130
抗微生物性　21
抗微生物ペプチドであるセクロピンB　21
交尾成立　211
交尾栓　130
交尾戦略　130
交尾地域　130
交尾対　130
交尾できる状態の雌　180
交尾テリトリー　130
交尾突起　21
交尾ナワバリ　130
交尾の　50
交尾のう　33, 205
交尾のう細胞の細胞核　149
交尾のう体　50
交尾の実際の相違　179
交尾の非対称性　130
交尾場所　129
硬皮板　165
交尾弁　228
交尾弁縁棘　28
交尾弁縁刺　28
交尾弁端部　53
交尾弁二尖　28
公表　175
公表された著作物　175
公表する　175
公表の日付　55
高品質な食料　103
口部　118, 153
後部　169
後ふ節　133
後部の　169
口部付属肢　140
高プロリン含有　173
口吻　100, 172, 219
高分散能力　103
口吻収縮　172
公文書史料保管　22
高分子量　103
口吻伸展反射(反応)　160, 172
広分布　233
広分布種　233
広分布属　233
広分布の属　233
口吻を伸ばしての食物探索　172
硬片　191
酵母　236
候補遺伝子　35
後方(の)眼状紋の縮小　181
合法的な関心　122
後方に　169
後方に伸長する　79
後方の　169
後方の大きな眼状紋　120
後方の眼状紋　169
後方の基翅節片　169
後方のフォーカス　169
後方部　169
後方部位　169
後方部分　169
候補基準化遺伝子　35
高木限界線　222
酵母種　236
酵母人工染色体　236
候補マーカー　35
肛脈　11
恒明　124
剛毛　33, 195
剛毛の　195
後模式指定　57
後模式標本　125
コウモリガ　95
コウモリからの逃避　27

コウモリセセリ　95
コウモリ探知　27
コウモリの超音波音の鳴き声　225
肛門　18
肛門褶　18
肛門部の脚　18
高薬量／保護区戦略　102
効用　228
硬葉樹林　191
広葉樹林地帯　33
高用量／保護区戦略　102
高力価のストック　103
高力価の貯蔵物　103
光量　18
香鱗　19, 191
小枝　224
超えた－　212
小枝に擬態　224
越えて－　220
ゴースト著者　95
コード　42
コードされた Dsx タンパク質　72
コードしている領域　42
コード配列　36, 42
コード領域　42
ゴール　90
コールマン法　43
コオロギ　51
古顎目　22
小型昆虫　200
小型鱗翅目　134
小型鱗翅類　134
個眼　152
語幹　207
個眼(面)　80
個眼の模式図　191
個眼ユニット　151
ゴキブリ目　30
コキマダラセセリ　120
呼吸　185
呼吸系　185
呼吸作用　185
呼吸量　185
国際イネ研究所　115, 117
国際稲研究所　115, 117
国際昆虫生理生態学センター　108
国際自然保護連合　117
国際植物防疫条約　114, 116
告示書　61
黒条　30
黒色化　132
黒色型　132
黒色系アゲハ　30
黒色色素　30
黒色の背側後翅　30
黒色の棒　83, 201
黒色のマーキングペン　30
黒色の蛹化台紙　30
黒色鱗粉　30
黒帯幅　30
国蝶　142
国定公園　177
黒点　30, 201
コクナーゼ　42
国内外来種　65
国内在来種　65
国内由来の外来種　65
コクヌストモドキ　85, 222
極保存配列　225
穀物　37
国立公園　142
国立生物工学情報センター　142, 143
国立バイオテクノロジーセンター(米国)　142, 143
国連環境計画　226
こげ茶色　54
語源学　76
午後のナワバリ制　15
心に描く　74
古細菌　22
誤差の程度　56
古参異名　194
古参同名　194
誇示　62
誇示行動　62
古植生の再構築　156
古植物学の　156
こずえ　222
コスト　50
こする　191
こする音　191
個性　161
個生態学　24
古生物学の場所　156
子世代　173
個体　110
個体間の移住パターン　114
個体間の近縁度　182
個体間の血縁度　182
個体間変異　114
個体群　97, 168
個体群圧の緩和　183
個体群遺伝解析　168
個体群遺伝的手法　168
個体群が発生してからの年数の効果(影響)　69
個体群間変異　115
個体群管理　168
個体群管理単位　140
個体群構造　168
個体群サイズ　168
個体群制御　168
個体群生態学　168
個体群成長　168
個体群成長率　168
個体群生物学　168
個体群絶滅の高まっているリスク　71
個体群全体の年間分布範囲　73
個体群調整　168
個体群動態　168
個体群動態モデル　168
個体群の安定性　168
個体群の遺伝学的研究法　168
個体群の遺伝的構造　168
個体群の入れ替わり率　224
個体群の永続性　161
個体群の回転率　224
個体群の孤立化　168
個体群の持続性　161
個体群の生存　168
個体群の保全　47
個体群変動　168
個体群密度　168

個体群レベルの変動 168
個体差 110
古第三紀 156
個体種数調査 138
個体水準の移動と個体群水準の分布との結合 47
個体水準の行動 111
個体数 12, 111, 150, 168
個体数あたりの種数の比 203
個体数減少傾向 226
個体数効果 69
個体数増加率 168
個体数調整 168
個体数に基づく希薄化 111
個体数に基づく希薄化の計算式 111
個体数に基づく曲線 111
個体数に基づくデータセット 111
個体数に基づくプロトコール 111
個体数に基づく分類サンプリングの曲線 111
個体数に基づく累積曲線 110
個体数の減少 56
個体数の変化 168
個体数密度 57
個体数量 12
個体数レベルの変化 168
個体多様性 110
古代 DNA 19
個体内の一致性指標 115
個体の空間分布 202
個体の微調整された反応 83
個体の母集団 168
個体発生(論) 152
コタイプ 50
個体変異 110
個体変動 110
個体密度 110
古代無性の(種) 19
個体レベルの行動 111
5' 側非翻訳領域 11
5' 非翻訳領域 11, 84
黒化 132
黒化異常型 30
黒化型 132
黒化擬態 132
骨格 200
骨格筋 200
黒化前駆物質 132
黒化の 132
黒化要素の優性 65
黒化鱗粉 132
黒環 30
国境 31
国境検査 31
国境通過後 169
国境通過前 170
コットンウッド 168
コツバメ 71
骨片 191
固定 84
固定遺伝子座 19
固定確率 84
固定効果の分析 18
固定剤 84
固定指数 84
固定した胴体 140
固定する 19
固定の齢数 84
固定分岐 84
誤適用する 136
古典仮説 40
古典的事例 40
古典的生物地理学 40
古典的な雑種不和合性 40
古典的な例 40
誤同定する 136
孤独性の 201
孤独性バチ 201
孤独相 161
異なった分岐群 62
異なる 59
異なる気候の地域 59
異なる基質 62
異なる機能 63
異なる行動 62
異なる色素で着色された鱗粉 60
異なるパターン 63
異なる表現型の遺伝子 59
子供を作る 199
コドラート 176
コドラート法 176
断り書き 36
コドン 42
コドン位置 42
コドン進化 42
コドンに基づく最尤解析 42
～後 24 時間 98
誤認 136
誤認する 110
コネクティビティ 47
コノハチョウ 110
好ましくない季節 226
好ましくない生息地 226
好みに合わない 227
好む 81
コバルトリジン 42
語尾 72
コピー数多型 42, 50
コヒオドシ 200
コピス 50
5- ヒドロキシトリプタミン 195
小瓶 229
こぶ 33, 146
瘤 33, 146
こぶ状の多様性曲線 107
コホート 42
コホート解析 43
細かい 83
ごまかしのない 208
鼓膜 67, 224
鼓膜器官 224
鼓膜耳 224
鼓膜室 224
鼓膜室の基部 26
鼓膜表面 224
鼓膜を持つ耳 224
ゴマシジミ 127
ゴマノハグサ 100
コマユバチ科 32
コマユバチ科の一種 50, 134
コマユバチファミリー 32

困らせる 176
込み合い 52
こみあいの程度 56
混み合う条件 52
混み具合 56
コミットメント 44
小麦粉につく甲虫 85
小麦畑 232
コムストック－ニードハム(ニーダム)の体系 46
コムラサキ亜科 72
コムラサキ属の蝶 72
米の害虫 161
米の水田 187
米畑 187
コメント 44
固有種 72, 142
固有周期 143
固有障壁 115
固有性 72
固有生物 72
固有の 72
固有の刺激 204
固有の対立遺伝子(座) 172
固有派生形質 24
子用語 39
ゴライアストリバネアゲハ 96
孤立化した個体群 117
孤立化していない個体群 47
孤立種 117
孤立発色団 117
コリドー 50
コリニアリティ 43
ゴルジ体 96
五齢後期 121
コレクション 43
コレクター 43
コレステロール 39
コレスポンディングオーサー 50
コロニー 43
コロニー成長 43
コロニーの健康状態 43
コロニスト 43
婚姻色 150
婚姻贈呈物 150
コンカテマー化したタンパク質 46
根茎 187
根元となる遺伝子 225
混合 14
混交一種 137
混合液 137
混合機能酸化酵素 134, 137
混合戦略 137
混合二次林 137
今後の研究 90
今後の方向性 89
混在 137
混雑条件 52
痕跡器官 229
痕跡的(なもの) 188
根絶 75
根絶キャンペーン 75
根絶した 79
コンセンサス配列 47
昆虫 112
昆虫－ 73
昆虫化学受容 112
昆虫学 73
昆虫学者 73
昆虫が宿主とする植物 165
昆虫館 112
昆虫寄生性(の) 73
昆虫綱 112
昆虫の口腔分泌物 112
昆虫(の)行動 112
昆虫行動制御剤 108, 112
昆虫行動制御物質 108, 112
昆虫飼育場 113
昆虫食塩水 112
昆虫植食者 112
昆虫生態学 112
昆虫成長制御剤 108, 112
昆虫成長制御物質 108, 112
昆虫生理学 112
昆虫－セリ科の関係 112
昆虫組織 112
昆虫追跡技術 112
昆虫の機能的な耳 89
昆虫の系統解析 112
昆虫の構成 46
昆虫の生物学的定量法 112
昆虫の生物検定 112
昆虫の適応度 112
昆虫のバイオアッセイ 112
昆虫の分散 112
昆虫の防御 112
昆虫の保護 112
昆虫の群れ 213
昆虫病原性(の) 73
昆虫目 112
昆虫類 112
コンティグ 48
コンティグ長 48
混同する 46
コントラスト 49
コントルタマツ 125
コントロールウイルス 49
コントロール(ポジコン／ネガコン)PCR 49
コントロール幼虫 49
コンパートメント 45
コンパクトカメラ 200
コンパスナビゲーション 45
コンピュータシミュレーション 46
コンピュータを用いて 109
棍棒 41
棍棒状の触角 41
棍棒状部 41
根本的な適応的差 225
根本的な適応的相違 225

サ

座 125
サーカディアン 40
サーペドビリン 190
サーモサイクラー 218
座位 125
再－ 179
再移住する 179
再移入 180
再移入する 180
催淫剤 21

催淫性フェロモン 21
再解析 179
サイカブト 187
細顆粒状の 97
細菌 26
最近縁種 41
細菌感染 26
最近記載された亜種 179
細菌細胞 26
細菌人工染色体 25
細菌のゲノム 26
最近の進歩 179
最近の調査 179
最近の発見 179
再現実験 184
再現する 184
再検討 186
再検討する 181
最高最低温度計 130
再構成されたアミノ酸 180
再構成された飛翔経路 180
再構成する 180
再構築する 180
最高点の個体数 159
最高濃度 103
最高用量 103
最後に 225
最後の著者 121
再コロニー化する 179
再サンプリング 179
採餌 86
採餌行動 86
採餌行動の効率 70
採餌成功(率) 86
採餌選好性 86
採餌戦略 86
鰓刺毛 171
採集 36
採集圧 43
採集期間 189
採集記録 180
採集効率 43
採集時間 189
採集者 107
採集する 43
在住する 185
採集地 43
採集地点 189
最終的に 225
最終投与 83
最終投与量 83
最終氷期最盛期 121, 123
採集旅行 43
採取期間 189
採取効果 189
採取時間 189
採取する 100
再受精 182
採取地点 189
採取地点の場所 125
再出現 180
最小閾値 136
最上位の遺伝子配列順序 140
最小可聴値 24
最小限の干渉 136
最小個体数 136
最小二乗法 122
最小偏差法 136
採食 86
最初の改訂者 83
採餌旅行 86
再水和する 182
サイズ関連形質 200
再スケール化 179
サイズ変異 200
再生 181
再生肢 181
再生する 184
採石場 177
最節約解析 131
最節約法 130
最節約法による祖先状態再構成 131
最節約モデル 140
最前線 86
再増強 179
採草地 131
最大吸収スペクトル 12
最大系統的情報性 130
最大系統発生学的情報量 130
最大限に優勢な 130
最大公約数 97
最大事後密度 106
最大種内分化 130
最大節約法 130
最大対数尤度 130
最大の著しい減少 140
最大変異の中心 37
最端 80
細短軟毛 175
再調査 179
再調査する 181
最低気温 126, 136
最低強度 126
最低致死温度 124, 126
再定着 179, 180
最適閾値 28
最適餌場滞在時間 153
最適餌選択モデル 153
最適解 153
最適化モデル 153
最適環境温度 153
最適個体群分散 153
最適採餌 153
最適採餌理論 151, 153
最適周囲温度 153
最適な閾値 28
最適な餌探索 153
最適な管理道具 153
最適な空間的分布 153
最適な探索戦略 153
最適な分割法 28
最適なランダム探索 153
最適(化)モデル 153
再導入 179, 182
再導入する 182
サイト枝解析 199
サイトカイン 53
再度見つける 179
栽培されたアブラナ科植物 53
栽培する 53
栽培品種 53
栽培変種 53
再発 180
再発見 181
サイパン 189
再評価 179
再分析 179

細分類　210
細分類学者　205
細分類数 - 分類数の比　210
再編集する　181
再訪　187
細胞　36
細胞アクチン　37
細胞応答　37
細胞外環境　79
細胞外基質　79
細胞外記録　79
細胞核　149
細胞学的観察　54
細胞学的に　54
細胞株　37
細胞間腔領域　36
細胞間相互作用　37
細胞間の間隔　114
細胞間の感染拡大　37
細胞機構　37
細胞(内)局在性　37
細胞区画　37
細胞計数　36
細胞系統　37
細胞骨格再配置　54
細胞骨格動態　54
細胞骨格の再構築　54
細胞骨格の再組織化　54
細胞骨格の細胞内動態　37
細胞骨格媒介性バキュロウイルスの運動性　54
細胞質　37
細胞質遺伝性遺伝子　54
細胞質因子による性比歪曲因子　54
細胞質性比歪曲因子　54
細胞質多角体病ウイルス　51, 53, 54
細胞質不和合　40, 54
細胞質不和合微生物　54
細胞質雄性不稔　42, 54
細胞周辺(部)　37, 160
細胞自律的　36
細胞自律的な方法　37
細胞侵入機構　36
細胞成分　37
細胞像　37
細胞相互作用　37
細胞増殖　37
細胞体　36
細胞内アクチン　37
細胞内器官　153
細胞内共生細菌　73
細胞内共生細菌ボルバキア　26
細胞内共生微生物　73
細胞内細菌　72
細胞内シグナル伝達系　115
細胞内装置　37
細胞内凍結　115
細胞の配列　37
細胞の配列パターン形成機構　37
細胞培養　36
細胞培養培地　36
細胞分裂頻度　137
細胞壁　37
細胞へのウイルス侵入　230
細胞磨砕物　36
細胞列の出現時期　72
最北個体群　149
最北端地域　148
最北端の個体群　149
細密なマッピング　83
材木　219
最尤(度)　130
最尤推定系統樹　131
最尤法　130
最尤法による系統樹　130, 137
最尤法による祖先状態再構成　131
採用　14
採用する　14
在来種　79, 125, 142
在来生物種　110
在来生物との競争に勝ち残る　154
在来の園芸植物　142
在来野生生物　142
サイラス型　53
細粒　83
最良線形不偏予測量　28
最良適合分割法　28
材料と方法　130
サイレンサー　198
サイレント置換　198
サイレント変異　198
サイン波　198
差延　59
坂　200
先のとがっていない　31
先へ　152
先細のピンセット　83
砂丘後背湿地　42
作業仮説　235
朔　35
索引　110
朔殻　35
酢酸エチル　76
搾取作用　79
サクセッション　211
作物組織　52
作物に被害を与える　100
作物畑　52
作物を傷つける　100
サケ科　152
裂け目　41
避けられない　111
避ける　27
叉甲孔　89
叉甲腹板　89
支える　213
さざ波状パターン　187
差込み図　113
差し込む　190
差し控える　235
差引き効率　210
差引きライブラリー　210
砂上の楼閣　106
誘う　51
撮影する　162
雑音閾値　146
雑音源　202
三角節片　25
雑記　136
殺菌する　208
雑種　52, 107, 137
雑種間の交配　107
雑種死滅　107
雑種種分化イベント　107

雑種第一代目　83
雑種致死　107
雑種の雄　107
雑種の機能不全　107
雑種のゲノム　107
雑種の生存不能　107
雑種の生存力実験　107
雑種の生存力低下　107
雑種の翅　107
雑種不妊性　107
雑種不稔　107
雑種崩壊　107
雑種卵の孵化　107
雑食性　91, 152
雑食(性)動物　152
雑草　232
殺虫剤　113, 161
殺虫剤抵抗性　113
殺虫剤抵抗性管理　113, 116
殺虫性結晶タンパク質　113
雑木林　33, 50, 191
雑木林の伐採　50
殺卵性捕食寄生者　70
殺卵の直接防御　70
殺卵防御　70
サテン地　190
作動筋　15
作動物質　15
砂糖溶液　211
作動様式　152
査読　186
査読者　118, 181, 186
里山　190
里山の蝶　51
蛹　39, 175
蛹クチクラ黒色化ホルモン　159, 175
蛹黒色化抑制因子　166, 175
蛹死亡率　175
蛹初期　236
蛹になる　176
蛹の重さ　175
蛹の後翅　175
蛹の色彩化　175
蛹の前翅芽　87
蛹の脱皮　175
蛹の表皮シート　175
蛹の表皮層　175
砂のう　95
サバイビン　212
砂漠　57
砂漠地帯　190
サハリン　189
サハリン島　189
サバンナ　190
サバンナ気候　190
サブクローン化した　209
サブ個体群　210
サブトラクション法　210
サブユニット　211
差分ライブラリー　210
サポートバリュー　212
サボテン好性ショウジョウバエ　34
妨げる　186
さまよい期　231
寒い冬　49
サムライコマユバチ亜科　134
サモア諸島　189
鞘　197
鞘翅　71
左右相称　29
左右非対称の色彩パターン　23
作用効果　13
作用スペクトル　13
ザラザラした　97
晒すこと　79
曝すこと　79
さらに悪化させる　15
さらに二回交雑する　52
サリチル酸　189
サリチル酸依存性生合成経路　189
サリチル酸依存(性)卵誘導性反応　189
サリチル酸応答性遺伝子　189
サリチル酸(SA)の生合成経路マーカー　189
サルチル酸メチル　134
サルペドビリン　190
サル免疫不全ウイルス　198, 200
3-　222
三-　222
三栄養　223
三角紙　73, 222
山岳性種　140
山岳地帯　140
三角板　25
三角板の　25
酸化状態　155
酸化／水酸化　155
酸化ストレス　155
三化性地域　223
三角形の　222
参考文献　28
サンゴ礁魚　50
サンゴ礁に生息する魚　50
三語名　223
残差　185
散在する　67
散在領域　190
残差項　185
残差値　185
残差 DNA　185
三叉に分かれる　222
三次元構造　11, 218
三次構造　217
産雌単為生殖　217
三枝に分かれる　222
三重交尾　223
35 サイクルでの PCR　11
産出された系統樹　236
参照　181
参照ゲノム　181
参照超計量系統樹　181
参照配列　181
三色型色覚　222
三色性の　222
酸素吸収　155
酸素呼吸　14
酸素消費(量)　155
酸素中毒　155
酸素毒性　155
残存個体群　183
残存種　183

残存生殖価　185
残存 DNA　185
残存繁殖価　185
残存物　183
3' 側非翻訳領域　11
3' 非翻訳領域　11, 218
山地　140
山地カシ林　139
山地(性)個体群　139, 140
山地生態型　140
山地性の　139
山頂　211
山頂占有行動　103
山頂占有性　103
散発的な　205
散発的に　205
3- ヒドロキシキヌレニン　11
散布体　173
残(留)物　183
サンプリング強度の指数　131
サンプリング曲線　189
サンプリング単位　190
サンプリング調査　189
サンプリング努力　189
サンプリングの手数　189
サンプリングプロトコール　189
サンプリング方法　189
サンプリング問題点　189
サンプル採集地　189
サンプルサイズ　200
サンプルサイズ依存性　189
サンプル数　189
サンプル数に基づく曲線　189
サンプル数に基づくデータセット　189
サンプル数に基づくプロトコール　189
サンプルの異種混交性　189
サンプルの大きさ　200
サンプルの相同サブセット　104
サンプルの配置　165
サンプルの不均質性　189
三峰性分布　223
三名式命名法　146
三名法　146
残余　183
残余項　185
産卵　70, 155
産卵活性　155
産卵管　155
産卵管鞘　155
産卵忌避物質　155
産卵孔　155
産卵行動　155
産卵された草　97
産卵刺激物質　155
産卵刺激物質受容系　155
産卵鞘　155
産卵植物　155, 165
産卵数　81, 150
産卵する　122, 155
産卵する雌　155
産卵前期　171
産卵選好　155
産卵選好性　155
産卵草選好性　155
産卵調節物質　155
産卵に適した葉の種類　86
産卵のバイオアッセイ　155
産卵分泌物　155
産卵弁　93
産卵弁片　228
産卵誘導性植物揮発性物質　151, 155
産卵抑制フェロモン　155
産卵抑制物質　155
産卵力　81
残留 DNA　185
残留物　185
３齢　11
三裂の　222

シ

肢　122
-翅　11
翅-　175
時-　39
シアノグルコシド　53
シアノバクテリア　53
ジアミジノフェニルインドール　54
シアン化物の放出　53
シアン化ベンジル　27, 28
シアン生成　53
CAPS 法　41
GAPDH 遺伝子　90
CFLP 法　41
GMYC モデル　91
glm 関数　95
glm コマンド　95
GO 用語　96
C 型レクチン遺伝子　34
C 型レクチンドメイン　34
飼育　179
シークエンシング技術　195
シークエンス　194
シークエンスアセンブリー　194
飼育温度　179
飼育環境　179
飼育環境下での繁殖　35
飼育環境で繁殖された希少種　35
飼育環境で繁殖されたストック　35
飼育ケージ複合体　34
飼育された雄　179
飼育されている　35
飼育実験　179
飼育場　32
飼育条件　35
飼育する　178, 179
GC 含量に偏りのある遺伝子変換　90
シーズン後期の成育条件　121
シータテハ　44
G タンパク質　90
G タンパク質共役型受容体　90
G タンパク質結合受容体　90, 96
C 値パラドックス　34
CDS コーディング　36
cDNA のサイズ画分の直接塩基配列決定法　61

cDNA のサイズ画分の直接シークエンシング法 61
Ct 値 52
シードポッド 193
GBS 法 94
C 末端部 34
C 脈 42, 50
CuP 脈 169
Cu 脈 20, 53
地色 97
Cytb 遺伝子 53
ジーンバンクアクセス番号 90
JH の循環濃度 40
JH の浄化 41
JH の生産 118
JC モデル 118
JTT モデルの(アミノ酸)置換 118
ジェネラリスト 91
ジェネラリストの蝶 91
ジェノタイピングシークエンシング 94
ジェル 90
シェルター 197
ジェンダー 90
塩水の滴下 66
シカ 56
翅芽 234
紫外域 225
紫外線 225, 228
紫外線反射 225
市街地 33
翅外中央帯 169
自家栄養 25
視覚 197, 230
視覚外光受容器 80
視覚化する 230
視覚器官 230
視覚系 230
視覚識別 230
視覚刺激 230
視覚信号 230
視覚通信 230
視覚的な接触 230
視覚による通信 230
自家受精 193
自家受粉 193
自家生殖 25
翅芽遅滅 133
自家不和合性 193
時間依存性 216
時間移動実験 41
時間がかかる作業 219
時間記憶 237
視環境 123
時間較正したゲノムベースの系統樹 219
時間尺度 216, 219
時間樹 219
時間スケール 216, 219
時間制限 219
時間生物学 39
時間制約 219
時間的異質性 216
時間的隔離 16, 216
時間的自己相関 216
時間的盛衰 216
時間的な景観動態 216
時間的不均一性 216
時間的分布 216
時間的変化 216
時間的変動 216
時間的予測可能性 216
時間と広がり説 107
時間に連続した 47
篩管部摂食性植食 162
時間偏移実験 41
時間を費やす作業 219
時期 191, 219
色覚 44
色覚用のスペクトル受容体の型 204
磁気感知 127
磁気コンパス 127
色彩 43
色彩化 44
色彩型 43
色彩恒常性 43
色彩情報 43
色彩選好地図 44
色彩多型 43
色彩適応 39
色彩のバリエーション 44
色彩パターン 43
色彩パターンの特異的発現 43
色彩斑紋 43
色紙モデル 156
磁気受容 127
時期尚早の停止 171
基翅節片筋 26
色素 164
色素拡散因子 159, 164
色素拡散ホルモン様ペプチド 164
色素形成 164
色素(生合成)経路 164
色素層 164
色素多型 164
色素沈着 164
色素沈着遺伝子 *ddc* 164
色素沈着過程 164
色素沈着制御遺伝子 90
色素沈着の遺伝学 93
色素分画 164
色帯 43
色調 44
時期はずれの発育 227
色斑 43
翅基部 234
翅基部の神経枝 234
識別 61, 108
識別形質 58
識別する 60, 61, 215
識別的遺伝子発現 59
識別点 63
翅基片 215
シキミ酸経路 197
支給 174
翅胸部神経球 175
翅胸部神経節 175
翅胸部側板 175
翅棘 88
至近機構 175
至近的基盤 174
至近的原因 174
至近要因 175
軸 25
時空 202
時空間 202

時空間尺度　190
時空間スケール　190
時空間の個体群動態　202
時空間の資源動態　203
時空間変化　203
時空間変動　203
時空情報　216
時空分離　216
軸索　25
軸索のコバルト染色法　42
ジグザグ飛翔　64
軸索末端　25
ジグザグ模様　237
軸性　25
軸節　36
軸側　25
シグナル因子　198
シグナル受容　198
シグナル生産　198
シグナル選好　198
シグナル伝達　198
シグナル伝達系　198
シグナル伝達系遺伝子　198
シグナル伝達経路　198
シグナル伝達の中心部　37
シグナル分子　197
シグナルペプチド　197
σ因子　197
シグマ因子　197
ジグリセリド　60
ジグリセリド運搬タンパク質　60
シクリッド　40
翅型　234, 235
翅形　234
翅型多型　234
時系列データ　219
刺激　208
刺激因子　208
刺激強度　208
刺激効果　208
刺激持続期間　208
刺激信号　53
刺激する　112, 208
刺激的な　174
刺激と選好の間の連鎖　124
止血膜　101
資源　185
資源環境　185
試験管内で　109
肢原基　122
翅原基　234
資源計画　185
資源景観　185
資源減少　185
資源勾配　185
資源枯渇　185
始原生殖細胞　161, 172
試験セット　217
資源選択　185
試験的研究　165
資源動態　185
資源の空間的不均一性　202
資源の空間的不均質性　202
資源の時空間動態　203
資源の相対量　182, 183
資源の多様性　64
資源の有用性　185
資源の利用可能性　185
資源の類似性　185
資源の類似度　185
資源パッチ　185
翅肩板　106
資源分布　185
資源防衛型の一夫多妻　185
資源保持力　185
資源密度　185
資源量　185
資源利用可能性の定性的尺度　176
自己-　24
翅鉤　99
指向移動　154
指向機構　154
視交叉上核　191, 212
施行水準　123
試行する　188
嗜好性　156
嗜好性の範囲　156
事後解析　121
自己回避　193
自己擬態　25
時刻信号　219
自己組織化　193
自己組織化マップ　193, 201
自己つぎはぎする　193
事後比較　168
自己複製　193
事後分析　121
自己防衛物質　56
視細胞　230
翅細胞　234
示唆する　211
示差走査熱量測定　59, 67
枝刺　191
翅刺　88
指示　110
支持構造　212
支持糸　95
支持値　212
脂質　124
翅室　202
脂質蓄積　124
脂質動員ホルモン　14
事実に基づいている　26
支持度　212
支持度の値　212
シジミタテハ　30
シジミタテハ科の蝶の総称　133
シジミチョウ亜科　50
シジミチョウ科　31, 96
シジミチョウ科の幼虫　127
シジミチョウ科のリスト　124
シジミチョウとアリとの関係　127
四重縮退サイト　87
自主的な交尾　110
示準化石　110
翅鞘　71
歯状感覚子　194, 206
糸状菌　89
糸状菌感染　89
矢状断面　189
指状突起　167
糸状の　83
矢状面　189
自殖　109
翅色選好の刺激　234

四肢類 176
私信 161
刺針 208
始新世 68
始新統 74
翅垂 118
視髄 132
シス因子 40
指数 177
シスエレメント 40
シス効果 40
シス-3-ヘキセノール 236
シス調節エレメント 40
シス調節転換 40
シス調節要素 40
システイン配置パターン 53
システム生物学 214
シスト 53
シスト細胞 53
姿勢 169
雌性化 82
雌性産生単為生殖 217
自生種 142
自生する植物 45
雌性生殖器 82
雌性先熟 174, 196
雌性先熟現象 174
始生代の 22
雌性単婚性 82
自生地 98
自生的な要素 142
自制のない採集 227
自生バキュロウイルス 142
磁性ビーズ 127
姿勢変更 169
次世代への餌の転化効率 70
自切 25
自切位置 25
糸節の 89
自然(界) 143
自然界の季節変異 143
自然界の規範 148
自然界の微生物 143
自然火災 234
自然環境 143
自然環境保全 143
事前期待 172
自然群 143
自然群集 143
事前研究 171
自然光 143
自然交雑帯 143
自然災害 143
自然史 143
自然死 143
自然条件 143
自然植生 143
自然植生の分断化 143
自然(の)生息地 143
自然選択 143
自然選択説 143
自然選択の分岐点モデル 32
自然(に起こる)単一遺伝子突然変異体 205
自然ではランダムに交尾する 130
自然淘汰 143
自然淘汰のサイン 198
自然淘汰のしるし 198
自然な花崗岩の露出部 143
自然な形の静止姿勢 143
自然な断片化 143
自然に起こる突然変異体 205
自然日長 143
自然の 142
事前の期待 172
自然の競争者 143
事前の許可 172
自然の構造 80
自然の障壁 142
自然の体系 214
自然の代物 143
自然のバランス 26
自然の捕食者 143
自然のままの森林地帯 142
自然のもの 143
自然発生 143
自然発生個体群 143
自然発生的な摂食反応 205
自然発生率 143
事前フィールド調査 171
自然分布域 143
自然保護 143
自然保護区 143, 234
自然保護者 47
自然無作為交配 130
自然免疫 112
自然免疫調節 182
自然誘発突然変異体 205
持続可能な開発 213
持続可能な森林地帯の管理 213
持続可能な生存 213
持続機構 213
持続収穫量 213
持続する交雑 161
持続性 160
持続性の 219
子孫 151, 173
子孫の質 151
子孫ファミリー 57
下 111
時代推定 219
時代的順序 39
次第に減る 214
下側 226
下草 225
親しい関係 115
下の 111
翅端 21, 69, 234
翅端部 21
枝長 32
翅長 234
翅頂室 21
翅頂部 21
翅頂部帯 21
室 36
10回につき1回 11
膝下器官 210
しっかりと組み合わせる 125
疾患粘液腫病 61
10km グリッド四方 97
湿気 137
湿原 31
実験アリーナ 78
実現遺伝率 179

実験期間　78
実験グループ　78
実験群　78
実験計画　78
実現系統樹　179
実験ケージ　119
実験研究　79
実験個体群　78
実験室研究　119
実験室条件　119
実験室で飼育された雌　119
実験室での交雑実験　119
湿原植生　232
湿原性蝶　31
実験地　78
実験的移動　72
実験的確証　78
実験的確認　78
実験的景観　72
実験的研究　79
実験的検定の不足　55
実験的証拠　78
実験的スクリーニング　78
実験的スクリーン　78
実験的操作　78
実験的導入　78
実験的な動物追跡データ　72
実験手順　78
実現ニッチ　179
実験に基づいた結果　72
実験の手順・観察(などの)記録　174
実験の標準化　78
実験プロトコル　174
しつこい追跡飛翔　113
実効塩基数　70
実効的交尾確率　13
漆黒色　118
実際の交尾確率　13
実時間追跡　179
櫛歯状の　159
実質的な差異　210
実質的な時間　210
実質的な変異　210
実質的な変化　210
湿重量　232
湿潤重量　232
湿潤森林　232
湿潤地帯　107
湿潤熱帯　107
櫛状器官　159
実証研究　72
実証試験　72
実証的関心　78
実証的研究　72
実証的証拠　72
実証的生産性勾配　78
実証的漸近　72
実証的相関解析　78
実証的相関研究　78
実証的な結果　72
櫛状突起　159
櫛状板　159
実線　48, 201
湿暖条件　137
湿地　54, 129, 213, 232
湿地性植生　232
湿地草　232
実地調査　83
質的確定要因　36
質的基準　177
質的固定要因　36
質的に同じ　177
質的物質　177
質的変化　177
湿度　107
湿度条件　107
室内条件　188
実の兄(弟)　89
シッフ塩基　191
疾病媒介生物　229
質量分析計　129, 140
指定　57
指定する　57, 204
ジデオキシターミネーター　59
ジデオキシ法　59
指摘しておく価値がある　236
自動-　24
自動運転ロボット　25
市(町)当局　141
自動車を利用した記録者　35
自動精製装置　24
自動性の　140
自動バーコードギャップ発見　12, 24
シトキン　53
指突起　53
肢内葉　120
ジナンドロモルフ　98
シニアオーサー　194
死肉　36
シニグリン　199
シネフリン　214
シネフリン受容体　214
シノニム　213, 214
死の病原体　55
シノモン　214
雌胚　82
翅背側　65
支配的信号　170
支配的で　171
支配的なシグナル　170
支配的な要因　170
支配要因　49
しばしば行く　88
しばしば伐採されている　88
芝土　224
自発的移動　13
自発的な摂食反応　205
自発飛翔　205
芝生　224
シバリング　197
翅斑紋パターン　129
地引網　213
ジヒドロキシフェニルアラニン　60
指標　110
翅表面　227
翅幅　235
刺腹板　205
視物質　162, 230
視物質感度　162
自分で動くことができる　140

自分と釣り合う(相手との交尾) 193
シベリア型 197
シベリア分布系統 197
子房 154
脂肪酸 81
脂肪酸 - アミノ酸縮合物 81
脂肪酸 - アミノ酸接合体 81
脂肪消費 48
脂肪体 81
死亡までの時間 219
死亡率 139
死亡率曲線 139
縞 209
姉妹群 199
姉妹種 199
姉妹属 199
姉妹ハプロタイプ 199
姉妹分類群 199
縞状バンド 26
縞状バンドの相称系 213
島の群集 117
島の生物地理学理論 218
島巡り 117
縞模様 44
しみ込む 160
シミ目 218
翅脈 229, 234
翅脈依存パターン 57
翅脈間隙 115
翅脈の 229
翅脈パターン 229
シミュレーションモデル 198
市民菜園 16
示す 78
閉め出される 206
ジメチルスルホキシド 60, 64
死滅 79
湿った砂 232
湿った土 232
湿った土壌 232
湿ったペーパータオル 137
翅(表)面 234
地面 97, 201
翅面荷重 234
刺毛 195, 208
刺毛式 38
刺毛状の触角 195
刺毛相 38
刺毛の 195
翅紋要素 234
シャーレ 161
シャイン・ダルガノ配列 197
社会機構 191
社会寄生 201
社会経済的な 201
社会コロニー 201
社会性 201
社会生態学 201
社会的コミュニケーション 201
社会的情報 201
社会的同調 201
社会的要因 201
シャクガモドキ科の蛾 100
シャクガモドキ上科 140
尺側の 225
弱体化した状態 232
若虫 120, 150
弱有害突然変異説 200
若齢期の発育 118
若齢幼虫 68, 236
ジャケツイバラ属の一種 34
ジャコウアゲハ 34
ジャコウアゲハ族 223
謝辞 13
写真乾板 162
写真撮影 162
写真図版 162, 165, 166
写真を撮る 162
ジャスモン酸 117, 118
ジャスモン酸依存性傷誘導性反応 117
ジャスモン酸メチル 134
射精 71
射精管 67, 71
射精する 71
射精のう棘 50
射精のう刺 50
尺骨(側)の 225
シャッフルされたカードの束 55
ジャノメチョウ 33
ジャノメチョウ(亜)科 33, 190, 235
ジャノメチョウ(亜)科の蝶 190
ジャノメチョウ科のタカの目と眼状紋 100
ジャノメチョウ科のタカの目と目玉模様 100
ジャノメチョウ族 190
蛇の目模様 151
シャノン - ウィナーの多様度指数 196
蛇腹 27
シャペロン 38
邪魔をする 99
斜面 200
蛇紋岩地の土壌環境 195
種 202, 203
鬚 156
主域 183
主遺伝子 128
朱色 229
主因 127, 171
縦 126
周囲 212
周囲 - 17, 40, 160
周囲温度 17
周囲条件 17
雌雄異体 60
11- シス型レチナール 11
11- シス - レチナール 11
収益 28
周縁効果 69
周縁植生 212
周縁に沿う(飛翔) 69
縦横比 23
十億年前 34
十億の 28
重回帰 141
重回帰式 141
重回帰分析 141
臭角 154, 191

収穫する 100
臭角分泌物 154
雌雄型 98
周期型 101, 160
従基準標本 157
周期性 160
周期ゼミ 53
周期的な 160
周期的な再発見 160
周期反応 160
獣脚類 218
褶曲 86
終局の結果 77
重鋸歯状の 30
重金属ストレス 100
ジュークス・カンターモデル 118
集群 15
集計する 214
自由継続周期 88
自由継続する 88
自由継続リズム 88
集合(体) 15
集合化 42
集合性 97
集合性があり 97
重合体 167
集合フェロモン 15
終止コドン 208
終止コドンの下流 66
収集家 43
収集する 43
収集フラスコ 43
重症度 195
修飾遺伝子 137
修飾因子 137
修飾－救済モデル 137
修飾された精子 137
重心 37
自由進行 88
修正 17, 71
習性 98, 99
習性に関して 98
習性の面では 98
修正名 71
雌雄選択 115
縦走脈 126
集束した表現型 49
収束する 49
従属的タクソン 210
収束的変化解析 49
従属変数 57
重大な革新の進化 77
集団 15, 23, 97, 168
集団遺伝学におけるボトルネック効果 168
集団帰塒 97
集団求愛場 122
集団行動 97
集団生態学 168
集団摂食 15
集団選択 97
集団で塒を形成する鳥 45
集団の遺伝的構造 168
集団の大きさ 168
集団の保全 47
集団繁殖地 43
雌雄短尾型 197
集団ボトルネック 168
縦断面 126
集中(的)サンプリング 113
集中的抽出法 113
集中的に刈られた 113
集中的に調べる 113
集中的に生草が食われた 113
自由度 54
雌雄同株 101
雌雄同体 101
周日本海地域 40, 156, 188, 191
周年経過 20
周年変化 19
周波数 88
周波数積算仮説 88
周波数弁別 88
臭物質 151
臭物質結合タンパク質 150, 151
獣糞 19
十分成長した幼虫 89
十分な解剖データ 18
十分な機会 18
十分なデータ 211
十分な飽和 210
十分に成長した草地 130
十分に定義された姉妹関係 232
十分に野焼きされた森 232
10 分の 1 11
周辺 212
周辺－ 40, 160
周辺域 212
周辺(的)種分化 160
周辺植生 212
周辺生息地 129
周辺地域 14
周辺投影様式 160
周辺投射パターン 160
充満させる 160
縦脈 126
雌雄モザイク 98, 195
雌雄モザイク現象 98
集約サンプリング 113
集約農業 113
自由蛹 78
重要害虫 119
重要な刺激 109
重要な進化的革新 77
重要な発見 109
重要要因 119
従来のアプローチ 220
従来のデシル硫酸ナトリウム-プロテイナーゼ K による消化 49
従来の報告 158
縦隆起 187
終了 72
重量 97
終了時刻 219
雌雄両方 31
重力 97
終令 209
終齢 89, 121, 209
終齢幼虫 121
終齢幼虫期 121
縦列出現 214
縦列重複 214
縦列発生 214
縦列(型)反復配列 197, 208

収斂 49
収斂現象 49
収斂した進化 49
収斂した表現型 49
収斂進化 49
収斂する 49
収斂的擬態 49
収斂的な結果 49
収斂的に 49
樹液 190, 222
受音波突起 191
シュガートランスポーター遺伝子 211
種階級群 203
種階級群名 203
種が多い 203
種が多い群集 204
種が少ない 203
種が少ない群集 203
種が乏しい 203
種が豊富な 203
種が豊富な場所 204
種間 115
樹冠 35, 222
樹幹 222
主観異名 210
種間干渉 115
種間競争 115
種間交雑 52, 107, 115
種間交配 115, 130
種間雑種 115
樹幹数 150
主観的 210
主観同物異名 210
種間ナワバリ 114
種間の遺伝子浸透を減少させる 181
種間の遺伝子流動 115
種間の交配 115
種間変異 115
樹幹密度 207
樹幹密度効果 69
種間要因 115
熟した果実 187
宿主 105
縮重フォワードプライマー 56
縮重プライマー 56
宿主個体群 105
宿主伸長因子 105
宿主世代 105
宿主節足動物 105
宿主組織 105
宿主ではない種 147
宿主特異性 105
宿主と病原体の軍拡競走 105
宿主に組み込まれたモチーフ 103
宿主認識タンパク質 105
宿主の体 105
宿主の間性形質 115
宿主の細胞機構 105
宿主の順化(順応) 105
宿主の生殖細胞系列細胞 105
宿主の適応度 105
宿主の適応度犠牲 78
宿主の発育停止 105
宿主の発生阻止 105
宿主の免疫防御 105
宿主の免疫抑制 105
宿主被子植物 19
宿主 - 病原体の軍拡競争 105
宿主 - 捕食寄生者相互作用 105
宿主由来の伸長因子 105
宿主幼虫 105
縮小(図) 181
樹形 219
樹形距離 219
樹形図 222
種系統樹 203
種構成 203
種交替 203
種子 193
種子が(農薬)処理された作物 193
種子形成 193
主刺激 175
種子食性甲虫 193
種子摂食甲虫 193
種子のさや 193
種子粉衣 193
種子粉衣用ネオニコチノイド 193
樹種 222
種鞘 193
樹状図 161
樹状突起 57
樹状突起先端部 57
手掌の 156
種小名 204
種水準の推定値 203
種数 150, 203
種数が多いこと 204
種数 - 科数の比 203
種数／個体数の比 179, 204
種数 - 個体数モデル 203
種数指数 187
種数推定 187
種数推定量 187
種数対個体数の比 204
種数対属数の比 204
種数対面積の比 179
種数の多い属 204
種数比較 187
種数累積曲線 203
種数を正規化することの危険 160
数珠玉状の 27
数珠玉状の触角 138
ジュズヒゲムシ目 237
受精 82, 113
授精 113
受精管 67
受精させる 82
受精した雌 113
種生態学 203
受精のう 179, 205
受精能力をもたない精子 147
主成分分析 159, 172
受精雌 113
受精卵 82
受精率 82
主席著者 194
首席著者 194
種選択 203

種族維持　127, 222
収束進化　49
種組成　203
受胎　97
受胎のう　205
種多様化　203
種多様性　203
種多様性指数の誤用　12
種多様性指標の誤用　12
種多様性の喪失　126
主張する　113
出芽ウイルス　33, 34
出芽型ウイルス　33, 34
出芽酵母　189
出現機構　131
出現時間　72
出現時刻　219
出現程度　56
熟考する　204
出生後の刺毛　169
出生地からの分散　142
出生島　142
出生率　142
出版する　175
出費　50
主働遺伝子　128
手動調整　129
種同定　203
受動的消失　158
受動発散　158
受動放散　158
種特異性　204
種特異的　203
種特異的な色選好性　204
種内(集団)間の遺伝的変異性　115
種内間の水平移動　115
種内間の水平伝播　115
種内擬態　25, 115
種内競争　115
種内交配(種)　130
種内多様性　115
種内ナワバリ　115
種内の臭跡　115
種内変異　115
種内捕食　35
種内要因　115
種認識　203
鬚の　156
種の起原　152
種の境界　203
種の境界設定　203
種の空間的集団　202
種の減少　203
種の個生態学　203
種の個体数　12
種の自然分布　143
種の出現(率)　203
種の衰亡　203
種の絶滅　203
種の喪失　203
種の総数　203
種の祖先　203
種の多様性　203
種の置換　203
種の転化　203
種の普通性と希少性　45
種の分布　203
種の分布と個体数　63
種の豊富さ　203
種の豊富さの空間的パターン　202
種のホットスポット　203
種の滅亡　203
樹皮食性　162
首尾よく羽化した成虫　211
種複合体　203
受粉　167
種分化　203
種分化機構　131
種分化した部分　203
種分化する　203
種分化速度　203
種分化の開始者　112
種分化の決定的な段階　52
種分化の最初期　68
種分化の時期　203
種分化の初期段階　109
種分化率　203
種分化(の)理論　218
受粉生態学　167
種分布圏　63
ジュベノイド　118
種密度　203
主脈　135
寿命　123, 126
種名　142, 203
樹木が茂った細長い土地　235
樹木に覆われた細長い土地　235
主要因　127, 171, 172
受容期間　180
主要経路　127
受容植物　179
受容性　12
主要組織適合遺伝子複合体　128, 134
受容体遺伝子　180
受容体個体群　180
主要な移行　128
主要なグループ　86
主要な目的　172
主要な理由　172
受容の　180
種よりも低位の学名　111
受理　12
主竜類　22
種類が多い単系統　203
種レベル　203
順位　178
春化　229
順化　12
馴化　12
巡回　159
巡回型　159
巡回行動　159
巡回飛翔　159
順化過程　12
馴化期間　12
馴化期間の効果　69
順化(順応)した *Ben* 遺伝子　65
春化処理　229
順化する　65
純化選択　176
純化淘汰　176
循環系　40
春季の再移動　206
順系相同　154
純高山性蝶　223

順次 224
順次戦略 195
遵守する 12
純粋型 176
純粋交配種 176
純粋種 176
純粋同腹仔 176
純粋な雌 176
純多様化速度 145
純熱帯性の種 223
順応 12, 13
順応学習 13
順応する 65
順応的管理 13
盾板 191
順番に 224
盾板の 191
準備行動 171
順不同 178
純分岐進化速度 145
順方向縮重プライマー 56
順方向プライマー及び逆方向プライマー 87
順方向または逆方向の読み込み 87
序 116
章 38
雌蛹 82
視葉 153
上 212
上- 74, 211
条 208
滋養(物) 149
上位(性) 75
上位異物同名 194
上位概念用語 157
上位シャッター対立遺伝子 75
上位性決定遺伝子座 75
上位同物異名 194
上位不和合性 75
小宇宙 134
消音装置 198
消化 60
上科 212
生涯 123, 126
上界 64
傷害関連分子構造 54
傷害関連分子パターン 54
上下位性 75
上外側 212
傷害で誘導される防御 236
障害物 103
傷害誘導防御 236
傷害由来分子パターン 54
消化管 16, 60
消化吸収率 60
小顎 130
小顎鬚 130
小顎の 130
消化系 60
浄化する 176
消化性 60
浄化選択 176
上科名 142, 212
小蛾類 134
小規模再配置 200
小球形にする 160
小球細胞 205
小球を投げつける 160
状況特異的な 48
状況特異的なやり方 48
状況に独立な 48
消極的移動 158
消去法による固定 84
消去力 50
上クチクラ 74
条件依存仮説 46
条件依存性 46
条件付きの 46
条件付きの提案(書) 46
条件的 80
条件的相利共生 80
条件的蛹休眠 80
正午 135
小腔 120
条項 23, 174, 188
症候群 214
上後側板 19
証拠標本 231
証拠物件 231
小根 178
小腮 130
小腮粒状体 130
小腮外葉 130
詳細空間スケール 83
小腮鬚 130
小腮節 130
詳細な遺伝子地図 83
詳細なマッピング 83
詳細に調べる 62
小腮の 130
孃細胞 55
娘細胞 55
蒸散 221
硝酸イオン 146
鞘翅 71
上翅 71, 86, 87
小翅室 22
小室 22
消失する 75
小室の 22
鞘翅目 43
照射 117
照射期間 160
照射強度 113
焼灼法 36
照射される光の期間 67
照射される光の強度 113
照射する 178
小縮尺の地図 200
小盾板 191
小盾板の 191
症状 213
上昇 75
鐘状感覚子 35
上昇した閾値 71
上昇した死亡率 71
ショウジョウバエ 89
ショウジョウバエ属 67
少食 151
証書番号 231
生じる 12, 71
上唇 120
小進化 134
上唇の 120
少数の 99
小生活圏 29
常染色体遺伝子 25
常染色体逆位 25
常染色体の遺伝子座 25

上前側板溝状線　19
上前腹板　19
上層の木　35
上層の広葉樹林　35
上層のマツ　35
掌側の　156
正体　108
正体が分からない　225
状態変数　207
承諾　47
小柱　220
消長　55
象徴グループ　84
象徴昆虫　84
冗長コンティグ　181
象徴種　84
象徴的なグループ　71
冗長複製　181
焦点　86
焦点信号　86
小転節　223
焦点の内訳　204
焦点をあてたグループ　86
照度　108, 123, 127
消毒　61
生得的な色選好性　112
照度差　108
衝突回避　46
小鈍鋸歯状の　51
承認手続き　24
小のう　189
乗馬道の拡大　233
視葉板　120
小斑点　66, 175
消費型競争　79
消費効果　48
上皮細胞　74, 75
上皮細胞成長因子　70, 74
上皮細胞増殖因子　70
消費者　48
消費者水準　48
消費者の多様性　64
消費する　48
上皮成長因子　70
証票　231
証憑　231
上表皮ワックス　74
上腹板　75
上腹板溝　75
上腹板線　75
上部の　212
小弁　228
小便　228
小胞　189, 228
情報交換　78
小胞子　135
上方制御遺伝子および下方制御遺伝子　227
上方制御する　227
上方選択　228
上方調節する　227
情報に基づく保全に関する意思決定　111
上方臨界温度　227
情報を持たないサイト　147
情報を持つサイト　111
漿膜　195
正味の競争効果　145
照明　123
承名の　147
消滅する　160
上面　54
上面の　65
消耗戦　231
上目　212
小葉　124
照葉－山地カシ林遺存型　122
将来の研究　90
将来の方向性　89
蒸留水　62
常緑広葉樹林　77
常緑樹林　77
常緑性　77
条例　34
女王アリ　177
女王蟻　177
女王(蜂)の産出　177
女王(蜂)の生産　177
除外された　78
初回実験　112
初回の継代時間　83
除外名　78
初夏型　68
初期型雄殺し　68
初期実験　112
初期種　109
初期図像学　68
初期探索　191
初期のウイルス侵入　112
除去　176
除去手術　12
除去する　183
－食　11
食塩水　189
食害　82
食細胞活動　161
食作用　161
食事　131
食餌供体植物　86
食餌資源　86
食餌植物　86
食餌性窒素　59
食餌摂取(量)　86
触刺毛　214
触手　83, 216
食樹　105
触手状器官　216
食餌要求　86
植食　101
植食者が誘導する植物の揮発成分　101, 103
植食者関連分子構造　101
植食者関連分子パターン　99, 101
植食者天敵の誘引　24
植食者に対する直接的な反発　61
植食者の産卵　101
植食者の天敵の誘引　24
植食者のパフォーマンス　101
植食者引き金性 JA 経路　101
植食者誘導性イベント　101
植食者誘導性植物揮発性物質　101, 103
植食者誘導性植物内シグナル伝達　101
植食者由来エリシター　101

植食者由来分子パターン 101
植食者由来誘発物 101
植食者を誘引する揮発性物質の放出 101
植食性 101, 164
植食性昆虫 101, 164
植食性昆虫と植物間の関係に関する生態学 69
植食性チョウ目昆虫 101
植食性の好蟻性昆虫 164
食事をする 81
食する 81
食性 82, 86
- 食性 11
植生 229
植生改変 229
植生型 229
植生帯 85, 229
植生調査 229
植生転換 229
食性転換 59
植生動態指標 110
植生の構造と(種)組成 229
食性分化 60, 82
植生分類 229
植生変化 229
植生変動 229
食草 86, 105
食草選択機構 105
食草転換 86
食草特化適応 105
食草認識 105
食草の移行帯 221
触知性刺毛 214
食虫性(の) 73
食虫性(の)鳥 113
食虫性の鳥種 113
食虫(性)動物 113
食道 75, 151
食道下神経球 209, 210
食道下神経節 209, 210
褥盤 175
植氷 108, 112
植氷過冷却点 112
植氷凍結回避 108
植物 165, 229
植物 - 164
植物エクジソン 164
植物科学 165
植物化学物質濃度 164
植物間のシグナル伝達 165
植物揮発性物質 165
植物群集 165
植物群落 165
植物検疫 165
植物香気 165
植物個体群 165
植物細胞 165
植物材料 165
植物色素 165
植物資源の転換 64
植物種 165
植物処理 165
植物数/面積 150
植物生産性指標 110
植物生態学者 165
植物生物量 165
植物相 85
植物 - 草食(性)動物の共同体 165
植物相的に 85
植物相的要素 85
植物帯 85
植物(の)多様性 165
植物多様性の保全 47
植物動態指標 110
植物と昆虫間の相互作用 165
植物と蝶の系譜 165
植物特化代謝物 165
植物内シグナル伝達 115
植物内情報伝達 115
植物内のシグナル伝達 235
植物二次代謝産物 165
植物の - 164
植物の異常値 165
植物の香り 165
植物の間接防御 110
植物の茎葉部組織 229
植物のゲノム 165
植物の種数 165
植物の種多様性 165
植物の生存 165
植物の食べられる部分 69
植物のバイオマス 165
植物の分泌物 165
植物の防御形質 165
植物の毛状突起 165
植物の誘導性防御応答 111
植物防疫情報総合ネットワーク 118
植物保護 165
植物ホルモン 164, 165
植物ホルモン経路間での相互作用 52
植物密度 165
植物量 165
植物量の密度 29
食物確保 86
食物供給 86
食物源 213
食物調達 86
食物の窒素 59
食物の豊富さ 86
食物不足 86
食物網 86
食物要求 59, 86
食物連鎖 86
食料 59
食料消費率 179
食料制限 86
食料制約 86
食料の質 59
食料の量 59
食料変換効率 70
食糧をあさる 86
食糧を入手する 86
食料を入手する場所 86
植林 165
植林地 15
植林プログラム 15
植林保護地域 87
初春 68
徐々に減少 96
除草剤 101
所蔵する 171
所属位置不明 109
初代培養 171
触覚 214, 220
触角 20, 83, 216

触角形状 20
触角形態 20
触覚刺激 214
触角先端部 41
触角挿入孔 20, 220
触角電図 67, 71
触角電図法 67, 71
触角の 20
触角の形状 20
触角の先端の太い棒 41
触角の鞭小節 20
触角鞭状部 89
触感 220
ショットガンシークエンス法 197
所定の野焼き 171
ショ糖 211
ショ糖含有溶液 201
ショ糖摂取 211
ショ糖トランスポーター遺伝子 211
ショ糖溶液 211
除脳 183
除脳休眠蛹 32
除脳動物 32
除脳蛹 32
書評 31
処分 176
初歩的研究 172
所有 168
所有する 168
処理 222
処理温度 222
処理区の幼虫および対照区(非処理区)の幼虫 222
処理グループ 222
処理群 222
処理された作物 222
処理された正常な母系群 222
処理時間 99
処理室 222
処理済みの幼虫および未(非)処理の幼虫 222
処理チェンバー 222
初齢 83
序論 116
ジョンストン器官 118
シラメスアカミドリシジミ 198
シリアゲムシ目 131
シリコーングリース 198
シリコングリス 198
シリコン内で 109
自律過程 25
自立持続可能性 193
自律振動 193
自律振動性 193
自律振動体 25
飼料 59
試料採集地 189
シルヴァーニ型の分岐群 198
しるし 194
シルビアシジミ 237
事例 36
事例研究 36
歯列 57
シロアリ 217
シロアリ目 117
シロアリモドキ目 71
白いすじ 233
白い線 233
白い大斑点 233
シロイチモジヨトウ 27, 205
シロイヌナズナ 22
シロイヌナズナ属の系統 22
シロイヌナズナのゲノム 217
シロオビアフリカアオバセセリ 209
シロオビドクチョウ 53, 100
白色素沈着 233
シロチョウ亜科 233
シロチョウ亜科の典型種 164
シロチョウ亜科の幼虫 164
シロチョウ科 233
シロチョウ科の蝶 164
白っぽい 233
シロヘリスミナガシ 208
シロベリセセリ 98
白丸 152
しわしわの翅 52
深 56
新亜種 145
新亜種記載 57
新亜種名 206
親暗相 191
人為構造 23
人為選択 23
人為的圧力 20
人為的攪乱 20, 106
人為的攪乱を受けた林分 63
人為的な影響 20
人為的な気候温暖化 20
人為的な景観 23
人為的な生息地移動 128
人為的な生息地変化 20
人為的に改変された生息地 20
人為導入種 23
深遠に 173
進化 77
進化過程 77
進化軌跡 77, 220
進化軌道 77, 220
進化距離 77
真核生物 76
真核生物型(翻訳)開始因子4α 70, 76
真核生物のゲノム 76
真核生物の生物多様性 76
新学名 145
進化傾向 77
進化ゲノム学 77
進行している衰亡 152
進化シナリオ 77
進化上の起点 77
進行する 173
進化生物学 77
進化生物地理学 77
進化戦略 77
進化速度 77
進化的安定戦略 75, 77
進化的意義 77
進化的意味 77

進化的影響 77
進化的応答 77
進化的可塑性 77
進化的関係 77
進化的慣性 77
進化的救済 77
進化的軍拡競争 77
進化的傾向 77
進化的結果 77
進化的最適条件 77
進化的視点 77
進化的収束 77
進化的重要単位 76, 77
進化的収斂 77
進化的新奇性 77
進化的スパイラル 77
進化的トラップ 78
進化的な観点 77
進化的に安定な最適収穫戦略 77
進化的に安定な戦略 75
進化的に重要な単位 76, 77
進化的に重要な単位の境界設定 76
進化的配列 77
進化的反応 77
進化的分岐 77
進化的変化 77
進化的放散 77
進化的見方 77
進化的類縁関係 77
進化的レスキュー 77
進化的罠 78
進化動態 77
進化の新規性 77
進化上のトラップ 78
進化の袋小路 77
進化の歴史 77
進化発生学 77
新科名 81
進化率 77
新規遺伝子の偶然的な獲得 12
審議会 44
審議会規則 48
審議会による裁定 188
新規形質 149
新規形質の獲得 13
新基準標本 144
新規の 149
新機能獲得 144
新規の未記載蝶 145
唇基部 42
真脚 223
新疆産遺伝子型 236
新記録 145
真菌類のゲノム 89
真空(状態) 228
シンク生息地 199
ジンクフィンガータンパク質 237
新組合せ 145
神経応答 145
神経活動 145
神経球 144
神経筋活動 145
神経結合 145
神経行動学 145
神経根 144
神経索 144
神経枝 144
神経シグナル 145
神経刺激 145
神経刺激伝達 112
神経支配 112
神経遮断の 55
神経終末 144
神経信号 145
神経スパイク 145
神経生物学的差異 145
神経生理的記録法 145
神経(的)漸増 145
神経内分泌制御 145
神経内分泌調節 145
神経内分泌機構 145
神経内分泌系 145
神経反応 145
神経分泌細胞 145
神経ペプチドホルモン 145
神経ホルモン的に 145
神経用染色 144
神経用染料 144
神経を刺激する 112
信号化学物質 194
人口学的確率性 56
人口学的関心 56
人口学的慣性 56
人口学的シナリオ 56
人口学的筋書き 56
人口学的パラメータ 56
人口学的母数 56
人工果汁 23
人工基質 23
人工気象室 41
人工構造物 128
人工産物 23
人工障壁 23
深紅色 127, 190
人工飼料 23
人工染色体 23
人工短日 23
人工知能的アプローチ 23
進行中の 226
人工的な環境 23
人工的な景観 23
人工的な転換 23
人工的な培養基 23
信号伝達 198
人口統計上の単位 56
人口統計的に 56
人工ニューラルネットワーク 19, 23
人工繁殖 35
信号物質 194
深刻な衰亡 195
新混棲地 145
審査 181
審査員 181
新参異名 118
新参同名 118
新翅群 144
新翅目 144
真社会性 76
新種 145
人種 177
真獣類 76
伸縮自在の触手状器官 186
伸縮突起 216, 223
伸縮突起器官 216
伸縮突起様蜜腺 216

伸縮蜜腺　216
新種名　202
針状　205
針状の　209
針状の形状　144
腎状紋　183
侵食する　75
新食草　145
真褥盤葉　76
親水性タンパク質　107
親水性部分　107
真正休眠　223
真性高山蝶　223
真正細菌　76
新生種　72
新生・消失の動態　30
真正染色質　76
真正相同遺伝子　154
新生代　37
新生代初頭　68
新生突然変異　55, 88
真正の移動記録　94
新生の型　72
真性メラニン　76
真正メラニン　76
新生幼虫　144
新世界熱帯　145
新世代シークエンサー　145
新世代シークエンシング　145
唇舌　123
新鮮重　88
新鮮な羽化したばかりの雌　88
新鮮な果汁　88
新鮮な切断部分　88
心臓　100
真足　223
迅速評価調査　178
新属名　90
新ダーウィン説　144
靭帯　123
身体の維持　201
身体の成長　201
シンタイプ　214
死んだ昆虫　55
診断　58
診断可能な　58
診断形質　58
診断 PCR 検出法　58
診断 PCR 法による検知　58
新置換名　145
新知見　145
伸長　71
伸長因子 1α　69, 71
伸長因子 1α 遺伝子　69
伸張した気管　73
伸張受容器　209
慎重に　36
伸長部　71
伸長要素　71
新訂版　145
シンテニー　214
伸展標本　233
浸透交雑　116
振動数　88
浸透する　111
振盪する　196
浸透性　160
浸透性交雑　116
浸透できる　160
振動特性　229
振動突起　229
浸透率　160
シンドローム　214
浸軟　127
侵入　116
侵入個体　116
侵入雑草の種子　116
侵入した　117
侵入した雌　110
侵入者　116
侵入種　116
侵入種の園芸植物　116
侵入種の野外放出　116
侵入する　116
侵入生物　116
侵入の初期　68
新熱帯　144
新熱帯区　144
新熱帯区に生息する蝶の属　94
新熱帯産の蝶　144
新熱帯種　144
心のう　160
真の系統樹　223
真の系統地理学　223
真の種数　223
深波状の　199
真皮　74, 107
新北区　143
新ホプキンス則　144
親密　115
新名　146
親明相　162
心門　168
深夜　121
親油性物質　124
針葉樹　47
信頼区間　46
信頼限界　46
信頼水準５％　122
信頼性の値　51
信頼性のある同定　183
信頼できる業者　185
信頼できる刺激　183
信頼できる手がかり　183
信頼できる同定　183
侵略性　116
侵略性ではない　147
侵略的　116
侵略的外来種　108, 116
森林　86, 87
森林遺存属　86
森林開拓地　87
森林火災　87, 234
森林型　87
森林ギャップ　87
森林限界　222
森林構造　87
森林サバンナ広域分布属　87
森林性広域分布蝶　86
森林性種　87, 235
森林性生息場所　235
森林生態系　87
森林性蝶　87
森林性の　235
森林性の生息地　87
森林性の蝶　235
森林地　87, 235

森林地域 235
森林地帯 235
森林的環境 87, 235
森林天幕毛虫 87
森林に覆われた景観 87
森林の周縁域 87
森林破壊 56
森林伐採 56
森林放牧地 235
森林マトリックス 87
親和性 15
親和性クロマトグラフィー 15

ス

巣 144
随意移動 80
水域 232
推移平衡 197
水銀灯 132
遂行 160
水酸化カリウム 169
水質汚染 48
水質浄化 232
随時的休眠 21
衰弱 57
衰弱化 55
衰弱する 43
水蒸気蒸留法 207
推進する 228
推進力 66
水生生物群集 208
水生無脊椎動物群集 22
髄層 132
水素炎イオン化型検出器 83, 84
推測 204
推測航法 158
推測する 204
衰退 55
水体 232
錐体視細胞 46
錐体視物質 46
垂直感染 229
垂直線 229
垂直層別 229
垂直伝達 229
垂直伝播 229
垂直に伝播する 221
垂直破線 229
垂直分布 229
スイッチ超遺伝子 213
スイッチを切る 213
推定アミノ酸配列 170
推定 ORF 176
推定オープンリーディングフレーム 176
推定確率 76
推定機能 176
推定距離 76
推定系統樹 76
推定結合タンパク質 176
推定された系統樹の信頼性 183
推定されるエクソン 176
推定される多様な役割 176
推定種 176
推定上のエクソン 176
推定上の ORF 176
推定上のゲノム規模水準 176
推定上のゲノムワイドな水準 176
推定(される)タンパク質 108
推定値 76
推定転写制御因子 176
推定ブレークポイント 176
推定密度 76
推定量の性能 160
水田 187
水田地帯 156
水道水 214
水分喪失 232
水分泌 85
水平感染 105
水平層 105
水平伝達 105
水平伝播 105
水平分布 105
衰亡 55
衰亡しつつある蝶 56
水胞(水疱)状の構造物 33
衰亡の危機にさらす 72
衰亡の空間的パターン 202
衰亡の空間分布パターン 202
ズイムシ(髄虫) 207
水溶性の 232
水溶性の細胞壁 22
水流 232
水路 232
推論にすぎない 204
水和 107
数学的表現 130
数日後 82
スーパー遺伝子 212
スーパーオキシドディスムターゼ 212
数量的反応 150
数量分類学 150
頭蓋骨 200
図解図 191
姿を消す時刻 219
姿を見せる時刻 219
スカヒュウム 190
スカラベ 190
図鑑 164
スキャンサンプリング法 190
スクリーニング 191
スクリーン付き蓋 191
スクロース 211
スクロース溶液 211
スケールバー 190
スケルトン 200
スケルトン光周期 200
少し楕円形状の眼状紋 200
図示 108
すす色の 112
すすぎ落とす 187
スズメガ 100
スズメガ科 205
スズメガ科の蛾 205
スズメバチ 105, 231
スタイロフォーム厚板 209
スタンダードサンプル 207
すっきりとした 208
スティグマジー 208
ステージ 207
ステージ別個体群解析 18

ステップ的環境 207
ステレオグラム 208
ステレオパワーアンプ 208
ステンレス鋼製フック電極 207
ステンレススチール製二次電極 192
ステンレススチール製フック電極 207
ストレス因子 209
ストレスが多い気象条件 209
ストレス誘導性揮発性物質 209
ストレス誘導性揮発物質応答 209
ストレス要因 209
ストレプトマイシン 208
巣仲間認識 144
砂時計型生物時計 106
砂時計型タイマー 106
すなわち 142
スニーカー 200
スパーのエポキシ樹脂 206
スパッタにて被覆した 206
スパッタリングコートした 206
素早く飛び回る 178
すばやく飛ぶ 81
素早く逃げる 64
スパルト遺伝子 189, 202
図版 165, 166
スピアマンの相関 203
スピーカー 203
スピロプラズマ 205
スプライシング 205
スプライシング型 205
スプライス型 205
スプライス信号 205
スフラギス 205
スペインヒョウモン 177
スペクトル感受性 204
スペクトル感度 204
スペクトル差 204
スペクトル多様化 204
スペクトル調整効果 204
スペクトル調整部位候補 35
スペクトル調節部位 204
スペクトル的にシフトされた系統 204
スペクトル同調 204
スペクトル反射率 204
スペクトル表現 204
スペクトル表現型 204
スペクトル偏移系統 204
スペクトル変化 204
スペクトル変動 204
スペシャリスト 203
スペシャリスト種 203
スペシャリスト植食者 203
スペシャリストである植食者 203
すべて雄が入っているケージ 16
すべて雄の同腹仔の産出 173
すべて雌の同腹仔 16
すべて雌を産生する母系群 16
SMART 法 213
住みかとなる 100
住みにくかった気候 112
すみわけ 98
住む 167, 185
スモールスキッパー 200
スモールセセリ 200
スライディングウィンド手法を用いた系統解析 200
スライドグラス 200
擦り切れた 236
刷り込み 109
すりつぶす 97
擦り減った 236
スリランカ 206
鋭い味覚 196
鋭くとがった 196
寸法 200

セ

背 25
性 90
世 75
生育期 97
生育条件 97
成育上の強い制約 209
成育達成度 160
生育地 98
性依存選択 195
斉一 75
斉一化 75, 226
斉一性 226
斉一的羽化 214
斉一的発生 214
斉一の 214
成因的相同 104
精液 71
精液タンパク質 194, 196
精液注入 194
精液中のタンパク質の量 18
精液物質 194
精液を貯蔵する 208
生化学 29
生化学的進化 29
生化学レベル 29
正確確率 78
正確な解答 13
正確なデータ 13
生活型 123
生活形 123
生活環 123
生活環多型 123
生活環特性 123
生活環の型 123
生活圏 123
生活現象 123
生活史 123
生活史形質 123
生活資源 123
生活史戦略 123
生活史多型 123
生活史特性 123
生活史の型 123
生活史の謎 142
生活の多様性 64
生活様式 123
性間の遺伝的相関 28
性間の非斉一化 24

性関連遺伝子 196
正規化 148
正規化された遺伝子発現（量） 148
正規化された計数データ 148
正基準標本 103
正規の宿主 182
正規分布した分散 148
正逆交配 180
正逆交配の F1 型 180
正逆交配の F1 子孫 180
正逆パターン 180
精球 205
制御 49
制御機構 49
制御筋 182
制御配列 182
生気論 230
整形パルス 196
性決定 195
性決定遺伝子 196
性決定カスケードの下流 66
性決定システムの下流 66
性決定と性分化 195
性決定の支配制御因子 195
性決定のマスター調節因子 195
制限酵素 186
制限酵素断片長多型 186, 187
制限酵素認識部位に関わる DNA 177, 186
制限断片長多型 186, 187
正弦波 198
制限(酵素切断)部位 186
生後 1 日 152
精孔 134
成功した増殖 211
生合成 29
生合成経路 29
生合成制御遺伝子 182
整合性のある枠組み 42
整合的な枠組み 42
精巧な仮説 71
精孔の 134
性語尾 90
正誤表 50
生痕化石タクソン 108
性差 195, 196
性差のある遺伝子発現 195
性差発現 90
生産者 173
生産する 129, 203
生産性勾配 173
青酸生成能 53
静止 177
精子 205
精子 - 205
静止位置 186
静止雄 160
精子(間)競争 205
精子形成 205
静止行動 160, 199
静止時間 160, 199
静止姿勢 186
精子修飾 137
精子数 205
性質 38
性質不明の 225
生死動態 30
精子の生産 205
精子の優先性 205
静止場所 160
脆弱線 123
精子優先度 205
成熟 130, 187
成熟期 130
成熟した卵巣 130
成熟卵 130
成熟卵母細胞 130
正常化選択 148
正常型 148
星状樹 207
正常でない位置に起こる形 87
正常な羽化率 148
正常な雌性生殖器官 148
正常な縞状バンド列 148
正常な精巣 148
正常な性比 148
正常な同腹仔 148
正常な母系群 148
正常な卵巣 148
正常に伸びた翅 148
正常蛹 148
生殖 184
生殖異常 184
生殖カースト 184
生殖核倍加型 90
生殖(的)隔離 184, 196
青色型の幼虫 30
生殖活動 184
生殖下板 107, 120, 152, 210
生殖器 93
生殖器官 184
生殖器官組織 184
生殖器官の発育 184
生殖器官複合体 184
生殖基節 96
生殖器付属腺 12
生殖休眠 184
生殖系 184
生殖系列細胞 95
生殖現象 184
青色光 30
生殖行動 184
生殖細胞 95
生殖細胞系列の細胞 95
生殖細胞で決定された性 95
生殖肢基節 96
青色色素 30
青色翅の雌モデル 30
青色シフト系統 30
青色スペクトルシフト 30
青色スペクトル範囲へのシフト 30
生殖制御による防除 184
生殖節 96
生殖腺の発育 96
生殖操作 184
生殖投資 184
生殖能 82, 184
生殖表現型 184
青色ビリン色素 30
青色ビリン色素の生成 30
青色ビリンの蓄積 30
青色ビリン量 18

生殖部位　93
生殖付属腺　12, 22
生殖付属腺上清　22
生殖付属腺の貯蔵器　22
生殖付属腺分泌物　12, 22
生殖弁　93
生殖瘤状突起　93
生殖力　81
精子を貯蔵する　208
精製　181
精製する　176
精製タンパク質　176
性染色質体　195
性染色体　195
性選択　196
性選択の副産物　34
整然とした研究　134
整然とした調査　134
精巣　217
製造する　129
清掃動物　190
生息域　98, 123, 178
生息域外保全　78
生息域内保全　109
生息域の拡大　178
生息域の縮小　178
生息域の変化　98
生息環境　98
生息環境依存性　48
生息環境適正　99
生息環境適正図　99
生息環境の悪化　98
生息環境要因　98
生息空間　146
生息圏　98
生息圏内の定住性　178
生息好適性地図　99
生息場適性　99
棲息する　67
生息する　112
生息生物　98, 112
生息帯　123
生息地　98, 181, 209
生息地隔離　98
生息地管理　98
生息地数　98
生息地適合性　99
生息地適正　99
生息地内の定住者　178
生息地の悪化　98
生息地の改変　98
生息地の開放性　152
生息地の季節的変動性　192
生息地の空間分布　202
生息地の減少　98
生息地の構造　99
生息地の好適性地図　99
生息地の個数　98
生息地の質　98
生息地の周縁　98
生息地の消失　98
生息地の喪失　98
生息地の造成　98
生息地の測定変数　98
生息地の多様性　228
生息地の特異性　99
生息地の特殊化　99
生息地の名残　98
生息地の比較測定項目　98
生息地の分断化　98
生息地の分布構造　99
生息地の利用可能性　25
生息地破壊　98
生息地パッチ　98
生息地パッチの占有率　99
生息地変化　98
生息場所　158
生息場所鋳型説　99
生息場所特定者　98
生息場所特定性の種　99
生息場所特定性の蝶　98
生息場所の操作　98
生存　212
生存価　212
生存可能な子孫　229
生存曲線　213
生存時間解析　212
生存上の利益　212
生存する　210
生存する子孫　212
生存度　229
生存分析　212
生存率　212, 229
生存力　229
生存力選択　229
生存力のある　229
成体　14
生態　69
生体　153
生体異物の相互作用　236
生体外で　109
生体外での試験　78
生態化学　68
成体が完成した幼虫　89
生態学　69
生態学的撹乱　68
生態学的結果　68
生態学的地域　69
生態学的変数　69
生態学的要求事項　69
生態学的予測　68
生態型　69
生態型コロンビア　43
生体機構　153
生態区域　123
生態系　69
生態系エンジニア　69
生態系(の)機能　69
生態系サービス　69
生態系主導のアプローチ　69
生態系の撹乱　68
生態系のバランス　68
生態系モデル　69
生態工学　68
生態写真　69
生体重　88
生態進化ダイナミクス　68
生態進化の動態　68
生態(学的)選択の副産物　34
生態地理学　68
生態的意義　69
生態的因子　68
生態的解放　69
生態的環境　68
生態的機会　69
生態的機能　68
生態的空間　69
生態(学)的群集　68
生態的結末　68

生態的個体群動態 69
生態的色彩変異 68
生態的条件 68
生態的多様化 68
生態的地位 69, 146
生態的特殊化 69
生態的特性 68, 69
生態的バランス 68
生態(学)的分岐 68
生態的平衡 68
生態的役割 69
生態的誘導多発生 69
生態的要因 68
生態的要求 69
生態で定義した個体群動態 69
生態転換の副産物 34
生体内エレクトロポレーション法 109
生体内で 109
生体内電気穿孔法 109
生態(学)モデル 69
性致死遺伝子 195
性致死的遺伝子 196, 213
成虫 14, 109
正中 132
成虫羽化時期 14
成虫羽化数 150
成虫羽化の時点 166
成虫越冬 14
成虫芽 108
成虫型 109
成虫期 14
成虫休眠 14
成虫形状 14
成虫形態 109
成虫原基 108
成虫個体群 14
成虫個体数 14
成虫翅 14
成虫翅の帯部分 26
成虫翅の帯領域 26
成虫出現期 14
成虫寿命 14
正中線 132, 135
正中線により近い 131
正中線により近い外側 121
成虫体重 14
正中断面 132
成虫での越冬 14
正中の 133
成虫の羽化 14
成虫の羽化時期 219
成虫の形態 14
成虫の好蟻性 108
成虫のサイズ 14
成虫の寿命 14
成虫の生活史段階 14
成虫の成熟 14
成虫の生殖能力 184
成虫の性特異的可塑性 14
成虫の体重 14
成虫の蝶 14
成虫の年齢 15
成虫の繁殖史 184
成虫の繁殖能力 184
成虫の頻度 88
成虫の離散世代 147
成虫盤 108
成虫表現型 109
成虫雌 14
正中面 133
成長 58, 97
生長 97
成長が停止する期間 108, 115
成長期 97
成長休止の時間間隔 108, 115
成長形質 98
成長条件 97
成長する植物組織 229
生長速度 58
成長速度 98
成長速度の差異 98
成長遅延 58
成長・発育を阻害する物質 98
成長率 58, 97, 98
成長量 97
生長力のある揮発性物質 229
生長力のある部分 229
精通性 81
制定 48
制定する 72
性的いやがらせ 196
性的隔離 196
性的隔離に関する形質の遺伝的連鎖 92
性的干渉 196
性的差異 196
性的受容可能な状態 196
性的受容可能なパートナー 196
性的体格二型 206
性的体長二型 196
性的対立 196
性的通信 196
性的二型(性) 196
性的二型指標 191, 196
性的二型の産出 173
性的二型の翅形態 196
性的二型の体色 60
性的に交尾可能な状態 196
性的に交尾可能なパートナー 196
性的に中間型の形質 196
性的に中間の交尾器 196
性的パートナー 196
性的発育時間二型 196
性的モザイク現象 196
晴天 83
制度 48
性淘汰 115, 196
正当な修正名 118
正当な目的 122
性特異的な mRNA スプライシング 196
性特異的な可塑性 196
性特異的なメラニン合成 196
性特異的反応 196
性の一致 90
正の環境要因 168
正の関連 168
正の機能 168
正の傾斜 168
正の自然選択 168
正の自然選択の継承 168
正の選択圧 168

正の相関　168
正の対照(区)　168
正の淘汰圧　168
正のフィードバック　168
青白色の透明斑　221
青白スクリーニング　233
性斑　32
性比　195
性非対称な近親交配　189, 195
性比淘汰　195
性標　32, 191, 195, 196
性標(斑)　195, 208
性フェロモン　195
性フェロモン形成　195
性フェロモン単位　195, 206
西部チベット　232
生物　51, 153
生物 -　29
生物学　29
生物学的因子　29
生物学的差異　29
生物学的種　29
生物学的種概念　29, 33
生物学的侵入　29
生物学的多様性　29
生物学的特徴　29
生物学的に　29
生物学的に重要な拡大　29
生物学的反復　29
生物学的反復実験　29
生物学的要因　29
生物学的理由　29
生物型　29
生物間相互関係　29
生物間相互作用　29
生物気象学　161
生物季節学　161
生物群系　29
生物経済学　29
生物系統学　29
生物圏　29
生物検定　29
生物個体　110
生物資源調査　29
生物資源目録　29
生物指標　29
生物種　29
生物情報解析　29
生物情報科学　29
生物情報研究解析　29
生物試料　29
生物相　29
生物相表現空間　69
生物多様性　29
生物多様性研究　29
生物多様性条約　36, 49
生物多様性の起源　154
生物多様性の代表的な構成要素　184
生物多様性の保全　29, 47
生物多様性評価　29
生物調査　29
生物地理学　29
生物地理学的研究　29
生物地理区　29
生物的隔離　29
生物的ストレス　29
生物的ストレスと非生物的ストレスとの共同効果　42
生物的相互関係　29
生物(学)的防除　29
生物的要因　29
生物時計　29
生物の多様性評価　29
生物の知覚　153
生物の表現型　153
生物の分布実態　153
生物表現型　153
生物標本　29
生物分布　63
生物分類学　29
生物目録　29
生物(体)量　29
正負電極　168
正文　151
性分化ホルモン　196
性別　90
性別特異的な発現　90
精包　205
精包タンパク質　205
青方偏移　30
精密な時計　209
生命の多様性　64
生命表　123
性毛　196
生毛　222
生毛細胞　222
性モザイク　195
性モザイク個体　196
性モザイクの発生　151
性紋　191
誓約する　166
セイヨウアブラナの種子　178
セイヨウイラクサ　208
セイヨウナシ形の　159
セイヨウナシ状の眼状紋　159
セイヨウワサビ　105
生来の多型系統　143
生理　164
生理学　164
生理学的アプローチ　164
生理学的因子　164
生理学的研究　164
生理学的・行動学的実験　163
生理学的差異　164
生理学的変数　164
生理学的要因　164
生理機構　164
生理(的)条件　164
生理状態　164
生理食塩水　164
成立要因　36
生理データ　164
生理(学)的記録　164
生理的形質　164
生理的コスト　164
生理的差異　164
生理(学)的時間　164
生理的調節　164
生理(学)的適応　163
生理的特性　164
生理的に調整された一連の行動　164
生理的に反応する　185
生理的媒介　164

生理的背景 164
生理的反応 164
生理的変化 164
生理的変数 164
生理的誘導多発生 164
生理的リサージェンス 164
生理特性 164
青緑色化 30
青緑色の 224
整列 16
整列群 48
整列させる 16
整列された曖昧領域 16
整列被覆(率) 16
整列網羅(率) 16
性連鎖 196
世界各地の分布 50
世界自然保全モニタリングセンター 232, 235
世界的規模 95
世界の生物多様性喪失 95
世界の名蝶 235
セカンドグロース林 192
赤外域 111
赤外光による生体イメージング 29
赤外光による生体センシング 29
赤外分光 111, 116
碩学 173
赤褐色の蛹 181
脊索 149
脊索動物 39
脊索動物門 39
積算温度 12, 53
積算温量 53
積算値 12
赤紫色 127
積翅目 166
赤色光 181
赤色色素濃縮ホルモン 181
脊椎動物 229
脊椎動物のゲノム 229
脊椎動物の生物学 229
脊椎動物の捕食者 229
脊椎動物網膜 229
赤道 75
赤道直下の 75
責任著者 50
蹠片 165
赤緑色覚 181
セクシャルハラスメント 196
セグメント 193
セクロピアサン 36
セクロピン 36
セスキテルペン 195
セセリチョウ亜科 97
セセリチョウ科の蝶 200
セセリチョウ上科 200
世代 91
世代数 150
世代を越える防御誘導 221
世代をわたる防御誘導 221
背丈の低い 126
節 192, 193
石灰岩台地 123
石灰岩の草地 123
石灰質土壌 34
石灰土壌の草地 34
積極的な追飛 113
接近 21
接近確率 22
節減 158
接合後隔離 169
接合子 237
接合子の死亡率 237
接合する 14
接合前隔離 170, 171
接合部 118
摂氏度 37
絶翅目 237
接写レンズ 127
接種 112
接種効果 69
摂取する 112
接種的放飼法 112
舌鞘 95
絶食 207
接触域 48
接触化学感覚器 48
接触化学受容器 38, 48
摂食活性 81
摂食活動 81
摂食期 82
摂食行動 81
接触作用を及ぼす 36
摂食刺激 82
摂食刺激物質 161
摂食傷害 82
摂食する植食者 82
摂食選好性 82
摂食阻害物質 21, 82
接触帯 48
摂食中 81
摂食停止 81
絶食の影響 69
摂食場所 82
接触フェロモン 48
摂食誘引物質 81
接する 31
節足動物 23
節足動物の転写開始点 23
節足動物門 23
絶対共生的関係 150
絶対項 12
切断型 223
切断状 223
切断する 62, 195, 197
接頭語 171
説得力のある証拠 161
(Z)-3-ヘキセニルビシアノシド 236
(Z)-3-ヘキセノール 236
Z染色体 236
Z染色体と連鎖した単コピー核遺伝子座 236
ZW型 237
接尾語 211
絶壁 41
切片 114, 193
節片 191
節片化 191
説明図 108
説明変数 79
絶滅 79
絶滅-移入動態 79
絶滅危惧種 72, 218
絶滅危惧の野生生物群 218
絶滅した 79
絶滅した台湾亜種 79

絶滅種　79
絶滅する傾向　173
絶滅 - 定着動態　79
絶滅の恐れのある野生生物群　218
絶滅のおそれのある野生動植物の種の国際取引に関する条約　40, 49
絶滅のおそれのある野生動植物の種の保存に関する法律　13, 122
絶滅の危機にさらす　188
絶滅の瀬戸際　33, 229
絶滅リスク　79
絶滅率　79
節約な系統樹　158
節約に関する情報　158
設立する　75
ゼニアオイ　128
ゼネラルプライマー　91
背の　65
背の高い草地　214
セバスチアナの種　118
セピアプテリン　194
ゼフィルス　237
ゼブラフィッシュ　54, 237
狭い接触域　142
狭い領域　142
セミ　40
セメント層　37
ゼラチン版印刷　100
セリン　195
セリンからアラニンへの置換　195
セルフリー　36
セルロース　37
セレンゲティ国立公園　195
ゼロ成長の等傾斜線　237
セロトニン　195
浅　212
前　51
前 -　86, 170, 172
全 -　103
全暗黒　55
遷移　211
前胃　174
繊維芽胞成長因子受容体タイプ3　82, 83
扇域　228
繊維質の　83
ゼン遺伝子　237
前縁　20, 50
全縁　73
前縁(部)　20
前縁 - 後縁軸　20, 21
前縁 - 後縁方向に取り巻く鱗粉リング　20
前縁 - 後縁方向へ　20
前縁溝状線　20
前縁室　50
前縁褶　50
前縁部　50, 122
前縁脈　34, 42, 50
前縁脈のひだ　50
前横脈　20, 178
線画　66
旋回角(度)　224
前外側盾板溝状線　20
旋回飛翔　40
旋回飛翔圏　40
旋回飛翔個体　40
旋回飛翔領域　40
前下胸板　171
前下唇　171
前基節溝状線　170
前記の　15
前脚　86, 88, 173
前脚の拡大　86
前脚ふ節　87
前脚ふ節感覚器　87
前脚ふ節受容器　87
前休眠期　170
前休眠期の交尾　170
前胸　174
前胸腺　174
前胸腺刺激ホルモン　174, 175
前胸腺ホルモン　174
前胸腺抑制ホルモン　175
前胸側斜架　75
前胸側板　173
前胸凸縁　173
前胸背襟部　173
前胸背側片　173
前胸背板　173
前胸背板溝状線　173
前胸背板の　173
前胸板　157, 166
前胸腹板　174
漸近種数　24
漸近種数推定量　24
漸近線　24
漸近的種数　24
漸近的推定量　24
漸近的に　24
漸近的に等価　24
先駆個体　165
先駆種　165
先駆樹種　165
先駆植物　165
先駆体　170
前駆体貯蔵庫　170
前駆体ドーパミン　170
前駆物質　170
前駆物質貯蔵体　170
前駆物質ドーパミン　170
前区脈　144
線形　124
線形一致　43
線形回帰　124
線形回帰分析　124
線形階層　124
線形混合効果モデル　124
前脛節　174
線形的形態測定学　124
線形の　124
扇形飛行　190
線形ランプ　124
全ゲノムサイズ　220
全ゲノムショットガン　232, 233
全ゲノムショットガンコンティグデータバンク　233
全ゲノムショットガンデータベース　232
全ゲノムショットガン配列　233
全ゲノム重複　232, 233
全ゲノム配列　233

全ゲノムバクテリア人工染色体ライブラリー 233
漸減 96
ゼンケンベルク博物館 194
選好 171
前口式 173
前後軸 20, 21
選好刺激 171
選好指数 171
選好指標 171
閃光色 84
先行する 170
選好性実験 171
選好性指標 171
選好に関する遺伝子座 171
前後方向に 20
潜在(的)害虫 169
全採集種リスト 124
先在情報 170
潜在的隠蔽種 169
潜在的拡大 169
潜在的機能分化 209
潜在的脅威 169
潜在的雑種 169
潜在的侵入種 169
潜在的増大 169
潜在的な逆位のサイズ 200
潜在的な候補(遺伝子) 169
潜在的な聴覚器官 169
潜在的な目標 169
潜在的に不和合な系統 169
潜在的にリスクの高い種 169
潜在的有効名 169
潜在繁殖成功度 169
全載標本 233
潜在力 169
センサス 37
潜時 121
前肢 86, 88
前翅 80, 86, 87
前翅縁毛色 87
前翅芽 86
前翅基骨 26
前翅基部 26
全色盲 13
前翅後縁 66
潜時帯 178
前翅長 87
前翅頂のとがり 166
全実験期間 73
前翅の色 87
前翅の帯 87
前翅の構造 87
前翅の最上端部 87
前翅の色彩遺伝子 87
前翅の大斑点 87
前翅の白色部分を黄色に変えた 233
前翅の発生 58
前翅背面 65
前翅表面 87, 227
全射域 75
前若虫 173
先住効果 69
先住者 185
先住者の雄 185
先取権 172
先取権の原則 172
先取権の原理 172
先取権の法則 172
戦術 214
前述の 15
前盾板 171
洗浄 41
線状観察路 123
全証拠解析 220
扇状の 84
線状の調査経路 123
扇状部 228
全照明 124
染色過程 164
染色糸 39
専食者 203
染色する 207
染色体 39
染色体位置 39
染色体外ゲノム 79
染色体間隔 39
染色体逆位 39
染色体ゲノム 39
染色体(の)構造 39
染色体再配置 39
染色体上で近接した 41
染色体進化 39
染色体切断 39
染色体地図 39
染色体調査 39
染色体並びが逆向き 39
染色体に組み込まれて伝達された内在性ウイルス 73
染色体の構成 39
染色体バンド法 39
染色体分染法 39
染色体マッピング 39
染色体融合 39
染色体レベル 39
前翅裏面 226
先進国 58
前進進化 18
漸進的進化 96
漸進的変化 96
全身の 214
前伸腹節 173
前伸腹節の 173
前伸腹節隆起線 173
センス鎖 194
潜性形質 180
先成現象 174
潜成虫 161
前成虫期 170
前前側板 107, 171
前前腹板 171
前前腹板線 171
漸増 180
前側板 75, 171
前側板の 75
前側片 74
前側片の 74
全組織標本 233
先祖状態再構築 19
先祖伝来のコウモリ探知器 19
全体が変化した斑紋 233
前腿節 173
全体像 233
全体の dN/dS 比 95
全体論的概念モデル 103
選択 193
選択圧 193

選択遺伝子　193
選択係数　193
選択交配　23
選択差　193
選択された部位　193
選択指数　171
選択的一掃　193
選択的恩恵　193
選択的干渉　193
選択的スプライシング　17
選択的中絶　193
選択的に　171
選択的ハラスメント　193
選択的表現型　17
選択的優勢　193
選択的利益　193
選択に対する反応　185
選択の起源　202
選択の分岐点テスト　32
選択(的)伐採　193
選択不利性　193
選択優位性　193
選択有利性　193
選択要因　193
先端　80, 122, 219
先端－　21
前端　20
先端の　21
先端の細い毛　21
先端部　122
全地球測位システム　95, 96
全地球測位網　95, 96
前膣片　120
前膣ラメラ　120
センチネル　194
センチモルガン　37, 42
線虫　34, 144
線虫のゲノム　144
前肘脈　20, 52, 53
前腸　86
全長　89
全長 ORF　89
全長オープンリーディングフレーム　89
全長配列　89
前低温馴化期間　170
選定期間　39
選定基準標本　122
前提条件　170
前適応　170
先天性免疫　112
前頭　88
蠕動運動　236
扇動する　113
前頭－頭盾縫合線　89
全同胞種　89
尖突起　53
前途有望な研究分野　173
セントラルドグマ　37
セントロメア　37
全日中期間　73
先入観　28
全能性　220
前培養をする　171
選抜　193
尖筆状の　209
前部－　20
潜伏期間　121
前腹板　74, 75, 171
前腹板の　74
前腹板隆縁　74
前腹板隆起線　74
前部伸縮突起　20
前ふ節　171, 174
前部の　20
全部の近縁種　89
前分枝　20
前報　171
前方眼状紋の切除　71
前方後側板　170
前方伸縮突起　20
前方に　152
前方に走行する　188
前方の　20
前方の位置　20
前方の眼状紋　20
前方の基翅甲　20
前方の基翅節片　20
全北区　103
繊毛　40
繊毛型　40
繊毛型オプシン　40
繊毛型光受容細胞　40
繊毛状の　40
専門家　203
占有空間　151
占有行動　217
占有個体　151
占有者　151
占有する　159
占有性　151
占有地　151
占有地域　151
占有パッチ　151
潜蛹　161
前蛹　171
前蛹期での幼若ホルモンの噴出　171
前蛹状態　171
潜幼虫　161
前蛹の　171
戦略　208
先例のない　227
先例のない解決策　227
前例のない解決策　227

ソ

相　161
層　208
双－　28, 58
双安定現象　30
相違　61, 63
走獲物性　171
騒音計　201
増加した体重　110
増加する　31
増加するタイター　110
増加する力価　110
走化性　38
相加的遺伝子型値　14
相加的遺伝分散　14
相加的樹形図　14
相加的な　14
相加的に　14
痩果の冠毛の長さ　157
痩果の分散距離　13
層化無作為抽出法　208
増加率　110
相関　50
相関因子　113

相関がある 50
相関角度 50
相関関係 50
相関関係がある 50
双眼鏡 28
相関係数 50
双眼顕微鏡 28
双眼実体顕微鏡 208
双眼実態顕微鏡 208
総乾重量 220
増感する 113
相関旋回 50
総乾燥重量 220
相関的な 50
爪間突起 72
壮観な配列 204
爪間盤(板) 17, 22
相関ランダム歩行 50
相関連多型 161
相関連変異 161
相関をもつランダム歩行 50
想起 180
早期発見 68
雑木林 33, 50, 191
雑木林の撤去 191
雑木林の伐採 50
双胸遺伝子 30, 34
増強された紫外線光に対する色覚 24
遭遇する 72
遭遇率 72
草原 97
草原遺存属 97
草原広域分布属 97
草原性生息地 97
草原性蝶 97
草原地帯 97
草原の蝶の衰退 56
草原的環境 97
相互 - 42
相互移植実験 180
相互遺伝子欠損 180
相互遺伝子消失 180
相互遺伝子喪失 180
草稿 129
総合 214
走光性 163
総合的害虫管理 113, 116
総合的生物多様性管理 108, 113
総合的理解 214
総行動範囲 220
総合防除 113
相互擬態種 42
相互検定 180
相互交雑 180
相互交配 114
相互作用 180
相互作用項 114
相互試験 180
相互相関因子 113
相互に - 113
相互の - 16
相互の追跡 141
相互(に)排他的 141
相互パターン 180
走根 208
操作 129
走査型高調波レーダー 190
走査型電子顕微鏡像 190
走査型レーザー振動計 190
操作基準 152
操作原理 152
走査サンプリング法 190
操作上の分類単位 152, 154
操作的判定基準 152
走査(型)電子顕微鏡 190
走査(型)電子顕微鏡検査 190
走査(型)電子顕微鏡法 190
操作能力 129
造山運動 154
創始者 87
創始者効果 87
創始者事象 87
創始者選択 87
走日性 100
双翅目 60
総翅目 218
双翅目の昆虫 60
双翅目の種 60
操縦性 129
総重量 220
早熟変態 174
早熟蛹 170
送受信機 221
早春 68
早春季生物 206
相称系の縞状バンド 213
相称系の正中線 135
相乗作用 214
双子葉植物 59
相称的樹形 213
相乗的に 214
相乗的に作用する 13
草食 101
装飾 229
増殖 62, 184
増殖効率 184
草食(性)昆虫 101
草食(性)動物 101
草食動物によって誘導された揮発性アルドキシム 101
草食(性)動物の群集 101
草食動物媒介による淘汰 101
増殖パターン 173
増殖率 184
走性 215
総説 186
想像する 74
創造的所産 23
想像できる 46
創造物 51
増大 24, 75
相対位置 183
相対確率 183
相対資源量 182
相対湿度 182, 187
相対尺度 183
相対種個体数 183
相対照度 182
増大する刺激強度 110
相対成長 16
相対成長速度 182, 187
相対成長率 187
相対速度テスト 183
相対的感受性 183

相対的交尾確率　182
相対的個体数分布　182
相対的個体数分布の均等度　77
相対的成長率　182
相対的生物量　182
相対的存在量　182
相対的定量　183
相対的投資量　182
相対的な資源量　183
相対的な豊富さ　182
相対的な用語　183
相対的に　182
相対的に少ない　183
相対的ニッチ空間の制御　182
相対的バイオマス　182
相対的不活性　182
相対的有効性　182
相対方向　182
相多型　161
草地　97, 213
早朝　68
想定された位置　170
想定上の縄張り　212
相同　104
相同異質形成　104
相同異質形成遺伝子　104, 106
相同遺伝経路　104
相同遺伝子　104, 154
相同化　104
相同関係　154
相同グループ　154
相同構造　104
相同(的)神経枝　104
相当数の調節要素　198
相同性　104
相同性検索　104
相同染色体　104
相同体モデル化　104
相同的構造　104
相同転写配列　104
相当な差異　210
相当な適応度費用　210
相同マーカー　104
相同(性)モデル　104
相同(性)領域　104
相同連鎖群　104
挿入　113
挿入器　14, 161
挿入器口　154
挿入器腹板　118
挿入欠失　110
挿入サイズ　113
挿入図　113
挿入名　115
創発　72
創発される個体群パターン　72
創発性　72
層板　120
相反交雑　180
総尾目　218
走風性　19
増幅産物　18
増幅断片　18
増幅断片長多型　15
増幅断片長多型マーカー　18
増幅 DNA 断片　18
増幅反応　18
送粉者　167
送粉者の誘引　24
送粉シンドローム　167
送粉生態学　167
相変異　161
双峰　28
双方向 SSH 法　28
双方向吸収スペクトル　224
双方向サブトラクションライブラリー　28
双峰性の活動期　28
相補的性決定　45, 52
相補的 DNA　36, 45
草本　101
草本植物　101
草本性つる植物　230
草本と低木との変化に富んだ混合　228
草本被覆率　101
造雄腺　19
造雄腺ホルモン　15, 19
総輸卵管　44
蔵卵数　81
相利型　141
相利共生　141
相利共生者のアリ　141
相利共生的相互作用　141
相利共生的　141
造林システム　198
造林施業システム　198
造林地　15
造林プログラム　15
双列型の鉤爪　30
ソース　201
ソース - シンクモデル　202
ソース生息地　201
ソースハビタット　201
ソーレー帯　201
ゾーン　237
阻害された胚形成　23
阻害条件　112
阻害要因　112
咀顎目　175
遡河性の太平洋サケ　18
ソキウス　201
属　90, 94
族　222
- 族　11
側 -　157
属あたりの種数　203
側圧した　46
側域刺毛板　121
属階級群　94
属階級群名　94
属格語尾　93
属間交雑　114
属間雑種　114
側基節溝状線　157
側棘　121
側系遺伝子　157
側系統群　157
側系統種　157
側系統の　157
側頸板　157
側鉤器　100
側後盾板　121
側後背板　121
側後方ニューロン　121, 126
側鰓　121

即座に現れる影響　109
側刺　121
測時機構　219
測時機能　219
側翅突　166, 176
測時反応　219
側刺毛　156
側所的　157
側所的亜種　157
側所的系統　157
側所的分布　157
側唇　93
促進因子　173
側心体　50
族水準　222
属数　91, 150
側舌　157
側単眼　121
測定結果　186
測定増幅器　131
測定の標準(単位)　137
速度　204
側突起　201
側突出縁　121
速度と方向　229
速度分布　63
属内競争　115
属内の亜属分化　115
側の　121
属の所属　91
側背板　121, 157
側背板筋　216
側板　166
側板溝　166
側板溝状線　166
側板翅突起　166, 176
側板の　166
側板の外葉　166
側板の翅突起　166
側板片　166
側板縫合線　166
側板葉　166
側部　121
側腹鰓　121
側腹板　166
属分布　63
属分布圏　63
側片　121, 157
側方に投影する　173
側方ニューロン　121, 124
側方の　121
側方部　121
側方部位　121
側方部分　121
側方抑制　121
俗名　44, 229
属名　91, 94, 142
族名　142, 222
側面日光浴　121
側面を接する　84
側輸卵管　121
側抑制　121
属よりも高位の　212
側稜　121
族レベル　222
ソケット　201
ソケット環節　220
ソケット形成細胞　201
ソケット細胞　220
ソケットなしの毛　147
ソケットを付けない毛　147
粗骨土壌　200
損なわれていないで　113
阻止因子　112
組織　219
組織学　103
組織学的断面　103
組織向性　219
組織親和性　219
組織損傷　219
組織(体系)的な　214
組織的な調査　134
組織特異的定量 PCR 法　219
組織特異的な発現　219
組織特有の定量的 PCR 法　219
組織粉砕懸濁液　104
阻止する　30, 50, 99, 114
咀嚼型昆虫　38
咀嚼口　38
疎水性化合物　107
疎水性分子　107
疎生植生　152
疎性連鎖　126
祖先(型)アミノ酸基質　19
祖先型化合物　19
祖先形質　166
祖先形質共有　213
祖先状態　19
祖先状態再構成　19
祖先状態復元　19
祖先的な幼虫の寄主植物　19
祖先伝来の化合物　19
祖先ノード　19
祖先のドクチョウグループ　19
育てる　178
外側に伸長する　79
外側の－　11
外側の着色フィルタリング色素　44
外側方向に大きくなる　173
外側方向に増殖する　173
外へ－　78
素のう　52
その後の解析　210
その後の指示　210
その後は　218
「その細胞が由来する生物個体内の本来あるべき場所」での交雑実験　109
その土地による選択　125
その中に　218
その日の時刻　219
そのままの　113
その溶液が希薄であることを示す接頭語　60
粗末な栄養　167
粗面小胞体　188
ソラーレン　175
そらす　56
ソラレン　175
粗粒面　188
疎林地帯　152
ソルビトール　201
それ自体が　160
それにもかかわらず　149
ソロモン諸島　201
存在　171

存在量　12
存在論　152
損失　50
損傷領域　54
存続可能個体群　229
存続する　210
存続する交雑　161
存続するメタ個体群　161
存続性　160, 229
存続能力　160

タ

多－　140, 167
ダーウィン　55
ダーウィンフィンチ　54
ダーウィン流淘汰　55
ターミナルフィラメント　216
態　87
代　75
退－　186
第○世代の　83
ダイアーの法則　67
ダイアナヒョウモン　58
帯域幅　26
第一胸節神経球　90
第一校訂者　83
第一校訂者の原理　172
第一次基準標本　172
第一世代　83
第一世代成虫　83
第一世代の雑種　84
第一代雑種　84
第一著者　83
第一腹部背板　193
第一節　83
大英博物館　31, 33
体液　30, 31
ダイオウショウ　126
対応する　55
対応する大きく有意なピーク　50
ダイオウマツ　126
体温　31
体温調節　218
体温調節機能　218
体温の低い巡回者　43
大腐　86
退化　182
タイガ　214
対外強硬主義　38
大顎　118, 128
体格依存の生存(率)　200
大顎の　128
体格の性的二型　206
対角破線　58
退化した器官　229
退化した相利共生的利益　229
退化する　56
退化的　56
耐寒性　43, 219
大気　24
大規模な移動パターン　120
大規模な研究　120
大規模な個体群分布　33
大規模な再編成　120
大規模な調査　79
大規模な地理的規模　120
大群　22, 41
体系学　214
台形状の音波　222
台形波のパルス音　222
退行　182
対抗手段　50
対抗戦術　51
体構造　31
退行脱皮　186
対抗適応　50
対差　156
対差異　156
大腮　118, 128
体サイズ　31
体サイズ指標　31
大腮の　128
対策　50
第三紀　217
第三紀周極要素　22
第三紀周北極要素　22
第三紀中期　135
第三紀中葉　135
第三紀北極要素　22
帯糸　26, 95, 198
台紙　156
帯糸形成　87
胎児死亡率　71
帯糸数　150
対峙する行動　46
代謝　133
代謝回路　133
代謝学　133
代謝活性　133
代謝活動　133
代謝経路　133
代謝(性)解毒　133
代謝源　133
代謝コスト　133
代謝産物レベル　133
代謝速度　133
代謝(上)の　133
代謝の起点　133
代謝費用　133
代謝変化　133
代謝変動　133
代謝レベル　133
体重　31
体重増加　110
体重の減少(率)　232
体重を量る　232
対照　49
対照遺伝子　181
対照ウイルス　49
対象外種　147
対象外の無脊椎動物　147
対象種　215
対照植物　49
対照処理　49
対称性のゆらぎ　85
対称的樹形　213
対称的な翅の損傷　213
対象とする害虫　119
対照場所　49
対照 PCR　49
対象物　150
対照幼虫　49
対照林分　49
体色　31
体色多型　164
体色変化　31, 43
対処する　55
帯糸をかける　95

大進化　127
大進化的結果　127
大進化の結果　127
対数オッズ　125
対数オッズ比　125
対数形式の雄の体サイズ　125
対数形式の照合指標　125
対数順位解析　125
対数順位検定　125
対数変換された距離　125
大豆サポニン　202
大スンダ列島　97
耐性　185, 219
大生態系　127
胎生の　230
耐性のある宿主　185
体成分　201
体制模式図　191
大西洋岸森林(南米)　24
堆積岩　193
堆積物　193
体節　193
腿節　82
腿節基部　26
腿節の　82
大草原に生息する種　170
タイター　219
代替の対立遺伝子　17
代替モデル　17
大多数　128
代置種　51
対地速度　97
タイ中部　37
体長　31
大腸菌　67, 75
大腸菌コロニー　67
体長の性的二型　196
多遺伝子　167
多遺伝子座で構成される遺伝子型の相違　141
多遺伝子座で構成されるカラーパターン構成　141
多遺伝子性形質　167
多遺伝子(性)の　167
大同小異　143
耐凍性　88
耐凍性種　88
大動脈　21
体内時計　29
体内時計の周期に影響を与える外的因子　237
ダイナミックプログラミング　67
ダイナミックレンジ　67
第二亜外縁室　192
第二化　192
第二仮説　192
第二後外縁室　192
第二次基準標本　192
第二世代　192
第二転節　223
第二の種　192
第二のピーク　192
大発生　117, 154
大発生した　117
大繁殖　117
大斑点　158
対比　49
対比較　156
対比較共祖係数　156
対比較コアンセストリー係数　156
対比較した遺伝距離　156
対比して目立つ飛翔経路　49
待避地　181
待避場所　181
代表画像　184
体表炭化水素　53
代表的な遺伝子座　184
代表的な画像　184
代表的なトレース　184
体表面積　31
タイプ　224
タイプ化　224
タイプ化の原理　172
タイプ固定　224
タイプ産地　224
タイプ種　224
タイプ宿主　224
タイプシリーズ　224
タイプ層準　224
タイプ属　224
体部の形態　31
体部の資源　31
タイプ標本　224
タイプ標本(レクトタイプ)の調査　78
大部分　128, 228
体分節　133
太平洋地域　156
大変革を起こす　187
大胞子　132
対捕食(者)回避行動　170
たいまつの火　220
帯蛹　16
太陽光　211
太陽光の強度　211
太陽光の強さ　113, 211
太陽黒点　211
太陽コンパス　201, 211
対様式遺伝的距離　156
太陽熱　201
太陽熱集熱器　201
代用品　175
太陽方位　211
太陽放射　201
代用名　210
代理　175
大陸規模　48
大陸規模での推定　48
大陸産　48
大陸性気候　48
大陸的気候　48
対立遺伝子　16
対立遺伝子転換　16
対立遺伝子特異的な発現　16
対立遺伝子のサイズ変異　16
対立遺伝子の代謝回転　16
対立遺伝子の違い　16
対立遺伝子の反復補充　183
対立遺伝子の変化　16
対立遺伝子頻度　16
対立遺伝子分岐　16
対立遺伝子変異　16
対立遺伝子変異体　16
対立遺伝子をヘテロで持つ個体　102

対立行動　50
滞留　195
大量絶滅　129
大量の　12
大量放飼法　116
大量捕獲　129
タイリンミヤコナズナ属　14
多雨の　166
多栄養段階(生物間)相互作用　141
耐える　225, 235
楕円体　71
楕円体状の眼状紋　71
多化　141
多寡　177
多回交尾　141
多回繁殖　117
多回繁殖性種　117
多回繁殖性の移住種　117
多角体タンパク質遺伝子　167
多角体タンパク質プロモーター　161
多角体病　167
多化性　141, 167
多化性集団　141
多化性の　140
他感作用物質　16
他感物質　16
多義性　17
たきつける　89
多義的電気泳動パターン　17
多義的な特性　17
妥協する　46
タギング　214
択一的スプライシング　17
卓越した　173
たくさんの草　12
たくさんの種　150
たくさんの努力　150
タクソン　215
タグ付け　214
択伐　193
択伐林業　193
竹　26
多型　167
多形型　167
多型擬態遺伝子座　167
多型系統　167
多型現象　167
多型個体群　167
多型性　167
多型的　167
多型的遺伝子座　167
多型的差異　167
多型的な違い　167
多系統　167
多系統群　167
多系統的　167
多型の　167
多型のベイツ型擬態(種)　167
多型部位　167
丈の短い芝　197
竹類植物　26
多国間環境協定　131, 140
多産な環境　89
確かめる　23, 46
多次元尺度解析　140
多次元尺度分析　140
多重侵入　141
多重整列　141
多重比較の誤り　75
多種群集　141
多少　201
他殖弱勢　154
多食性　167
多数　12
多数回繁殖　117
多数決原理総意樹　128
多数決合意樹　128
多数合意樹　128
多数著者　141
多数派支配型コンセンサス樹　128
タスキシジミ　120
多精　167
他生的な種　16
他生的な要素　16
多世代　141
唾(液)腺　189
多相同染色体の擬態構成　104
たそがれ　67
正しい原綴り　50
但し書き　36
正しく認識する　21
ただの虫　145
立ち上がり／立ち下がり　187
立木　219
立ち去る　235
立ち戻る　180
脱外被　225
脱殻　225
脱字を施す　71
脱水　56
脱水する　56
達成する　24
脱皮　68, 138, 140
脱皮液　138
脱皮殻　80
脱皮顆粒　68
脱皮ゲル　138
脱皮刺激ホルモン　68, 76
脱皮ステロイド力価　68
脱皮線　68
脱皮腺　68
脱皮ホルモン　68, 138, 174
脱皮齢期　138
たっぷりな人工果汁　18
脱糞　98
脱蛹　68
脱抑制する　61
脱抑制をきたす　61
脱落化　67
多 DNA ウイルス　167
建て込んだ地区　33
縦縞模様　209
縦の　126
タテハチョウ亜科　223
タテハチョウ科　33
タテハチョウ科の蝶　54
タテハチョウ科の分岐　150
縦方向に　122
縦方向に切断する　192
縦横比　23
妥当性　183
他動的移動　158
妥当な説明　179

ダナイドン 54
ダナエツマアカシロチョウ 51
タナグモ 106
ダニ 136
多肉植物 211
ダネット(の)検定 67
種本 177
多年生の草 160
多胚形成 167
多倍数体分岐細胞核 103
タバコ 146
タバコスズメガ 129, 219
ダビ 54
wsp 遺伝子断片 236
WntA スイッチ遺伝子 235
WGS データベース 232
W 字状の斑点 231
ダブルトン 66
多分岐 140
たぶんテリトリーを持たない種 171
たぶん二倍 224
タペータム 214
タペート 214
食べ尽くす 48
多変量点のパターン解析 141
タマオシコガネ 190
卵 70
卵が産み付けられた植物 70
卵形 154
卵形の 154
卵から成虫までの発育時間 70
卵寄生蜂を誘引する揮発性物質 70
タマゴコバチ属 *Trichogramma* のハチ 222
タマゴコバチ属 *Trichogramma* の卵寄生蜂 222
卵殺し防御 70
卵死亡数 70
卵死亡率 70
卵付きの草 97
卵で誘導された植物応答 70
卵なしの草 97
卵の一群 70
卵の生産 70
卵の共食い 70
卵の二倍体化 70
卵の孵化 70
卵の孵化成功率 70
卵の包囲化 70
卵孵化率 70
卵捕食寄生者 70
卵捕食寄生者 - 幼虫捕食寄生者 70
卵を生む 155
卵をたくさん付けた芽キャベツの葉 70
だまし花 55
騙す 86
玉虫色の 116
だめにする 119
多面発現 166
多面発現効果 166
多様化 63
多様化期間 63
多様化選択 64
多様化速度 64
多様化速度の変化 197
多様化の正味速度 145
多様化爆発 33
多様化率 64
多様性 64, 228, 229
多様性指数 64, 131
多様性指数の推定 76
多様性指標 64
多様性指標の推定 76
多様性の維持 127
多様性の起源 154
多様性の進化 77
多様体 228
多様度指数 64
多様な機能 63
多様なグループ 63
多様な昆虫目 63
多様な種 63
多様な種のホットスポット 64
多様な樹齢構造 63
多様な草本群落 63
多様な中間型の存在 78
多様なテンプレートソース 63
多様な配偶者選好 63
多様な翅パターン 63
多様な分岐 63
多様な変異 63
多様な幼生期食性 63
多様な列挙 63
多量 12, 177
多量体 167
ダルトン 54
単 - 99, 138, 226
単閾値的 GMYC 解析 206
単為生殖 138, 158
単為生殖による女王位継承システム 22, 23
単為生殖誘導 158
単位生物量 226
単位地域 226
単一 - 99
単一(の)遺伝子型 199
単一遺伝子座の遺伝 199
単一遺伝子の突然変異 199
単一遺伝子の突然変異体 199
単一 S(短波長型)オプシン変異体 199
単一塩基多型 199, 200
単一型 138
単一感覚子記録法 199
単一起源 199
単一局所性 199
単一クローン 198
単一形質の推定値 198
単一個体(種) 199
単一種属 138
単一生化学的機構 198
単一単眼優性ニューロン 138
単一の常染色体遺伝子座 198
単一のヌクレオキャプシドを含む NPV 199, 200
単一のヌクレオキャプシドを含む核多角体病ウイルス 199, 200

単一ピーク　199
単一ポジティブクローン　199
単一陽性クローン　199
単位バイオマス　226
単位複製配列　18
暖温帯性の祖先型　231
暖温帯の種　231
段階　207
段階的重回帰　208
段階的探索法　208
段階的変化　96
短角型　197
炭化水素　107
単眼　151, 198, 207
単感染の雌　199
単眼体　53
単眼内縁隆起線　171
単眼の　151
短期間　197
短期の移動　197
短期の飛翔バウト　197
探求　177
短距離シグナルの中継　183
単型　138
単系統　40, 138, 198
単系統群　138
単系統姉妹群　138
単系統種　138
単系統種概念　138
単系統性　138
単系統に感染した正常な同腹仔　199
単系統の　138
単型の　138
単型の雄　138
単系列　40
端鉤　72
暖候期の黒化表現型　132
単光センシング構造　199
単コピー核遺伝子　199
単コピー核遺伝子座　199
単コピーマイクロサテライト　199
単婚　138
探索型　159
探索行動　191
探索時間　222
探索戦略　191
探索像　191
探索的歩行　79
短鎖散在(型)核内反復配列　197
短鎖散在反復配列　197, 198
短鎖縦列反復配列　197, 208
端三角部　22
炭酸ソーダ　201
炭酸ナトリウム　201
短翅　32
端指　60
端刺　206
端糸　216
短翅型　32, 134
短時間　197
短時間の光パルス　32
短時間の飛翔　32
単雌系統　117
端室　21
短日型　197
短日型反応　197
短日効果　197
短日サイクル数　197
短日処理　197
短日日長　197
単雌の子孫　117
探雌のための行動　130
探雌のためのナワバリ　130
探雌反応　130
単純家族　198
単純形状の触角　198
単純配列長多型　198, 206
単純反復配列　198, 206
単純反復配列多型　198, 206
誕生　30
単子葉植物　138
短焦点の望遠鏡　197
単女王性　138
単食　138
単食性　138
単食性の　138
炭水化物　35
炭水化物結合タンパク質　35
炭水化物代謝　35
淡水魚　88
淡水産無脊椎動物　88
単数回交尾　138
単数回交尾の雌　138
単数二倍体の性決定様式　99
単生性の　201
単性不妊性　226
単性不稔性　226
断絶する　195
短潜時　197
単相の　99
暖帯林上部　227
短断片の増殖　18
短断片の増幅　18
だんだん細くなる　214
暖地性種　231
断頭　55
短刀状の　54
単刀直入な　208
単独記録者　199
単独植物　199
単独スイッチ遺伝子　199
単独性　201
単独性バチ　201
タンニン含量　214
端背板　13
端背片　13
タンパク質　174
タンパク質架橋　174
タンパク質結合　174
タンパク質合成　174
タンパク質構造データバンク　159
タンパク質コード遺伝子　174
タンパク質生合成　174
タンパク質相同グループ　174
タンパク質のアミノ酸配列　174
タンパク質の補給　184
タンパク質配列データベース　174

タンパク質非コード領域 174
タンパク質の豊富な食餌植物 174
短波長感受性錐体視物質 197
単発休眠期 199
単板綱 138
端部 21
端部の 62
単分子膜細胞層 138
断片化された個体群 88
短報 197
探訪旅行 185
単木 110
担名タイプ 142
短命な昆虫 197
短命な餌源 74
断面 52
断面図 52
短毛 197
単門亜目 138

チ

チアクロプリド 218
チアメトキサム 218
地域擬態多型 125
地域系統 125
地域個体群 125
地域個体群動態 125
値域ごとにまとめた生データ 28
地域集団 56, 125
地域多様性 125, 182
地域的に絶滅した 182
地域的にまれな代替寄主植物 182
地域動物相 182
地域の生物多様性調査 182
地域変異 94, 95
地域変動 125
小さい黒色幼虫の突然変異体 60
小さくなる 197
小さな比率を占める 225
チーター 38
地位変更 207
地衣類 123
遅延 186
遅延出現 56
地縁性 182
地縁的 182
遅延発生 56
知覚刺激 160
知覚情報 160
近くの自然生息地 143
近づける 22
力強く飛ぶ 169
置換 210
置換生殖虫 184
置換速度 210
置換速度の不均質性 179
置換速度の不均質性のガンマモデル 90
置換／代用 184
置換の飽和 210
置換パターン 210
置換名 184
置換モデル 210
置換率 210
地球 68, 95
地球温暖化 95
地球規模の環境問題 95
地球上の生物群系 217
地球上の生物多様性 165
地球上の多様性 64
地球生物 217
地球の磁場 68
逐次 AIC 手順 207
逐次重回帰 208
逐次線形回帰 208
逐次的逸脱度分析 208
蓄積 12, 195
地形 120, 219
地形図 219
地形図モデル 219
地形的 219
致死 122
地磁気 68
地磁気座標 95
致死限界 122
致死性微生物 55
致死相当量 122
地質時代 95
致死的 122
地史的遷移 95
地史的背景 94, 95
地誌的背景 94
致死的病原体 55
地史的変異 95
地上の生息地 217
致死量寸前の効果 210
致死量に近い効果 210
致死(限界)レベル 122
地図距離 92
地図の全長 220
父親性発現遺伝子 158, 160
父方の PA 158
父方のピロリジジンアルカロイド 158
縮む 197
地中海植物 132
地中海地域 132
地中海地域の植物 132
チッカディ 39
腔小葉 124
腔前室 229
窒素 142, 146
窒素化合物 146
窒素含有物 142
窒素含有無機イオン 146
窒素欠乏食料 146
窒素固定食餌植物 146
窒素固定食草 146
窒素物質 146
窒素要求量 146
チトクロ(ー)ム 53
チトクロムオキシターゼ I と II 53
チトクロム酸化酵素 53
チトクロム酸化酵素サブユニット 1 43, 53
チトクロム b 遺伝子 53
チトクロム P450 53
地の利 228
遅発幼虫 56
地表水 212
地表面 201
チベット西部 232
地方擬態多型 125

地方系統　125
地方自治体　141
地方種　125
地方の蝶愛好家　124
地方変異　125
致命的に損傷する　81
チャールズ・ダーウィン　55
チャイロイチモンジ　230
チャイロドクチョウ　118
チャオビタテハ　96
茶褐色の小さな翅　66
着色　44
着色過程　164
着色された縞状バンドの相称系　213
着色する　207
着色パターン　164
着色紋様　164
着地　120
着陸　16, 120
チャノコカクモンハマキ　200
チャパラル　38
チャマダラセセリ亜科　206
茶葉　215
柱　220
中－　109, 115, 132, 133, 135
注意点　165
虫癭　90
中央－　132
中央閾値　132
中央欠刻　132
中央鰓　132
中央裂目　132
中央神経分泌細胞　37
中横線　132
中央帯　61, 131, 132
中央値　132
中央に近い　174
中央の　131, 132, 133, 135
中央の帯糸　37
中央部　37, 61, 132
中央部の青白い縞状バンド　156
中央部の縞状バンド　37
中央部の縞状バンドの発育不全　109
中央分節　146
中央片　132
中横脈　131
肘横脈　53
中央融合型　37
中央輸卵管　44
注解　19
中核的生合成経路　50
中間　114
中間温帯　114
中間温度　114
中間型　114
中間型個体群　114
中間型種　114
中間的措置　114
中間的段階　114
中間的な温度　114
中間的な好適生息地　114
中間的な性表現型　196
中間的な日長　114
中間的な率　114
昼間の　63
昼間の強い自然光　209
中間表現型　114, 162
チューキー型 LSD 法の比較　224
チューキー型最小有意差法の比較　224
チューキー検定　224
チューキーの HSD 検定　224
チューキーの HSD 事後比較法　224
チューキー法　224
中脚　133, 135
中胸　133
中胸後側板　132
中胸翅基部　133
中胸盾板　133
中胸小盾板　133
中胸前側板　132
中胸前腹板　132
中胸側線　208
中胸側板　133
中胸側板溝　132
中胸側腹板　133
中胸背縦斜溝　149
中胸背盾板　149
中胸背板　133
中胸腹板　133
中脛節　133
昼行活動　63
昼行性蛾　63
昼行性蝶　63
昼行性で音を出さない　63
昼行性で音を出さない蝶　45
昼行性の　55, 63
昼行性の森林性若虫　63
中後背板翅突起　169
中国型　39
中国山地　39
中国山地（日本）　40
中国西部系要素　232
中国大陸　127
中国中部　37
中止　11
中枝　135
中肢　135
中軸　135
中止する　36
中室　36, 61
忠実性　83
中室端斑　21, 36
中室の棒状紋　36
注射　112
注釈　19
注釈付きの遺伝子　19
中縦線　132
中終齢幼虫　135
抽出 DNA　80
抽出物　80
中心窩　81
中心教義　37
中新世　136
中心線　25
中心的な役割　155
中心点採餌　37, 51
中新統　136
中心フォーカス　37
中心複合体　37
抽水植物　72

中枢種 119
中枢神経系 37
中枢の属 165
中節 193
中舌 95
中線 135
中層の広葉樹 135
中層のマツ 135
中体節 133
中腿節 133
中体節の 133
中断 37
中肘横脈 131, 132
中腸 135
中腸の腸管微生物叢の組成 135
中腸の腸内細菌叢の組成 135
肘臂横脈 144
柱頭 208
注入 112
注入する 221
注入バッファー 112
虫媒花 73, 205
中背翅突 132
中背板翅突起 132
中波長感受性視物質 135
中波長感受性錐体視物質 135
中腹側 135
中腹板溝状線 135
中部サハリン 37
中ふ節 133
中部ラオス 37
中片 132
中胞 189
稠密な 57
稠密な種 57
稠密な属 57
稠密な日陰 57
中脈 20, 131, 132
肘脈 52, 53
肘脈後枝 53
肘脈前枝 52, 53
肘脈中断 53
肘脈中断遺伝子 40, 90
肘脈の 52
中脈分岐 169
注目すべきクライン(的)変異 183
注目すべき蝶種 183
注目の期間 86
注目の蝶 86
中立遺伝子 145
中立(的)進化 145
中立進化説 108
中立進化論 145
中立性 145
中立説 145
中立対立遺伝子 145
中立的景観 145
中立的突然変異の累積 13
中立的に進化する 78
中立突然変異 145
中立(的)部位 145
チューリングモデル 224
中肋 135
中和剤 145
中和物 145
蝶類調査 37
蝶 33
蝶(蛾)の研究者 122
超- 107
超遺伝子 211, 212
超遺伝子構成 212
超遺伝子族 212
蝶園 34
長円形の 71
長円状またはらせん状の飛翔 122
超音波音感受性の 225
超音波音感受性の耳 225
超音波クリック音 225
超音波の反響定位音 225
超音波の反響定位鳴声 225
超科 212
聴覚 24, 100
聴覚閾 24
聴覚閾値 24
聴覚応答 24
長角型 125
聴覚器官 100
聴覚機能 100
聴覚神経 24
聴覚の 13, 24
聴覚反応 24
蝶化石 34
超過分 78
チョウ・ガ類 122
長期 125
超幾何学的希薄化曲線 107
超幾何学的サンプリング分布 107
長期間 125
長期間にわたる衰亡 126
長期間にわたる生息範囲の拡大 126
長期間の移動 126
長期間の交尾 173
長期間の進化 126
長期間の成功 126
長期間暴露 173
長期休眠 126
長期的進化 126
長期動態 125
長期の産卵前期間 173
長期防除 126
長距離 120, 125
長距離移動 125
長距離シグナル 126
長距離の刺激 125
長距離のフェロモン 125
蝶群集間の非類似性 62
蝶群集の安定性 33
蝶系統群が定着した植物 34
超計量系統樹 225
超計量的樹形図 225
徴候 208
超高純度の水 225
蝶咬節 208
蝶咬節の 208
調査 37, 185
蝶採取 34
蝶採集 34, 43
蝶採集記 33
調査員 116
調査回数 150
調査季節 116
調査計画 212
調査個体数 150

調査材料　116
調査されていない　226
長鎖散在(型)反復配列　123, 125
調査する　155
調査対象の種　155
調査地域　209
調査データ　37
長鎖ノンコーディングRNA　125
長鎖非コードRNA　125
長鎖末端反復配列　125, 127
調査旅行　185
調査を行う　37
超酸化物アニオン　19
超酸化物イオン分解酵素　212
超酸化物不均化酵素　212
長散在型反復配列　123
長翅　127
弔辞　46, 132
長翅型　127
長翅型個体　126
長軸　125
頂室　21
長日型　125
長日型反応　125
長日植物　125
長日処理　125
長日にさらすこと　125
長日日長　125
長日養生　125
長翅目　131
超純水　225
調製　171
調整する　35
蝶成虫　14
聴性反応　24
調節　182
調節遺伝子　182
調節機構　49, 182
調節筋　182
調節効果　137
調節信号　182
調節スイッチ遺伝子　182
調節する　35, 132, 137
調節配列　182
調節要素の拡大　182
調節領域　182
チョウセンコムラサキ　176
挑戦的な質問　38
チョウセンヒョウモンモドキ　129
朝鮮民主主義人民共和国　54
チョウ相　33
蝶相　33
超双胸遺伝子　225
蝶相の変遷　211
長楕円形の　155
超多様な分類　107
頂端側　21
頂端の　21
蝶道　86
蝶道型　152
蝶と植物の軍拡競争　34
蝶と花との相互作用　34
腸内細菌叢　98
蝶の維持　34
蝶の羽化　33
蝶の活動　33
蝶の観察地　34
蝶の観察路　34
蝶の木　33
蝶の共回転　34
蝶の来る庭作り　34
蝶の形態(学)　34
蝶の解毒機構　33
蝶のゲノム　34
蝶の個体群　34
蝶のこと　187
蝶の色素　34
蝶の出現　33
蝶の状態　34
蝶の食性の進化　77
蝶の進化過程　77
蝶の旅　34
蝶の単一属　198
蝶の地図帳　33
蝶の分布　33
蝶の眼の異質性　101
蝶の眼の多様性　101
蝶の紋様研究入門　116
蝶の幼虫の餌となる部分　69
蝶の楽園　157
長波長感受性視物質　119, 126
長波長感受性錐体　126
挑発する　174
超微細構造　107
頂部　21
重複　67
重複遺伝子　67
重複型紫外線オプシン　67
重複感染した雌　66
重複感染しているスピロプラズマ系統　43
重複感染する　43
重複感染の昆虫系統　66
重複されたサンプル　184
重複しない機能　148
重複植物　184
重複する区域　22
重複断片　155
重複地域　22
重複なしの大量サンプル　227
重複保持　186
長方形　180
長命　126
チョウ目　122
超優性　154
超優性淘汰　154
聴力検査手順　24
聴力図　24
聴力図作成手順　24
聴力図実験　24
蝶類　33
鳥類学者　154
蝶類多様性の隠蔽的な部分　52
蝶類多様性の保全　47
蝶類多様性のホットスポット　33
調和した　46
調和的な生活　100
調和平均　100
蝶を呼ぶ花　85
直翅目　154

直射日光　61
直進飛翔　208
直接クローニング法　60
直接シークエンス法　61
直接接触　60
直接的な共進化の相互作用　60
直接的な選択　61
直接の影響要因　61
直接の刺激　175
直接反復結合　61, 66
直接防衛　60
直線状分子の円形化　40
直線状分子の環状化　40
直線的な衰亡　124
直前の刺激　175
直線飛翔　124
直腸　180
著作物　235
著作リスト　124
著者　24
著者たちの　24
貯精のう　179, 194
貯蔵タンパク質　208
直系遺伝子　154
直系遺伝子対　154
直交三元計画　154
チリ　39
地理型　182
地理区　94
地理的亜種　94
地理的位置　94
地理的解像度　94
地理的隔離　94
地理的規模　94
地理的距離　94
地理的空間　94
地理的クライン　94
地理的系統　94
地理的勾配　94
地理的個体群　94
地理的産地　94
地理的種　94
地理的種分化　94
地理的制限　94
地理的制約　94
地理的地域　94
地理的な場所　94
地理的な分化　94
地理的に近傍な　94
地理的に狭い地域　142
地理的背景　94
地理的場所　94
地理的パターン　94
地理的範囲　94
地理的品種　94
地理的分化　94
地理的分布　94
地理的分布域　94
地理的分布型　94
地理的変異　94
地理的放散　94
地理的モザイク　94
チロシン　224
チロシン組換え酵素遺伝子　225
チロシン水酸化酵素　217, 224
チロシンヒドロキシラーゼ　217, 224
チロシンリコンビナーゼ遺伝子　225
沈下　210
珍奇なこと　53
沈降　170, 210
珍蝶　178
沈殿　170
チンパンジー　39

ツ

追憶　183
追加記録　14
追加研究　14
追加された知見　14
追加実験　89
追加調査　14
追加要因　14
追求　177
追跡　176
追跡飛翔　38
追悼　132
追悼文　132
対になったt検定　156
追飛　38
追飛行動　38
費やした時間に相当する　236
通過移動　220
通過移動飛翔　221
通過雌　158
通気孔　15
通気性　229
通時性　58
通常　45
通称名　44
通信担当者　50
通俗名　229
ツェアクヌルト遺伝子　237
使い古した　236
使う　78
接ぎ木　96
月降水量　139
月コンパス　127
突き刺して吸汁する植食者　164
突き出る　174
突きとめる　23
次の作製手順(書)　86
次の手順　86
次の日　86
次のプロトコール　86
月平均降水量　139
作り上げる　164
造り出す　43
ツタ類などの植物種　41
筒状の　223
包み　73
包み込む　109
綴り　205
つの　173
つのがある頭部　105
ツバメガ科オオツバメガ亜科の蛾　228
坪枯れ　105
褄　21
ツマキチョウ族　20
ツマジロクサヨトウ　80, 205
ツマジロクサヨトウの卵巣細胞由来の樹立培養細胞株　196
つまり　142

罪 151
爪 41, 51, 226
冷たい 39
強い発生的制約説 108
強く関係する 209
強く示唆する 209
強める 113
釣糸 84
つる植物 41
つる性の植物 41
ツンドラ 224
ツンドラ的環境 224

テ

出会う 72
〜であろうとも 15
提案書 173
提案する 211
定位 153
Tre 系統 222
TE バッファー 223
定位移動 154
DHPLC 法 57, 58
*dsx*遺伝子 66, 67
*dsx*遺伝子から生じた雌特異的なアイソフォーム 82
dsDNA の環 67
DNase 処置 64
DNase 処理 64
DNA 結合のモチーフ・二量体形成の領域 64
DNA 交換 64
DNA 修復 64
DNA 増幅 64
DNA 多型 64
DNA 断片長多型 64
DNA チップ 64
DNA 抽出 64
DNA トランスポゾン 64
DNA に基づく同定 64
DNA に基づく評価 64
DNA に基づく標本同定 64
DNA の短配列 197
DNA の部分劣化 158
DNA バーコーディング 64
DNA バーコーディングの性能 64
DNA バーコード 64
DNA バーコード記録 64
DNA バーコード参照ライブラリー 64
DNA バーコード法 64
DNA バーコード法の性能 64
DNA バーコードリファレンスライブラリー 64
DNA フィンガープリント法 64
DNA 分解 64
DNA マーカー 64
DNA マーカー利用選抜 64
定位機構 154
t 検定 214
定位性 61
t 統計量 80
D 統計量 54
DV 軸 65, 66, 67
ディーム 56
低温 49, 126
低温化 39
低温乾燥期 126
低温期 52, 126
定温器 110
低温期間 126
低温系列 43
低温刺激 126
低温障害防御物質 52
低温ショック 43, 126
低温処理 49, 126
低温耐性 126
低温短日 126
低温致死温度 124, 126
定温の 104, 231
低温の感受期 126
低下 56, 75
低下した生殖能力 181
低下した発現 181
低下する 56
低下する力価 80
低下する類似度 56
定花性 85
定義 56
提供側生息地 65
提供する 65
定型的でない 24
抵抗(性)機構 185
抵抗する 185
抵抗性 185
抵抗力がある 185
抵抗力のある宿主 185
定時休眠 84
停止コドン 208
停止する 36
提示する 78
底質 210
定日型 84
低湿地の草 232
定時的休眠 84
低次分類数 - 高次分類数の比 126
定住(性) 185
定住域 193
定住圏 185, 193
定住種 185
定住する 195
定住性の動物 193
定住戦略 185
定住動物 193
低周波音 126
低周波数音 126
提唱 173
定常脱皮 207
泥水の吸水行動 140
ディスクリーメン 61
訂正 186
定棲性種 185
定性的に異なる手法 177
定性的に異なる入力情報 177
訂正と追加の情報 50
底側 165
低体温巡回型 43
泥炭 159
低地 126
低地広葉樹林 31
低地性の蝶 31
低地性の花 31
低窒素試料 126
低地熱帯雨林 126
定着 43, 76
定着イベント 43

定着確率 43
定着からの経過時間 219
定着剤 84
定着しつつある個体群 43
定着してしまった侵略的外来種 75
定着者 43
定着性 76
定着動態 43
定着能力 43
定着の歴史 43
定着ボトルネック 43
定着率 43
定着力 43, 76
ティッシュペーパー 219
低品質飼料 126
低風 126
ディフェンシン 56
低分子干渉 RNA 197, 199
低分子代謝産物 126
低分子代謝物 126
定方向性選択 61
定方向選択 61
低木 197
低木層 226
低木の列管理 100
低木林 33
低木林地帯 191
ティリャード表記法 219
定量 177
定量 RT-PCR 177
定量化 177
定量化可能な指標 177
定量化可能な尺度 177
定量化する 177
定量逆転写 PCR 176
定量的 RT-PCR 解析 177
定量的逆転写 PCR 177
定量的手法 177
定量的に継承された 177
定量的 PCR 176
定量的 PCR 解析 177
定量的 PCR 法 177
定量的予測 177
定量 PCR 177
データ計測器 55
データ準備 55
データの質 55
データの適切性 55
データベース 55
データマイニング 55
テーダマツ 124
データロガー 55
デオキシリボ核酸 57, 64
デカペンタプレジック遺伝子 55, 66
手軽な同定法 46
敵 86
適応 13
適応学習 13
適応形質 13
適応しない 226
適応上の意義 13
適応進化 13
適応性の季節的相違 191
適応性の相違 59
適応戦略 13
適応ダイナミクス 13
適応的意義 13
適応的遺伝子移入 13
適応的仮説 13
適応的季節多型 13
適応的収斂 13
適応的新奇性 13
適応的新規性 13
適応的探索行動 13
適応的突然変異 13
適応的反応 13
適応的分化 13
適応的分散 13
適応度 84
適応度因子 84
適応度基準 84
適応度向上の利益 84
適応度パラメータ 84
適応パラメータ 84
適応放散 13
適応様式 13
適格性 25
適格な著作物 25
適格な命名法的行為 25
適格名 25
適合性 211
適合度 83
適合法 45
適合溶質 45
摘出する 100
適切性 211
敵対者 86
敵対する 106
滴注式給餌法 66
滴定実験 219
摘要 12, 211
適用名 25
テグメン 215
凸凹のある 190
凸凹のある縁 190
デジタルオシロスコープ 60
デジタルカメラ 60
デジタルシグナルプロセッサー 60
デジタル写真 60
デジタル信号処理装置 60
デジタルビデオ 60
デシベル 55
デスクランプ 57
手操作 99
テチス型 217
テチス区 217
撤回 235
手つかずの林 113
鉄線柵 235
徹底的な研究 113
徹底的な分析 109
でっぱり 124
テトラサイクリン 217
テトラサイクリン入り人工飼料 217
テトラサイクリン塩酸塩 217
テトラサイクリンヒドロクロリド 217
テトラサイクリン補給飼料 217
テトラピロール系色素 217
でない- 225
手に入れる 12
テネラル 216
テネラル(羽化)以後の移動飛翔 168

デノボゲノムアセンブリー　55
デノボシークエンシング　55
デノボトランスクリプトーム生成　55
〜ではあるが　15
テリトリー　217
テリトリー行動　217
テリトリー接触域　217
テリトリー占有雄　217
テリトリーの境界　32
テリトリーの変動　217
テリトリーの防衛　217
テリトリーの保有　217
テリトリーの保有期間　217
テリトリーマーカー　217
出る　71
テルペノイド防御分泌物　217
テルペン　217
テルペン合成酵素　217
テレメーターデータ　215
テレメトリデータ　215
テロメア　215
転位　221
転移　222
転移 RNA　221, 223
転移因子　215, 222
転移因子の水平伝播　105
転移型置換　221
臀域　18
転位した眼状紋　69
転位した眼状紋の形成　87
転位した白色紋様　69
転移する　222
転移相　161
臀縁　18
田園地域　188
田園地帯　51
臀扇　18
天蓋　35
天蓋に覆われている木　35
転換　222
点眼　207
転換型置換　222
電気泳動(法)　71
電気泳動解析　71
電気泳動型特異性　71
電気泳動図の検査　113
電気泳動パターン　71
電気泳動分析　71
電気泳動法の検査　113
電気計量器　71
電気生理学的測定　71
電気生理学的反応　71
電気生理学的方法　71
電気穿孔処理された　71
電気穿孔法　71
電気穿孔法を用いた低分子干渉 RNA の取込み実験　71
電気バランス　71
電極　71
天空コンパス　200
天空光コンパス　200
テングチョウ　200
テングチョウ亜科　200
デング熱　57
典型種　78
典型的な　224
典型的な秋型　224
典型的な遺伝子座　184
典型的な系統樹形　184
典型的な痕跡　184
典型的な対数傾向　38
典型的な標本　184
点刻　175
転座　221
点在している森林地帯　190
点在する　202
点在する植生　202
点在領域　190
転座した発生　221
展翅　140, 206
展翅板　140, 206
撚翅目　208
転写　220
転写因子　221
転写開始点　221, 223
転写酵素　220
転写産物抑制因子　221
転写制御因子　221
転写リプレッサー　221
転写(の)リプログラミング　221
天山山地　218
天山山脈　218
転節　223
伝染　221
伝染病　74
伝染病学　75
伝染病菌　111
天体航法　36
天体コンパス　36
伝達　221
転地　221
天敵　143
天敵真空空間　73
天敵低密度空間　73
天敵の誘引　73
天敵不在空間　73
点々を付ける　208
伝統的生物的防除　40
伝統的なアプローチ　220
伝統的な形態分類学　220
テントウムシ　120
天然記念物　143
天然色版　44
天然の病原体　143
天然バキュロウイルス　142
伝播　62, 221
電波探知機　177
伝播の方向　61
臀部　18
テンプレート　216
デンプン　207
臀脈　11, 18
臀葉　18

ト

問い合わせ(る)　177
問い合わせ配列　177
という意味での　109
と一致して　109
胴　31
同 -　104
銅／亜鉛超酸化物不均化酵素領域　49
糖アルコール　211
同意　47

同位体 117
等位置価で描かれた線 75
統一 214
同一環境実験 44
同一環境条件 44
同一性 108
同一性閾値 108
同一相対位置 189
同一祖先種 44
同一地方 189
同一庭園で飼育した蝶による標識 - 再捕獲研究 44
統一的な枠組み 226
同一のハプロタイプの除去 183
同一の累積種数 108
同一場所 189
同一方向 189
同位の原理 172
同位表現型 157
動員 180
動員する 180
糖液 104
冬芽 235
同化 18, 23
透過域 221
頭蓋骨 200
頭蓋縫合線 74
等価基準標本 214
頭殻 100
頭額溝 75
透過光 160
同化作用 23
同化する 23
透過性 160
透過性のある 160
同化速度 23
統括著者 121
等価な染色体位置 75
トウガラシ 160
同化率 23
同化量 18
同期化した二年生の蝶 214
冬季気温 235
同義サイト 214
同義性固定置換 84
頭基節縫合線 132
同義置換 214
同義置換率 67
同義の 214
冬期の気温 235
冬期の気温の上昇 187
冬期の厳しさの度合い 127
冬期の寒さ 235
冬期の生存率 235
東京近郊 211
統計(学) 207
同系化 104
統計解析 207
統計学的研究 207
同系交配 109
同系交配の 109
統計集団 207
同型性 104
同型接合体 104
統計的期待値 207
統計的に有意な 207
統計的に有意な関係 207
統計的比較 207
統計的有意性 207
同形配偶 117
同型配偶子をもつ性 104
統計母集団 207
凍結感受性 88
凍結防止物質 20
動原体 37, 119
統合 214
投稿中 210
統合的アプローチ 113
頭骨 200
同語反復 215
動作擬態 125
等差数列 22
同時攻撃 198
同時出生集団 42
糖質 35
等質性 104
等質性のχ二乗検定 39
等質性のカイ二乗検定 39
同質倍数体 25
同時に - 213
同時に起こる 42
同時(性)の 214
同時発生 214
同時分割法 198
等翅目 117
謄写版印刷 135
同種 46
同種 - 24
同種異亜種 189
同種間コミュニケーション 47
同種間通信 47
同種間の交配 47
同種他個体に対する誘引 47
同種的誘引 47
同種認知 48
同種の 47
同種の個体 46, 47
同種の翅 48
同種の表現型 47
同種の雌 47
同種表現型 47
同種類 47
頭盾 42
頭盾上区 212
頭盾の 42
頭盾板 42
島嶼 117
凍傷 88
凍傷以外の低温障害 147
筒状の 223
橙色で囲まれた黒色 30
同色の翅内の求愛 235
同所(的)種 213
同所性 213
同所性／異所性の関係 213
島嶼生息地 98
同所性の 213
同所的擬態分類単位 213
同所的形態 213
同所的系統 213
同所的種分化 213
同所的相互擬態型のベニモンドクチョウ 213
同所的対立遺伝子 213
同所的に 213
同所的に共存する 42
同所的分布 213
同所的変異型 213

投資量　116
同心円　46
同心円状の環　46
同心円状の眼状紋パターン　46
銅(重金属)ストレス　52
統制　49
動性　119
統制群の場所　49
統制群の林分　49
統制処理　49
同性内性選択　115
同性内選択　115
闘争　27
同属種　46
同属の　46
同属の共存　46
淘汰　193
淘汰圧　193
胴体　31
動態覚　140
動体視　140
動体視覚　140
淘汰の不利性　193
淘汰の有利性　193
淘汰不利益　193
淘汰利益　193
冬虫夏草　165
同調　73
同調因子　237
同調した進化　49
頭頂部　229
同定　108
同定困難な種　60
同定システム　108
同定成功率　108
同定ための捕獲　35
動的環境　67
動的気流　67
動的空気流　67
動的景観　67
動的状態　67
動的増大過程　67
動的発現プロファイル　67
動的平衡　67
同等者による査読　160
同動態的　104
島内効果　235
東南アジア　202
東南アジア森林広域分布属　86
東南アジア島嶼　117
東南アジア島嶼の蝶　34
導入　98, 116
導入された遺伝子　221
導入種　116
導入する　116
同輩審査　160
逃避擬態　75, 77
逃避行動　75
逃避点　84
逃避の飛翔行動作戦　76
逃避反応　77
頭部　100
頭幅　100
同腹　33
同腹仔　33
同腹仔依存の変動　33
同腹仔の比較　33
同腹の仔　33
頭部切断　100
動物　19
同物異名　213, 214
同物異名の検討　214
動物学者　237
動物学的タクソン　237
動物学的名称　237
動物学における適格名リスト　124
動物学における適格名リストの分冊　158
動物性カロチノイド　237
動物相　81
動物相の多様性　81
動物地理区　81, 237
動物定型名　237
動物の移動　19
動物の移動軌跡　19
動物のゲノム　19
動物の現在位置　19
動物の行動に関する意思決定　19
動物の種分化　19
動物の仕業　235
動物のナビゲーション　19
動物命名法　237
動物命名法紀要　33
動物命名法国際審議会　108, 114
同父母の兄(弟)　89
同方向　189
同胞交配　33, 197
同胞種　197
等方性光条件　117
倒木ギャップ　222
糖蜜　222
冬眠　102, 235
冬眠以後　88
冬眠以降　88
冬眠状態　220
冬眠場所　102
同名　104
同名異物　104
同名関係　104
同名状態　104
透明な斑点　107
透明な斑紋部分　221
どうも進化しているらしい　171
トウモロコシ　127
トウモロコシの市販交雑種　44
トウモロコシ畑　50
動揺　161
等容(変化)　117
東洋区　153
同様に　123
東洋熱帯起源　153
東洋熱帯地域　153
同裏面　63
糖類　189
同類　42
同類交配　23
同類交配選好　23
同類的に交配する　129
登録する　182
道路で車にひき殺された動物の死体　188
当惑する　176
トウワタ　135
遠縁　62

遠縁種 62
遠縁生物 62
遠縁の 62
遠縁の相互擬態種 62
遠くに 15
遠くの目標 62
遠く離れた目標 62
通して－ 58, 160
トータル RNA 抽出試薬 222
トータル RNA 分離用試薬 223
ドーパ 65
ドーパ脱炭酸酵素 55, 65
ドーパデカルボキシラーゼ 55, 65
ドーパミン 65
ドーパミンキノン 65
ドーパミン誘導体 65
ドーム状の形 65
トカゲ 124
溶かす 62
とがった 19
とがった先端 166
と合致して 109
時折 151
時々 151
と境界を成す 31
毒 166
特異性 159
特異的エリシター 204
特異的組織 204
特異的な 108, 159
特異的な有無相関 204
特異的なエフェクター 204
特異的な交尾場所 204
特異的な生息地 204
特異的な(微)生物 204
特異的プライマー 204
特異的誘発物質 204
特異点 199
特異なタテハチョウ種 226
ドクガ科の蛾の総称 224
特質のない 225
毒蛇 166
特殊な 159
特殊部位 204
特性 38, 159
毒性 220, 230
毒性遺伝子 230
毒性因子 230
毒性化合物 220
毒性－解毒モデル 166
毒性の 220
特性のない 225
毒素 220
毒素遺伝子 230
毒素因子 230
独創性に富んだ考え 194
毒素分解産物 220
特徴 38, 81
ドクチョウ 100
ドクチョウ亜科 126
ドクチョウ属の蝶 100
特徴的な保護色 38
特徴的分布 38
特定音周波数 204
特定外来生物 108, 116
特定する 204
特定の閾値体重 204
特定の花蜜への集中(化) 158
特定の交尾場所 204
特定の状態変数 37
特定の生息地 204
特定の(微)生物 204
特定の臨界体重 204
特定部位 204
毒－毒消しモデル 166
特に 149, 158
毒物 220
特別保護地区 203
匿名の 20
匿名の査読者 20
匿名の著作物 20
特有性 72
特有の 224
独立栄養 25
独立した遺伝子座 110
独立種 110, 194
独立の観測機器 110
独立の進化系統 110
独立の進化の系譜 110
独立変数 110
独立モデル 150
トゲ 205
刺 205
棘 205
とげ 218
棘瘤 191
棘のある粘着性剛毛の粘着力 14
都市型土地利用 228
都市型被覆 228
吐糸管 205
年ごとの予測可能性 114
都市(的)地域 228
都市的土地利用 228
～として知られている 119
土壌 201
土壌形成 201
土壌食性 95
土壌水 201
土壌水分 201
土壌摂取 201
土壌表面 201
土食 95
土台 166
土地管理 120
土地被覆 120
土着種 110, 142
土着植物 142
土着の 110
土地利用パターン 120
突縁 84
凸凹のある 190
凸凹のある縁 190
特化 203
突起 124, 172, 187
突起した 173
突起したドーム形の内膜 173
突起部 173
突出部 79, 173
凸状の泡 49
凸状の気泡 49
凸状の内膜 49
突然の変化 12
突然変異 141
突然変異型眼状紋表現型 141

突然変異種 141
突然変異説 141
突然変異体 141
突然変異体表現型 141
突然変異率 141
突然変異を起こさせる 141
突発的な個体群パターン 72
トップダウン効果 219
とどまる 12
ドナー 65
ドナーサークル 65
ドナー種 65
ドナーストック 65
トビイロウンカ 33, 187
トビケラ目 222
飛び散る 86
トビバッタ 125
飛ぶ 85
途方もない個数 80
徒歩記録者 159
トポタイプ 219
トポロジー 219
トマト黄化葉巻ウイルス 224
トマト黄化葉巻病 224
とまり木 188
ドメイン 64
止める 207, 213
共食い 35
共倒れ 80
共に－ 213
トラウマを被った受精 222
捉えどころのない数量 71
トラップ 222
トラップ用道づくり 222
トラップ覆工 222
トラップを仕掛ける 222
トラップを敷くこと 222
トラフアゲハ 219
トラフタイマイ 237
ドラフトゲノムのアセンブリー 66
トラベキュラ 220
ドラミング 67
トラ模様の形態 219
トランスエレメント 220
トランスクリプターゼ 220
トランスクリプトーム 221
トランスクリプトミクス 221
トランス効果 220
トランスジェニック花粉 221
トランスジェニック植物 221
トランスジェニックマウス 217
トランスジェニックマウスの作出 221
トランスジェニックマウスの作製 221
トランスファーRNA 221, 223
トランスフェクション 221
トランスポーター 222
トランスポゾン 222
トランスポンダー 221
トランスメンブラン受容体 220
トランセクト 221
トランセクト調査 221
トランセクトの記録者 221
トランセクト法 221
鳥 29
取り扱い 99, 222
トリートメント 222
トリオースリン酸イソメラーゼ 223
取り込み 110
トリシン 222
TRIzol 試薬 223
トリトン X-100 223
鳥の嘴型 27
鳥のさえずり 30
取り除く 71, 183
鳥の鳴き声 30
鳥の糞 30
鳥の呼ぶ音 35
取り外し可能なガーゼ布 183
取り外す 57
トリバネアゲハ 30
トリプトファン 223
トリプレット 223
鳥捕食者 25
鳥捕食者の飛翔音 85
トルイジンブルー 219
トルコ 224
ドルトン 54
トレードオフ 220
トレハラーゼ 222
トレハロース 222
泥 61, 140
泥の水溜まり 140
泥水の吸水行動 140
トンガ王国 119
鈍角 151
トンボ 66
トンボシロチョウ亜科 136
トンボマダラ亜科 41
トンボマダラ科の蝶 117
トンボマダラ科の蝶の群集 117
トンボ目 151
蜻蛉目 151
貪欲アルゴリズム 97
貪欲法 97

ナ

内 114
内－ 72, 109, 115
内陰茎端部 229
内因性 150
内因性移動 150
内因性休眠 150
内因性刺激 73
内因性リズム 73
内因的 115
内因的な概日時計 72
内因的な発育最適温度 116
内縁 112
内横線 20, 174, 209
内温性の 73
内外軸 132, 137
内規 34
内クチクラ 72
内原表皮 72
内在性ウイルス 73
内在性ウイルスの統合的な性質 113

内在性ウイルス由来配列 73, 77
内在性コントロール遺伝子 114
内在性のオクトパミン受容体 73
内在的 115
内在的性質 116
内斜 110
内集団 112
内鞘 229
内食性の 73
内挿 115
内挿する 115
内側 131
内側 - 132
内側基部 132
内側中央帯 132
内側中央部 20
内側部位の切除 131
内側部位の微細焼灼 131
内的自然増加率 116
内的障壁 115
内的増殖率 116
内的繁殖力 116
内的符合モデル 114
内胚乳 73
内板 114
内表皮 72
内部寄生 73
内部寄生者 73
内部共生微生物 73
内部形態 114
内部結節 114
内部構造 114
内部コントロール 114
内部生殖器官 114
内部転写スペーサー 117
内部の 114
内部標準 114
内部表皮の袋 114
内部捕食寄生者の卵 73
内部捕食寄生性の蜂 73
内部捕食性寄生蜂 73
内部捕食性寄生(蜂の)卵 73
内分泌器官 72
内分泌機構 72
内分泌系 72
内分泌腺 72
内分泌調節 72
内分泌物質 72
内分泌要因 72
内膜 112
内膜表面 112
内面 112
内面層 214
内葉 120
内葉片 112, 114
内陸性気候 48
内陸地方 114
内陸部 114
ナイロン膜 150
苗木 150
長生き 126
長い日長 125
長い領域 125
長く伸びる 71
中に入れる - 71, 72
長年にわたる論争 126
長年の疑問 126
中ほどにある 125
流れ 208, 218
名残り 183
ナスビ形の 215
謎 142
謎の種類 142
ナタネ 151
夏型 211
夏型成虫 211
夏型の誘導阻害 112
夏型雌 211
夏型誘導 211
夏型誘導因子 111
夏休眠 14, 76, 211
夏世代 211
夏の気温 211
夏の雑木林の帝王 119
夏場の気温 211
ナトリウム 201
ナトリウム塩の摂取 189
ナトリウムが補給された泥水 201
7 回膜貫通型受容体 11
7 回膜貫通型タンパク質 11, 195
72 穴ローター 11
ナナフシ目 161
竹節虫目 161
斜め線 58
斜めの条線 58
ナノグラム 146
名のない 20
ナビゲーション技能 143
ナビゲーション障害 109
ナビゲーション能力 143
名前 146
生データ 179
生展翅 140, 206
鉛化合物 122
ナメクジ形の 200
ナメクジのような 200
滑らかな希薄化曲線 200
滑らかな曲線 200
滑らかなクライン 200
並び方 22
並べる 16
並んで 17
なりすまし 129
ナワバリ 217
縄張り 217
縄張り活動 217
ナワバリ行動 217
縄張り行動 217
ナワバリ制 217
ナワバリ制の存在 78
なわばりを防御する 56
南縁部 202
軟化処理 127
軟化展翅 140, 206
南極区 20
南限のコロニー 202
軟骨異栄養症 13
軟骨発育［形成］不全症 13
南西諸島の個体群 202
南西方向 202
ナンセンス変異 148
軟体動物 138
ナンダ - ハムナー実験 142

ナンダ-ハムナープロトコール　142
なんでも屋　117
南方性の種　202
南方的な種　202
南方へ飛翔　202
軟毛　175
軟毛のある　164
難溶性　113

ニ

２-　28, 58
二-　28, 58
におい　200
匂い源　151
匂い付け行動　191
臭物質　151
臭物質結合タンパク質　150, 151
二化　224
２回洗浄する　231
二化性　30
二化性集団　30
二化性地域　30
二化性の　30
二化地帯　30
ニカメイチュウ　187, 207
肉眼で観察した　132
肉質の　84
肉質の突起　84
肉食性天敵の誘引　24
肉食性動物　36
肉食性の　36
肉食(性)動物　170
肉食(性)哺乳類　36
ニクバエ　84
二型　60
二形性　60
二系統に感染したすべて雌の同腹仔　66
二型の産出　173
二型の体色　60
二元配置分散分析　11
二元配置分散分析法　224
二項回帰モデル　29
二項誤差分布　28
二項対立　59
二項データ　28
二項反応　29
二項分布回帰モデル　29
２個体だけ現れた種　66
ニコチンアミドアデニンジヌクレオチド還元型　181
ニコチンアミドアデニンジヌクレオチドの還元型　142
ニコチン性アセチルコリン受容体　142, 146
二語名　28, 29
２・３日後　82
二次異物同名　192
虹色の　116
西から東にかけての明確な勾配　62
ニシキオオツバメガ　127, 211, 228
二次項　176
二次構造　192
二次刺毛　192
二次植物成分　192
二次森林　192
二次成長林　192
二次代謝産物　192
二次的接触　192
二次同名　192
二次文献　192
滲み出た樹液　80
二者択一の表現型　17
二次誘引物質　192
二重感染のすべて雌の同腹仔　66
二重交叉　66
25 塩基オーバーラップ　11
二重鎖 RNA　66, 67
二重鎖 DNA　66
二重線　152
二重に切断した　66
20-ハイドロキシエクダイソン　11
20-ヒドロキシエクジソン　11
24 時間後　98
24 時間まで　98
二重らせん　66
二種の間での安定平衡　224
二種の拮抗的な生産経路　20
２種類の制限酵素で切断した　66
二色型色覚の　59
二色性の　59
二次林　192
にする-　71, 72
偽-　175
にせの近縁種　80
偽の高山蝶　80
二相安定現象　30
二足動物　29
二足を有する　29
日常行動　223
日常的移動　54
日常的に利用される植物　45
日常飛翔　54, 223
日度　55, 56
二値変数　28
日没　211
日没の方位　211
日没の方位角　211
日齢　55
日齢に関係する交尾受容性　15
日齢に関係する出生場所固執性　15
日齢に関係する生息場所執着性　15
日光　211
日光性変温動物　100
日光に恵まれた地点　211
日光浴　27, 211
日光浴好きな蝶　211
日光浴に出かける場所　211
日光を浴びて体温を上げる変温動物　100
日射　201
日周活動　54, 63
日周行動　54, 63
日周サイクル　54
日周性　63
日周の個体数パターン　54

日周変化　38
日周リズム　40, 63
日照時間　55, 113
ニッチ　146
ニッチ空間　146
ニッチ空間の相対量　182
ニッチシフト　146
ニッチ集合モデル　146
ニッチ説　146
ニッチ分化　146
ニッチ分配モデル　146
日中型　199
日中区　199
日中区系　199
日中系統　199
日中に活動する　63
日中に活動的な　63
日長　55, 162
日長効果　162
日長時間　55
日長調節　162
日長変化　38
似ている　185
似てない行動　62
ニトリル形成能　146
ニトリル指定子タンパク質　146, 149
ニトリルスペシファイアータンパク質　146, 149
二年ごとの　28
二年生の　28
二年生の化性　101
二倍-　28
二倍体胚　60
二番目に多い個数　192
二番目の最高点　192
鈍い　31
鈍い色の　67
二分　59
二分割系統樹　29
二分岐　28
二分岐のキー　59
二分法　59
2変量散布図　30
二峰的に　28
日本応用動物昆虫学会　118
日本型　118
日本固有種　72
日本昆虫学会　73
二本鎖 RNA　66, 67
二本鎖 DNA　66
日本産蝶　118
日本産蝶種名一覧　203
日本産蝶類標準図鑑　207
日本産蝶類リスト　124
日本蝶類科学学会　34
日本蝶類学会　34
日本 DNA データバンク　55, 64
日本特産　72
日本特産種　72, 203
日本とその周辺の東アジア諸国　118
日本(産)の　118
日本の国蝶　142
日本の南西地方の低地　202
日本の秘蝶　178
日本本土　118, 127
日本鱗翅学会　122, 126
日本列島　118
二名式命名法　28, 146
二名式用語体系　28
二名法　28, 29, 146
二名法の原理　172
～にもかかわらず　15
にもかかわらず　149
二門亜目　63
乳液　121
乳液の流れ　121
乳液の質　121
ニューカレドニア　145
ニューギニア(島)　145
乳酸酢酸オルセイン　120
入射光　109
乳状液　135
入植　43
乳濁液　135
入念な調査　134
乳白色　51
乳棒と乳鉢で磨砕した　97
尿　228
尿酸　228
尿酸誘導体　228
二様長短交互型の鉤爪　29
二様長短交互型の鉤爪　29
二齢　192
ニレを食餌植物とする　82
庭造り　90
ニワトリのゲノム　39
ニワムシクイ　90
任意閾　22
任意共生的関係　80
任意交配　156, 178
任意交配集団　156
任意交配単位　156
任意的相利共生　80
任意に設定された　22
任意の　22, 80
任意の閾値　22
人間活動　106
人間の住居　106
人間の人口密度　106
人間の生息地　106
人間の繁栄　106
人間の福祉　106
認識可能な成虫形態形成　180
妊性　81, 82
認知地図　42
ニンフ　150

ヌ

ぬかるみ　140
抜き刷り　151, 194
ヌクレオキャプシド伝播　149
ヌクレオソーム　149
ヌクレオチド位置　149
ヌクレオチド混合溶液　149
ヌクレオチド伸長　149
ヌクレオチド水準　149
ヌクレオチドストレッチ　149
ヌクレオチド多様性　149
ヌクレオチドデータ　149
ヌクレオチドの位置番号　149
ヌクレオチド配列　149
ヌクレオチド部位　149
ヌクレオチド分岐　149
ヌクレオチドモデル　149

ヌクレオチドレベル　149
ヌクレオモルフ　149
脱け殻　68, 72, 80, 225
ヌディウイルス　150
沼地　31, 129, 213
ヌルモデル　150

ネ

根　188
ネアンデルタール人　143
ネイティブアメリカン　142
ネイの遺伝(学)的距離　144
ネオダーウィニズム　144
ネオタイプ　144
ネオテニー　144
ネオテニック生殖虫　144
ネオニコチノイド　144
ネオニコチノイド汚染　144
ネオニコチノイド(系)殺虫剤　144
ネオニコチノイドの使用　144
ネオニコチノイドの利用　144
ネオプテロビリン　144
ネオ - ホプキンス則　144
ネオリグノイド摂食抑制剤　144
ネガティブコントロール　144
ネガティブ鎖 RNA ウイルス　144
ねじれ　220
ネジレバネ目　208
ネズミ　140
熱慣性　218
熱吸収　100
熱取得効率　100
熱ショックタンパク質　100, 106
熱ショックタンパク質遺伝子　100, 106
熱心な　68
熱ストレス　100
熱帯　220, 223
熱帯アフリカ区　15
熱帯遺存　223
熱帯雨林　223
熱帯型　223
熱帯気候　223
熱帯降雨林　223
熱帯産甲虫　223
熱帯産節足動物　223
熱帯産節足動物データセット　223
熱帯産の蝶　223
熱帯産の薄明活動性蝶　223
熱帯産無脊椎動物群集　223
熱帯収束帯　115, 117
熱帯種子銀行データセット　223
熱帯性甲虫　223
熱帯性の蝶　223
熱帯地方　223
熱帯蝶　223
ネッタイツメガエル　236
熱帯の赤道収斂域　115, 117
熱帯の二次植生林　223
熱帯の老齢林　223
熱帯モンスーン　223
熱帯モンスーン地域　223
熱帯林　223
熱電対　218
ネットワーク　145
ネットワーク樹　145
ネットワーク内変異　235
熱麻痺　100
熱烈な吸蜜者　25
ネナシカズラ　64
ネマトーダ　144
眠(り)　138, 200
年一回発生　152
年一世代　152
粘液腫病　142
粘液腺　43
粘液のう　140
年間および年内　28
年間の予測可能性　114
年間変動　113
年三世代　218
年次変動　20, 113
年中　16, 218
稔性　81, 82
稔性の　82
粘性のある　230
年代(推定学)　39
年代領域説　107
粘着テープ　14
粘土処理　41
年表　39
年譜　219
年変動　113, 236
年齢別生命表　15

ノ

ノイズ閾値　146
の意味で　194
のう　229
脳　32
農家が選択した農産物　81
脳間部　158
脳間部背面　65
農業　15
農業害虫　15
農業環境計画　15
農業・環境スキーム　15
農業景観　15
農業集約化　15
農業用殺虫剤　15
農耕　15
濃縮水準　123
脳除去　183
脳 - 前胸腺 - アラタ体 - 側心体の複合体　32
農村地域　81
農村地帯　81
農地　81
農地性生息場所　81
農地の場所　81
濃度　46
濃度依存的な様式　66
のう胞　53
脳ホルモン　32
濃密な　57
濃密な地帯　57
農薬　161
農林水産省(日本)　136
ノーザンブロッティング　148
芒形の触角　22

の近似 15
のこぎり歯のような形 117
残っている森林 87
残りの系統樹 183
～の真価(性質、差異)を認める 21
～の造成 58
覗き見る 65
ノックアウトする 119
ノックアウト・ノックイン解析 119
ノックダウン実験 119
ノッチ 149
能登半島 149
野火 83
上り傾斜 168
ノミ目 199
野焼き 83
野焼き強度 113
野焼きされたダイオウショウ 33
野焼きされた林分 33
野焼き処理 33
野焼き体制 83
野焼きで維持されている群落 83
野焼きと生息地との相互作用 114
野焼きの影響 69
野焼きの管理体制 83
野焼きの効果 69
野焼きの強さ 113
のり付けする 96
ノンコーディング配列 147
ノンパラメトリック推定量 148

ハ

葉 86, 122
パー遺伝子 160
把握器 40, 157
バーコード・オブ・ライフ・データ 31
バーコード解析 26
バーコードギャップの自動発見 12, 24
バーコード索引番号システム 26, 28
パージ 176
パーツパービリオン 169
ハーディー‐ワインベルグ期待値 100
ハーディワインベルグ平衡 100
ハーブ 101
バーモント州の河川 229
ハーレム 100
バーンイン 33
バイアス 28
配位子 123
灰色の破線 55
背縁 65
バイオアッセイ 29
バイオインフォマティクス解析 29
バイオインフォマティクス研究 29
バイオーム 29
バイオタイプ 29
パイオニア個体 165
バイオマス 29
バイオミメティク 29
バイオミメティックス 29
徘徊移動 231
徘徊行動 231
媒介昆虫 112
媒介者 228
媒介者によって媒介される疾病 229
媒介者防除 229
媒介する 132, 221, 228
背管 65
背眼状斑点 65
背棘 65
配偶競争 129
配偶行動 130
配偶行動戦術 130
配偶行動を延期する 169
配偶子 90
配偶子隔離 90
配偶子合体 214
配偶システム 130
配偶子単離 90
配偶者位置探索行動 130
配偶者位置探索戦略 130
配偶者競争 129
配偶者選好の刺激 130
配偶者選好の臨界刺激 51
配偶者選択 129, 130
配偶者選択実験 129
配偶者選択の組合せ実験 44
配偶者争奪戦 129
配偶者探索 130
配偶者認識 130
配偶者認知 130
配偶者認知シグナル 130
配偶様式 130
背景依存性 48
背景活動 25
背景雑音 25
胚形成 71
背景絶滅 25
背腔 65
胚細胞 95
背刺 65
背軸側の 11
背翅突 149, 150
胚死亡率 71
胚珠 155
背縦走筋 64, 65
杯状細胞 96
排除確率 78
排水 66
排水法 66
排水路 66
倍数関係 166
倍数性 166
倍数性レベル 166
倍数体化 167
倍数変化の差 86
排泄 78
排泄系 78
倍増時間 66
背側 54, 65
背側縁 65
背側解剖法 65
背側から見た図(写真) 65
背側縦走筋 64, 65
背側突起 65

背側に　66
背側の前翅　65
背側の側線　65
背側の非可塑性の眼状紋　148
背側板筋　216
背側板溝　18
背側板溝状線　19
背側網膜　65
胚帯　95
媒体となる　132
排他的分野　78
背断面　65
培地　53
配置換えをする　181
配置に付かせる　57
配置パターン　22
配糖体　96
胚のう細胞　132
胚の初期発生イベント　68
胚の性決定　71
胚発育　71
胚発生　71
胚発生の初期段階　68
胚発達停止　23
背板　149, 216
背板溝　216
背板翅突起　149, 150
背板の　149, 216
背板の翅突起　149
背板の割目　216
胚盤葉　30
背板裂　216
背部眼状斑点　65
背腹軸　65, 66, 67
背腹飛翔筋　66
背腹方向に走る　67
背腹方向に走る飛翔筋　66
背腹方向に走る飛翔筋組織　66
背部中央　135
背部と体側　25
背部の　65
背部の剛毛　65
背部の刺毛　65
背部の付属肢　65
背部の模様パターンの地理的変異　94
背部蜜腺　65
ハイブリダイゼーション捕獲法　35
ハイブリッド米　107
ハイフン　107
背方ニューロン　64, 65
背脈管　65, 100
背面　25
背面図　65
背面日光浴　65
背面の　65
背面方向　65
背毛　65
培養液の小分け　16
培養基　53
培養細胞　53
培養室　97
培養する　110
背瘤　37
背粒　37
配列　22, 153, 194
配列アセンブリー　194
配列アライメント　194
配列アラインメント　194
配列間の差異　195
配列多様性　195
配列追跡ファイル　195
配列同一性　195
配列統計　195
配列のアラインメント　16
配列の一致度　195
配列比較　194
配列変異　195
配列保存　195
パイロシークエンシング　176
ハウスキーピング遺伝子　106
ハエ　85
ハエ目　60
ハエ目の昆虫　60
ハエ目の種　60
歯形の　219
はかまの裾のような後翅(尾端)　66
吐き戻し　182
バキュロウイルス　26
バキュロウイルスが宿っている幼虫　120
バキュロウイルス感染　26
バキュロウイルス感染に対する防御　174
バキュロウイルス増殖　26
バキュロウイルスに対する耐性　185
バキュロウイルスに対する抵抗性　185
バキュロウイルスに対する防御　174
バキュロウイルスの運動性　26
バキュロウイルスの感染性　26
白亜紀　51
白色素沈着　233
白色／黄色スイッチ　233
白色／黄色の色彩転換　233
白色対立遺伝子のヘテロ接合　102
白色の帯　232
白色の前翅と後翅のシャッター対立遺伝子　233
白色の斑紋　233
白色のまま　225
白色の蛹化台紙　233
白色抑制遺伝子　119
白色鱗粉　233
白青スクリーニング　233
白帯型と黄帯型　232
バクテリア　26
バクテリア細胞　26
バクテリア人工染色体　25, 26
博物館の標本　141
薄片　120
薄片を作る　192
薄暮　67
薄暮活動性　51
薄暮時の　51
薄膜　120, 218
薄膜の鞘　19
バクミド　25

バクミドDNA　26
バクミド由来ウイルス　26
薄明　51, 55
はく裂　197
暴露　79
暴露時の　79
刷毛　191
刷毛足の　33
波形縁の　73
激しい競争　195
～はさておき　23
ハサミムシ目　57
初めから　55
はじめに　116
播種　62
場所　125
波状の線　232
場所固執性　199
走る　222
パスカル　156
パスツール　158
はずみ　138
ハスモンヨトウ　44, 205
派生グループ　57
派生形質　21
派生形質共有　214
破線　33, 55
破線の灰色の線　55
葉組織　122
バター色のハエ・アブ　33
パターソンのD統計量　159
パターン化遺伝子　159
パターン形成機構　159
パターン指向モデリング　159
パターン認識受容体　159
パターン変化　159
裸　150
ハダニ　205
バタバタさせている翅　84
バタフライガーデニング　34
バタフライガーデン　34
働かせる　78
働きアリ　235
鉢　169
ハチ　231
ハチクイモドキ　140
蜂の遺伝子セット　231
蜂の共通祖先　44
ハチの尻振りダンス　231
蜂の卵　231
蜂の幼虫　231
蜂の卵巣　231
パチパチという音　51
ハチミツ溶液　104
ハチ目　107
ハチ目の昆虫　107
ハチ目のレクチン　107
爬虫類　184
波長特異的な行動　232
波長領域　232
バチルスチューリンゲンシス　25
発育　58
発育安定性　58
発育異時性　58
発育が阻害された蛹　58
発育過程　58
発育が抑制された蛹　58
発育期　58
発育期間　58
発育季節　58
発育継続の効果　69
発育限界温度　218
発育恒常性　58
発育時間　58
発育時間の性的二型　196
発育ステージ　58
発育相　58
発育速度　58, 179
発育段階　58
発育段階の差　59
発育遅延　23, 58
発育遅滞　22
発育調節因子アケロン　58
発育停止　22
発育日数　98
発育の比率　179
発育プログラム　58
発育誘導　58
発育率　58
発育零点　58
発音行動　201
白化異常型　11
白化型　15, 233
発芽型ウイルス　33, 34
白化型形成　233
白化型生産　233
白化型誘起　233
発芽する　95
発芽成功　95
白化の　15
はっきりした　31
はっきりしない　150
はっきりわかる　77
初記録　83
罰金　83
バックグラウンド的絶滅　25
バックフィル　25
発見　61
発現　21, 79
発現安定性　79
発現遺伝子配列断片　75, 79
発現遺伝子配列断片データベース　75
発現系　79
発現傾向　79
発現検証　79
発現調節　79
発見的探索法　102
発見的方法　102
発現配列タグ　75
発現パターン　79
発現比の差　86
発現プロファイル　79
発現プロフィール　79
発現変動遺伝子　60
発現量　79
発現量が異なる遺伝子　60
発現量の倍数差　86
発現レベル　79
発香器官　19, 191
発香器官系　19
発香器官分泌物　19
発光クラゲ　29
発酵した　82
発酵した果実　82

発香腺　190
発香総　50
発香鱗　19, 191
伐採　125
伐採された生息地　125
伐採された森　125
伐採されなかった森　227
発祥地　165
発色結合ポケット　39
発疹　179
発する　71
発生　58, 71, 91, 154
発生遺伝機構　58
発生遺伝子　58
発生回数　150
発生学　71
発生期間　85
発生機構　58
発生休止状態の休眠　58
発生経路　58
発生拘束　58
発生時期の斉一化　72
発生時刻　72
発生上のスイッチ機構　58
発生上の制約　58
発生上の強い制約　209
発生する　151
発生生物学　58
発生相同　58
発生地　165
発生調節因子アケロン　58
発生的恒常性　58
発生的選択　58
発生的な立場　207
発生の異時性　58
発生の中心　37
発生の不斉一性　24
発生パターン　72
発生プログラミング　58
発生プログラム　58
発生ホモロジー　58
発生率　58
罰則　160
バッタ　97, 125
発達時間　58
発達の悪い目玉模様　232
発達率　58
バッタ目　154
パッチ　158
パッチ間移動　115
パッチ(状)構造　158
パッチ状分布　158
パッチ占有率　158
パッチネス　158
パッチネス指数　131
パッチネットワーク　158
パッチへの移住　158
初通過時間　83
発展史　58
発展性のある考え　194
発展中の研究分野　152
発展的固有　58
初到達時間　83
初発見　83
発毛　222
派手な　209
派手な眼状紋　209
鼻　200
花化学　85
花が放散する揮発性ベンジルアセトン　85
花形　85
ハナグモ　85
花植物　85
花選好性　85
花園　85
花に関する広範な局所的知識　79
花の香りの発散　85
花の香りの放散　85
花の揮発性成分　85
花の揮発性物質　85
花の群生　42
花の形状　85
花の香気成分　85
花の構成　85
花の個体数　85
花の選択　39
花の発散香気成分　85
花の訪問者　85
ハナハッカ　129
パナマの森　156
花由来の揮発性ベンジルアセトン　85
離れた系統　62
離れた分岐群　62
離れて－　11, 21
離れているらしい　176
はにかみ　51
葉に擬態　122
翅　234
－翅　11
翅色選好の刺激　234
翅裏面　126
翅鉤　88, 99
翅荷重　234
翅型　234, 235
翅形　234
翅関連形質　234
翅基部　234
翅形態　234
翅原基　234
翅細胞　234
翅組織　234
翅多型　234
翅棘　88
翅二型　234
翅二型のキリギリス　234
翅によるフリック音　234
翅による摩擦音　234
翅の一次神経枝　172
翅の裏面　229
翅の折りたたみ機構　131
翅の折りたたみ系　234
翅の外縁部　234
翅の開張　234
翅の硬さ　234
翅の擬態紋様　136
翅の強度　234
翅の形態形成　234
翅の黒化の程度　79
翅の色素　234
翅の状態　234
翅の先端　234
翅の損傷度　234
翅のパターン形成候補遺伝子　234
翅の斑紋　234
翅の表皮細胞　234
翅の表面　234
翅の部位　234

翅の辺縁部の延長部　173
翅の膜部分　158
翅の脈系　234
翅の脈相　234
翅の紋様形成　234
翅の紋様要素　234
翅の鱗粉　234
翅パターン化遺伝子　234
翅パターンの多様性　234
翅パターンの表現型多型　234
翅発生　58
翅表面　227
翅変異　234
翅面積　234
翅をパタパタと開閉する　85
翅を開いた　154
歯のある　57
葉の硬さ　122
葉の硬化　220
葉の光合成率　122
葉のシェルター　122
葉の中央脈　135
葉の表面　122, 227
母親　82
母親種　130
母株　140
羽ばたき　234
はばたき反応　85
羽ばたきをする　84, 85
はばたく　235
ハバチ　190
ハバロフスク　119
ハパントタイプ　99
ハビタット　98
パピリオクローム　157
パプアコムラサキ　21
パプアニューギニア　157
パブリックドメインソフトウェア　175
ハプログループ　99
ハプロタイプ　99
ハプロタイプ解析　99
ハプロタイプ多様性　99
ハプロタイプ地図　99
ハプロタイプネットワーク　99
葉粉末　122
～は別にして　23
ハマキムシ　122
ハミルトン則　99
ハミルトンの法則　99
葉物　86
早いものと遅いものを両極に持つ連続体　81
早い幼虫成長　81
パラオ諸島　156
ハラスメント　99
パラタイプ　157
パラダイムシフト　157
パラビオーシス　157
パラフィルム　157
パラホルムアルデヒド　157, 161
パラメータ推定値　157
パラレクトタイプ　157
パラログ　157
バランス変動　197
バリアー抵抗　26
針状　205
針状の　209
針状の形状　144
葉リター　122
ハリネズミ状紋　100
春型　206
春型成虫　206
春型雌　206
春型誘導　206
パルス音　201
春世代　206
春の　229
バルバ　228
ハルペ　100
葉を出す　122
反－　20, 109
半－　100, 194
範囲　22, 178
範囲外の－　79
範囲境界　178
範囲の端部　178
範囲の断片化　178
範囲の分断化　178
繁栄　173
繁栄する　218
繁栄の段階　173
バンカー法　26
半化性　101, 194
半環状　133
半乾燥地　193
半乾燥地域　193
半径の増分　178
半月紋　127
半減期　99
パン酵母　189
パン酵母のゲノム　93
犯罪　221
犯罪行為　51
反軸側の　11
半自然草地　194
半自然な生息地　194
半自然の生息地　194
半翅目　100
半社会性ルート　194
反射光　181
反射層板　214
半種　194
晩秋　121
反証　81
繁盛　173
板状感覚器　165
板状感覚子　165
反証をあげる　62
繁殖　184
繁殖価　184
繁殖隔離　184
繁殖干渉　184, 187
繁殖期　32
繁殖期間　184
繁殖場　32
繁殖する　32, 141, 184, 218
繁殖する地域　32
繁殖成功度　184
繁殖潜在力　184
繁殖戦略　184
繁殖体　173
繁殖地　32
繁殖地間の距離　43
繁殖適応度　184

繁殖投資　184
繁殖努力　184
繁殖能力　184
繁殖の偏り　184
繁殖の公平さ　184
繁殖場所　32
繁殖への投資量　116
繁殖領域　32
繁殖力　81, 184
繁殖力のある　82
半数体　99
半数体集団の有効な大きさ　99
半数体の遺伝子型　99
伴性　196
伴性遺伝　196
伴性遺伝子　196
伴性遺伝の　196
汎生殖個体群　156
伴性の休眠反応　196
汎存種　50
反対側の－　11
反対色反応　44
反対に－　225
反対の極　153
反転　116
斑点　205
反転性伸縮突起　77
斑点のある　205
斑点米　159
半島　160
半透明の　221
バンドパターン　26
ハンドペアリング　99
ハンドペアリング法　99
パントラップ　156
汎熱帯型　156
万能家　91
反応基準　179
反応規準　179
反応系　185
反応継続期間　67
反応する　186
反応潜時　121
反応特性　185
反応変数　186
反応様式　185
反応率　185
半倍数性　99
半倍数性の性決定システム　99
反発行動　185
板部　120
反復　184
反復関与　183
反復実験　184
反復配列　184
反復補充　183
反復名　215
反復要素　184
反復リサンプリング　183
ハンプソン式　99
ハンプソンの分類　99
板部の　120
繁茂している緑色の葉　127
繁茂する　85
斑紋　129, 159
斑紋プレパターン　129
斑紋変異　129, 234
斑紋変化　204
半優性突然変異　194
汎用プロトコール　91
半落葉性中生林　194
半落葉性の中生植物の森　194
範例シフト　157
反論　48
反論する　62

ヒ

比　177
非－　11, 61, 78
ピアソンの相関係数　159
日当たり条件　211
日当たりのよい地域　211
ピアレビュー　160
PR 遺伝子　158, 170
PI 分布　164
BI 法による系統樹　27, 28
BI 法の解析　28
Ben 遺伝子　27
BEN タンパク質　27
pH　169
BAC クローン同定　25
PAUP　163
pFBD-多角体タンパク質プロモーター導入ベクター　161
pFBD-pH 導入ベクター　161
pFBD-ポリヘドリンプロモーター導入ベクター　161
BLL 遺伝子　30, 32
p 距離　155
ピーク感度　159
ピーク吸収　159
ピーク個体数　159
ピーク方向　159
ビークマーク　27
PCR アンプリコン　159
PCR 混合物　159
PCR 産物　159
PCR 断片　159
PCR に基づくマーカー　159
PCR の温度プロファイル　159
PCR 法　159
PCR 由来プローブ　159
ヒース地帯　100
ヒース地帯の植生　100
ヒースランド　100
非依存性感染　110
p 値　155
PD 軸　159, 174, 175
PDV レクチン　159
p10 プロモーター　156
ピート　159
非移動型　147
非移動性種　147
非移動性蝶種　147
非移動性の　147
ビート仮説　27
非意図的導入　226
PBS 洗浄　159
BV 遺伝子　34
緋色　190
非ウイルス性病原体　148
ピエリシン　164
ビオトープ　29
控えめな説明　158
比較　45

尾角 37
比較可能な密度 45
比較形態学的研究 45
比較形態学的方法 45, 134
比較ゲノム解析 45
比較ゲノム学 45
比較研究 45
比較 Ct 法 45
比較的少ない 183
比較的低い温度 183
比較的弱々しい昆虫 183
尾角の 37
比較マッピング 45
日陰生息地 196
日陰生息地の許容性 219
日陰生息地の耐性 219
日陰になった生息地 196
日陰の植生 196
東アジア 68
非加重結合法 227
非加重平均結合法 227
光- 162
光異性化する 162
光異性化閃光 162
光エネルギー量 18
光り輝く虫 205
光感覚器 162
光環境 123
光感受性介在神経 162
光感受性発色団 123
光色素 162
光色素含有微絨毛膜 162
光刺激 153
光受容器 162
光受容器の可能性をもつ小器官 176
光受容器のスペクトル的に異なる型 204
光受容器の単一スペクトル型 199
光受容細胞 162
光受容細胞核 162
光受容体 162
光受容体物質 123
光条件 123
光照射 123
光中断効果 69
光中断反応 123
光の波長領域 123
光パルス 123
光を当てる 197
微環境 135
非感染細胞 147
非感染雌 226
引き起こす 32, 36, 78, 174
引き下がる 235
引き裂く 215
非寄主の種 147
微気象 134
微気象制約条件 134
微気象の影響 111
非季節的移住 23
非擬態型組み換え種 148
非擬態中間型種 148
非擬態的表現型 148
非擬態の 147
引きちぎる 215
非機能性遺伝子 147
尾脚 18
非求愛行動 146
非休眠個体 147
非休眠性幼虫発育期間 147
非休眠になる場合の幼虫発育期間 147
非休眠蛹 147
非休眠幼虫 147
非休眠卵 147
ビクトリア朝 230
日暮 224
日暮信号 224
鬚 156
非系統樹ネットワーク 147
非血縁者との交配 154
鬚の 156
ビコイド遺伝子 27
尾鉤 51
非公開データ 227
飛行境界層 84
飛行経路 86
微高地 135
飛行能力 84
飛行方向の機構 131
非コード RNA 143, 147
非コード配列 147
非コード領域 148
腓骨の 83
ひこばえ 211
ピコモル 166
膝 94
尾鰓 36
微細空間スケール 83
微細構造 83, 134
微細焼灼 134
微細スケールの空間(的)モザイク 83
微細な 83
微細な鉗子 83
微細なハサミ 83
微細ブラシ 83
非在来種 147
非在来生物種 148
尾索動物 228
非作物植物 147
非作物植物の汚染 48
非雑種の雌 176
微刺 13
ひし形 58
ひし形記号 58
被子植物 19
被子植物の寄主植物 19
被子植物の花 19
被子植物門 20
微視的な細胞内共生生物 134
微視的な標本 134
非社会性昆虫 147
微絨毛 135
非宿主種への刺入 208
非宿主種を刺す行動 208
飛翔 84
飛翔音 85
飛翔型 86
飛翔活動 84
飛翔関連形態 85
飛翔関連行動 85
飛翔期間の長さ 122
飛翔期間の頻度 88
飛翔筋 85
飛翔筋重量 85
飛翔筋多型 85

飛翔空間　85
飛翔経路の形状　95
飛翔経路のジオメトリー　95
微小結節　134
飛翔行動多型　84
飛翔時間　85
飛翔写真　164
非ショウジョウバエ様式　147
微小針状体　13
飛翔する　85
飛翔する領域　84
飛翔制御　49, 85
微小生態系　134
飛翔速度　85, 204
微小注射　134
飛翔調節　49, 85
非冗長なタンパク質データベース　149
尾状突起　103, 214
尾状突起付きアゲハチョウ　213
尾状突起を欠く　214
微小な蜂　136
非常に－　160
非常に大きな個数　80
非常に大きな割合　80
非常に局在化した種　103
非常に詳細　97
尾状の　36
飛翔能力　84
飛翔の解析　18
飛翔の季節　85
飛翔バウト回数　88
飛翔バウト(一続きの行動期間)の距離　62
飛翔パターン　86
非消費効果　147
飛翔様式　137
飛翔領域　84
被食者　171
被食植物　111
非自律　147
非神経支配の微毛　147
披針形の　209
非侵略性　147
ヒステロテリー　108
ヒストン　103
微生息場所　134
微生物　134, 158
微生物感染　134, 158
微生物群集　134
微生物集団　134
非生物的環境　12
微生物的殺虫剤　134
非生物的ストレス　12
非生物的ストレス応答　12
非生物的制約　12
微生物的防除　134
非生物的要因　12
尾節　18, 107, 215
尾節の　107
非漸近的種数　148
非線形性　147
非線形の　148
非選好性　147
非占有場所　227
非相称スケルトン光周期　23
非相称的樹形　24
非相称枠光周期　23
腓側　83
ひそむ　100
ひだ　86
ひたい　88
非対称遺伝子流動　23
非対称的樹形　24
非対称の　23
非耐凍型昆虫　88
非耐凍型種　88
非耐凍性　88
非耐凍性昆虫　88
非耐凍性種　88
ヒタキ　86
ひだ状の触角　120
日溜り　211
ビタミン　230
ひだ模様　51
尾端　219
尾端光受容器　93, 214
微地形　135
日付　55
引っ込む　235
ヒツジ　197
羊　197
必須条件　170
必須の分類群　150
ヒッチハイキング　103
ヒッチハイキング効果　103
筆頭著者　83
ヒットした登録番号　103
ピットフォール　165
必要条件　170
否定－　55
否定する　181
非適応的　148
非適応的組み換え種　148
日照り　67
ビテリン　230
ビテロジェニン　230
非同義性固定置換　84
非同義置換　147, 148
非同義置換と同義置換の比　118
非同義置換率　64
非同義・同義塩基置換数比　64
ひとかたまりの種　127
ヒトゲノム　106
ヒトゲノム計画　106
ヒトゲノム中の遺伝子総数　220
ヒトゲノムの塩基総数　220
人里　106
等しく　75
尾突　201
一つしか含まれない分類　138
ヒト免疫不全ウイルス　103, 106
一文字状白帯　208
ヒトリガ　90, 219
ヒトリガ科の蛾　22
一人の記録者　199
避難区域　181
避難場所　181
非任意分散　147
非野焼き　225
火花　123
非パラロガス遺伝子　148

日々の訪問 54
日々(の)変化 55
批評 44
非標的種 147
非標的無脊椎動物 147
皮膚 113, 215
被覆 51
非復元抽出 179
被覆されている領域の深度 51
皮膚腺 57
皮膚表面 53
ビブリスハレギチョウ 181
非分散移動 148
微胞子虫 135
微胞子虫病 135
非放出性遺伝形質転換タバコ植物 147
備忘録 132
非翻訳領域 148, 228
ヒマラヤ 103
ヒマラヤ型 103
ヒマラヤ蝶紀行 184
ヒマラヤの蝶類レポート 184
ヒマラヤ北西部 149
ヒマワリ 211
肥満体 81
美味 156
美味性 156
微妙な局所的変異 200
ヒメアカタテハ 156
ヒメギフチョウ 127
ヒメシジミ亜科 31
ヒメシルビアシジミ 237
ヒメフンバエ 236
ヒメホリカワコウモリ 14
被毛 229
微毛 13, 135, 175
尾毛 37
微毛長 122
微毛の長さ 122
微毛の密度 57
微毛密度 57
非モデル昆虫種 148
100 万 135
100 万年 141
100 万年前 141
100 万分の 1 170
百万分率のこと 170
白夜の太陽 135
冷やす 181
ビューフォート風力階級 27
ヒューリスティックス 102
被蛹 151
費用 50
尾葉 18, 37
費用がかかる 50
氷河期 95
評価基準 23
氷河期の退避地 95
氷核形成 108
評価計画案 23
氷河時代 95
評価指標 76
評価プレート 23
評価プロトコール 23
氷期 95
病気の雑種同腹仔 61
病気の症状 61
病気の大流行 61
病気の発生 61
表形学 161
氷結 89
表現 79
表現 - 161
表現可塑性 79, 162
表現機構 162
病原菌 134
病原菌認識 158
病原菌媒介生物 229
表現型 162
表現形 162
表現型可塑性 162
表現型可変性 162
表現型効果 162
表現型酷似 162
表現形質 162
表現型進化 162
表現型多型 167
表現型多型の発生 167
表現型多様性 162
表現型的に異なる個体群 162
表現型の可塑的反応 166
表現型の効果 162
表現型の転換 162
表現型の平行進化 157
表現型発現 162
表現型分化 162
表現型分岐 162
表現型発散 162
表現型変異 162
表現図 161
病原性遺伝子 230
病原性因子 230
病原性ウイルス 158
病原体 158
病原体感染 158
病原体に対する反応 185
病原体認識 158
病原体の遺伝子資源 158
病原力 230
標高 17, 71, 191
標高クライン 17
標高傾度 17
標高変動 71
氷山の一角 219
表翅 227
表示 62
標識 194
標識擬態 136
標識再捕獲統計 129
標識再捕獲統計学 129
標識再捕獲法 129
標識再捕実験 129
標識再捕法 129
標識的擬態 136
標識物質 129
標識 - 放出 - 再捕獲の研究 129
標準化された回帰係数 207
標準化された調査 207
標準化された比較 207
標準化石 110
標準曲線 207
標準ゲノム 181
標準誤差 191
標準データセット 27
標準的な飼育温度 207
標準的な実験手順書 207

標準的に通用している 207
標準的比較 207
標準的標本 207
標準手順書 207
標準のチョウ目のプライマー 207
標準の鱗翅類のプライマー 207
標準プライマー 207
標準プロトコル 207
標準偏差 207
標準マーカー 207
氷晶核 108
表題 219
標徴 38
病徴 61, 213
標的遺伝子 214
標的器官 214
氷点下 89
氷点下の気温 211
氷点下の日数 150
病毒性 230
表皮 53, 74, 200
表皮器官 74
表皮形成 53
表皮形成組織 53
表皮細胞 74
表皮細胞シート 74
表皮細胞層 74
表皮成長因子 70, 74
表皮タンパク質 53
表皮タンパク質の架橋 52
表皮の厚みのある隆起 218
表皮の強固さ 53
表皮の超微細構造 53
表皮反応 74
表皮鱗粉 53
標本 171, 204
標本採集地 189
標本抽出率 189
標本の証拠書 204
標本の命名 142
表面 54, 227
表面形状 212
表面構造 212
表面上の分類学的(意思)決定 212
表面積 212
表面積／質量比 212
表面積／体重比 212
表面的な分類学的意思決定 212
表面の 65
表面(上)の 212
表面領域 212
ヒョウモンチョウ 88
ヒョウモンドクチョウ 158
ヒョウモンモドキ 38
肥沃にする 82
日よけの場所 22
ひょっとすると 46
日和見休眠 153
日和見的な生活史戦略 153
飛来個体数 150
飛来してくる無交尾の雌 110
飛来雌 110
開けた狭い草地 152
開けた地域 152
開けた場所 152
開けた陽のあたる生息地 152
開けた落葉樹林 152
平皿トラップ 156
ヒラタカメムシの一種 22
平べったい先端 84
ピラミッド形のもの 176
非ランダム分散 147
ピリオド遺伝子 160
ビリオン 230
比率 177
比率的体重消失 173
比率的体重喪失 173
ビリベルジン IX 28
ビリベルジンタイプ 28
微粒 83
微量遠心管 134
微量遠心チューブ 134
微量定量プレート 135
ビリン 28
ビリン結合タンパク質 28
非類似度行列 62
ヒルトッピング 103
昼間の 63
美麗種 96
非連鎖カラーパターン遺伝子座 227
広い意味で 33, 188, 194
広い変異能力 233
広がる 161, 205
広く受け入れられた見方 233
広く行われている方法 171
広く分散する能力 103
広く分布し連続域に生息する個体群 120
広く分布する種 233
広げる 209
ピロシークエンシング 176
ピロリジジンアルカロイド 156, 176
火を付ける 108
敏感期 194
敏感な気候 210
ビンクルム 230
貧困 169
品種 80, 177, 228
品種間の 114
頻度 88
頻度依存性捕食 88
頻度依存選択 88
頻度依存的な競争 88
頻度依存的な捕食 88
頻度の中心 37
頻度分布 88
頻発する落し穴 180
品目 23

フ

不－ 20, 61
ファイトアレキシン 164
ファウナ 81
ファラデーケージ 81
ファラデー箱 81
ファルス 161
ファレート蛹 161
ファレート状態 161
ファレート成虫 161
ファレート幼虫 161
ファロイジン 161
不安定な夏の天候 227

不安定な防御脂質　119
ファントホッフの法則　228
フィールド　83
フィールド遠征　83
フィールド観察　83
フィールド研究　83
フィールド支援　83
フィールド実験　83
フィールド周縁部　83
フィールド状況　83
フィールド証拠　83
フィールドメモリーレコーダー　83
V 字形の斑紋　228
フィッシャーの正確確率検定　84
不一致　61, 110
不一致率　179
フィットネス　84
部位特異的組み換え　199
フィブリノペプチド　83
フィラメント　83
フィルタリング手順　83
フィンチ　83
風変わりな　30
封じ込め　48
封じ込め地域　48
風速　234
ブータンシボリアゲハ　127
ブータン東部　68
風潮　218
フウチョウソウ科　35, 41
風洞実験法　234
ブートストラップ　31
ブートストラップ確率　31
ブートストラップサポート　31
ブートストラップ値　31, 34
ブートストラップ複製　31
ブートストラップ法　31
封入された細胞　140
封入体　108, 110
封入体症　110
封入体病　110
プールされた個体　167
プールされた方形区　167
フェナントリジンアルカロイド　161
フェニルアラニン　162
フェノールオキシダーゼ　167
フェノール - クロロホルム　161
フェノール - クロロホルム抽出法　161
フェノール酸化酵素前駆体活性化因子　173
フェノール赤色染料　161
フェノールレッド染料　161
フェノロジー　161
フェラムリン　161
フェロモン　162
フェロモン生合成活性化神経ペプチド　159, 162
フェロモン前駆体　162
フェロモン前駆物質　162
フェロモンの運搬　162
フェロモンの輸送　162
フォイルゲン反応　82
不応期　181
フォーカス　86
フォーカス以外の部位切除　147
フォーカスシグナル　86
フォーカスの移植　86
フォーカスの仕様　204
フォーカスの切除　86
フォーゲル器官　231
フォーゲル器官の存否　171
フォスミド　87
フォスミドクローン　87
フォスミドシークエンシング　87
フォスミド配列　87
フォスミドライブラリー　87
フォルカビリン　162
フォワードプライマーとリバースプライマー　87
腐果　55, 56
孵化　71, 100
深い関係　115
不快性　227
深い森の景観　100
孵化器　110
深く分岐した系統　56
不可欠な　110
孵化サイズ　100
孵化した卵　100
孵化したての雌　145
富化水準　123
孵化する　71, 100
孵化成功　100
不活性代謝物　111
付加的な　14
不可分　113
深みのある　56
不完全変態　100, 110
不完全変態する昆虫類　100
不規則な休眠に関連する現象　117
不規則な突起部　117
不揮発性化合物　148
腐朽　56
普及啓発　175
不休眠条件　147
不休眠成虫　147
不休眠世代　147
不休眠選択　147
不休眠で発育した個体　60, 61
不休眠蛹　147
不休眠幼虫　147
不休眠卵　147
不均一な混み具合　226
不均一な生息地　102
不均一に発現したフィルタリング色素　102
不均一パラメータ　158
不均衡　61
不均質性　101
不均質選択　102
不均等性　19
副花冠　50
複眼　46
副基準標本　50
腹脚　11, 173
腹胸側線　208
複合(の)　46
複合活動電位　46

副後基準標本　157
複合区間マッピング法　46
複合顕微鏡　46
複合種　203
複合体　45
複合ピーク　45
複合名　46
複雑な局所的オログラフィー　45
複雑な局所的山地地形　45
複雑な状況　45
複雑な統計的補正　45
複雑な論争点　45
副作用　197
副産物　34
複式顕微鏡　46
副視髄　12
輻射熱　178
腹上側板溝状線　208
複女王性　167
複数回繁殖の種　117
複数箇所営巣性種　141
複数(の)季節　141
複数原綴り　141
複数個の新種名　205
複数個の非同義置換率と同義置換率の比　95
複数世代　141
複数単眼優性ニューロン　141
複数年期間　140
複数のヌクレオキャプシドを含むNPV　137, 141
複数のヌクレオキャプシドを含む核多角体病ウイルス　137, 141
複数の豊富な資源のある地域　141
複製　184
複製単位　226
腹性防御腺　11
副(次)専門分野　209
腹側　228, 229
腹側からの光　229
腹側から見た図　229
腹側の　229
腹側の青白いベージュ色の後翅　156
腹側の後翅　229
腹側の翅面　229
腹側の伸長した眼状紋列　73
腹側の翅の斑紋　229
腹側の翅の目立つ斑紋　48
腹側表面　229
腹側網膜　229
腹中線　135
副転節　223
腹板　208
副(次)標本(を取る)　210
腹部　11, 228
腹部重量　11
腹部神経節　11
腹部の　11, 229
腹部の脚　11
腹部背板　11
腹部腹板　229
腹部防御腺　11
腹部末端　219
腹部末端節　217
腹柄　161
覆片　215
副模式標本　157, 192
ふくれた　213
フクロウ　155
フクロウチョウ　155
父系染色体　158
父系分析　158
不顕性　180
符合する　214
布告書　55
房　224
負鎖RNAウイルス　144
ふ先　21
ふさぐ　151
塞ぐ　151
房状　224
フジウツギ　33
不思議なことに　53
不十分な予測因子　167
腐食性　58, 190
腐食性昆虫　190
腐食(性)生物　58
付随研究　14
付随コメント　12
付随論評　12
父性　158
不斉一化　147
不斉一性　147
不斉一発生　24
不斉休眠　147
不整形な突起部　117
不整成虫　216
不正な後綴り　110
不正な原綴り　110
不正評価する　110
父性分析　158
防ぐ　57
跗節　215
ふ節　215
ふ節亜節　215
ふ節式　215
ふ節小節　215
ふ節端突起　165
ふ節端葉片　165
ふ節の　215
ふ節の剛毛列　215
ふ節の接触化学感覚毛　215
ふ節盤　76
付属器　21
付属肢　21
付属肢の先端　62
付属腺　12
付属突起　21
ブタ　164
フタオチョウ　38
フタオチョウ亜科　38, 122
不確かさ　17
再び居住させる　184
二つの顕著に区別できる表現型　224
二つの役割　67
二方式に　28
二股賭け戦略　28
ふ端中片　165
縁どり　31
縁どりのある斑紋　187
付着した外来種　103
不注意　227
不調和　158

普通種 45
普通種と希少種 45
普通地域 153
普通にみられる微生物 44
普通鱗(片) 153
伏角 60
復帰 186
復帰突然変異 25
復権 182
不都合な条件 226
不都合な非生物的条件 226
物質循環 150
物質代謝 133
プッシュ・プル法 176
沸点 31
ブッドレア 33
不釣り合いなほど多くの 62
不釣合いな面積 110
物理的距離 163
物理的性格 163
物理的相違 163
物理的特性 163
物理的防除 163
物理的要因 163
物理的連鎖 163
仏領ギアナ 88
不定芽 14
腐泥食性 58
腐泥食(性)生物 58
不適応 128
不適格性 225
不適格な著作物 225
不適格な命名法的行為 225
不適格名 225
不適切名 109
不適当な組み合わせ 136
プテリジン環 175
プテリジン系色素 175
プテリン系色素 175
浮動 66
ぶどう園 230
不動化 109
不等交叉 226
浮動選択 66
不凍タンパク質 21, 218
不当な修正名 226
ブドウの木 230
不等分散 226
フトオビルリツバメガ 237
太線 31
太らす 81
ブナ帯 27
腐肉 36
腐肉食(性)動物 190
不妊カースト 208
不妊性 208
不妊虫放飼法 199, 208
不稔性 208
負の影響を与える 144
負の関連 144
負の効果 144
負の自然選択 144
負の自然淘汰 144
負の相関 144
負の対照(区) 144
負の淘汰 144
負の二項(分布) 144
負のフィードバック 144
腐敗 56
腐敗しかけた 55
腐敗している果実 55, 56
部分 158
部分化性 158
部分個体群 210
部分精製された前胸腺刺激ホルモン 158
部分的な解決(策) 158
部分的に未知 158
部分的保護種 158
部分母集団 210
不変 116
普遍遺伝暗号 226
普遍化 91
普遍種 50
不偏性相利共生 80
普遍性の種 91
普遍的 226
普遍的な形質 226
普遍的な発生 226
普遍的なプライマー 91
普遍的パターン 226
不変部位 116
普遍プライマー 91
踏み石の生息地 207
踏みつける 220
不明瞭な 150
不明瞭な褐色の外側の環 110
冬枯 235
冬休眠 235
冬の生存率 235
冬日 235
冬芽 235
腐葉土 107
プライオリティ 172
プライベートアリル 172
プライマー 172
プライマー設計 172
プライマー対 172
プライマーの組み合わせ 172
プライマーフェロモン 172
ブラコウイルス遺伝子 32, 34
ブラコウイルス遺伝子配列 32
ブラコウイルス関連の蜂 32
ブラコウイルスタンパク質 32
ブラコウイルス毒性タンパク質 32
ブラコウイルスの生活環 32
ブラコウイルスの挿入 32
ブラコウイルスの注入 32
ブラコウイルス配列の順化 65
ブラコウイルス様 *Ben*(遺伝子)配列 32
ブラコウイルス様レクチン遺伝子 30, 32
ブラコウイルス粒子の侵入 32
ブラコウイルス - レクチン様タンパク質 32
プラスチックカップ容器 166
プラスチック瓶 166
プラスチック袋 166
プラスチックボトル 166

BLAST 検索　26, 30
BLAST 比較　30
プラズマ細胞　166
プラスミド　166
プラスミド精製　166
プラスミド精製法　166
プラスミド調製　166
ブラックリスト　30
フラビンアデニンジヌクレオチド　80, 84
フラボノイド　84
フラボノイド系色素　84
フラボノイドの蓄積　195
フラボノール　84
フラボノールグリコシド　84
フラボノール配糖体　84
フラボン　84
フランキング配列　84
フランキング領域　84
不利　111
ブリアムストリバネアゲハ　34
プリカーサー　170
ブリコラージュ　32
プリセット　171
不利な条件　226
振り回し式乾湿計　200
振り回し湿度計　200
不良な生息地　226
不慮の過誤　109
プリン環　176
フリンジ　88
プリンターの間違い　172
プリン誘導体　176
ふるい操作　197
ふるい分け　191, 197
ブルーシフト　30
ブルード　33
ブルード比較　33
ブルーモルフォ　139
ブルシコン　33
ブルジョワ戦略　32
プレインキュベートする　171
フレームシフト　84
フレームシフト突然変異　88
プレパターン　171
プレパラート　171
触れること　220
不連続な帯　33
不連続な翅のパターン要素　61
不連続分布　61
プロウイルスが組み込まれたサークル(環)配列　174
プロウイルス型　174
ブロードキャスト　33
プローブ　172
フローラ　85
プログラム細胞死　21
プログラムされた細胞死　173
プログラム(化)できる　173
プロスペロ遺伝子　174
プロセス　172
プロセテリー　174
プロテオーム　174
プロテオミクス　174
プロトコール　174
プロフェノールオキシダーゼ活性化因子　173
プロボースィペディア遺伝子　159
プロモーター　173
プロモーター p10　156
プロルサ型　173
不和合性　110
不和合性交配　110
不和合性誘発微生物　110
糞　67, 88, 190
雰囲気　24
粉衣された種子の条播き　66
分化　60, 63, 203
分解　56
分解者　56
分解生物　56
分隔甲　163
分化した　203
分化する　60
文化庁(日本)　15
分割　158, 205
分割特異的な置換モデル　158
分割頻度　205
分化の中心　37
分化の中心地　37
分化の歴史　203
分岐　63, 64
分岐解析　40
分岐学　40
分岐関係図　40
分岐群　40
分岐鎖アミノ酸　27, 32
分岐時間　63
分岐した個体群　63
分岐進化　40, 63
分岐図　40
分岐する　63, 205
分岐選択　40
分岐年代　63
分岐の先験的選択　11
分岐配列　63
分岐パターン　63
分岐分類主義　40
糞球　160
分極化された　166
分岐論　40
文献探索技能　200
分光応答度　204
分光学的特徴　204
分光学的に　204
分光学的微調整　204
分光感度　204
分光光度計　204
分光反射率　204
粉砕　197
分散　61, 135, 228
分散移動　62
分散型　61
分散傾向　61
分散形質　61
分散行動　61
分散剤　61
分散する　63
分散性向　61
分散多型　61
分散能力　61, 62
分散の進化　77

分散の潜在能力 61
分散分析 18, 20
分散分布 62
分散率 61
分散力 61
分子 138
分子遺伝学 138, 221
分枝過程 32
分子機構 138
分子技法 138
分子系統(学) 138
分子系統解析 99
分子系統学的解析 138
分子系統学的観点 138
分子系統学的研究 138
分子系統学的証拠 137
分子系統樹 138
分子系統分類学 138
分子サイン 138
分子シグネチャー 138
分子指標 138
分子署名 138
分子(的)進化 137
分子進化(学)的解析 137
分子進化(学)的研究法 137
分子生物学的証拠 137
分子成分 137
分子データ 137
分子的証拠 137
分子的対抗適応 137
分子時計 137
分子時計の単位 226
分子分散分析 18
分子マーカー 138
噴射 205
分集団 210
分子レベル 138
分節 193
扮装擬態 129
吻側 188
吻足遺伝子 159
分断 62
分断化された景観 88
分断化された個体群 88
分断化生態選択 62
分断された地形 88
分断自然選択 62
分断自然淘汰 62
分断種分化 229
分断色 62
分断性生態選択 62
分断性選択 62
分断性淘汰 62
分断選択 62
分断的生態選択 62
分断淘汰 62
分断分布 229
分断分布を示す種分化 229
分配する 63
糞蝿 67
分泌(物) 192
分泌顆粒 192
分泌状態 192
分泌する 192
分泌タンパク質 192
分泌物の特異的混合物 204
分泌物の特定混合物 204
分布 63
分布域 63
分布拡大 63, 78, 94
分布型 63, 224
分布系統 224
分布経路 63
分布する 63
分布調査 63
分布南限地 202
分布の断絶 117
分布の中心 37
分布の変化 63
分布パターン 63
分布頻度 63
分布辺縁生息地 129
分布変化 63
分布北限 148
分布要素 63
分布様相 63
文脈依存性 48
噴門弁 36, 208
分野 83
分離 117, 193
分離 - 61
分離した焦点 61
分離する 57, 193
分離線 61
分離帯 194
分離点 194
分離発色団 117
分離分布 61
糞粒 88
分類 40, 215
分類階級 215
分類階級名 36
分類学 214, 215
分類学上の位置 215
分類学上の名称 215
分類学(的)情報 215
分類学的位置 215
分類学的位置づけ 215
分類学的階級 215
分類学的境界 215
分類学的研究 215
分類学的検証 215
分類学的タクソン 215
分類学的単位 215
分類学的な構成 215
分類学的範囲 215
分類学的分布 215
分類学的名称 215
分類群 215
分類群が多いグループ 215
分類群間の成分 18
分類系列 215
分類項目 - 細分類項目の比 36
分類された対象群 40
分類される 80
分類サンプリング 215
分類サンプリングの曲線 215
分類出現頻度 215
分類順位 215
分類数 215
分類組成 215
分類単位 40, 215
分類の階層構造 215
分類配列 215
分類比 215
分類分野 215
分類法 215
分類本位の調査 215
分類密度 215

分類名　36
分裂　62, 64
分裂酵母　191
分裂させる　62
分裂促進因子活性化タンパク質キナーゼ　129, 137

へ

ペアエンドのリード　156
ペア整列　156
ヘアペンシル　99, 106
ヘアペンシルから分泌されるジヒドロピロリジン化合物　99
平圧される　57
平滑な面　200
平均　25, 131
平均音圧レベル　25
平均過冷却点　131
平均距離法　227
平均混み合い度　131
平均棍　99
平均最低気温　131
平均種数値　131
平均深度　131
平均世代時間　131
平均体重　131
平均値　25
平均長(塩基の)　25
平均発現水準　131
平均発現レベル　131
平均分散傾向　131
平均密度　25
平均累積個体数　131
並行進化　157
平行進化　157
平衡推移　197
平衡遷移　197
平衡選択　26
平行的表現型進化　157
平行的変化解析　157
平行発散　157
併合派分類学者　127
平行表現型　157
平行分岐進化　157
平衡密度　75
閉鎖個体群　41
閉翅日光浴　121
瓶首　31
瓶首効果　31
平常活動　148
平常な行動範囲　148
ベイズ系統解析　27
ベイズ情報量基準　27, 28
ベイズ推定法による系統樹　27, 28
ベイズのアルゴリズム　27
ベイズの事後確率　27
ベイズの問題解決手法　27
ベイズ法　27
ベイズ法による系統解析　27
柄節　190
柄側刻　96
併体結合　157
平坦状の　84
平地個体群　126
平地性種　126
平地と高標高地間を垂直に移動する　85
ベイツ型擬態　27
ベイトマンの原理　27
平年　153
平板　166
柄部　161
平面位置表示器　165, 170
並列比較　197
ペースメーカー　156
β-アラニン　28
ベータアラニン　28
β-カロチン　28
ベータカロチン　28
β-グルコシダーゼ　28
β-グロビン　28
β-多様性　28
ベータ多様性　28
ベータ多様性上での種の入れ替わりの影響　111
ペーハー　169
ベールを取る　227
ペガラオオモンヒカゲ　235
ヘキサマープライマー　102
北京市郊外　230
ベクター　228
ベクタークリッピング　228
ベクター媒介性疾患　229
ヘクタール　100
凹む　46
隔たり　90
別-　16
別亜種　62
ヘッジホッグ　100, 102
ヘッジホッグシグナル伝達　100
別種　62
別刷り　151, 184
別属　59
ヘッドスペース法　100
別の状態へ-　220
別の目的で用いる　181
別模式標本　16
ヘディストスジャコウアゲハ　34
ヘテロクロマチン　101
ヘテロ接合雄　102
ヘテロ接合個体　102
ヘテロ接合性　102
ヘテロ接合体　102
ヘテロ接合体遺伝子型　102
ヘテロ接合体個体の分布　102
ヘテロ接合体の　102
ヘテロ接合体の個体　102
ヘテロ接合体領域　102
ヘテロ接合度　102
ヘテロ接合度の期待値　78
ヘテロダイナミック　101
ヘテロダイマー受容体　101
ヘテロな性の染色体　101
ヘテロ二量体受容体　101
ヘテロ二量体を形成する受容体　101
ベトナム中部　37
ペトリ皿　161
ペトロラタム　161
ベニオビイナズマ　76
ベニオビコバネシロチョウ　219
ベニシジミ　200
ペニシリン　160
ベニシロチョウ　153

ベニヒカゲ　17
ベニボシイナズマ　181
ベニモンオオキチョウ　153
ベニモンクロヒカゲ　213
ベニモンドクチョウ　169
ペプチド　160
ペプチドグリカン　160, 161
ペプチドグリカン認識タンパク質　160, 161
ヘモグロビン　99, 101
ヘモフィラス・インフルエンザ菌　99
ヘモリン　101
減らす　60
ペリット　160
ヘリトロン　100
ペルー奥地　161
ヘルパープラスミド　100
ペルム紀　160
ペレイデスモルフォ　44
ペレット化する　160
弁　228
変異　141, 228
変域　64
変異系列　228
変異個体群　228
変異集団　228
変異性　228
変異速度　141, 228
変異体　228
ベン遺伝子　27
変異の最小頻度　136
変異のないサイト　116
変異の幅　178
変異幅　178
変異様式　228
変異率　141
辺縁　212
辺縁の生息地　129
辺縁部　50
辺縁部の単眼　31
変温処理　85
変温性脊椎動物　166
変温動物　166
変温の　43, 69
片害共生　17
片害作用　17
変化させるもの　221
変化した調節　17
変化する　228
変化置換　184
変化程度　56
変化の駆動因子　66
変化の早期警報　68
変換距離法　221
変換する　221
偏光　166
変更　17
変更遺伝子　137
偏向コンパス　166
偏在する　225
遍在発現転写体　225, 228
変種　228
編集距離　69
編集者　69
鞭小節　84
鞭状の　232
ベンジル　28
ペンシルベニア紀　160
ベン図(表)　229
変成岩　133
変成高速液体クロマトグラフィー分析法　57, 58
変性種　12
鞭節　84, 89
鞭節の　84
変体　12
変態　133, 221
変態期　133
変態機構　221
変態する　133, 221
変態脱皮　133
変体綴り　228
変態の開始　152
変態の阻害　112
変態の抑制　112
ベンタンパク質　27
ベンチマークデータセット　27
偏重主義　38
変動　85, 228
変動係数　42, 53
変動主要因　119
変動主要因分析　119
変動する　85
変動性　75, 228
変動体　228
変動非対称性　80, 85
ベンドレス遺伝子　27
ベンドレスタンパク質　27
偏波依存色覚　166
扁平な楕円体状の翅表面　84
扁平な袋　84
片利共生　44
片利共生的　44
片利共生的相互作用　44

ホ

ポアソン過程　166
ポアソン過程に従って起きると仮定した系統樹　166, 175
ポアソン系統樹過程　166, 175
ポアソン分布　166
包囲作用　72
防衛遺伝子　56
防衛行動　56
防衛的グルコシノレート　56
防衛的芳香　56
防衛特性　56
防衛の最前線　83
貿易相手　220
ホウオウボク属の一種　56
崩壊　62
崩壊させる　62
妨害する　109
訪花行動　85
包括的アプローチ　110
包括適応　110
包括適応度　110
包括的なモデリング枠組み　46
包括度　51
放棄　11
胞器　33
紡脚目　71
防御遺伝子　56

防御応答　56
防御化合物の下流の産出機構　66
防御機能　174
防御行動　56
防御腺　184
防御的効果　174
防御的吐き戻し　56
防御的役割　174
防御反応　56
防御物質　56
方形区　176
方形区当りの種数　187
方形区当りの平均樹幹数　131
方形区数に基づく種数　176
方形区法　176
傍系相同　157
方形枠　176
抱鉤　186
膀胱　229
方向移動能力　61
方向移動飛翔　61
芳香(性)化合物　151
方向感覚を失う　61
芳香源　151
方向性効果　61
方向性選択　61
方向性淘汰　61
縫合線　213
芳香組成物　88
方向比　61
報告の続編　48
ホウ砂溶液　31
放散　60, 61, 178
放散過程　60
放散する　178
放飼　183
放飼増強法　24
胞子体　205
ホウジャク　107
放射状型　179
放射状の　177
放射線　178
放射(線)免疫測定法　178, 187
防除　49
帽鞘　35
防除措置　49
防水効果　232
紡錘糸　205
紡錘状の　89
防水性　232
放送　33
膨大部　18
放逐　183
放逐区域　183
放逐実験　183
放逐地点　183
放蝶　183
法的保護種　122
包のう　53
胞のう　189
胞胚葉　30
防備を強化する　87
豊富　12
豊富な　12
豊富な草　12
豊富な証拠　12
豊富な生物多様性　187
豊富にある　12
方法論的進歩　134
放牧　97
放牧パターン　146
包埋型ビリオン　151
包埋体　150, 151
包埋体由来のウイルス粒子　151
訪問回数　150
ホウライカガミ族　158
ホウライカガミ属　158
抱卵　97
法令　48
飽和度検査　190
ホーフバウアー‐ブフナーアイレット　103
ホームベース　104
ホームレンジ　104
ホールデインの法則　99
ホールデンセンチモルガン　99
ホールデンの法則　99
ボールのようなかたちをした毛束　26
ホールマウント　233
捕獲下条件　35
他の科　154
他の側へ‐　220
他の種　154
補間　115
補給する　184
補強　73, 212
補強する　50, 182
保菌昆虫　112
北縁部　148
北欧　148
北限地域　148
牧草地　97, 131, 158
北端　148
ホクベイオオギンモンセセリ　198
撲滅　75
撲滅する　75
牧養力　36
保健休養の目的　180
保護　47
歩行　231
歩行速度　231
保護外被　174
保護機関　47
保護区　174, 181
保護支出　47
保護種　174
保護色　52, 174
保護色効果　174
保護色の色彩多型　174
保護団体　47
保護地　181
保護地域　174
保護地の個体群　181
保護努力　47
保護に対する懸念　47
保護費用　47
保護方針　47
保護用鉄線柵　174
保護鱗粉　174
母細胞　140
星形記号　207
星コンパス　207
保持する　186

ポジティブコントロール 168
ポジティブ細胞 168
ホシホウジャク 107
母株 140
補充 180
補充女王 212
補充する 184
母集団 168
保守的仮説 47
保守的選択 47
保守的に 47
圃場 83
圃場縁 83
保証する 231
補償成長 45
捕食 170
捕食圧 170
捕食寄生(の) 157
捕食寄生者 157
捕食寄生者を誘引する揮発性物質 157
捕食寄生者を誘引する刺激 157
捕食寄生者を誘引する植物刺激 157
捕食寄生虫 157
捕食寄生蜂 157
捕食者 24, 170
捕食者回避 170
捕食者回避反応 170
捕食者攻撃のぶれ 56
捕食者逃避 170
捕食する 171
捕食性のカスミカメムシ 170
捕食性のカブリダニ 170
捕食虫 170
捕食(性)鳥 170
捕食鳥のさえずり 230
捕食鳥の鳴き声 230
捕食(性)動物 170
捕食の本性 143
母植物 140
捕食率 170
補助色素細胞 12
補助的摂食 212
母数効果の分析 18
ポストマン型 169
ホストレース 105
母性効果 130
母性効果優勢胚発育停止因子 131
母性発現遺伝子 130
保全 47
保全決定 47
保全された著作物 47
保全する 47
保全生態学 47
保全生物学 47
保全生物地理学 47
保全戦略 47
保全措置 47
保全対策 47
保全的導入 28, 47
保全努力 47
保全方針 47
保全名 47
細い 200
細い毛 200
舗装道路 159
補足 187
ホソチョウ亜科 13
細長い 124
細長い光受容細胞 71
ホソバセセリ 198
保存 47
保存されたスーパー遺伝子座 47
保存されたタンパク質非コード領域 42, 47
保存された発生経路 47
保存された PCR プライマー 47
保存されたポリメラーゼ連鎖反応用プライマー 47
保存されているウイルス性部位 47
保存されている領域 47
保存する 171
保体 186
母体型の影響 130
補注 212
捕虫網 144, 213
母蝶 82
北寒帯 22
北極圏種 22
北極帯 22
北極(圏)の 22
Hox 遺伝子群 106
発端種 109
ポット 169
ホットスポット 106
北方起源 148
北方系 148
北方系種 31
北方高山性の分類群 31
北方針葉樹林 214
北方大陸 148
北方的性格 148
北方的な種 148
北方に拡大する個体群 78
北方へと広がる 206
北方への侵入 149
北方への広がり 149
北方へ飛行 148
ボディプランの発育 31
ボトムアップ効果 31
ボトルネック 31
ボトルネック効果 31
ほとんどすべての 143
ほとんど致死量に近い用量 210
哺乳類 128
哺乳類のゲノム 128
哺乳類の草食動物 128
骨 31
骨組 190
母年齢 130
ホプキンスの寄主選択則 105
匍匐茎 208
匍匐枝 208
ポプラ 167
頬 38, 90
ほぼ同じ大きさの 209
ほぼ完全な擬態の局所的な収束 143
ほぼ同一なカラーパターンモザイク 143
ほぼ等長の 209

類の　90
ほぼ間違いなく　22
ホメオシス　104
ホメオティック遺伝子　104, 106
ホメオティック(突然)変異　104
ホメオボックス　104
ホモイオロジー　104
ホモエイシス　104
ホモジネート　104
ホモ接合型　104
ホモ接合型対立遺伝子　104
ホモ接合体　104
ホモダイナミック　104
ポモナ型　167
ホモニム　104
ホモログ　104
ホモロジー　104
ホモロジー検索　104
ホモロジーモデリング　104
ホヤのゲノム　23
保有(菌)株　27
ポリアデニル化　167
ポリガミー　167
ポリジーン　167
ポリジーン支配　167
ポリジーンの　167
ポリ(ュ)テス型　167
ポリドナウイルス　167
ポリドナウイルス由来レクチン　159
ポリネーター　167
ポリプロピレン管　167
ポリプロピレンチューブ　167
ポリヘドリン遺伝子　167
ポリヘドリンプロモーター　161
ポリペプチド因子　167
ポリマー　167
ポリメラーゼ連鎖反応　159, 167
保留する　235
ボルネオ　31
ボルバキア　235
ボルバキアによる雄殺し　235
ボルバキアによる生殖操作　235
ボルバキアによる単為生殖　235
ボルバキアによる雌化　235
ボルバキア媒介による中毒　235
ボルバキア媒介による中毒機構　235
ボルバキア媒介による中毒説　235
ボルバキア密度　235
ホルマリン水溶液　87
ホルマリン溶液　87
ホルモン　105
ホルモン機構　105
ホルモンシグナル伝達　105
ホルモン支配　105
ホルモン制御　105
ホルモン調節　105
ホルモン動態　105
ホルモンの差異　105
ホルモン力価　105
ホロタイプ　103
ホワイトリスト　233
本気の　68
本質的な性格　112
本質的な洞察力　75
本質的に　160
本質において　109
本州個体群　105
本州北限地域　148
本数　150
本数密度　207
本数密度効果　69
本体論　152
盆地　26
本土 - 島モデル　202
本土の群集　127
ボンビキシン　31
ボンビコール　31
ポンピング　175
ボンフェローニの調整　31
ボンフェローニ(の)補正　31
本物の科学的目的　31
本来的性格　142
本来の在来野生生物　154
本来の生息地　142
本来の場所での保全　109

マ

マーカー　129
マーカー遺伝子　129
マーカー遺伝子座　129
マーカー組換え率の解析　18
マーカー候補　35
マーラースペース　128
マイアミブルーシジミ　134
マイオティックドライブ　132
マイクロ RNA　134, 136
マイクロアレイ　134
マイクロインジェクション　134
マイクロ遠心チューブ　134
マイクロサイト　135
マイクロサテライト　134
マイクロサテライト多型　134
マイクロサテライトマーカー　134
マイクロタイタープレート　135
マイクロハビタット　134
マイクロホン　134
毎年　77
マイマイガ　98
マウス　140
前　20
前側の　20
前刷り　171
前に -　20
前に突き出た　168
前の -　20
前もって学習した　171
前もって決まっている翅の紋様の色彩　44
磨縁部　128
間置き　202
曲がった　53

曲がり　200
まき散らす　115
マキバジャノメ　131
膜　132
膜貫通受容体　220
膜貫通ドメイン　221
膜貫通領域　221
膜構造　132
膜質表皮層　132
膜翅目　107
膜翅目の昆虫　107
膜翅目のレクチン　107
膜状の構造　132
膜性表皮　132
膜タンパク質　132
膜チャンバー　224
マクドナルド-クレイトマンテスト　131, 137
マグニチュードオーダー　153
膜の高速振動　178
膜(の)幅　233
膜輸送体　222
マクロレンズ　127
磨砕する　97
摩擦発音　209
真下に　27
真下に置かれる　123
まずい　62, 227
まずい味の蝶　227
麻酔をかける　19
マスター遺伝子　129
マスター制御遺伝子　129
マスタード　141
マスタード油　141
マスタード油グリコシド　141
マスタード油配糖体　141
マゼンタ　127
マダラチョウ亜科　135
マダラチョウ亜族　54
マダラチョウ科　135
まだら模様　38
間違いを証明する　181
待ち構える　25
待ち伏せ　98
待ち伏せ型　160, 231
待つ　12
マツが優占した林　165
マツキリガ　165
マッククレイド　127
まっ黒の　118
マツ群落　165
まつ毛がある眼　121
末梢側で結合する　24
末梢側に　62
末梢神経　160
末梢の　62
末梢領域　62
末端　216, 217, 219
末端側で結合する　24
末端側に　62
末端側の縞状バンド　62
末端機構　217
末端細胞　21
末端小粒　215
末端突起　217
末端の　62
末端の枝　216
末端の距　217
末端配列　72
末端部位の切除　62
末端部のない遺伝子　62
末端部のない遺伝子の発現　62
末端部のない遺伝子発現の消失　126
末端部のないタンパク質　62
末端分岐　216
末端分類群　217
末端融合型　216
末端領域　62
マツと被子植物との異なる構成　60
マツノキシロチョウ　165
松林　165
マツ林　165
マッピング　129
マッピング手順　129
マッピング分解能　185
末ふ節　215
窓　234
窓(の)網戸　234
マトリックス生息地　130
惑わせる　64
真夏　135
まばらに植えた　202
麻痺させる　19
目蓋　80
マメ科　80
マメゾウムシ　27, 193
マメゾウムシ科　33
守る　12
繭　42
迷い種　208
真夜中　135
真夜中の太陽　135
マヨラナ　129
マリナー様転移因子　129, 137
マルコフ過程　129
マルコフ連鎖モンテカルロ法　131
マルチクローニングサイト　140
マルチプレックスPCR　141
丸で囲った領域　40
丸点　152
マルハキバガ科　151
マルハナバチ　33
マルピーギ管　128
マルピーギ管細胞　128
マルピーギ管細胞の細胞核　150
丸まった葉　188
丸みをおびた　188
マレー型　128
まれな　112, 190
まれなこと　190
稀な例　178
まれに発現する　79
まれに有意となる減少　26
まん延する　161
マングローブ　129
卍ともえ飛翔　205
慢性の　39
マンテル検定　129
マンテル-コックス検定　129
マントファスマ目　129

マン‐ホイットニーのU検定　129

ミ

見出す　61
未解決な問題　152
未改良の草地　226
味覚　98, 215
味覚感覚子　215
味覚感覚毛　98
味覚器官　98
味覚受容体遺伝子候補　35
味覚受容体　96, 98
味覚受容体遺伝子の一つ　96
味覚受容体候補遺伝子　35
味覚受容能力　98
味覚神経細胞　98
味覚増強効果　215
未確定種　203
味覚ニューロン　98
未攪乱環境　226
見かけ固定しているミトコンドリアDNAの遺伝子浸透　21
見かけ上複雑な行列　193
みかけの　212
見かけの競争　21
三日月形　127
三日月形の　51
ミカン科　188
ミカン科植物　188
幹　207
右上がり斜線　168
右鼓膜　187
幹数　150
ミクロコスモス　134
未経験の土地への侵入・定着　116
未結合の下流遺伝子　227
未結合領域　227
未交尾雌　227, 230
未交尾雌の分泌物　192
ミコニウム　134
見込み　123
短い再帰期間　197
短い再発期間　197
短い挿入　197
短い突起　124
短い日長　197
MISHIMA法　134, 136
未熟果　236
未受精の半数体卵　226
未受精卵　226
未出版　227
未照射の　148
実生の苗木　193
未使用の培養液　88
未使用培地　88
未処理　136
ミス　75
ミズアオシロチョウ　15
水清き国　41
水先案内　165
水資源　232
ミスセンス変異　136
水たまり　167, 175
水たまりの蝶　175
水に溶ける　232
水の汚染　48
水のパン　156
ミスマッチ　136
水を張った皿　156
未成熟期　109
見せかけ　98
みせかけ　129
溝　41, 97, 211
ミゾホオズキ　138
見たところ存続できる個体群　21
道しるべ　220
道しるべフェロモン　220
道に迷う　208
未知の　226
未知の生態学的要因　226
道ばたの蝶　188
蜜　104, 143
ミツオシジミタテハ　96
三つ組　223
見つける　61
蜜源の量　12
密集した植生　57
密集した低地雨林　57
密集したパッチ　57
密集した日陰の多い森　57
密集した緑に覆われた地域　57
蜜食動物　144
ミッションヒメシジミ　136
密接な因果関係　41
密接な関係　15, 115
密接な生態関係　41
密接な相互作用　115
密接に相互関連した生活史形質　41
蜜腺　104, 144
三つの対立遺伝子システム　218
密度依存的分散　57
密度効果　57
密度に依存しない過程　57
密度に依存する過程　57
密度の変化を介する間接効果　57
密度板　57
密に分散されたマーカー　57
密に連鎖した遺伝子座　219
蜜の愛飲者　25
ミツバチ　104
蜜蜂　27
ミツバチ飼育　21
ミツバチ女王蜂の大顎腺フェロモン　177
ミツバチのコロニー　104
蜜標　143
蜜を摂取する昆虫　144
ミトコンドリア遺伝子　136
ミトコンドリア遺伝子の符号化　136
ミトコンドリア遺伝子配列　137
ミトコンドリアゲノム　137
ミトコンドリアDNA　136, 140
ミトコンドリアDNAの塩基配列　195
ミトコンドリアの遺伝的多様性　137
認められた亜種　180
認める　14

ミドリタテハ 128
緑の革命 97
緑髭効果 97
緑豊かな植生 127
見なす 56
ミナミイヌモンキチョウ 202
南西日本の低地 202
南ベトナム 202
ミネラル 136
ミノムシ 26
未野焼き 225
実り豊かな環境 89
ミバエ 89
未発育の卵巣小管 226
未発表 227
未発表データ 227
未発表の著作物 227
みはり 160
見張り 160
未舗装道路の路端 69
耳 67
ミミウイルス 136
ミミクリー 136
ミミック 136
耳の 24
ミメシス 135
脈 229
脈管系(統) 228
脈系 229
脈翅目 145
脈相 229, 234
ミヤマシロチョウ族 21
ミュラー型擬態 140
ミュラー擬態 140
ミリアミド 136
ミリモーラー 137
ミリモル濃度 137
魅力たっぷりな趣味 35
見分ける 61
眠 138
眠性 138, 140

ム

無- 11, 61, 78
無遺伝子移入の帰無仮説 150
無栄養型卵巣小管 156
無縁種 227
無害種 100
無害の 112
無核精子 21, 22
無攪乱の対照 226
無関係遺伝子 227
無機態窒素イオン 112
無機物 136
無極性化合物 148
向け直す 181
向ける 35
無限サイトモデル 111
無限対立遺伝子モデル 111
無効な 116
無効な命令法的行為 116
無効名 116
無根系統樹 227
無細胞 36
無作為順序 178
無差別な交換 173
むさぼり食う 58
無翅遺伝子 91, 232, 235
無翅型 22, 235
無指向機構 147
虫こぶ 90
無翅種 235
無視できる程度の個体数 62
無翅の 22
虫眼鏡 127
無周期型 104, 147
矛盾 61
無照射で 235
無傷植物 113
無処理 136
無処理の昆虫 227
無処理の昆虫系統 148
無処理の雌化した母系 227
無数の菜園 51
無数の場所 51
結びついた 114
娘細胞 55
無性芽 173
無精子タンパク質 147
無脊椎動物 116
無脊椎動物の保全努力 116
無操作の対照 227
無定位運動性 119
無定位機構 147
無毒種 147
無毒にする 58
無反応期 181
無尾型雌 214
無柄 195
無変更の黄色 225
無変更の白色 225
無変態 17
無矛盾の連鎖関係 146
無毛の光沢のある葉表面 95
無毛の眼 99
無目標移動 147
無紋型 52
無有意差 146
無有意相関 146
無融合生殖 21
むらがある 158
群がり 52
無力化する 114
群れ 22, 177
ムレキシド反応 141
群れて 97

メ

眼 80
明暗サイクル 123
明暗視 13
明暗周期 122, 123
明暗対比型視覚 32
明黄緑色の蛹 32
明快に実証する 225
明快に同定された 225
メイガ科の蛾 176
明確な色彩選好 173
明確に区別できる対立遺伝子 63
明確に証明された 41
明確に定義された姉妹関係 232
明確に特定された 225
メイガ成虫 207
メイガの幼虫 207
明期 119, 123, 162
明期開始 27

名義科階級群タクソン　147
明期終了　72
名義属　147
名義タイプ亜種　147
名義タイプタクソン　147
名義タイプの　147
名義タクソン　147
名義タクソン名　147
鳴禽類の種　201
迷光　208
名詞句　149
名称　142
明相　123
メイチュウ　207
迷蝶　208
明度　32
明度処理　32
明白な　129, 173
明白な衰亡　21
明白にする　225
命名者　24
命名法　142, 147
命名法的　146
命名法的行為　147
命名法的行為の先取権　172
命名法的地位　147
名誉会長　105
名誉議長　105
名誉著者　104
名誉のオーサーシップ　104
明瞭なカラーパターン　62
目がくらむ　55
メガネトリバネアゲハ　34
メガピクセル　132
メガBLAST解析　132
芽キャベツ　33
メゲラツマジロウラジャノメ　231
目覚め　25
眼－触角原基　80
目じるし　120
♀　82
メス　82
雌　82
雌異型配偶子型　82
雌化　82
雌化する　82
雌型幼虫　82
メスキートヒメシジミ　179
メスグロトラフアゲハ　219
雌探し行動　130
雌特異的染色体　82
雌特有の器官　82
雌における妊性選択　81
雌に偏ったSSD　82
雌に限定されたアイソフォーム　82
雌に限定された擬態　82
雌に限定された多型　82
雌に限定された多型擬態　82
雌に限定されたベイツ型擬態　82
雌に対する妊性選択　81
雌に特異的な　204
雌に特有の分子機構　82
雌に有益な　82
雌による密かな性選択　38, 52
雌の羽化　82
雌の羽化スケジュール　82
雌の羽化場所での交尾可能性　168
雌の体　31
雌の交尾拒否行動　27
雌の子孫　82
雌の収入　82
雌の出現予定表　82
雌の生殖器官　82
雌の性的交尾受容力　82
雌の性分化　82
雌の選択　82
雌の胚　82
雌の表現型　82
雌の表現型発現　79
雌のほうが雄よりも体格が大きい性的二型　82
雌発育の有利点　58
雌ヘテロ型　82
雌ヘテロ型性染色体構成　82
雌ヘテロ型性染色体システム　82
雌ヘテロ型性染色体様式　82
雌防衛型の一夫多妻　82
雌モデル　82
雌有尾型　214
珍しい　190
珍しい種　225
珍しい例　227
雌を見つけるためのナワバリ　130
メタ解析　133
メタゲノム　133
メタゲノム解析　133
メタゲノム学　133
メタ個体群　133
メタ個体群効果　133
メタ個体群生存率　133
メタ個体群生態学　133
メタ個体群動態　133
メタ個体群動態解析　133
メタ個体群動態モデル　133
メタ個体群の遺伝的形質　92
メタ個体群の人口学的特性　56
メタ個体群の生存能力　133
メタ個体群の存続性　133
メタ種　133
メタセテリー　133
目立たないくすんだ茶色　148
目立つ　48
目立つ色　48
目立つ眼状紋　48, 173, 209
目立つこと　48
目立った　48
目立つテリトリー行動　48
メタノール酢酸　134
メタ分析　133
メタボロミクス　133
目玉模様　80
目玉模様の痕跡　188
メチオニン　134
メチオニン型貯蔵タンパク質　134
メチオニン由来化合物　134
メチオニン由来グルコシノレート　134

メチルアルキルピラジン　134
メチル化　134
メチル化 DNA 免疫沈降法　132, 134
メチルトランスフェラーゼ　134
滅菌蒸留水　191, 208
滅菌する　208
メッセンジャー RNA　133, 140
めったに起こらない　112
滅亡の段階　56
メディア因子　131
メディアン　132
メディアン閾値　132
メディアン値　132
メネラウスモルフォ　30
目の後ろあたりで　25
メバロナート　134
メバロン酸　134
メモ　132
目盛り棒　190
メラトニン　132
メラニン　132
メラニン色素　132
メラニン前駆体　132
メリッサミヤマシジミ　119
メルポメーネベニモンドクチョウ　169
芽を出す　95
目を見張る違い　204
免疫　109
免疫関連因子　109
免疫系因子　109
免疫系関連遺伝子ファミリー　109
免疫障害　109
免疫組織化学的局在　109
免疫反応　109
面会場所　183
面積に基づく累積曲線　22
メンデル　132
メンデル遺伝子座　132
メンデル因子　132

モ

毛縁　40
毛管　35
毛細管　35
毛細気管層　220
毛式　38
網室　191
網翅目　30
毛翅目　222
網状　186
毛状感覚子　222
毛状突起　222
毛状の　99
網状の斑点　186
毛状の眼　99
毛生　222
毛束　99, 106
毛束状の　81
盲のう　73
毛母　222
毛母細胞　222
網膜　186
網膜外光受容器　80
網膜の光異性化酵素　186
網膜の領域化　186
網膜光受容器　186
網膜部域性　186
網羅的探索法　78
網羅的な解析　109
網羅度合　56
毛隆　38
目　153
目撃(例)　197
目撃情報　197
木材燃料　235
目視検査　230
目視識別　230
木質燃料　235
木炭　38
目的の PCR 産物　214
目標位置　214
目標移動　35
目標地点　120
木本寄生　235
木本植生の侵食　72
木本の林床被覆(率)　235
モグラ　137
模型用粘土　137
模糊脈　143
モザイク解析　139
モザイク状　139
もしあれば　108
模式種　224
模式図　191
模式標本　224
文字の任意組合せ　22
モジュール　137
もたらす　32
モチーフ　140
持つ　168
最も近縁な　140
最も健康な個体群　100
最も厳密にいえば　194
最も近い原因　174
最も早く飛来した個体　83
最も豊富な　139
最もよく研究された分野　28
もっともらしい説明　166
もつれをほどく　61
モデリングの包括的枠組み　46
モデル　137
モデル系　137
モデル係数　137
モデル種　137
モデル選択　137
モデルで予想した寄主植物選好性　137
モデル分類群　137
戻し交雑　25
戻し交雑の結果の雄の選好性　25
戻し交配　25, 27
戻し交配の家系　25
戻し交配の同胞　25
戻り移動　186
戻る　186
モニタリング　138
モニタリングスキーム　138
モノグラフ　138
モノテルペン　138
物珍しさ　53

モミジの木　129
模様　26, 129
最寄りの近縁　143
盛り上がった翅脈　173
森に覆われた景観　87
森の残部　87
森の住人　87
森の蝶　34
モルフ　139
モルフォゲン　139
モルフォゲン勾配　139
モルフォゲンの円錐形濃度勾配　47
モルフォゲンの隆起線　187
モルフォチョウ　139
モルフォチョウ亜科　139
門　163
モンキチョウ亜科　211
モンゴル　138
モンシロチョウ　34, 200
モンスーン　138
モンスーンの季節　138
問題解決のための段階的手法　15
問題の侵入種　116
紋様形成　159
紋様形成の反応拡散モデル　179
紋様の変換則　159

ヤ

ヤガ　155
野外　83
野外遠征　83
野外から採集された蛹　175
野外観察　83
野外観察者　83
野外研究　83
野外検定　83
野外採集雄　234
野外採集個体　83, 234
野外採集した雌　83
野外飼育　154
野外支援　83
野外実験　83
野外条件　83
野外成虫　233
野外で採集した雄　234
野外で採集した動物　83
野外で採取した動物　83
野外での手順　83
野外では　234
野外の昆虫飼育場　154
ヤガ科　146
ヤガ科の蛾　146
ヤガ科の種　146
焼かれていない　225
夜間活動する　146
夜間の　146
焼きなまし温度　19
焼畑　213
焼く　108
約　22
約1日後　12
約50%を含んでいる　46
薬物食性　161
薬物摂食　161
約60%　34
夜行活動　146
夜行性　146
夜行性蛾のような祖先　146
夜行性原猿　146
夜行性の　146
夜行性の蛾　146
夜行性の生活様式　146
野蚕　234
矢印状の斑紋　23
休む　186
野生型　234
野生型個体群　234
野生型幼虫　234
野生個体　233
野生個体群　233
野生植物　233
野生のアブラナ科植物　233
野生の十字花科植物　233
野生のショウジョウバエ　233
野生の花　233
野生のミツバチ　233
野草　233
夜長時間　146
ヤドリギ　136
宿る　100
ヤヌスグリーンB　117
矢のようにすばやく飛ぶ　54
やぶ　33
夜分　121
山火事　234
山形紋　38
山積み　164
山積みにする　164
ヤマトスジグロシロチョウ　97
やめさせる　57, 213
やや　201
やや白い　233
やりがいのある　236
槍状の　120
槍の穂先形の　120
柔らかい葉　201
ヤンキーコヒオドシ　135

ユ

優位系列　65
有意鎖　194
有意性閾値　198
有意な閾値　198
有意な正のD統計量　198
有意な相違　198
有意な変異　210
有意な偏差　198
有意に好む　198
有意により少ない　198
有意に悪い　198
誘引機能　71
誘引効果　71
誘引する　24
誘引物質　24
有益　228
有益な形質　27
有益な効果　27
有益な植物　187
UXT　225, 228
融解曲線　132
有害効果　56
有害作用　56
有害対立遺伝子　56
有害な遺伝的影響　56
有害な侵入種　54

有害な真昼の温度 58
有害な虫 161
有害な劣性種 56
有核精子 76, 149
有桿細胞 191
有桿体細胞 191
有桿体帽鞘 191
有桿帽鞘 191
有機(従属)栄養生物 102
有機化学者 153
有機質肥料 129
誘起する 71, 174
有機体 153
有棘毛瘤 165
ユークロマチン 76
有効アリル数 70
融合遺伝子 89
有効温度 70
有効温量 220
有効個体群サイズ 70
有効個体数 70
融合した微絨毛膜 89
融合して一つになった眼状紋の対 89
有効集団サイズ 70
有効スペクトル 70
融合する 89
有効性 228
有効成分 13
有効積算温度 220
有効積算温度定数 218
有効積算温量 70, 185, 220
有効積算温量定数 218
有効対立遺伝子数 70
有効中立突然変異モデル 137
有効な集団の大きさ 70
有効な命名法的行為 228
有効にされた 228
有孔の鐘状感覚器 168
有効波長 70
有効名 228
有根系統樹 188
有翅型 15, 235
有翅胸節側板 175
有翅種 15
U3X 遺伝子 225, 227
優性 65
優性関係 65
優勢形質 65
雄性交尾器 128
有性生殖 196
雄性生殖器 128
雄性先熟 174, 196
優性突然変異 65
優性の 65
優性白色／劣性黄色の対立遺伝子 65
有精卵 82
優先権 170
優占種 65
優先的に 171
優占度 65
優先度を下げる 66
優占な広域分布種 65
遊走期 231
有胎盤哺乳類 165
有蹄類 226
尤度 123
尤度アプローチ 123
誘導する 208
誘導する要因 219
誘導性植物揮発性物質の放出 111
誘導多発生 186
誘導フェロモン 172
誘導防衛 111
誘導防御 111
尤度関数 123
有毒蝶 100
有毒な 166, 220
有毒のモデル 220
尤度比検定 123
雄胚 128
誘発 71
誘発機能 71
誘発効果 71
有尾型雌 214
UV-C 228
有柄眼のハエ 207
有柄の 161
有望 173
遊牧 146
遊牧(的)移動 146
有毛の 164
幽門弁 172, 176
有用 228
ユーラシア起源 76
ユーラシア起源草原属 97
ユーラシア山岳起源 154
ユーラシア大陸 76
遊離型グリシン 88
遊離グリシン 88
有利な点 228
有利な繁殖条件 81
優良遺伝子モデル 96
有力で 171
有力な源泉 169
有力な要因 170
ユール過程 236
ゆがみ 63, 200
ゆがめる 63
ユクスタ 118
癒合 89
輸精管 228
豊かにする 73
油断 227
ユッカガ 236
ゆっくり歩行 200
ユニークな求愛器官 226
ユニバーサルパターン 226
指寸法で計る 202
夢物語(的な) 228
ゆらぎ 85
ゆらぎの非対称性 85
輸卵管 155
ゆるやかな 96

ヨ

夜明 55
良い天候 83
陽イオン 36
要因配置計画 80
蛹越冬 175
葉円板 122
蛹化 176
溶解したミネラル 62
溶解した無機物 62
溶解度 201
蛹殻 176
蛹化時期 176, 219

蛹化した個体　176
蛹化する　176
蛹化場所　176
蛹化率　176
蛹期　175
蛹期間　175
蛹期後期　121
蛹期初期　68
蛹期前期の後翅　68
蛹期中期　135
蛹期の一部期間　158
蛹期の後翅　175
蛹期の前翅原基　87
蛹期の翅　175
蛹休眠　175
蛹休眠種　175, 203
蛹休眠の発生　109
蛹休眠の誘起　109
蛹休眠率　109
幼形成熟　144
幼形成熟の生殖虫　144
蛹後期　121
擁護名　146
幼根　178
蛹翅　175
要旨　12
幼若成虫期　170
幼若ホルモン　118
幼若ホルモン(JH)の血中濃度　40
幼若ホルモンの産出　118
幼若ホルモン(JH)の除去　41
幼若ホルモン(JH)の分解速度　179
幼若ホルモンの力価　118
幼若ホルモン類似体メトプレン　118
蛹重　175
蛹鞘　95
葉状の　124
葉状片　75
蛹初期　68, 236
蛹初期の後翅　68
蛹初期の翅　68
蛹初期の表皮　68
養殖　81
葉食者　122
養殖する　53
葉食性　163
葉食動物　86
用心深く　36
要するに　109
幼生　120
幼生期　68, 171
幼生期調査　191
幼生期の発育　118
幼生期100日をこえる　68
陽性クローン　168
陽性細胞　168
幼生生殖　156
幼生の生育　118
葉積層　122
様相　159
葉組織　122
幼体　118
蛹態配偶行動　175
幼虫　36, 98, 120, 150
幼虫越冬　121
蛹中期　135
幼虫期　36, 109, 121
幼虫期間　121
幼虫寄生蜂　121
幼虫期の成虫原基　120
幼虫期の脱皮　121
幼虫期の中頃　135
幼虫期の翅原基　121
幼虫期の翅成虫原基　121
幼虫期の表皮器官　171
幼虫期の齢期間　67
幼虫休眠　120
幼虫時間　121
幼虫死亡数　121
幼虫死亡率　121
幼虫宿主の免疫反応(応答)　36
幼虫宿主の免疫防御　36
幼虫生育　120
幼虫の生存率　121
幼虫生存力　121
幼虫組織　121
幼虫体　120
幼虫脱皮　121
幼虫(体)段階　118
幼虫の新たな色彩型　46
幼虫の過剰な発育時間　78
幼虫の感染　121
幼虫の客棲性　121
幼虫の休眠誘起　120
幼虫の飼育温度　121
幼虫の時期　121
幼虫の色素沈着　121
幼虫の集合性　120
幼虫の終齢期　121
幼虫の食餌植物　120
幼虫の食草　120
幼虫の食草問題　117
幼虫の巣単位での頭数　121
幼虫の成育　121
幼虫の摂食効率　120
幼虫の中腸　121
幼虫の肉食性　120
幼虫の発育　121
幼虫の防御　121
幼虫の保護　121
幼虫の蛹化　176
幼虫の齢期　121
幼虫の齢数　150
幼虫発達　120
幼虫皮膚　121
幼虫ホルモン　118
幼虫末期　217
容認可能な限度　12
溶媒　201
溶媒処理済みの対照植物　201
溶媒抽出　201
葉柄　207
葉片　18, 120, 124
蛹便　131, 134
葉片円盤　122
葉片状の　120
蛹便を排泄する　78
養蜂業　21
要防除密度　49, 52
羊膜　18
羊膜類　18
羊膜類のゲノム　18
葉脈　122, 229
葉脈切断行動　229
要約　211

葉翼 122
用量依存的手法 66
葉緑体 39
葉緑体ゲノム 39
葉緑体 DNA 39, 51
ヨーロッパアカタテハ 180
ヨーロッパ諸国による植民地化 76
ヨーロッパシロジャノメ 129
ヨーロッパ動物相データベース 81
ヨーロッパの蝶 76
ヨーロッパの北極圏の高山性鱗翅類 76
予期されなかった機構 226
抑圧 212
抑圧された 57
抑圧名 212
翼(翅)荷重 234
良く研究された寄主植物と草食生物とのシステム 232
抑止する 57
翌日 86
抑止力 57
抑制 24, 212
抑制因子 57, 112
抑制解除する 61
抑制差引ハイブリダイゼーション法 212
抑制サブトラクティブ・ハイブリダイゼーション法 206, 212
抑制された感染密度 212
抑制された著作物 212
抑制条件 112
抑制する 12, 184, 186
抑制的な欲求 186
抑制ホルモン 112
抑制名 212
抑制要因 112, 212
良くなる 17
良く発育した大草原植生 232
よく分岐した系統 232
よく訪問する 45
翼葉 122
よく理解された分子機構 232
予見可能性の欠如 120
予見不可能な環境急変 227
横(の) 222
横風ドリフト 52
横風偏流 52
横梁 52
横方向に切断する 192
四次構造 177
四次の 177
余剰ストック 212
よじれた 86
寄集群 43
寄せ付けない 183
予想されていなかった機構 226
予想する 74
予測 170
予測因子 170
予測可能性の欠如 120
予測可能な 170
予測可能な環境 170
予測可能な季節変動 170
予測可能な差異 170
予測可能な方向 170
予測可能な方法 170
予測された位置 170
予測的 170
予測不可能な移動 227
予測不能な 227
予測不能な環境 227
予測変数 170
四つ組 177
4つに枝分かれした櫛歯状の 176
ヨツモンマメゾウムシ 33
ヨナクニサン 24
予備的観察 151
予備的研究 171
予備的フィールド調査 171
予備凍結 170
予備パターン 171
予防 171
予防的アプローチ 170
予防的手法 170
読み取り枠 152, 153
読み枠 153
ヨモギギク抽出 214
より暖かい夏の気温 231
より暖かい熱帯雨林 231
より後で現れる影響 121
より薄暗い場所 54
より大きすぎる影響 62
より温暖な条件 231
より暗く 54
より厳密に言えば 194
より濃い 57
より低品質な食料 126
より遠く離れた 89
選り分ける 61
夜に活動的な 146
よろず屋 117
弱い休眠 151
弱い競争相手 167
弱い負の関係 232
弱く支持された結節点 232
弱められた選好性 181
弱める 24, 60, 109
4種類のデオキシリボヌクレオチド三リン酸(dATP, dCTP, dGTP, dTTP)を混合したものを表記する 64
四足類 176
四分裂の 176
4-メチルスルフィニルブチル 11
四裂片の櫛歯状の 176

ラ

裸 150
らい菌 141
ライ菌 141
ライトトラップ 123
ライト-フィッシャーモデル 236
ライバル 188
ライフサイクル 123
ライブラリー 123
ライマビーン 123
ライマメ 123
ライム病 127
ライントランセクト 123

ラオス　120
ラオス人民民主共和国　120
ラオス中部　37
ラオス北部　148
ラクシャリー遺伝子　127
ラグジュアリー遺伝子　127
ラクナ　120
落葉広葉樹林　55
落葉樹林　55
落葉性　55
落葉性の　55
落葉性の種　55
落葉層　122
落葉林の経済収支　69
ラジオイムノアッセイ　178, 187
裸子植物　98
らせん飛翔　205
裸地　26
落果　80
ラテックス　121
ラテックスの質　121
ラテックスの流出　121
ラテックスを分泌する葉　121
ラテン語　121
ラテン語化する　121
ラトランクリン　122
ラミート　178
ラムサール条約　178
裸名　146
ラメート　178
裸蛹　78
卵　70, 155
卵越冬　70
卵越冬一化型　226
卵円形の　154, 155
卵黄形成　230
卵黄膜　230
卵塊　27, 42, 70
卵蓋　152
卵回収　70
ランカウイ島　120
卵殻　35, 39, 70
卵期　70
卵寄生蜂を誘引する揮発性物質　70
卵休眠　70
卵休眠種　70, 203
卵菌　152
卵形
卵形成　70, 152
卵形成 - 飛翔形質群　152
卵形成 - 飛翔症候群　152
卵形の　154
卵形の外膜　154
乱婚　173
卵細胞　155
卵子　155
卵死亡数　70
卵死亡率　70
卵収集　70
卵食　70
卵数の評価　70
卵生の　155
卵巣　154
卵巣休眠　154
卵巣細胞培養　154
卵巣小管　154
卵巣成熟　154
卵巣動態　154
卵巣の動態　154
卵巣の発達停止　212
卵巣発育　154
卵巣由来細胞株　154
卵胎生の　155
卵堆積　70
ランタナ種の切り花　53
ランダムウォーク　178
ランダム傾きを加味したモデル　178
ランダム交配　178
ランダム順　178
ランダム増幅多型 DNA 法　178
ランダムな遺伝的浮動　178
ランダムなサンプルの累積曲線　178
ランダムに採集する　43
ランダムに採取する　43
ランダムに生成された景観　178
ランダムに捕獲した個体　178
ランダム配置曲線　178
ランダム浮動　178
ランダムヘキサマープライマー　178
ランダム方向　178
ランダム歩行　178
卵タンパク質　70
ランデブーサイト　183
ランドマーク点　120
ランナウェイ過程　188
ランナウェイ淘汰　188
卵認識　70
廊坊市　120
ランプ応答　178
卵母細胞の存在　152
卵捕食寄生者　70
卵捕食寄生者 - 幼虫捕食寄生者　70
卵門　134
卵門の　134
卵由来エリシター　70
卵由来分泌物　70

リ

リアルタイム蛍光定量的 PCR 法　179
リード　179
利益　28
リオ宣言　187
リガンド　123
力価　219
力説する　226
陸域甲殻類　217
陸域生態系　217
陸域節足動物　217
陸産甲殻類　217
陸産節足動物　217
陸上甲殻類　217
陸上生態系　217
陸上生物　217
陸生節足動物　217
陸生動物の生物多様性　217
陸地　48
陸のバイオーム　217
陸標　120
陸標情報　120
リクルートメント　180

リケッチア 187
利己的遺伝因子 193
利己的因子 193
利己的な細胞内共生微生物 193
リサージェンス 186
離散サンプリング 61
離散的な選択的形態 61
リサンプリング 179
リシークエンシング 185
リジェクト 182
リスク管理 188
リスク評価 187
リスク分析 187
リスケールする 179
理想化されたスペクトル 108
理想郷の 228
理想自由経路 108
理想自由分布 108
理想的体系 108
理想的な個体群分布 108
理想的なシステム 108
理想的なスペクトル 108
利他(的)行動 17
立体画像 208
立方根 52
離島 183
リニアランプ 124
理にかなった説明 179
利尿 63
利尿ホルモン 63
リノール酸 124
リノレン酸 124
リピッド 124
リファレンス遺伝子 181
リファレンスとなるべき配列データベース 181
リポカリン 124
リポキシゲナーゼ 124
リボ酵素 188
リボザイム 188
リボソーム 187
リボソーム RNA 187, 188
リボソーム遺伝子 187
リボソームタンパク質 187
リボソームタンパク質 L3 遺伝子 79
リボソーム DNA 179
リポフォリン 124
裏面 27, 226, 228
リモートセンシング技法 183
略記 11
流域 26, 188
隆起 172, 187, 227
隆起した翅脈 173
隆起線 36
琉球列島 188
流行している 171
流行性 75
流行病 74, 75
流行病学 75
竜骨 190
流産する 12
粒子 158
粒子産出 158
粒子生成 158
粒状の 97
留蝶の雄 185
流動食 124
流動体 85
流入した種 111
量 177
領域 45, 64
領域外の－ 79
領域シークエンス 181
両掛け戦略 28
両掛け戦略適応 28
両掛け適応 28
利用価値 25
両極電極 168
両鋸歯状の 30
良好な繁殖条件 81
両翅 83
両翅裏面 226
両櫛歯状の 29
良質な分散した生息地 103
両翅表面 227
利用種 228
両親媒性アミノ酸 18
両性 31, 66
両性遺伝子 66, 67
両性遺伝子の相同体 104
両性動物 101
両性のアイソフォーム 66
両性の相同遺伝子 104
両性のホモログ遺伝子 104
両生類 18
両生類のゲノム 18
稜線 36, 187
両側 Z 検定 224
両側の隣接マーカー 31
量的遺伝子 167
量的形質遺伝子座 176, 177
量的形質座位 176, 177
量的測時モデル 177
量的物質 177
量的変化 177
両輪精管 31
両立しない要求 110
緑色蛍光 97
緑色蛍光タンパク質 95, 97
緑色光 97
緑色葉の植食(性) 97
緑色葉の組織 97
緑青色斑紋 97
リリーサー 183
リリーサーフェロモン 183
リリース 183
リリースした地区 183
リリース実験 183
リリースするマーキング調査 129
リリース地点 183
理論家 217
理論的基礎 217
理論的研究 217
理論的根拠 217
理論的進展 217
理論的調査 217
理論的展開 217
理論的取り扱い 217
理論的な 204
理論的モデル 217
理論的論争 217
林縁 87
林縁効果 69

林縁部　87
臨界閾値　51
臨界光周期　51
臨界値　51
臨界日数　51
臨界日長　51
臨界夜長　51
林冠　35, 87
林冠性　87
林冠層の広葉樹林　35
林冠層のマツ　35
林冠の間引き　218
林冠閉鎖　35
林業　87
リング状の　187
リンケージマッピング　124
リンケージマッピング研究　124
リンゴミバエ　21
リン酸緩衝食塩水　159, 162
リン酸緩衝生理食塩水　159, 162
リン酸緩衝生理食塩水 - ウシ血清アルブミン　159
リン酸緩衝生理食塩水で処理された対照植物　159
リン酸緩衝生理食塩水での洗浄　159
リン脂質　162
鱗翅目　122
鱗翅目昆虫の生理学　122
鱗翅目の昆虫　122
鱗翅目の種　122
林床　87
林床性　87
鱗翅類　122
鱗翅類学　122
鱗翅類昆虫　122
鱗翅類宿主の体　31
鱗翅類に特異的な塩基配列　122
鱗翅類のゲノム　122
鱗翅類の個体数　122
鱗翅類の宿主細胞 DNA　122
鱗翅類群集　122
鱗翅類系統　122
隣接遺伝子　144
隣接する遺伝子　144
隣接する競合植物　144
隣接する個体群　144
隣接する同種植物　144
隣接分布地域　14
隣接抑制　121
隣接レーン　14
林地　235
リンネ　124
リンネウス　124
リンネ式同語反復　124
リンネ式動植物分類(命名)法　124
林分　207
鱗粉　190
鱗粉形成細胞　190
鱗粉細胞　190
鱗粉細胞の発生開始　112
鱗粉の形態　190
鱗粉の脱落　126
鱗粉の配列　190
鱗粉の母細胞　190
鱗粉の列　190
鱗紛列　190
鱗粉列リング　190
鱗片　190
鱗毛　190
倫理的　76

ル

類縁化合物　182
類縁関係的　182
類縁種　182
類形　104
類似　15, 198
類似 -　213
類似化合物　198
類似性　198
類似度　198
類似の進化面の急速変化　18
類人猿　21
累積荷重　53
累積曲線　13
累積死亡率　53
累積重量　53
累積種数　53
累代　74
ルイ・パスツール　158
類別　40
ルードルフィア線　127
ルテイン　127
ルテキシン　127
ルベンチーナグループ　127
ルリオビトガリバワモンチョウ　190
ルリボシスミナガシ　208
ルリモンアゲハ　157

レ

令　113
齢　113, 207
冷温　49
冷温帯　49
冷温帯域　49
冷温帯性　49
冷温耐性　126
冷温帯性の種　49
冷温帯性落葉広葉樹林　49
冷温帯性落葉樹林　49
零下の気温　211
冷却処理　49
冷却する　181
冷却蛹　49
例証する　78
齢数　113, 150
冷蔵期間　39
霊長類　172
霊長類の共通祖先　44
霊長類の系統　172
霊長類の色覚系　172
霊長類の色素　172
冷凍で付随して発生する潜熱の放出　183
冷凍標本　89
冷凍保存　52
齢別生命表　15
○齢幼虫　113
冷涼な場所　49
レヴァナ型　122
レヴィ飛行　123
レース　177

レース編みの 120
レース形成 177
レース状 120
レーダー(装置) 177
歴史(的)生物地理学 103
歴史的生物地理学的仮説の検定 217
歴史的に 103
歴史的分布 103
歴史年表 219
歴代のコレクション 211
レクチン 122
レクトタイプ 122
レクリエーションの目的 180
レシリン 185
レスキュー効果 185
レチナール光異性化酵素 186
レチナール領域化 186
劣化する 57
レック 122
劣勢 111
劣勢形質 180
劣性致死遺伝子 180
劣性致死対立遺伝子 180
劣性突然変異体 180
劣性の 180
劣性有害遺伝子 180
劣性有害対立遺伝子 180
劣等 111
レッドデータブック 180
レッドリスト 181
レッドリスト基準 181
レッドリスト区分 181
レッドリスト部門 181
裂片 41, 124
レテノールモルフォ 51
レトロウイルス受容体遺伝子 186
レトロウイルスレセプター遺伝子 186
レトロトランスポゾン関連コーディング領域 186
レトロポゾン 186
レフェリー 181
レフュジア 181
レプリケート 184
レリーサー 183
レリーフパターン 183
連環型タンパク質 46
連結 114
連結器官 234
連結したタンパク質 46
連結性 47
連結断片 48
連結飛行 36
連結飛翔 36
連合学習 23
連合定位法 23
連鎖 124
連鎖群 124
連鎖群関係 124
連鎖群の割り当て 124
連鎖状タンパク質 46
連鎖する遺伝子座 124
連鎖地図 124
連鎖の順序 124
連鎖不平衡 124
連鎖分析研究 124
連鎖を徐々に密にして行く 96
レンシュ則 183
レンシュの法則 183
連続環節体 214
連続 30 分間 47
連続した塩基 48
連続した生息地 48
連続したヌクレオチド 48
連続した変異 48
連続したリング 48
連続する塩基３個の１組 223
連続性 47
連続性繁殖動態的 104
連続線 48
連続的進化過程 211
連続的な齢期 211
連続的変異 48
連続飛翔 48
連続飛翔型 48
連続飛翔行動 48
連続分布 48
連続変異 41, 48
連打 67
連邦政府によって絶滅危惧種に指定されている蝶 81
連絡先となる著者 50
連絡著者 50

ロ

ロイコプテリン 122
ロイシン 122
ロイド - ゲラルディの J 指標 124
老衰死 143
ロウ層 232
老廃物 231
老齢林分 151
ロードキル 188
ロードマチン 187
濾過して取り除く 83
録音の再生 166
六脚亜門 102
ログランク検定 125
濾紙 12, 83
ロシア・マガダン地域 127
ロジスティック回帰 125
ロジスティック回帰モデル 125
露出 154
露出した場所 79
六角形 102
ロッド 125
六本脚 102
ロドプシンタイプ 187
ロドプシンタイプ受容体 187
ロドプテリン 187
濾胞 86
路傍 188
濾胞細胞 86
路傍の植込み 188
- 論 11
論説 23, 222
論争 49
ロンドン自然史博物館 31, 33
論破する 181
論評 44